Lecture Notes in Computer Science 3418

Commenced Publication in 1973
Founding and Former Series Editors:
Gerhard Goos, Juris Hartmanis, and Jan van Leeuwen

Ulrik Brandes   Thomas Erlebach (Eds.)

Network
Analysis

Methodological Foundations

 Springer

Volume Editors

Ulrik Brandes
University of Konstanz
Department of Computer and Information Science
Box D 67, 78457 Konstanz, Germany
E-mail: ulrik.brandes@uni-konstanz.de

Thomas Erlebach
University of Leicester
Department of Computer Science
University Road, Leicester, LE1 7RH, U.K.
E-mail: t.erlebach@mcs.le.ac.uk

Library of Congress Control Number: 2005920456

CR Subject Classification (1998): G.2, F.2.2, E.1, G.1, C.2

ISSN 0302-9743
ISBN 3-540-24979-6 Springer Berlin Heidelberg New York

Springer is a part of Springer Science+Business Media

springeronline.com

© Springer-Verlag Berlin Heidelberg 2005
Printed in Germany

Typesetting: Camera-ready by author, data conversion by Markus Richter, Heidelberg
Printed on acid-free paper   SPIN: 11394051   06/3142   5 4 3 2 1 0

Preface

The present book is the outcome of a seminar organized by the editors, sponsored by the *Gesellschaft für Informatik e.V.* (GI) and held in Dagstuhl, 13–16 April 2004.

GI-Dagstuhl-Seminars are organized on current topics in computer science that are not yet well covered in textbooks. Most importantly, this gives young researchers an opportunity to become actively involved in such topics, and to produce a book that can provide an introduction for others as well.

The participants of this seminar were assigned subtopics on which they did half a year of research prior to the meeting. After a week of presentations and discussion at Schloss Dagstuhl, slightly more than another half-year was spent on writing the chapters. These were cross-reviewed internally and blind-reviewed by external experts. Since we anticipate that readers will come from various disciplines, we would like to emphasize that it is customary in our field to order authors alphabetically.

The intended audience consists of everyone interested in formal aspects of network analysis, though a background in computer science on, roughly, the undergraduate level is assumed. No prior knowledge about network analysis is required. Ideally, this book will be used as an introduction to the field, a reference and a basis for graduate-level courses in applied graph theory.

First and foremost, we would like to thank all participants of the seminar and thus the authors of this book. We were blessed with a focused and determined group of people that worked professionally throughout. We are grateful to the GI and Schloss Dagstuhl for granting us the opportunity to organize the seminar, and we are happy to acknowledge that we were actually talked into doing so by Dorothea Wagner who was then chairing the *GI-Beirat der Universitätsprofessor(inn)en*. We received much appreciated chapter reviews from Vladimir Batagelj, Stephen P. Borgatti, Carter Butts, Petros Drineas, Robert Elsässer, Martin G. Everett, Ove Frank, Seokhee Hong, David Hunter, Sven O. Krumke, Ulrich Meyer, Haiko Müller, Philippa Pattison and Dieter Rautenbach. We thank Barny Martin for proof-reading several chapters and Daniel Fleischer, Martin Hoefer and Christian Pich for preparing the index.

December 2004

Ulrik Brandes
Thomas Erlebach

List of Contributors

Andreas Baltz
Mathematisches Seminar
Christian-Albrechts-Platz 4
University of Kiel
24118 Kiel, Germany

Nadine Baumann
Department of Mathematics
University of Dortmund
44221 Dortmund, Germany

Michael Baur
Faculty of Informatics
University of Karlsruhe
Box D 6980
76128 Karlsruhe, Germany

Marc Benkert
Faculty of Informatics
University of Karlsruhe
Box D 6980
76128 Karlsruhe, Germany

Ulrik Brandes
Computer & Information Science
University of Konstanz
Box D 67
78457 Konstanz, Germany

Michael Brinkmeier
Automation & Computer Science
Technical University of Ilmenau
98684 Ilmenau, Germany

Thomas Erlebach
Department of Computer Science
University of Leicester
University Road
Leicester LE1 7RH, U.K.

Marco Gaertler
Faculty of Informatics
University of Karlsruhe
Box D 6980
76128 Karlsruhe, Germany

Riko Jacob
Theoretical Computer Science
Swiss Federal Institute
of Technology Zürich
8092 Zürich, Switzerland

Frank Kammer
Theoretical Computer Science
Faculty of Informatics
University of Augsburg
86135 Augsburg, Germany

Gunnar W. Klau
Computer Graphics & Algorithms
Vienna University of Technology
1040 Vienna, Austria

Lasse Kliemann
Mathematisches Seminar
Christian-Albrechts-Platz 4
University of Kiel
24118 Kiel, Germany

Dirk Koschützki
IPK Gatersleben
Corrensstraße 3
06466 Gatersleben, Germany

Sven Kosub
Department of Computer Science
Technische Universität München
Boltzmannstraße 3
D-85748 Garching, Germany

Katharina A. Lehmann
Wilhelm-Schickard-Institut
für Informatik
Universität Tübingen
Sand 14, C108
72076 Tübingen, Germany

Jürgen Lerner
Computer & Information Science
University of Konstanz
Box D 67
78457 Konstanz, Germany

Marc Nunkesser
Theoretical Computer Science
Swiss Federal Institute
of Technology Zürich
8092 Zürich, Switzerland

Leon Peeters
Theoretical Computer Science
Swiss Federal Institute
of Technology Zürich
8092 Zürich, Switzerland

Stefan Richter
Theoretical Computer Science
RWTH Aachen
Ahornstraße 55
52056 aachen, Germany

Daniel Sawitzki
Computer Science 2
University of Dortmund
44221 Dortmund, Germany

Thomas Schank
Faculty of Informatics
University of Karlsruhe
Box D 6980
76128 Karlsruhe, Germany

Sebastian Stiller
Institute of Mathematics
Technische Universität Berlin
10623 Berlin, Germany

Hanjo Täubig
Department of Computer Science
Technische Universität München
Boltzmannstraße 3
85748 Garching, Germany

Dagmar Tenfelde-Podehl
Department of Mathematics
Technische Universität
Kaiserslautern
67653 Kaiserslautern, Germany

René Weiskircher
Computer Graphics & Algorithms
Vienna University of Technology
1040 Vienna, Austria

Oliver Zlotowski
Algorithms and Data Structures
Univeristät Trier
54296 Trier, Germany

Table of Contents

1 Introduction

Ulrik Brandes and Thomas Erlebach

Many readers will find the title of this book misleading – at least, at first sight. This is because 'network' is a heavily overloaded term used to denote relational data in so vast a number of applications that it is far from surprising that 'network analysis' means different things to different people.

To name but a few examples, 'network analysis' is carried out in areas such as project planning, complex systems, electrical circuits, social networks, transportation systems, communication networks, epidemiology, bioinformatics, hypertext systems, text analysis, bibliometrics, organization theory, genealogical research and event analysis.

Most of these application areas, however, rely on a formal basis that is fairly coherent. While many approaches have been developed in isolation, quite a few have been re-invented several times or proven useful in other contexts as well. It therefore seems adequate to treat network analysis as a field of its own. From a computer science point of view, it might well be subsumed under 'applied graph theory,' since structural and algorithmic aspects of abstract graphs are the prevalent methodological determinants in many applications, no matter which type of networks are being modeled.

There is an especially long tradition of network analysis in the social sciences [228], but a dramatically increased visibility of the field is owed to recent interest of physicists, who discovered the usefulness of methods developed in statistical mechanics for the analysis of large-scale networks [15]. However, there seem to be some fundamental differences in how to approach the topic. For computer scientists and mathematicians a statement like, e.g., the following is somewhat problematic.

> "Also, we follow the hierarchy of values in Western science: an experiment and empirical data are more valuable than an estimate; an estimate is more valuable than an approximate calculation; an approximate calculation is more valuable than a rigorous result." [165, Preface]

Since the focus of this book is on structure theory and methods, the content is organized by level of analysis rather than, e.g., domain of application or formal concept used. If at all, applications are mentioned only for motivation or to explain the origins of a particular method. The following three examples stand in for the wide range of applications and at the same time serve to illustrate what is meant by level of analysis.

U. Brandes and T. Erlebach (Eds.): Network Analysis, LNCS 3418, pp. 1–6, 2005.
© Springer-Verlag Berlin Heidelberg 2005

Element-Level Analysis (Google's PageRank)

Standard Web search engines index large numbers of documents from the Web in order to answer keyword queries by returning documents that appear relevant to the query. Aside from scaling issues due to the incredible, yet still growing size of the Web, the large number of hits (documents containing the required combination of keywords) generated by typical queries poses a serious problem. When results are returned, they are therefore ordered by their relevance with respect to the query.

The success of a search engine is thus crucially dependent on its definition of relevance. Contemporary search engines use a weighted combination of several criteria. Besides straightforward components such as the number, position, and markup of keyword occurrences, their distance and order in the text, or the creation date of the document, a structural measure of relevance employed by market leader Google turned out to be most successful.

Consider the graph consisting of a vertex for each indexed document, and a directed edge from a vertex to another vertex, if the corresponding document contains a hyperlink to the other one. This graph is called the Web graph and represents the link structure of documents on the Web. Since a link corresponds to a referral from one document to another, it embodies the idea that the second document contains relevant information. It is thus reasonable to assume that a document that is often referred to is a relevant document, and even more so, if the referring documents are relevant themselves. Technically, this (structural) relevance of a document is expressed by a positive real number, and the particular definition used by Google [101] is called the PageRank of the document. Figure 1.1 shows the PageRank of documents in a network of some 5,000 Web pages and 15,000 links. Section 3.9.3 contains are more detailed description of PageRank and some close relatives.

Note that the PageRank of a document is completely determined by the structure of (the indexed part of) the Web graph and independent of any query. It is thus an example of a structural vertex index, i.e. an assignment of real numbers to vertices of a graph that is not influenced by anything but the adjacency relation.

Similar valuations of vertices and also of edges of a graph have been proposed in many application domains, and "Which is the most important element?" or, more specifically, "How important is this element?" is the fundamental question in element-level analysis. It is typically addressed using concepts of structural centrality, but while a plethora of definitions have been proposed, no general, comprehensive, and accepted theory is available.

This is precisely what made the organization of the first part of the book most difficult. Together with the authors, the editor's original division into themes and topics was revised substantially towards the end of the seminar from which this book arose. A particular consequence is that subtopics prepared by different participants may now be spread throughout the three chapters. This naturally led to a larger number of authors for each chapter, though potentially with heavily

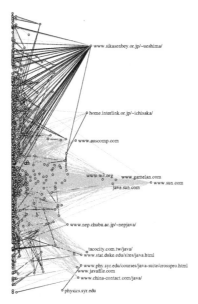

Fig. 1.1. PageRank in a network of some 5,000 Web pages containing the keyword 'java' (documents with higher value are further to the right; from [93])

skewed workload. To counterbalance this effect, leading authors are identified in such chapters.

Chapter 3 provides an overview of centrality measures for network elements. The authors have organized the material from a conceptual point of view, which is very different from how it is covered in the literature. Algorithms are rarely discussed in the application-oriented literature, but of central interest in computer science. The underdeveloped field of algorithmic approaches to centrality is therefore reviewed in Chapter 4. Advanced issues related to centrality are treated in Chapter 5. It is remarkable that some of the original contributions contained in this chapter have been developed independently by established researchers [85].

Group-Level Analysis (Political Ties)

Doreian and Albert [161] is an illustrative example of network analysis on the level of groups. The network in question is made up of influential local politicians and their strong political ties. This is by definition a difficult network to measure, because personal variations in perception and political incentives may affect the outcome of direct questioning. Therefore, not the politicians themselves, but staff members of the local daily newspaper who regularly report on political affairs were asked to provide the data shown in Figure 1.2.

Black nodes represent politicians who are members of the city council and had to vote on the proposed construction of a new jail. The County Executive,

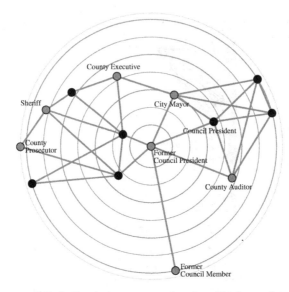

Fig. 1.2. Strong political ties between prominent politicians of a county; the two apparent groups predict the voting pattern of City Council members (black nodes) on a crucial issue (data from [161])

who was in favor of building the new jail, and the County Auditor were in strong personal opposition, so that the latter publicly opposed the construction. While the diagram indicates that the former Council President is structurally most important (closeness to the center reflects a vertex index called closeness centrality), it is the group structure which is of interest here.

The voting pattern on the jail issue is predicted precisely by the membership to one of two apparent groups of strong internal bonds. Members of the group containing the County Executive voted for the new jail, and those of the group containing the County Auditor voted against. Note that the entire network is very homogeneous with respect to gender, race, and political affiliation, so that these variables are of no influence.

Note also that two council members in the upper right have ties to exactly the same other actors. Similar patterns of relationships suggest that actors have similar (structural) 'roles' in the network. In fact, the network could roughly be reduced to two internally tied parties that are linked by the former Council President.

Methods for defining and finding groups are treated extensively in the second part of the book. Generally speaking, there are two major perspectives on what constitutes a group in a network, namely strong or similar linkages.

In the first three chapters on group-level analysis, a group is identified by strong linkages among its members. These may be based on relatively heavy induced subgraphs (Chapters 6) or relatively high connectivity between each

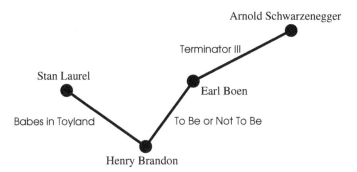

Fig. 1.3. Actors appearing jointly (proving that the co-starring distance of S. Laurel and A. Schwarzenegger is no larger than 3)

pair of members (Chapter 7). Methods for splitting a network into groups based on strong linkage are then reviewed in Chapter 8.

Chapters 9 and 10 focus on groups defined by the pattern of relations that members have. While such groups need not be connected at all, strong internal combined with weak external linkage can be seen as a special case.

Network-Level Analysis (Oracle of Bacon)

Empirical networks representing diverse relations such as linkages among Web pages, gene regulation in primitive organisms, sexual contacts among Swedes, or the power grid of the western United States appear to have, maybe surprisingly, some statistical properties in common.

A very popular example of a network that evolves over time is the movie actor collaboration graph feeding the Oracle of Bacon at Virginia.[1] From all movies stored in the Internet Movie Database[2] it is determined which pairs of actors co-appeared in which movie. The 'Oracle' can be queried to determine (an upper bound on) the co-starring distance of an actor from Kevin Bacon, or in a variant game between any two actors. Except for fun and anecdotal purposes (exemplified in Figure 1.3), actual links between actors are not of primary interest. The fascinating characteristics of this data are on the aggregate level. It turns out, for instance, that Kevin Bacon is on average only three movies apart from any of the more than half a million actors in the database, and that there are more than a thousand actors who have the same property.

Many more properties of this data can be studied. A particularly pertinent observation is, for instance, that in many empirical networks the distribution of at least some statistic obeys a power-law. But the network could also be compared to other empirical networks from related domains (like science collaboration) or fabricated networks for which a suitable model would be required.

---

[1] www.oracleofbacon.org
[2] www.imdb.com

The focus of network-level analysis in general is on properties of networks as a whole. These may reflect, e.g., typical or atypical traits relative to an application domain or similarities occuring in networks of entirely different origin.

Network statistics, reviewed in Chapter 11, are a first indicator of network similarity, often employed in complex systems analysis. In Chapter 12, more rigorous methods for detailed structure comparison of equally (or at least comparatively) sized networks are discussed. A different line of research is the attempt to understand the governing principles of network formation. Chapter 13 is therefore devoted to models for networks with certain properties. A particularly powerful approach to global network analysis is the utilization of spectral properties of matrices defined describing the network. These are described in detail in Chapter 14. The final chapter of this book is devoted to the important question of how sensitive a network is to the loss of some of its elements.

Despite the wealth of material covered, the scope of this book is necessarily limited. No matter which personal background, the reader will easily identify gems from the repertoire of network analysis that have been consciously omitted or woefully overlooked. We nevertheless hope that the book will serve as a useful introduction and handy reference for everyone interested in the methods that drive network analysis.

2 Fundamentals

Ulrik Brandes and Thomas Erlebach

In this chapter we discuss basic terminology and notation for graphs, some fundamental algorithms, and a few other mathematical preliminaries.

We denote the set of integers by $\mathbb{Z}$, the set of real numbers by $\mathbb{R}$, the set of complex numbers by $\mathbb{C}$, and the set of rationals by $\mathbb{Q}$. For a set X of numbers, X^+ denotes the subset of positive numbers in X, and X_0^+ the subset of non-negative numbers. The set of positive integers is denoted by $\mathbb{N} = \mathbb{Z}^+$ and the set of non-negative integers by $\mathbb{N}_0 = \mathbb{Z}_0^+$.

We use $\mathbb{R}^{n \times m}$ to denote the set of all real-valued matrices with n rows and m columns. If the entries of the matrix can be complex numbers, we write $\mathbb{C}^{n \times m}$. The n-dimensional identity matrix is denoted by I_n. The n-dimensional vector with all entries equal to 1 (equal to 0) is denoted by $\mathbf{1}_n$ (by $\mathbf{0}_n$).

For two functions $f : \mathbb{N} \to \mathbb{N}$ and $g : \mathbb{N} \to \mathbb{N}$, we say that f is in $\mathcal{O}(g)$ if there are positive constants $n_0 \in \mathbb{N}$ and $c \in \mathbb{R}^+$ such that $f(n) \leq c \cdot g(n)$ holds for all $n \geq n_0$. Furthermore, we say that f is in $\Omega(g)$ if g is in $\mathcal{O}(f)$. This notation is useful to estimate the asymptotic growth of functions. In particular, running-times of algorithms are usually specified using this notation.

2.1 Graph Theory

We take the term *network* to refer to the informal concept describing an object composed of elements and interactions or connections between these elements. For example, the Internet is a network composed of nodes (routers, hosts) and connections between these nodes (e.g. fiber cables). The natural means to model networks mathematically is provided by the notion of graphs.

A *graph* $G = (V, E)$ is an abstract object formed by a set V of *vertices* (nodes) and a set E of edges (links) that join (connect) pairs of vertices. The vertex set and edge set of a graph G are denoted by $V(G)$ and $E(G)$, respectively. The cardinality of V is usually denoted by n, the cardinality of E by m. The two vertices joined by an edge are called its *endvertices*. If two vertices are joined by an edge, they are *adjacent* and we call them *neighbors*. Graphs can be *undirected* or *directed*. In undirected graphs, the order of the endvertices of an edge is immaterial. An undirected edge joining vertices $u, v \in V$ is denoted by $\{u, v\}$. In directed graphs, each directed edge (arc) has an *origin* (*tail*) and a *destination* (*head*). An edge with origin $u \in V$ and destination $v \in V$ is represented by an ordered pair (u, v). As a shorthand notation, an edge $\{u, v\}$ or (u, v) can also be

U. Brandes and T. Erlebach (Eds.): Network Analysis, LNCS 3418, pp. 7–15, 2005.

denoted by uv. In a directed graph, uv is short for (u, v), while in an undirected graph, uv and vu are the same and both stand for $\{u, v\}$. For a directed graph $G = (V, E)$, the *underlying undirected graph* is the undirected graph with vertex set V that has an undirected edge between two vertices $u, v \in V$ if (u, v) or (v, u) is in E. Graphs that can have directed edges as well as undirected edges are called *mixed graphs*, but such graphs are encountered rarely and we will not discuss them explicitly in the following.

Multigraphs. In both undirected and directed graphs, we may allow the edge set E to contain the same edge several times, i.e., E can be a multiset. If an edge occurs several times in E, the copies of that edge are called *parallel edges*. Graphs with parallel edges are also called *multigraphs*. A graph is called *simple*, if each of its edges is contained in E only once, i.e., if the graph does not have parallel edges. An edge joining a vertex to itself, i.e., an edge whose endvertices are identical, is called a *loop*. A graph is called *loop-free* if it has no loops. We will assume all graphs to be loop-free unless specified otherwise.

Weighted graphs. Often it is useful to associate numerical values (weights) with the edges or vertices of a graph $G = (V, E)$. Here we discuss only edge weights. Edge weights can be represented as a function $\omega : E \to \mathbb{R}$ that assigns each edge $e \in E$ a weight $\omega(e)$. Depending on the context, edge weights can describe various properties such as cost (e.g. travel time or distance), capacity, strength of interaction, or similarity. One usually tries to indicate the characteristics of the edge weights by the choice of the name for the function. In particular, a function assigning (upper) capacities to edges is often denoted by u, especially in the context of network flow problems (see below). In general, we will mostly use ω to denote edge weights that express costs and other letters to denote edge weights that express capacities or interaction strengths. For most purposes, an unweighted graph $G = (V, E)$ is equivalent to a weighted graph with *unit edge weights* $\omega(e) = 1$ for all $e \in E$.

Degrees. The *degree* of a vertex v in an undirected graph $G = (V, E)$, denoted by $d(v)$, is the number of edges in E that have v as an endvertex. If G is a multigraph, parallel edges are counted according to their multiplicity in E. The set of edges that have v as an endvertex is denoted by $\Gamma(v)$. The set of neighbors of v is denoted by $N(v)$. In a directed graph $G = (V, E)$, the *out-degree* of $v \in V$, denoted by $d^+(v)$, is the number of edges in E that have origin v. The *in-degree* of $v \in V$, denoted by $d^-(v)$, is the number of edges with destination v. For weighted graphs, all these notions are generalized by summing over edge weights rather than taking their number. The set of edges with origin v is denoted by $\Gamma^+(v)$, the set of edges with destination v by $\Gamma^-(v)$. The set of destinations of edges in $\Gamma^+(v)$ is denoted by $N^+(v)$, the set of origins of edges in $\Gamma^-(v)$ by $N^-(v)$. If the graph under consideration is not clear from the context, these notations can be augmented by specifying the graph as an index. For example, $d_G(v)$ denotes the degree of v in G. The maximum and minimum degree of an undirected graph $G = (V, E)$ are denoted by $\Delta(G)$ and $\delta(G)$, respectively.

The average degree is denoted by $\bar{d}(G) = \frac{1}{|V|} \sum_{v \in V} d(v)$. An undirected graph $G = (V, E)$ is called *regular* if all its vertices have the same degree, and *r-regular* if that degree is equal to r.

Subgraphs. A graph $G' = (V', E')$ is a *subgraph* of the graph $G = (V, E)$ if $V' \subseteq V$ and $E' \subseteq E$. It is a *(vertex-)induced subgraph* if E' contains all edges $e \in E$ that join vertices in V'. The induced subgraph of $G = (V, E)$ with vertex set $V' \subseteq V$ is denoted by $G[V']$. The *(edge-)induced subgraph* with edge set $E' \subseteq E$, denoted by $G[E']$, is the subgraph $G' = (V', E')$ of G, where V' is the set of all vertices in V that are endvertices of at least one edge in E', If C is a proper subset of V, then $G - C$ denotes the graph obtained from G by deleting all vertices in C and their incident edges. If F is a subset of E, $G - F$ denotes the graph obtained from G by deleting all edges in F.

Walks, paths and cycles. A *walk* from x_0 to x_k in a graph $G = (V, E)$ is an alternating sequence $x_0, e_1, x_1, e_2, x_2, \ldots, x_{k-1}, e_k, x_k$ of vertices and edges, where $e_i = \{x_{i-1}, x_i\}$ in the undirected case and $e_i = (x_{i-1}, x_i)$ in the directed case. The length of the walk is defined as the number of edges on the walk. The walk is called a *path*, if $e_i \neq e_j$ for $i \neq j$, and a path is a *simple path* if $x_i \neq x_j$ for $i \neq j$. A path with $x_0 = x_k$ is a *cycle*. A cycle is a *simple cycle* if $x_i \neq x_j$ for $0 \leq i < j \leq k - 1$.

2.2 Essential Problems and Algorithms

2.2.1 Connected Components

An undirected graph $G = (V, E)$ is *connected* if every vertex can be reached from every other vertex, i.e., if there is a path from every vertex to every other vertex. A graph consisting of a single vertex is also taken to be connected. Graphs that are not connected are called *disconnected*. For a given undirected graph $G = (V, E)$, a *connected component* of G is an induced subgraph $G' = (V', E')$ that is connected and maximal (i.e., there is no connected subgraph $G'' = (V'', E'')$ with $V'' \supset V'$). Checking whether a graph is connected and finding all its connected components can be done in time $\mathcal{O}(n + m)$ using depth-first search (DFS) or breadth-first search (BFS).

A directed graph $G = (V, E)$ is *strongly connected* if there is a directed path from every vertex to every other vertex. A *strongly connected component* of a directed graph G is an induced subgraph that is strongly connected and maximal. The strongly connected components of a directed graph can be computed in time $\mathcal{O}(n+m)$ using a modified DFS [542]. A directed graph is called *weakly connected* if its underlying undirected graph is connected.

2.2.2 Distances and Shortest Paths

For a path p in a graph $G = (V, E)$ with edge weights ω, the weight of the path, denoted by $\omega(p)$, is defined as the sum of the weights of the edges on p. A path from u to v in G is a *shortest path* (with respect to ω) if its weight is the smallest possible among all paths from u to v. The length of a shortest path from u to v, also called the shortest-path distance between u and v, is denoted by $d_{G,\omega}(u,v)$, where the subscripts G and/or ω are usually dropped if no confusion can arise.

The single-source shortest paths problem (SSSP) is defined as follows: Given a graph $G = (V, E)$ with edge weights $\omega : E \to \mathbb{R}$ and a vertex $s \in V$ (the source), compute shortest paths from s to all other vertices in the graph. The problem is only well-defined if the graph does not contain a cycle of negative weight. If the edge weights are non-negative, SSSP can be solved in time $\mathcal{O}(m + n \log n)$ using an efficient implementation of Dijkstra's algorithm [133]. If the edge weights are arbitrary, the Bellman-Ford algorithm uses time $\mathcal{O}(mn)$ to detect a cycle of negative length or, if no such cycle exists, solve the problem. For the special case of unit edge weights, BFS solves the problem in linear time $\mathcal{O}(n + m)$.

In the all-pairs shortest paths problem (APSP), one is given a graph $G = (V, E)$ with edge weights $\omega : E \to \mathbb{R}$ and wants to compute the shortest-path distances for all pairs of nodes. Provided that G does not contain a cycle of negative length, this problem can be solved by the Floyd-Warshall algorithm in time $\mathcal{O}(n^3)$, or by n SSSP computations in time $\mathcal{O}(nm + n^2 \log n)$.

These algorithms work for both directed and undirected graphs.

2.2.3 Network Flow

A *flow network* is given by a directed graph $G = (V, E)$, a function $u : E \to \mathbb{R}$ assigning non-negative capacities to the edges, and two distinct vertices $s, t \in V$ designated as the *source* and the *sink*, respectively. A flow f from s to t, or an *s-t-flow* for short, is a function $f : E \to \mathbb{R}$ satisfying the following constraints:

- Capacity constraints: $\forall e \in E : 0 \le f(e) \le u(e)$
- Balance conditions: $\forall v \in V \setminus \{s,t\} : \sum_{e \in \Gamma^-(v)} f(e) = \sum_{e \in \Gamma^+(v)} f(e)$

The *value* of the flow f is defined as

$$\sum_{e \in \Gamma^+(s)} f(e) - \sum_{e \in \Gamma^-(s)} f(e).$$

The problem of computing a flow of maximum value is called the *max-flow problem*. The max-flow problem can be solved in time $\mathcal{O}(nm \log(n^2/m))$ using the algorithm of Goldberg and Tarjan [252], for example.

For a given graph $G = (V, E)$, a *cut* is a partition $(S, \bar{S})$ of V into two non-empty subsets S and $\bar{S}$. A cut $(S, \bar{S})$ is an *s-t-cut*, for $s, t \in V$, if $s \in S$ and $t \in \bar{S}$. The capacity of a cut $(S, \bar{S})$ is defined as the sum of the capacities of the edges with origin in S and destination in $\bar{S}$. A *minimum s-t-cut* is an s-t-cut whose capacity is minimum among all s-t-cuts. It is easy to see that the value of an s-t-flow can never be larger than the capacity of a s-t-cut. A classical result in the theory of network flows states that the maximum value and the minimum capacity are in fact the same.

Theorem 2.2.1 (Ford and Fulkerson [218]). *The value of a maximum s-t-flow is equal to the capacity of a minimum s-t-cut.*

Algorithms for the max-flow problem can also be used to compute a minimum s-t-cut efficiently. A *minimum cut* in an undirected graph $G = (V, E)$ with edge capacities $u : E \to \mathbb{R}$ is a cut that is an s-t-cut for some vertices $s, t \in V$ and has minimum capacity.

In the *min-cost flow problem*, one is given a directed graph $G = (V, E)$, a non-negative capacity function $u : E \to \mathbb{R}$, a cost function $c : E \to \mathbb{R}$, and a function $b : V \to \mathbb{R}$ assigning each vertex a demand/supply value. Here, a flow is a function $f : E \to \mathbb{R}$ that satisfies the capacity constraints and, in addition, the following version of the balance conditions:

$$\forall v \in V : \sum_{e \in \Gamma^+(v)} f(e) - \sum_{e \in \Gamma^-(v)} f(e) = b(v)$$

The cost of a flow f is defined as $c(f) = \sum_{e \in E} f(e)c(e)$. The problem of computing a flow of minimum cost can be solved in polynomial time.

2.2.4 Graph k-Connectivity

An undirected graph $G = (V, E)$ is called *k-vertex-connected* if $|V| > k$ and $G - X$ is connected for every $X \subset V$ with $|X| < k$. Note that every (non-empty) graph is 0-vertex-connected, and the 1-vertex-connected graphs are precisely the connected graphs on at least two vertices. Furthermore, a graph consisting of a single vertex is connected and 0-vertex-connected, but not 1-vertex-connected. The *vertex-connectivity* of G is the largest integer k such that G is k-vertex-connected. Similarly, G is called *k-edge-connected* if $|V| \geq 2$ and $G - Y$ is connected for every $Y \subseteq E$ with $|Y| < k$. The *edge-connectivity* of G is the largest integer k such that G is k-edge-connected. The edge-connectivity of a disconnected graph and of a graph consisting of a single vertex is 0.

The notions of vertex-connectivity and edge-connectivity can be adapted to directed graphs by requiring in the definitions above that $G - X$ and $G - Y$, respectively, be strongly connected.

Consider an undirected graph $G = (V, E)$. A subset $C \subset V$ is called a *vertex-separator* (or *vertex cutset*) if the number of connected components of $G - C$ is larger than that of G. If two vertices s and t are in the same connected component of G, but in different connected components of $G - C$, then C is called an *s-t-vertex-separator*. *Edge-separators* (*edge cutsets*) and *s-t-edge-separators* are defined analogously. The notion of s-t-separators can be adapted to directed graphs in the natural way: a set of vertices or edges is an s-t-separator if there is no more path from s to t after deleting the set from the graph.

Let $G = (V, E)$ be an undirected or directed graph. Two (directed or undirected) paths p_1 and p_2 from $s \in V$ to $t \in V$ are called *vertex-disjoint* if they do not share any vertices except s and t. They are called *edge-disjoint* if they do not share any edges. By Menger's Theorem (see Chapter 7 for further details), a graph G with at least $k + 1$ vertices is k-vertex-connected if and only if there are k vertex-disjoint paths between any pair of distinct vertices, and a graph G with at least 2 vertices is k-edge-connected if and only if there are at least k edge-disjoint paths between any pair of distinct vertices.

The number of vertex- or edge-disjoint paths between two given vertices in a graph can be computed in polynomial time using network flow algorithms. Therefore, the vertex- and edge-connectivity of a graph can be determined in polynomial time as well. Special algorithms for these problems will be discussed in Chapter 7.

2.2.5   Linear Programming

Let A be a real $m \times n$ matrix, b a real m-dimensional vector, and c a real n-dimensional vector. Furthermore, let $x = (x_1, \ldots, x_n)$ be a vector of n real variables. The optimization problem

$$\max \ c^T x$$
$$\text{s.t.} \ \ Ax \leq b$$
$$x \geq 0$$

is called a *linear program*. It asks to find a real vector x that satisfies the constraints $Ax \leq b$ and $x \geq 0$ (where $\leq$ is to be understood component-wise) and maximizes the objective function $c^T x = \sum_{i=1}^n c_i x_i$. Linear programs with rational coefficients can be solved in time polynomial in the size of the input.

If the variables of a linear program are constrained to be integers, the program is called an *integer linear program*. Computing optimal solutions to integer linear programs is an $\mathcal{NP}$-hard problem (see the next section), and no polynomial-time algorithm is known for this problem.

2.2.6 $\mathcal{NP}$-Completeness

It is important to consider the running-time of an algorithm for a given problem. Usually, one wants to give an upper bound on the running time of the algorithm for inputs of a certain size. If the running-time of an algorithm is $n^{\mathcal{O}(1)}$ for inputs of size n, we say that the algorithm runs in polynomial time. (For graph problems, the running-time is usually specified as a function of n and m, the number of edges and vertices of the graph, respectively.) For many problems, however, no polynomial-time algorithm has been discovered. Although one cannot rule out the possible existence of polynomial-time algorithms for such problems, the theory of $\mathcal{NP}$-completeness provides means to give evidence for the computational intractability of a problem. A decision problem is in the complexity class $\mathcal{NP}$ if there is a non-deterministic Turing machine that solves the problem in polynomial time. Equivalently, for every yes-instance of the problem there is a proof of polynomial size that can be verified in polynomial time. A decision problem is $\mathcal{NP}$-hard if every decision problem in $\mathcal{NP}$ can be reduced to it by a polynomial many-one reduction. Problems that are in $\mathcal{NP}$ and $\mathcal{NP}$-hard are called

$\mathcal{NP}$-complete. An example of an $\mathcal{NP}$-complete problem is SATISFIABILITY, i.e., checking whether a given Boolean formula in conjunctive normal form has a satisfying truth assignment. A polynomial-time algorithm for an $\mathcal{NP}$-hard problem would imply a polynomial-time algorithm for all problems in $\mathcal{NP}$—something that is considered very unlikely. Therefore, $\mathcal{NP}$-hardness of a problem is considered substantial evidence for the computational difficulty of the problem. For optimization problems (where the goal is to compute a feasible solution that maximizes or minimizes some objective function), we say that the problem is $\mathcal{NP}$-hard if the corresponding decision problem (checking whether a solution with objective value better than a given value k exists) is $\mathcal{NP}$-hard. In order to solve $\mathcal{NP}$-hard optimization problems, the only known approaches either settle with approximate solutions or incur a potentially exponential running-time.

2.3   Algebraic Graph Theory

Two directed graphs $G_1 = (V_1, E_1)$ and $G_2 = (V_2, E_2)$ are *isomorphic* (written as $G_1 \simeq G_2$) if there is a bijection $\phi : V_1 \to V_2$ with

$$\forall u, v \in V : (u, v) \in E_1 \Leftrightarrow (\phi(u), \phi(v)) \in E_2 \,.$$

Such a bijection is called an *isomorphism*. An isomorphism that maps a graph onto itself is called an *automorphism*. Usually we consider two graphs to be the same if they are isomorphic. Isomorphism and automorphism for undirected graphs are defined analogously.

The *incidence matrix* (or *node-arc incidence matrix*) of a directed graph $G = (V, E)$ with $V = \{v_1, \ldots, v_n\}$ and $E = \{e_1, \ldots, e_m\}$ is a matrix B with n rows and m columns that has entries $b_{i,j}$ satisfying

$$b_{i,j} = \begin{cases} -1, & \text{if } v_i \text{ is the origin of } e_j \\ 1, & \text{if } v_i \text{ is the destination of } e_j \\ 0, & \text{otherwise} \end{cases}$$

The *adjacency matrix* of a simple directed graph $G = (V, E)$ with $V = \{v_1, v_2, \ldots, v_n\}$ is an $n \times n$ matrix $A(G) = (a_{i,j})_{1 \le i,j \le n}$ with

$$a_{i,j} = \begin{cases} 1, & \text{if } (v_i, v_j) \in E \\ 0, & \text{otherwise} \end{cases}$$

If G is an undirected graph, its adjacency matrix is symmetric and has $a_{i,j} = 1$ if v_i and v_j are adjacent. For weighted graphs, the non-zero entries are $\omega(v_i, v_j)$ rather than 1.

The *Laplacian* of an undirected graph $G = (V, E)$ is an $n \times n$ matrix defined by $L(G) = D(G) - A(G)$, where $D(G)$ is the diagonal matrix that has its i-th diagonal entry equal to $d_G(v_i)$. Note that $L(G) = BB^T$ for any fixed orientation of the edges of G. The *normalized Laplacian* of G is the $n \times n$ matrix defined by $\mathcal{L}(G) = D(G)^{-1/2} L(G) D(G)^{-1/2}$, where $D(G)^{-1/2}$ is the diagonal matrix where the i-th diagonal entry is 0 if $d_G(v_i) = 0$ and $1/\sqrt{d_G(v_i)}$ otherwise.

Let $A \in \mathbb{C}^{n \times n}$ be a matrix. A value $\lambda \in \mathbb{C}$ is called an *eigenvalue* of A if there is a non-zero n-dimensional vector x such that $Ax = \lambda x$. Such a vector x is then called an *eigenvector* of A (with eigenvalue λ). The (multi-)set of all eigenvalues of a matrix is called its *spectrum*. It is equal to the set of the roots of the *characteristic polynomial* of A, where the characteristic polynomial of A is defined as the determinant of $A - \lambda \cdot I_n$.

If A is a real symmetric matrix, all eigenvalues are real. Therefore, the spectra of the adjacency matrix, the Laplacian, and the normalized Laplacian of an undirected graph $G = (V, E)$ are multisets containing n real values. The spectrum of the adjacency matrix $A(G)$ of a graph G is also called the spectrum of G. The spectra of the Laplacian and the normalized Laplacian of G are called the Laplacian spectrum and the normalized Laplacian spectrum of G.

2.4   Probability and Random Walks

A *discrete probability space* is a pair (Ω, Pr), where Ω is a non-empty, finite or countably infinite set and Pr is a mapping from the power set $\mathcal{P}(\Omega)$ of Ω to the real numbers satisfying the following:

- $\mathrm{Pr}[A] \geq 0$, for all $A \subseteq \Omega$
- $\mathrm{Pr}[\Omega] = 1$
- $\mathrm{Pr}\left[\bigcup_{i \in \mathbb{N}} A_i\right] = \sum_{i \in \mathbb{N}} \mathrm{Pr}[A_i]$, for every sequence $(A_i)_{i \in \mathbb{N}}$ of pairwise disjoint sets from $\mathcal{P}(\Omega)$.

We call Ω a *sample space*. Subsets of Ω are called *events*. Note that we write the probability of an event A as $\mathrm{Pr}[A]$ (and not as $\mathrm{Pr}(A)$). The conditional probability of event A given the occurrence of event B is written as $\mathrm{Pr}[A \mid B]$ and is well-defined by $\mathrm{Pr}[A \cap B]/\mathrm{Pr}[B]$ whenever $\mathrm{Pr}[B] \neq 0$.

A random variable X is a mapping from the sample space to the real numbers. The image of X is denoted by $I_X = X(\Omega)$. The expected value of a random variable X is defined as $\mathbb{E}[X] = \sum_{\omega \in \Omega} X(\omega) \mathrm{Pr}[\omega]$. Note that this definition implies $\mathbb{E}[X] = \sum_{x \in X(\Omega)} x \mathrm{Pr}[X = x]$.

A *Markov chain* on state set S, where S can be finite or countably infinite, is given by a sequence $(X_t)_{t \in \mathbb{N}_0}$ of random variables X_t with $I_{X_t} \subseteq S$ and an initial distribution q_0 that maps S to $\mathbb{R}_0^+$ and satisfies $\sum_{s \in S} q_0(s) = 1$. It must satisfy the *Markov condition*, i.e. for all $t > 0$ and all $I \subseteq \{0, 1, \ldots, t-1\}$ and all $i, j, s_k \in S$ we must have:

$$\mathrm{Pr}[X_{t+1} = j \mid X_t = i, \forall k \in I : \ X_k = s_k] = \mathrm{Pr}[X_{t+1} = j \mid X_t = i]$$

In words, the probability distribution of the successor state X_{t+1} depends only on the current state X_t, not on the history of how the chain has arrived in the current state. We interpret X_t as the state of the Markov chain at *time t*. By q_t we denote the probability distribution on the state set S at time t, i.e., q_t is a vector whose i-th entry, for $i \in S$, is defined by $q_t(i) = \mathrm{Pr}[X_t = i]$.

If $\mathrm{Pr}[X_{t+1} = j \mid X_t = i]$ is independent of t for all states $i, j \in S$, the Markov chain is called *homogeneous*. We consider only homogeneous Markov chains with finite state set S in the following. For such Markov chains, the *transition matrix* is defined as the $|S| \times |S|$ matrix $T = (t_{i,j})$ with $t_{i,j} = \Pr[X_{t+1} = j \mid X_t = i]$. The transition matrix is a *stochastic matrix*, i.e., a non-negative matrix in which the entries in each row sum up to 1. Note that the probability distribution q_{t+1} on the state set S at time $t + 1$, viewed as a row vector, can be computed from the probability distribution q_t at time t by $q_{t+1} = q_t \cdot T$, for all $t \geq 0$. This implies that $q_t = q_0 \cdot T^t$ holds for all $t \geq 0$.

A Markov chain is called *irreducible* if for every pair (i, j) of states there exists a $k > 0$ such that $\Pr[X_k = j \mid X_0 = i] > 0$. In other words, a Markov chain is irreducible if every state can be reached from any given state with positive probability. The graph of a Markov chain is defined as the directed graph with vertex set S and edges (i, j) for all i, j with $\Pr[X_{t+1} = j \mid X_t = i] > 0$. A Markov chain is irreducible if and only if its graph is strongly connected.

The *period* of a state $s \in S$ of an irreducible Markov chain is the greatest common divisor of all $k > 0$ such that $\Pr[X_k = s \mid X_0 = s] > 0$. A Markov chain is *aperiodic* if all its states have period 1.

For a given Markov chain with state set S and transition matrix T, a non-negative row vector $\pi = (\pi_s)_{s \in S}$ is called a *stationary distribution* if $\sum_{s \in S} \pi_s = 1$ and $\pi \cdot T = \pi$. Every irreducible Markov chain with finite state set S has a unique stationary distribution. If, in addition, the Markov chain is aperiodic, the probability distribution on the states converges to the stationary distibution independently of the initial distribution, i.e., $\lim_{t \to \infty} q_t = \pi$.

The *hitting time* of state j starting at state i is the expected number of steps the Markov chain makes if it starts in state i at time 0 until it first arrives in state j at some time $t \geq 1$.

A *random walk* in a simple directed graph $G = (V, E)$ is a Markov chain with $S = V$ and:

$$\Pr[X_{t+1} = v \mid X_t = u] = \begin{cases} \frac{1}{d^+(u)}, & \text{if } (u, v) \in E \\ 0, & \text{otherwise} \end{cases}$$

In every step, the random walk picks a random edge leaving the current vertex and follows it to the destination of that edge. The random walk is well-defined only if $d^+(v) \geq 1$ for all $v \in V$. In this case, the transition matrix of the random walk is the stochastic $|V| \times |V|$ matrix $T = (t_{i,j})$ with $t_{i,j} = 1/d^+(i)$ if $(i, j) \in E$ and $t_{i,j} = 0$ otherwise. Note that the Markov chain given by a random walk in a directed graph G is irreducible if and only if G is strongly connected.

Random walks in undirected graphs can be defined analogously.

2.5   Chapter Notes

There are many good textbooks for the topics discussed in this chapter. Graph theory is treated in [145, 67]. An introduction to algorithms can be found in [133]. Network flows are treated in [6]. Linear programming is covered extensively in [505]. The standard reference for the theory of $\mathcal{NP}$-completeness is [240]. A textbook about algebraic graph theory is [247]. An introduction to probability theory is provided by [498].

3 Centrality Indices

Dirk Koschützki,* Katharina Anna Lehmann,* Leon Peeters, Stefan Richter,
Dagmar Tenfelde-Podehl,* and Oliver Zlotowski*

Centrality indices are to quantify an intuitive feeling that in most networks some vertices or edges are more central than others. Many vertex centrality indices were introduced for the first time in the 1950s: e.g., the Bavelas index [50, 51], degree centrality [483] or a first feedback centrality, introduced by Seeley [510]. These early centralities raised a rush of research in which manifold applications were found. However, not every centrality index was suitable to every application, so with time, dozens of new centrality indices were published. This chapter will present some of the more influential, 'classic' centrality indices. We do not strive for completeness, but hope to give a catalog of basic centrality indices with some of their main applications.

In Section 3.1 we will begin with two simple examples to show how centrality indices can help to analyze networks and the situation these networks represent. In Section 3.2 we discuss the properties that are minimally required for a real-valued function on the set of vertices or edges of a graph to be a centrality index for vertices and edges, respectively.

In subsequent Sections 3.3–3.9, various families of vertex and edge centralities are presented. First, centrality indices based on distance and neighborhood are discussed in Section 3.3. Additionally, this section presents in detail some instances of facility location problems as a possible application for centrality indices. Next we discuss the centrality indices based on shortest paths in Section 3.4. These are naturally defined for both, vertices and edges. We decided to present both, vertex and edge centrality indices, in one chapter together since many families of centrality indices are naturally defined for both and many indices can be easily transformed from a vertex centrality to an edge centrality, and vice versa. Up to date there have been proposed many more centrality indices for vertices than for edges. Therefore, we discuss general methods to derive an edge centrality out of the definition of a vertex centrality in Section 3.5. The general approach of vitality measures is also applicable to edges and vertices. We will describe this family in Section 3.6. In Section 3.7, a family of centrality indices is presented that is derived from a certain analogy between information flow and current flow. In Section 3.8 centrality indices based on random processes are presented. In Section 3.9 we present some of the more prominent feedback centralities that evaluate the importance of a vertex by evaluating the centrality of its surrounding vertices.

---

* Lead authors

U. Brandes and T. Erlebach (Eds.): Network Analysis, LNCS 3418, pp. 16–61, 2005.

For many centrality indices it is required that the network at hand be connected. If this is not the case, computing these centralities might be a problem. As an example, shortest paths based centralities encounter the problem that certain vertices are not reachable from vertices in a different component of the network. This yields infinite distances for closeness centrality, and zero shortest-path counts for betweenness centrality. Section 3.10 of this chapter discusses how to deal with these problems in disconnected graphs.

Before we close the chapter we want to discuss a topic that spans the bridge between the analysis of networks on the level of elements and the level of the whole graph. In Section 3.11, we propose a very general method with which a structural index for vertices can be transformed into a structural index for graphs. This is helpful, e.g., in the design of new centrality indices which will be explained on a simple example. We close this chapter with some remarks on the history of centrality indices in Section 3.12.

3.1 Introductory Examples

Election of a leader is a frequent event in many social groups and intuitively, some persons in such an event are more important or 'central' than others, e.g. the candidates. The question is now how centrality indices can help to derive a measure of this intuitive observation. On this first example we want to illustrate that different kind of networks can be abstracted from such a social interaction and we want to show how network analysis with centrality indices may help to identify important vertices of these networks. A second example illustrates how the application of an edge centrality index may help to figure out important edges in a network. Both illustrations underline that there is no centrality index that fits all applications and that the same network may be meaningfully analyzed with different centrality indices depending on the question to be answered.

Before we begin the discussion on the examples, it should be noted that the term 'centrality' is by no means clearly defined. What is it that makes a vertex central and another vertex peripheral? In the course of time there have been different answers to this question. Each of them serves another intuition about the notion of centrality. Centrality can be interpreted as - among other things - 'influence', as 'prestige' or as 'control'. For example, a vertex can be regarded as central if it is heavily required for the transport of information within the network or if it is connected to other important vertices. These few examples from a set of dozens other possibilities show that the interpretation of 'centrality' is heavily dependent on the context.

We will demonstrate the application of three different interpretations on the following example: A school class of 30 students has to elect a class representative and every student is allowed to vote for one other student. We can derive different graph abstractions from this situation that can later be analyzed with different centrality indices. We will first look at a network that represents the voting results directly. In this network vertices represent students and an edge from student A to student B is established if A has voted for B. In such a situation a student could be said to be the more 'central' the more people have voted for him or her. This kind of centrality is directly represented by the number of edges pointing to the corresponding vertex. The so called 'in-degree centrality' is presented in Section 3.3.1.

Another view on the same situation results in another network: In this network an edge between A and B represents that student A has convinced student B to vote for his or her favorite candidate. We will call this network an 'influence network'. Let us assume that the class is mainly split into two big groups X and Y. Let some person have social relationships to members from both groups. If this person has a favorite candidate from group X and convinces a big part of group Y to vote for this candidate, he or she is 'central' because he or she mediates the most information between both groups. With this argument we can say that a vertex in the given influence network is the more central the more it is needed to transport the opinion of others. A family of centrality indices that tries to capture this intuition of 'being between groups' is the family of betweenness centrality indices, presented in Sections 3.4.2, 3.6.1 and 3.8.2.

In yet another perspective we could view the general social network of the class: Who is friends with whom? Someone who is a friend of an important person could be regarded as more important than someone having friends with low social prestige. The centrality of a vertex in this kind of network is therefore given by the centrality of adjacent vertices. This kind of 'feedback centrality' is captured by many centrality indices that are presented in Section 3.9.

In analogy to the centrality of vertices, some of the edges in a network can be viewed as being more important than others. We will illustrate this on a commonly used network, the Internet. Looking at the backbone of the Internet it is clear that the cables between servers on different continents are few and thus very important for the functionality of the system. This importance stems from the enormous data flow through the intercontinental cables that had to be redirected if one of these cables was out of service. There are mainly two different approaches to measure the centrality of an edge in a network: The first counts the number of substructures like traversal sets or the set of shortest paths in the graph on which an edge participates. An example for this approach is the betweenness centrality of edges, presented in Section 3.4.2. The second approach is based on the idea of measuring how much a certain network parameter is changed if the edge is removed. An example for this approach is the flow betweenness vitality, presented in Section 3.6.1.

We have shown for two examples that very different ideas of centrality can lead to centrality indices that help to analyze the situation represented by the given network. It is important to note that none of these measures is superior to the others. Every one is appropriate for some but not all questions in network analysis.

3.2   A Loose Definition

Before presenting any centrality indices, we first have to give a definition for centrality indices.[1] Historically there is no commonly accepted definition of what a centrality index is, and almost everybody introduced his or her centrality without giving a strict definition for centrality in general. Thus, here we will just state the least common ground for all centralities presented in the following sections. In Section 5.4 we will give some classes of centralities that follow much stricter definitions.

The intuition about a centrality is that it denotes an order of importance on the vertices or edges of a graph by assigning real values to them. As we have pointed out in the introduction to this chapter, the notion of 'importance' is by no means unambiguous. Nonetheless, as a minimal requirement we demand that the result of a centrality index is only depending on the structure of the graph. This demand is stated in the following definition of a structural index. Every of the centrality indices presented here is a structural index but it is important to note that not every structural index will be accepted as a centrality index. Section 5.4 will also show that to date there is no stricter definition that captures all of the introduced centrality indices.

Recall, that two graphs $G_1 = (V_1, E_1)$ and $G_2 = (V_2, E_2)$ are isomorphic $(G_1 \simeq G_2)$ if there exists a one-to-one mapping $\phi\colon V_1 \to V_2$ such that (u, v) is an edge in E_1 if and only if $(\phi(u), \phi(v))$ is an edge in E_2 (cf. Section 2.3).

Definition 3.2.1 (Structural Index). *Let $G = (V, E)$ be a weighted, directed or undirected multigraph and let X represent the set of vertices or edges of G, respectively. A real-valued function s is called a structural index if and only if the following condition is satisfied: $\forall x \in X\colon G \simeq H \implies s_G(x) = s_H(\phi(x))$, where $s_G(x)$ denotes the value of $s(x)$ in G.*

A centrality index c is required to be a structural index and thus induces at least a semi-order on the set of vertices or edges, respectively. By this order we can say that $x \in X$ is at least as central as $y \in X$ with respect to a given centrality c if $c(x) \geq c(y)$. Note that, in general, the difference or ratio of two centrality values cannot be interpreted as a quantification of how much more central one element is than the other.

The definition of a structural index expresses the natural requirement that a centrality measure must be invariant under isomorphisms. In particular, this condition implies that a centrality measure is also invariant under automorphisms.

3.3   Distances and Neighborhoods

In this section we will present centrality indices that evaluate the 'reachability' of a vertex. Given any network these measures rank the vertices according to the

---

[1] Centrality index will be used synonymously with centrality measure and, shortly, centrality.

number of neighbors or to the cost it takes to reach all other vertices from it. These centralities are directly based on the notion of distances within a graph, or on the notion of neighborhood, as in the case of the degree centrality. We start with this very basic index, the degree centrality. Other centralities, like eccentricity or closeness, will be presented in the light of a special application, the facility location problem.

3.3.1 Degree

The most simple centrality is the degree centrality $c_D(v)$ of a vertex v that is simply defined as the degree $d(v)$ of v if the considered graph is undirected. In directed networks two variants of the degree centrality may be appropriate: the in-degree centrality $c_{iD}(v) = d^-(v)$ and the out-degree centrality $c_{oD}(v) = d^+(v)$. The degree centrality is, e.g., applicable whenever the graph represents something like a voting result. These networks represent a static situation and we are interested in the vertex that has the most direct votes or that can reach most other vertices directly. The degree centrality is a local measure, because the centrality value of a vertex is only determined by the number of its neighbors. In the next section we investigate global centrality measures and consider their applications in a special set of problems, namely Facility Location Problems.

3.3.2 Facility Location Problems

Facility location analysis deals with the problem of finding optimal locations for one or more facilities in a given environment. Location problems are classical optimization problems with many applications in industry and economy. The spatial location of facilities often take place in the context of a given transportation, communication, or transmission system, which may be represented as a network for analytic purposes.

A first paradigm for location based on the minimization of transportation costs was introduced by Weber [575] in 1909. However, a significant progress was not made before 1960 when facility location emerged as a research field.

There exist several ways to classify location problems. According to Hakami [271] who considered two families of location problems we categorize them with respect to their objective function. The first family consists of those problems that use a minimax criterion. As an example, consider the problem of determining the location for an emergency facility such as a hospital. The main objective of such an emergency facility location problem is to find a site that minimizes the maximum response time between the facility and the site of a possible emergency. The second family of location problems considered by Hakimi optimizes a minisum criterion which is used in determining the location for a service facility like a shopping mall. The aim here is to minimize the total travel time. A third family of location problems described for example in [524, 527] deals with the location of commercial facilities which operate in a competitive environment. The goal of a competitive location problem is to estimate the market share captured by each competing facility in order to optimize its location.

Our focus here is not to treat all facility location problems. The interested reader is referred to a bibliography devoted to facility location analysis [158]. The aim of this section is to introduce three important vertex centralities by examining location problems. In the subsequent section we investigate some structural properties of the sets of most central indices that are given by these centrality indices.

The definition of different objectives leads to different centrality measures. A common feature, however, is that each objective function depends on the distance between the vertices of a graph. In the following we assume that $G = (V, E)$ is a connected undirected graph with at least two vertices and we suppose that the distance $d(u, v)$ between two vertices u and v is defined as the length of the shortest path from u to v (cf. in Section 2.2.2). These assumptions ensure that the following centrality indices are well defined. Moreover, for reasons of simplicity we consider G to be an unweighted graph, i.e., all edge weights are equal to one. Of course, all indices presented here can equally well be applied to weighted graphs.

Eccentricity. The aim of the first problem family is to determine a location that minimizes the maximum distance to any other location in the network. Suppose that a hospital is located at a vertex $u \in V$. We denote the maximum distance from u to a random vertex v in the network, representing a possible incident, as the eccentricity $e(u)$ of u, where $e(u) = \max\{d(u, v): v \in V\}$. The problem of finding an optimal location can be solved by determining the minimum over all $e(u)$ with $u \in V$. In graph theory, the set of vertices with minimal eccentricity is denoted as the center of G (cf. Section 3.3.3). The concept is illustrated in Figure 3.1. The eccentricity values are shown and the most central vertices are highlighted.

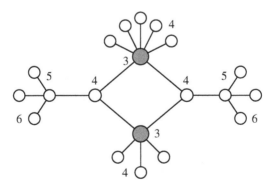

Fig. 3.1. Eccentricity values of a graph. Vertices in the center are colored in grey

Hage and Harary [278] proposed a centrality measure based on the eccentricity

$$c_E(u) = \frac{1}{e(u)} = \frac{1}{\max\{d(u,v)\colon v \in V\}}.$$ (3.1)

This measure is consistent with our general notion of vertex centrality, since $e(u)^{-1}$ grows if the maximal distance of u decreases. Thus, for all vertices $u \in V$ of the center of G: $c_E(u) \geq c_E(v)$ for all $v \in V$.

Closeness. Next we consider the second type of location problems – the minisum location problem, often also called the median problem or service facility location problem. Suppose we want to place a service facility, e.g., a shopping mall, such that the total distance to all customers in the region is minimal. This would make traveling to the mall as convenient as possible for most customers.

We denote the sum of the distances from a vertex $u \in V$ to any other vertex in a graph $G = (V, E)$ as the total distance[2] $\sum_{v \in V} d(u, v)$. The problem of finding an appropriate location can be solved by computing the set of vertices with minimum total distance. In Figure 3.2 the total distances for all vertices are shown and the vertices with minimal total distance are highlighted.

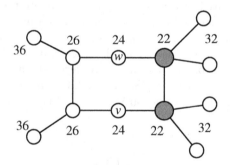

Fig. 3.2. Total distances of a graph. Lowest valued vertices are colored in grey. Note, the vertices v and w are more important with respect to the eccentricity

In social network analysis a centrality index based on this concept is called closeness. The focus lies here, for example, on measuring the closeness of a person to all other people in the network. People with a small total distance are considered as more important as those with a high total distance. Various closeness-based measures have been developed, see for example [500, 51, 52, 433, 558, 451, 88]. In Section 3.10 we outline a measures developed for digraphs. The most commonly employed definition of closeness is the reciprocal of the total distance

---

[2] In [273], Harary used the term status to describe a status of a person in an organization or a group. In the context of communication networks this sum is also called transmission number.

$$c_C(u) = \frac{1}{\sum_{v \in V} d(u,v)}. \tag{3.2}$$

In our sense this definition is a vertex centrality, since $c_C(u)$ grows with decreasing total distance of u and it is clearly a structural index.

Before we discuss the competitive location problem, we want to mention the radiality measure and integration measure proposed by Valente and Foreman [558]. These measures can also be viewed as closeness-based indices. They were developed for digraphs but an undirected version is applicable to undirected connected graphs, too. This variant is defined as

$$c_R(u) = \frac{\sum_{v \in V} (\Delta_G + 1 - d(u,v))}{n - 1} \tag{3.3}$$

where Δ_G and n denote the diameter of the graph and the number of vertices, respectively. The index measures how well a vertex is integrated in a network. The better a vertex is integrated the closer the vertex must be to other vertices. The primary difference between c_C and c_R is that c_R reverses the distances to get a closeness-based measure and then averages these values for each vertex.

Centroid Values. The last centrality index presented here is used in competitive settings: Suppose each vertex represents a customer in a graph. The service location problem considered above assumes a single store in a region. In reality, however, this is usually not the case. There is often at least one competitor offering the same products or services. Competitive location problems deal with the planning of commercial facilities which operate in such a competitive environment. For reasons of simplicity, we assume that the competing facilities are equally attractive and that customers prefer the facility closest to them. Consider now the following situation: A salesman selects a location for his store knowing that a competitor can observe the selection process and decide afterwards which location to select for her shop. Which vertex should the salesman choose?

Given a connected undirected graph G of n vertices. For a pair of vertices u and v, $\gamma_u(v)$ denotes the number of vertices which are closer to u than to v, that is $\gamma_u(v) = |\{w \in V : d(u,w) < d(v,w)\}|$. If the salesman selects a vertex u and his competitor selects a vertex v, then he will have $\gamma_u(v) + \frac{1}{2}(n - \gamma_u(v) - \gamma_v(u)) = \frac{1}{2}n + \frac{1}{2}(\gamma_u(v) - \gamma_v(u))$ customers. Thus, letting $f(u,v) = \gamma_u(v) - \gamma_v(u)$, the competitor will choose a vertex v which minimizes $f(u,v)$. The salesman knows this strategy and calculates for each vertex u the worst case, that is

$$c_F(u) = \min\{f(u,v) : v \in V - u\}. \tag{3.4}$$

$c_F(u)$ is called the centroid value and measures the advantage of the location u compared to other locations, that is the minimal difference of the number of customers which the salesman gains or loses if he selects u and a competitor chooses an appropriate vertex v different from u.

In Figure 3.3 an example is shown where the centroid vertex is highlighted. Notice that for each vertex $u \in V$ in graph shown in Figure 3.4 $c_F(u) \leq -1$.

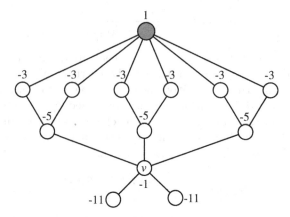

Fig. 3.3. A graph with one centroid vertex. Note that v is the vertex with maximal closeness centrality

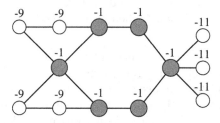

Fig. 3.4. All centroid values are negative. There is no profitable location for the salesman

Here, the salesman loses his advantage to choose as first. The strategy "choose after the leader has chosen" would be optimal.

Also the centroid measure is a structural index according to Definition 3.2.1. But in contrast to eccentricity and closeness, centroid values can be negative as well.

3.3.3   Structural Properties

In this section we will investigate several structural properties for the distance-based vertex centralities introduced in Section 3.3.2. Using Definition 3.2.1 the set of maximum centrality vertices $\mathcal{S}_c(G)$ of G with respect to a given vertex centrality c is given by

$$\mathcal{S}_c(G) = \{u \in V : \forall v \in V \; c(u) \geq c(v)\}. \tag{3.5}$$

Center of a Graph. In Section 3.3.2 the eccentricity of a vertex $u \in G$ was defined as $e(u) = \max\{d(u,v) : v \in V\}$. Recall, that by taking the minimum over all $e(u)$ we solve the emergency location problem. In graph theory, this minimum is called the radius $r(G) = \min\{e(u)\colon u \in V\}$. Using the radius of G the center $\mathcal{C}(G)$ of a graph G is

$$\mathcal{C}(G) = \{u \in V : r(G) = e(u)\}. \tag{3.6}$$

It is easy to show that $\mathcal{S}_{c_E}(G) = \mathcal{C}(G)$. Clearly, every undirected connected graph has a non-empty center. But where are the vertices of the center located? A basic result concerning the center of a tree is due to Jordan [336]

Theorem 3.3.1. *For any tree, the center of a tree consists of at most two adjacent vertices.*

Proof. The result is trivial if the tree consists of at most two vertices. We show that any other tree T has the same center as the tree T' which is obtained from T by removing all leaves. For each vertex u of T, only a leaf can be an eccentric vertex. Aa vertex u is an eccentric vertex of a vertex v if $d(u,v) = e(v)$. Since the eccentricity of each $u \in T'$ is one less than its eccentricity in T, T and T' have the same center. If the process of removing leaves is continued, we successively obtain trees having the same center as T. Finally, we obtain a subtree of T which consists of either one vertex or a pair of adjacent vertices. $\qquad\square$

The proof shows that it is possible to determine the center without computing the vertex eccentricities. The following generalization of Theorem 3.3.1 due to Harary and Norman [281] deals with the location of the center in a connected separable graph, i.e., a graph which contains a cut-vertex. Recall, a cut-vertex of a graph is a vertex whose removal increases the number of components, i.e., if u is a cut-vertex of a connected graph G, then $G - u$ is disconnected. We call a graph 2-vertex-connected if G contains no cut-vertices (cf. Section 2.2.4). Note, each vertex of a graph distinct from a cut-vertex lies in exactly one 2-vertex-connected subgraph, and each cut-vertex lies in more than one.

Theorem 3.3.2. *Let G be a connected undirected graph. There exists a 2-vertex-connected subgraph in G containing all vertices of $\mathcal{C}(G)$.*

Proof. Suppose there is no 2-vertex-connected subgraph in G containing all the vertices of $\mathcal{C}(G)$. Then G has a cut-vertex u such that $G - u$ decomposes into the subgraphs G_1 and G_2 each of them containing at least one vertex of $\mathcal{C}(G)$. Let v be an eccentric vertex of u and P the corresponding shortest path between u and v of length $e(u)$. Then v must lie in G_1 or G_2, say G_2. Furthermore there exists at least one vertex w in G_1 which does not belong to P. Now, let $w \in \mathcal{C}(G)$ and let P' be a shortest path in G between w and u. Then $e(w) \geq d(w,u) + d(u,v) \geq 1 + e(u)$. So w does not belong to the center of G, a contradiction. Thus, there must be a 2-vertex-connected subgraph containing all vertices of center of G. $\quad\square$

Figure 3.1 in Section 3.3.2 shows a graph consisting of fourteen 2-vertex-connected subgraphs consisting of two vertices and one 2-vertex-connected subgraph in the middle containing the two central vertices.

Median of a Graph. The service facility problem presented in Sect. 3.3.2 was solved by determining the set of vertices with minimum total distance. If the minimum total distance of G is denoted by $s(G) = \min\{s(u)\colon u \in V\}$, the median $\mathcal{M}(G)$ of G is given by

$$\mathcal{M}(G) = \{u \in V \colon s(G) = s(u)\} \,. \tag{3.7}$$

Clearly $\mathcal{S}_{cc}(G) = \mathcal{M}(G)$. Truszczyński [552] studied the location of the median in a connected undirected graph.

Theorem 3.3.3. *The median of a connected undirected graph G lies within a 2-vertex-connected subgraph of G.*

Similar to the center of a tree Theorem 3.3.3 implies the existence of at least one 2-vertex-connected subgraph containing the median of a tree.

Corollary 3.3.4. *The median of a tree consists of either a single vertex or a pair of adjacent vertices.*

The graph in Figure 3.2 contains a 2-vertex-connected subgraph of six vertices containing the median. Moreover, the example illustrates that $\mathcal{C}(G) \cap \mathcal{M}(G) = \emptyset$ is possible. Let $\langle \mathcal{M}(G) \rangle$ and $\langle \mathcal{C}(G) \rangle$ denote the subgraphs induced by $\mathcal{M}(G)$ and $\mathcal{C}(G)$, respectively. The results due to Hendry [293] and Holbert [300] show that the center and median can be arbitrarily far apart.

Theorem 3.3.5. *Let H_1 and H_2 be two connected undirected graphs. For any integer $k > 0$, there exists a connected undirected graph G, such that $\langle \mathcal{M}(G) \rangle \simeq H_1$, $\langle \mathcal{C}(G) \rangle \simeq H_2$, and the distance between $\mathcal{M}(G)$ and $\mathcal{C}(G)$ is at least k.*

This result is not surprising, because the center and the median represent solution sets of distinct objective functions.

Centroid of a Graph. The computation of the centroid of a graph is a maximin optimization problem. In Sect. 3.3.2 we have shown the relation to a competitive location problem. We defined the centroid value for a given vertex u by $c_F(u) = \min\{f(u,v)\colon v \in V - u\}$. In addition we call the objective function value $f(G) = \max\{c_F(u)\colon u \in V\}$ the centroid value of G and denote by

$$\mathcal{Z}(G) = \{u \in V \colon f(G) = c_F(u)\} \tag{3.8}$$

the set of vertices representing the centroid of G. With it the set $\mathcal{Z}(G)$ consists of all appropriate locations for the competitive location problem considered in Section 3.3.2.

We now focus on the location of the centroid in a graph. First we assume the graph is an undirected tree $T = (V, E)$. Let u be vertex of T. A branch of u is a maximal subtree containing u as a leaf. The number of branches at u is equal to the degree of u. The branch weight of u is the maximum number of edges among all branches of u. The vertex u is called a branch weight centroid vertex if u has minimum branch weight and the branch weight centroid of T consists of all such vertices. Zenlinka [594] has shown that the branch weight centroid of T is identical with its median. Slater [524] used this result to show

Theorem 3.3.6. *For any tree the centroid and the median are identical.*

Theorem 3.3.6 and Corollary 3.3.4 together imply that the centroid of a tree consists of either a single vertex or a pair of adjacent vertices. Smart and Slater [527] also studied the relative location of the centroid in a connected undirected graph. The following Theorem is a generalization of Theorem 3.3.6.

Theorem 3.3.7. *For any connected undirected graph, the median and the centroid lie in the same 2-vertex-connected subgraph.*

Reconsider the graph in Fig. 3.3. The median and the centroid lie within the subgraph but $\mathcal{Z}(G) \cap \mathcal{M}(G) = \emptyset$. Let $\langle \mathcal{Z}(G) \rangle$ be the graph induced by $\mathcal{Z}(G)$. Smart and Slater [527] have shown the following.

Theorem 3.3.8. *Let H_1 and H_2 be to connected undirected graphs. For any integer $k \geq 4$, there exists a connected undirected graph G, such that $\langle \mathcal{Z}(G) \rangle \simeq H_1$, $\langle \mathcal{M}(G) \rangle \simeq H_2$, and the distance between $\mathcal{Z}(G)$ and $\mathcal{M}(G)$ is at least k.*

Furthermore, Smart and Slater [527] proved that the center, the median, and the centroid can be arbitrarily far apart in a connected undirected graph. In Fig. 3.5 an example is given where all sets are pairwise distinct. The following result summarizes Theorems 3.3.5 and 3.3.8.

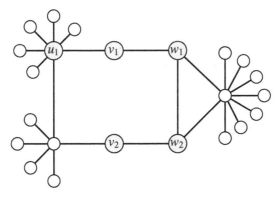

Fig. 3.5. $\mathcal{C}(G) = \{v_1, v_2\}$, $\mathcal{M}(G) = \{u_1\}$, and $\mathcal{Z}(G) = \{w_1, w_2\}$ are pairwise distinct

Theorem 3.3.9. *For three connected undirected graphs H_1, H_2, and H_3, and any integer $k \geq 4$, there exists an undirected connected graph G such that $\langle \mathcal{C}(G) \rangle \simeq H_1$, $\langle \mathcal{M}(G) \rangle \simeq H_2$, $\langle \mathcal{Z}(G) \rangle \simeq H_3$, and the distances between any two of them is at least k.*

Some of concepts presented here can be extended to digraphs. Chartrand et al. [115] showed that the result of Theorem 3.3.5 also holds for digraphs.

3.4 Shortest Paths

This section presents centrality indices that are based on the set of shortest paths in a graph. Shortest paths are defined on vertices as well as on edges and such, some centrality indices were first introduced as vertex centralities and later adapted as edge centralities. In the following, we will sometimes make a general statement regarding vertices and edges equally. We will call a vertex v or an edge e (graph) 'element' and denote the centrality of an element in general by x. The first two indices, stress and betweenness centrality of an element x, are based on the (relative) number of shortest paths that contain x. The last centrality index is only defined on edges and based on traversal sets. All three centrality indices can be defined on weighted or unweighted and directed or undirected and simple or multi graphs. For simplification we will discard any information about the underlying graph in the notation for a given centrality. Thus, c_X might denote the centrality indices of a weighted, undirected graph or any other combination of weights, direction and edge multiplicity. Note that the set of all shortest paths has to be computed in a preprocessing step with the appropriate algorithm, depending on the combination of these graph properties.

3.4.1 Stress Centrality

The first centrality index based on enumeration of shortest paths is stress centrality $c_S(x)$, introduced in [519]. The author was concerned with the question how much 'work' is done by each vertex in a communication network. It is clear that communication or transport of goods will follow different kinds of paths in a social network. Nonetheless, the author of [519] models the set of paths used for communication as the set of shortest paths. The assumption is that counting the number of shortest path that contain an element x gives an approximation of the amount of 'work' or 'stress' the element has to sustain in the network. With this, an element is the more central the more shortest paths run through it. The formal definition is given by:

$$c_S(v) = \sum_{s \neq v \in V} \sum_{t \neq v \in V} \sigma_{st}(v) \tag{3.9}$$

where $\sigma_{st}(v)$ denotes the number of shortest paths containing v. The definition given in [519] is not rigorous, but in analogy to the betweenness centrality all shortest paths that either start or end in v are not accounted for this centrality index. The calculation of this centrality index is given by a variant of a simple all-pairs shortest-paths algorithm that not only calculates one shortest path but all shortest paths between any pair of vertices. More about the algorithm for this centrality can be found in Section 4.2.1.

Although this centrality was designed to measure stress on vertices, the same definition can be applied for edges:

$$c_S(e) = \sum_{s \in V} \sum_{t \in V} \sigma_{st}(e) \tag{3.10}$$

where $\sigma_{st}(e)$ denotes the number of shortest paths containing edge e. In both cases stress centrality measures the amount of communication that passes an element in an all-to-all scenario. More precisely, it is not only an all-to-all scenario but every vertex sends as many goods or information units to every other vertex as there are shortest paths between them and stress centrality measures the according stress.

We next want to show how the stress centrality value of a vertex v is related to the stress centrality indices of the edges incident to v.

Lemma 3.4.1 (Relation between $c_S(v)$ and $c_S(e)$). *In a directed graph $G = (V, E)$, stress centrality on vertices and edges are related by*

$$c_S(v) = \frac{1}{2} \sum_{e \in \Gamma(v)} c_S(e) - \sum_{v \neq s \in V} \sigma_{sv} - \sum_{v \neq t \in V} \sigma_{vt} \tag{3.11}$$

for all $v \in V$.

Proof. Consider any shortest path connecting a pair $s \neq t \in V$. It contributes a value of 1 to the stress of each of its vertices and edges. Summing the contribution of a path over all edges that are incident to a vertex v thus yields twice its contribution to v itself if $v \in V \backslash \{s, t\}$, and 1 otherwise. The sum of contributions of all shortest paths to edges incident to a common vertex v hence satisfies the above relation, since v is $\sum_{v \neq s \in V} \sigma_{sv} + \sum_{v \neq t \in V} \sigma_{vt}$ times an endvertex of any shortest path. □

3.4.2 Shortest-Path Betweenness Centrality

Shortest-path betweenness centrality can be viewed as some kind of relative stress centrality. Here, we will first define it and then discuss the motivation behind this centrality index: Let $\delta_{st}(v)$ denote the fraction of shortest paths between s and t that contain vertex v:

$$\delta_{st}(v) = \frac{\sigma_{st}(v)}{\sigma_{st}} \tag{3.12}$$

where σ_{st} denotes the number of all shortest-path between s and t. Ratios $\delta_{st}(v)$ can be interpreted as the probability that vertex v is involved into any communication between s and t. Note, that the index implicitly assumes that all communication is conducted along shortest paths. Then the betweenness centrality $c_B(v)$ of a vertex v is given by:

$$c_B(v) = \sum_{s \neq v \in V} \sum_{t \neq v \in V} \delta_{st}(v) \qquad (3.13)$$

As for stress centrality, the shortest paths ending or starting in v are explicitly excluded. The motivation for this is that the betweenness centrality of a vertex measures the control over communication between others.

The betweenness centrality index was introduced in [32, 226] and has found a wide field of applications. In [226] this new centrality index was introduced because it is problematic to apply the closeness centrality to a disconnected graph: the distance between two vertices in different components is usually set to infinity. With this, the closeness centrality (see subsection 3.2) in disconnected graphs will give no information because each vertex is assigned the same centrality value, namely $1/\infty$. We will discuss some resorts to this problem in Section 3.10.

The betweenness centrality does not suffer from this problem: Any pair of vertices s and t without any shortest path from s to t just will add zero to the betweenness centrality of every other vertex in the network.

Betweenness centrality is similar to stress centrality introduced in [519], but instead of counting the absolute number of shortest paths, the shortest-path betweenness centrality sums up the relative number of shortest paths for each pair of endvertices. These are interpreted as the extent to which a vertex v controls the communication between such pairs. Figure 3.6 gives an example why this might be more interesting than the absolute number of shortest paths. It shows two tripartite graphs in which the middle layer mediates all communication between the upper and the lower layer. The stress centrality of vertices in the middle layer is the same in both graphs but the removal of the middle vertex on the right would disconnect the whole system whereas in the right graph the removal of a single vertex would not. This is because the former has full responsibility for the communication in its graph whereas on the left side every vertex just bears one third of it.

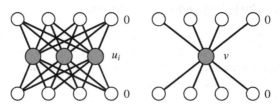

Fig. 3.6. $c_S(u_i) = 16$ and $c_B(u_i) = \frac{1}{3}$, $i = 1, 2, 3$ and $c_S(v) = 16$ but $c_B(v) = 1$. The graph shows on an example that stress centrality is not designed to evaluate how much communication control a vertex has

In [32] the shortest-path betweenness centrality – here called 'rush' – is viewed as a flow centrality: "The rush in an element is the total flow through that element, resulting from a flow between each pair of vertices". In this sense,

$\delta_{st}(v)$ is interpreted as the amount of flow that passes if one unit of flow is sent from s to t along shortest paths, and with a special division rule. In [32] the 'rush' is also defined for edges with $\delta_{st}(e)$ as the flow over edge e:

$$\delta_{st}(e) = \frac{\sigma_{st}(e)}{\sigma_{st}} \tag{3.14}$$

For reasons of consistency we will denote the resulting centrality not as 'rush on edges' but as the betweenness centrality $c_B(e)$ of edge e:

$$c_B(e) = \sum_{s \in V} \sum_{t \in V} \delta_{st}(e) . \tag{3.15}$$

Variants of Shortest-Path Betweenness Centrality. In [111, 580] some variants of the shortest-path betweenness centrality have been introduced. The authors generalize the approach of betweenness centrality by changing the set of paths $P(s,t)$ on which the betweenness centrality is evaluated. Instead of just using the set of all shortest paths between s and t any other set can be used for this variant. The general pattern is always the same: For each node pair s and t compute the fraction of paths in $P(s,t)$ that contain an element from the sum of all paths between s and t. To get the betweenness centrality $c_B(P(s,t))$ on a specified path set p sum over the terms for all node pairs. In [580], a number of possible path sets $P(s,t)$ was defined, as e.g. the set of k-shortest paths, i.e. the set of all paths not longer than $k \in \mathbb{N}$ or the set of k-shortest, node-disjoint paths. The according betweenness centralities did not get any special name but for reasons of consistency we will denote them as k-shortest paths and k-shortest vertex-disjoint paths betweenness centrality.

The authors in [111] were motivated by the fact that the betweenness centrality is not very stable in dynamic graphs (see also our discussion of the stability and sensitivity of centrality indices in Section 5.5). The removal or addition of an edge might cause great perturbations in the betweenness centrality values. To eliminate this, $P(s,t)$ was defined to contain all paths between a node pair s and t that are not longer than $(1+\epsilon)d(s,t)$. The resulting betweenness centrality for nodes and edges has been named ϵ-betweenness centrality. The idea behind this centrality index seems reasonable but analytical or empirical results on the stability of this index were not given.

Other variants of the general betweenness centrality concept are fundamentally different in their approach and calculation. We will discuss the flow betweenness centrality in Section 3.6.1 and the random-walk betweenness centrality in Section 3.8.2.

In the following theorem we state the relation between the edge and vertex betweenness centrality $c_B(e)$ and $c_B(v)$ of vertices and edges incident to each other:

Lemma 3.4.2 (Relation between $c_B(v)$ and $c_B(e)$). *In a directed graph $G = (V, E)$, shortest-path betweenness on vertices and edges are related by*

$$c_B(v) = \sum_{e \in \Gamma^+(v)} c_B(e) - (n-1) = \sum_{e \in \Gamma^-(v)} c_B(e) - (n-1) \qquad (3.16)$$

for all $v \in V$.

Proof. Consider any shortest path connecting a pair $s \neq t \in V$. It contributes exactly $\frac{1}{\sigma_{st}}$ to the betweenness of its vertices and edges. Summing the contribution of a path over all incoming (or outgoing) edges of a vertex v thus equals its contribution to v itself if $v \in V \setminus \{s, t\}$, and $\frac{1}{\sigma_{st}}$ otherwise. The sum of contributions of all shortest paths to edges incident to a common vertex v hence satisfies the above relation, since v is $(n-1)$ times the first (last) vertex of paths to some vertex t (from some vertex s).   □

3.4.3   Reach

In 2004, Ron Gutman [266] published a new approach to shortest path computation in hierarchical networks like road maps, for example. It is based on employing Dijkstras algorithm or the A* algorithm alternatively on a select subset of nodes. More specifically, only nodes having a high *reach* are considered. The concept is defined as follows:

Definition 3.4.3. *Given*

- *a directed graph $G = (V, E)$ with a nonnegative distance function $m : E \to \mathbb{R}^+$, which is called* reach metric
- *a path P in G starting at node s and ending at node t*
- *a node v on P*

the reach *of v on P is defined as $r(v, P) := \min\{m(s, v, P), m(v, t, P)\}$, the minimum of the distance from s to v and the distance from v to t, following path P according to the reach metric. The* reach *of v in G, $r(v, G)$ is the maximum value of $r(v, Q)$ over all least-cost paths Q in G containing v.*

When performing a Dijkstra-like shortest-path search towards a target t, nodes are only enqueued if they pass $test(v)$, where $test(v) := r(v, G) \geq m(P) \vee r(v, G) \geq d(v, t)$. That is v is only disregarded if its reach is too small for it to lie on a least-cost path a distance $m(P)$ – denoting the length of the computed path from the origin s to v at the time v is to be inserted into the priority queue – from s and at a straight-line distance $d(v, t)$ from the destination. Note that this requires a distance function that is consistent with reach metric, such that on a path P from u to v, the path length $m(P) = m(u, v, P)$ must be at least $d(u, v)$.

At first, this reach centrality does not seem to make sense in order to simplify computation of shortest paths, since there is no obvious way of computing $r(v, G)$ for all nodes without solving an all pairs shortest path problem in the first place. However, Gutman goes on to show that in the above algorithm, even an upper bound for $r(v, G)$ suffices to preserve guaranteed shortest paths. Naturally, using an upper bound increases the number of nodes that need to be enqueued. The author gives a sophisticated algorithm that yields practically useful bounds in a more feasible time complexity. Unfortunately, both quality and complexity are only empirically analyzed.

3.4.4 Traversal Sets

For $G = (V, E)$ and an edge $e \in E$ we call

$$T_e = \{(a, b) \in V \times V \mid \exists p. \; p \text{ is a shortest path from } a \text{ to } b \text{ and contains } e\}$$

the edge's *traversal set* – the set of source-destination pairs where for every pair some shortest path contains this edge. Now, the size of the traversal set would be an obvious measure for the importance of the edge. As claimed by Tangmunarunkit et al. [540], this simple method may not yield the desired result in some cases, so they propose the following different counting scheme.[3]

The traversal set T_e can be seen as a set of new edges, connecting those pairs of vertices that have shortest paths along e in the original graph. These edges (together with the vertices they connect) naturally constitute a graph, which is bipartite as we will now see.

Fig. 3.7. The traversal set graph is bipartite

Let (a, b) be any edge in the traversal set graph T_e of edge $e = (y, z)$. This means that there is a shortest path p connecting a and b via e (cf. Figure 3.7). Without loss of generality, assume that p has the form $a - \cdots - y - z - \cdots - b$. Then, there cannot be an $a - z$ path shorter than the $a - y$ prefix of p, for else the resulting path along $a - \cdots - z - \cdots - b$ would be shorter than our shortest path p. In the other direction, no $y - b$ path may be shorter than our $z - b$ suffix of p. To summarize, a is closer to y, and b is closer to z. Let $\mathcal{Y}$ denote the set of all vertices closer to y than to z and let $\mathcal{Z}$ denote the set of all vertices closer to z. Thus, $\mathcal{Y}$ and $\mathcal{Z}$ form a partition of V. No two vertices belonging to the same set can be connected by an edge in this graph since the shortest path connecting them can never contain e. Thus, T_e is naturally bipartite with regard to $\mathcal{Y}$ and $\mathcal{Z}$.

---

[3] Both ways of counting yield values of different orders of magnitude for certain example graphs. However, we have not been able to identify a case where one scheme differentiates between two situations while the other does not. That is why we can only rely on the experience of Tangmunarunkit et al (ibid.).

An edge's *value* is then defined as the size of a minimum vertex cover on the bipartite graph formed by the traversal set:

$$C_{ts}(e) = \min\{|H| \mid H \text{ is a vertex cover for } T_e\}$$

Unlike the non-bipartite case, this is computable in polynomial time (less than $\Theta(n^3)$) using a theorem by Kőnig and Egerváry [366, 173], which states that the minimum size of a vertex cover equals the size of a maximum matching on bipartite graphs.

In [540] this centrality index has been used to characterize a graph with regard to its hierarchical organization. The authors determine the edge value pattern of sample paths in the original graph. If a high fraction of paths shows an up-down pattern of edge values, i.e., a paths begins with edges having a small value, the value raises along the path and then drops again to low values, the authors assume that this shows a high level of hierarchical organization of the underlying graph. An example on which this assumption is intuitively true is the graph of streets in a country: Some of them are only within cities, others are connecting smaller suburbs and some are high-speed freeways. Most paths from one location to another will follow streets that have low values at the beginning, then the driver will use a freeway and at last will use inner-city streets again at the end. This example shows that hierarchically organized networks may show an up-down pattern in the edge value distribution on many paths but the reverse will be hard to prove. This empirical finding should thus be treated with care.

3.5   Derived Edge Centralities

Historically, centrality indices were developed to analyze social networks. From this application, the emphasis lay on the analysis of the most central persons in social networks. This lead to a great number of different centrality indices for vertices. Most centrality indices for edges, e.g., the shortest path betweenness centrality, were only developed as a variant of the centrality index for vertices. Here, we want to discuss two methods with which every given centrality index for vertices can be transformed into a centrality index for edges.

3.5.1   Edge Centralities Derived from Vertex Centralities

One intuitive idea to derive an edge centrality from a vertex centrality is to apply the vertex centrality to the *edge graph* that is corresponding to the network to be analyzed:

Definition 3.5.1. *The edge graph of $G = (V, E)$ is $G' = (E, K)$ where K is the set of all edges $e = ((x, y), (y, z))$ where $(x, y), (y, z) \in E$. That is, two edges have a connection if they are adjacent to the same vertex y (with the first one in- and the second outbound for directed graphs).*

There are biased and unbiased centralities for vertices. Note that methods that incorporate previous knowledge usually do this by assuming that a subset of 'root vertices' is especially important. For details on personalization see Section 5.2. Unlike the approaches described in there, an application on the edge graph then needs a description of central *edges*.

The size of the edge graph may be quadratic in the size of the original graph. For large graphs and computationally expensive methods this might well be a hindrance.

There is another caveat. Some of the more advanced techniques for vertices incorporate weighted edges, a feature that allows for more detailed models. However, in the edge graph these become weighted vertices, and there is no canonical way to use this data.

Finally, there is a philosophical point to be made against this approach: The vertex centralities described so far fall into the categories of degree, closeness and betweenness centrality. On the edge graph, these concepts translate into counting incident edges, closeness to other edges and position on paths between pairs of edges. However, when modeling phenomena using networks, we tend to have vertices representing entities, while edges describe relationships between these. Most of the time, these relationships are meaningless without the entities they connect. Therefore, none of the three mentioned categories seems to make a lot of sense as a centrality measure for edges.

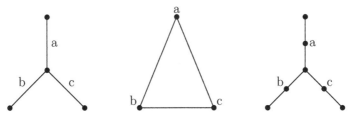

Fig. 3.8. Edge graph example

As an illustrative instance, look at the evaluation of the stress centrality on the left example graph in Figure 3.8. For a vertex x it is defined as the number of shortest paths that use x and do not end in x. The straightforward translation for an edge, say a, would be the number of shortest paths that use a, adding up to three in this example. In the middle, you find the corresponding edge graph. In contrast to the above, no shortest path (except those that end in a) crosses the vertex a. Obviously, the edge graph does not lead to the natural edge generalization of stress centrality. However, this natural generalization may be attained using a different graph translation. We will call this construction the *incidence graph*, and there is an illustrative instance on the right hand side of Figure 3.8: Each edge e is split by a new 'edge vertex' that receives the link's name.

Definition 3.5.2. *The* incidence graph *of* $G = (V, E)$ *is*

$$G'' = (V \cup E, \{(v, e) \mid \exists w : e = (v, w) \in E\} \cup \{(e, w) \mid \exists v : e = (v, w) \in E\}.$$

That is, a 'real vertex' and an 'edge vertex' become linked if they are incident in the original graph.

We can now use a biased version of stress vertex betweenness (see Section 5.2 for details on how to personalize measures), which only takes into account 'real vertex' pairs to measure the importance of 'edge vertices'. This way, most vertex measures may be translated into edge measures. As with the original centralities, it remains to check if the measure we achieve does have a sensible semantics with respect to the function of the network.

3.6 Vitality

Vitality measures are commonly used to determine the importance of vertices or edges in a graph. Given an arbitrary real-valued function on G a vitality measure quantifies the difference between the value on G with and without the vertex or edge. The main motivation behind this idea is that most networks have some quality that can be evaluated by a function on G: Imagine a transport network with different capacities on the edges in which the goal is to transport as much as possible of some good from some vertex s to another vertex t. The functionality of a network for this goal can be described by the maximal possible flow in it (see Section 2.2.3). The degree to which this quality is impaired by the loss of an edge or vertex can be viewed as the extent to which this edge or vertex is 'central' for the network. A second example is a graph representing a mobile communication network in which every vertex should be indirectly connected to all others over as few switching points as possible. The quality of this graph could be evaluated by its Wiener index, the sum over all distances in the graph (see Section 3.6.2). Then, the vitality of a vertex or edge x denotes the loss of this quality if x was removed from the network. More formally:

Definition 3.6.1 (Vitality Index). *Let $\mathcal{G}$ be the set of all simple, undirected and unweighted graphs $G = (V, E)$ and $f : \mathcal{G} \to \mathbb{R}$ be any real-valued function on $G \in \mathcal{G}$. A vitality index $\mathcal{V}(G, x)$ is then defined as the difference of the values of f on G and on G without element x: $\mathcal{V}(G, x) = f(G) - f(G \backslash \{x\})$.*

We will begin with a centrality index that is derived from the field of network flow problems. After that, a new centrality index, the closeness vitality, is presented that might be useful for some applications. The next subsection presents a new centrality index that is not a vitality index in the strict sense but the relationship to vitality indices is strong. The last subsection presents a discussion in how far the stress centrality presented in Section 3.4.1 can be interpreted as a vitality index.

3.6.1 Flow Betweenness Vitality

In this subsection we present a vertex centrality based on network flows. More precisely a measure for max-flow networks is presented which is similar to the shortest-path betweenness described in Section 3.4.2 and makes the measure proposed in Freeman et al. [229] concrete.[4] As Stephenson and Zelen [533] observed, there is no reason to believe that information in a communication network between a pair of vertices takes place only on the shortest path. Obviously, there are applications where the centrality values computed by shortest path betweenness leads to misleading results. Thus other paths have to be considered instead.

Taking up the example of communication networks, Freeman et al. assumed information as flow and assigned with each edge a non-negative value representing the maximum of information that can be passed between its endpoints. In extending the betweenness model to flow networks, a vertex u will be seen as standing between other vertices. The goal is to measure the degree that the maximum flow between those vertices depends on u.

Based on this idea we provide a concise definition of a vertex centrality based on maximum flows. We call this centrality the max-flow betweenness vitality. Note that the maximum-flow problem between a source vertex s and a target vertex t was introduced in Section 2.2.3. For reasons of simplicity we further assume $G = (V, E)$ as a connected undirected network with non-negative edge capacities. By f_{st} we denote the objective function value of a maximum s-t-flow. The value f_{st} represents the maximal flow between s and t in G with respect to the capacity constraints and the balance conditions. As indicated above, we are now interested in the answer of the questions: How much flow must go over a vertex u in order to obtain the maximum flow value? And how does the objective function value change if we remove u from the network?

According to the betweenness centrality for shortest paths we define the max-flow betweenness for a vertex $u \in V$ by

$$c_{mf}(u) = \sum_{\substack{s,t \in V \\ u \neq s, u \neq t \\ f_{st} > 0}} \frac{f_{st}(u)}{f_{st}} \tag{3.17}$$

where $f_{st}(u)$ is the amount of flow which must go through u. We determine $f_{st}(u)$ by $f_{st}(u) = f_{st} - \tilde{f}_{st}$ where $\tilde{f}_{st}(u)$ is the maximal s-t-flow in $G \setminus u$. That is, $\tilde{f}_{st}(u)$ is determined by removing u form G and computing the maximal s-t-flow in the resulting network $G \setminus u$.

It is important to note, that this concept may also be applied to other network flow problems, e.g., the minimum-cost maximum-flow problem (MCMF) which may be viewed as a generalization of the max-flow problem. In a MCMF network each edge has a non-negative cost value and a non-negative upper capacity bound. The objective is to find a maximum flow of minimum cost between two designated vertices s and t. Applying the idea of measuring the vitality of each vertex to MCMF networks yields a new meaningful vitality measure. For further details relating to the MCMF problem see [6].

---

[4] Note that the original definition in [229] is ambiguous, because it neglects that a max-flow is not unique in general.

3.6.2   Closeness Vitality

In analogy to the closeness centrality index presented in Section 3.3.2, we will introduce a new centrality, based on the Wiener Index[5] [583]. The Wiener Index $I_W(G)$ of a graph G is defined as the sum over the distances of all vertex pairs:

$$I_W(G) = \sum_{v \in V} \sum_{w \in V} d(v, w) \qquad (3.18)$$

It is easy to see that the Wiener Index can also be written as the sum of the closeness centrality values $c_C(v)$ (see Section 3.2) of all vertices v:

$$I_W(G) = \sum_{v \in V} \frac{1}{c_C(v)} \qquad (3.19)$$

We will now define a new centrality called closeness vitality $c_{CV}(x)$, defined on both vertices and edges:

$$c_{CV}(x) = I_W(G) - I_W(G \setminus \{x\}) \qquad (3.20)$$

Clearly, this new centrality is a vitality, with $f(G) = I_W(G)$. What does this centrality index measure? Let the distance between two vertices represent the costs to send a message from s to t. Then the closeness vitality denotes how much the transport costs in an all-to-all communication will increase if the corresponding element x is removed from the graph. With a small modification we can also calculate the average distance $d_{\varnothing}(G)$ between two vertices:

$$d_{\varnothing}(G) = \frac{I_W(G)}{n(n-1)} \qquad (3.21)$$

This variant computes how much the costs are increased on average if the element x is removed from the graph.

There is one pitfall in the general idea of a closeness vitality: If x is a cut-vertex or a bridge, respectively, the graph will be disconnected after the removal. Then $c_{CV}(x)$ is $-\infty$ for this element. We will discuss some ideas to deal with the calculation of distance based centrality indices in Section 3.10.

3.6.3   Shortcut Values as a Vitality-Like Index

Although shortcut values are not a vitality index in the sense of Definition 3.6.1, they are nevertheless based on the concept of vitality. Thus, we present shortcut values here as a vitality-like index.

The shortcut value for edge e is defined by the maximum increase in distance between any two vertices if $e = (u, v)$ is removed from the graph. It is clear that this maximum increase can only be found between vertices that use e for all of

---

[5] Wiener itself named it 'path number' which is misleading. Subsequent articles quoted it as 'Wiener Index' [592]

their shortest paths. We claim that the increase in path length is maximized for the pair (u, v). This can easily be seen as follows. Clearly, the increase in distance for the pair (u, v) equals the difference between the length of e and the length of the shortest path p from u to v that does not use e. Further, other pair of vertices will either use their old path with e replaced by p, or use an alternative that is shorter than that.

Alternatively, the shortcut value can also be defined as the maximum relative increase in distance when all edge lengths are non-negative. In this case, the length of a shortest path using e is larger than the length of e, such that the relative increase is also maximized for the pair (u, v).

The shortcut values for all edges can be computed naïvely by $m = |E|$ many calls to a single-source shortest-path routine. Section 4.2.2 introduces a more efficient algorithm that is as efficient as computing $|V|$ single-source-shortest paths trees.

The notion of a shortcut value for an edge can be directly generalized to vertices, as the maximum increase in distance if the vertex is deleted.

3.6.4 Stress Centrality as a Vitality-Like Index

Stress centrality can be viewed as a vitality-like measure: Stress centrality (Section 3.4.1) counts the number of shortest paths containing a vertex or an edge and can thus be interpreted as the number of shortest paths that are lost if the vertex or edge is removed from the graph.

This sounds like a vitality measure but there is a crucial difference to the definition of vitality: The number of lost shortest paths has to be measured relatively to the number of shortest paths in the original graph. This is important, because a simple example shows that the total number of shortest paths can actually increase if a vertex or edge is removed from a graph (see Figure 3.9).

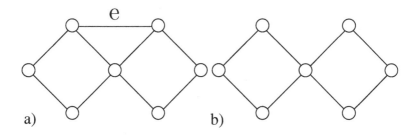

Fig. 3.9. The figure shows that the removal of an edge can actually increase the number of shortest paths in a graph

On the left side of Figure 3.9 (a) a small graph is shown with a total number of 54 shortest paths, 8 of them containing edge e. After the removal of e we find 64 shortest paths in the resulting graph. Of course, 18 of them are now longer than before. When will the removal of an edge lead to an increase in the edge number? In this example, edge e is a shortcut for some of the paths from or to the two outermost vertices. As an example, we will take the path from the left outermost vertex to the right outermost vertex. As soon as e is removed, the distance between these nodes increases by one. Additionally, the number of shortest paths between them increases by three because now there are four paths with length 4 instead of only one with length 3 as before.

To interpret the stress centrality as a vitality measure we have to disregard shortest paths that have an increased length after the removal of an element. To formalize this idea we will give a definition of $f(G \setminus \{x\})$ that allows us to interpret the stress centrality of a vertex or an edge as vitality.

Let $f(G)$ be the number of all shortest paths in G and $f(G\setminus\{v\})$ be defined as following:

$$f(G\setminus\{v\}) = \sum_{s\in V}\sum_{t\in V} \sigma_{st}[d_G(s,t) = d_{G\setminus\{v\}}(s,t)] \qquad (3.22)$$

The definition is given in Iverson-Notation, first described in [322], adapted by Knuth in [365]. The term inside the parentheses can be any logical statement. If the statement is true the term evaluates to 1, if it is false the term is 0. This notation makes the summation much easier to read than the classical notation in which logical statements are combined with the index of the sum. The definition of $f(G \setminus \{v\})$ is thus defined as the sum over the number of all those shortest paths that have the same length as the distance of s and t in G.

Analogously, let $f(G\setminus\{e\})$ be defined as following:

$$f(G\setminus\{e\}) = \sum_{s\in V}\sum_{t\in V} \sigma_{st}(e)[d_G(s,t) = d_{G\setminus\{e\}}(s,t)] \qquad (3.23)$$

Defined in this way, the stress centrality $C_S(x)$ of an element x is exactly the difference between $f(G)$ and $f(G\setminus\{x\})$. It is important to note that the definition of $f(G \setminus \{x\})$ does not match the formal definition for a vitality measure. Nonetheless, the similarity of both is evident and thus we will denote the stress centrality as a vitality-like centrality index.

3.7 Current Flow

Shortest paths centralities rely on the crucial assumption that the flow of information, or the transport of goods in general, takes place along shortest paths. This section describes the current flow centralities, which are appropriate when the flow of information or transport does not adhere to this shortest paths assumption, but rather follows the behavior of an electrical current flowing through an electrical network.

3.7.1 Electrical Networks

Current flow centralities are based on the flow of electrical current in a network. We briefly describe currents in electrical networks below, and refer to [67] for an extensive discussion. An electrical network is defined by an undirected, connected, and simple graph $G = (V, E)$, together with a conductance function $c : E \rightarrow \mathbb{R}$. External electrical current enters and leaves this network, which is specified by a supply function $b : V \rightarrow \mathbb{R}$. Positive values of b represent entering current, negative values represent current that leaves the network, and the amounts of entering and leaving currents are required to be equal: $\sum_{v \in V} b(v) = 0$. Since it is useful to talk about the direction of a current in the undirected graph, each edge $e \in E$ is arbitrarily oriented to obtain an oriented edge $\vec{e}$, which results in an oriented edge set $\vec{E}$.

A function $x : \vec{E} \rightarrow \mathbb{R}$ is called a (electrical) current in $N = (G = (V, E), c)$ if

$$\sum_{(v,w) \in \vec{E}} x(v, w) - \sum_{(w,v) \in \vec{E}} x(w, v) = b(v) \text{ for all } v \in V$$

and

$$\sum_{e \in C} x(\vec{e}) = 0$$

for every cycle $C \subseteq E$, that is, for every cycle in the undirected graph G. The former equation is known as Kirchoff's current law, and the latter as Kirchoff's potential law. Negative values of x are to be interpreted as current flowing against the direction of an oriented edge.

Alternatively to the current x, an electrical flow can also be represented by potentials. A function $p : V \rightarrow \mathbb{R}$ is a (electrical) potential if $p(v) - p(w) = x(v, w)/c(v, w)$ for all $(v, w) \in \vec{E}$. As an electrical network $N = (G, c)$ has a unique current x for any supply b, it also has a potential p that is unique up to an additive factor [67].

Define the Laplacian matrix $L = L(N)$ of the electrical network N to be

$$L_{vw} = \begin{cases} \sum_{e \ni v} c(e) & \text{if } v = w \\ -c(e) & \text{if } e = \{v, w\} \\ 0 & \text{otherwise} \end{cases}$$

for $v, w \in V$. Then, a potential p for an electrical network $N = (G, c)$ and a supply b can be found by solving the linear system $Lp = b$.

Finally, for the purpose of stating centralities based on electrical currents, define a unit s-t-supply b_{st} as a supply of one unit that enters the network at s and leaves it at t, that is, $b_{st}(s) = 1, b_{st}(t) = -1$, and $b_{st}(v) = 0$ for all $v \in V \setminus \{s, t\}$.

3.7.2   Current-Flow Betweenness Centrality

Newman [443] first considered centrality measures based on electrical currents. The current-flow betweenness of a vertex represents the fraction of unit s-t-supplies that passes through that vertex, just as shortest paths betweenness counts the fraction of shortest s-t-paths through a vertex. For a fixed s-t pair, the so-called throughput of a vertex v forms the current-flow equivalent of the number of shortest paths $\sigma_{st}(v)$ through v. More precisely, the throughput of a vertex $v \in V$ with respect to a unit s-t-supply b_{st} is defined as

$$\tau_{st}(v) = \frac{1}{2} \left(-|b_{st}(v)| + \sum_{e \ni v} |x(\vec{e})| \right).$$

Here, the term $-|b_{st}(v)|$ sets the throughput of a vertex with non-zero supply equal to zero. For the current-flow betweenness, this ensures that a given unit s-t-supply does not count for the throughput of its source and sink nodes s and t. Further, the term $\frac{1}{2}$ adjusts for the fact that the summation counts both the current into and out of the vertex v.

Using the throughput definition, the current-flow betweenness centrality $c_{CB} : V \rightarrow \mathbb{R}$ for an electrical network $N = (G = (V, E), c)$ is defined as

$$c_{CB}(v) = \frac{1}{(n-1)(n-2)} \sum_{s,t \in V} \tau_{st}(v),$$

for all $v \in V$, where $1/(n-1)(n-2)$ is a normalizing constant. Thus, current-flow betweenness measures the fraction of throughput through vertex v, taken over all possible s-t pairs. Since an electrical network has a unique current for a given supply, current-flow betweenness is well defined.

3.7.3   Current-Flow Closeness Centrality

As with betweenness, the concept of closeness can also be extended from shortest paths to electrical current. For shortest paths, closeness is a measure of the shortest path distance from a certain vertex to all other vertices. For electrical current, Brandes and Fleischer [94] propose a closeness centrality that measures the distance between two vertices v and w as the difference of their potentials $p(v) - p(w)$. Their current-flow closeness centrality $c_{CC}(v) : V \rightarrow \mathbb{R}$ is defined as

$$c_{CC}(v) = \frac{n-1}{\sum_{t \neq v} p_{vt}(v) - p_{vt}(t)}$$

for all $v \in V$, where $(n-1)$ is again a normalizing factor. Here, the subscript vt on the potentials means that the potential stems from a unit v-t-supply b_{vt}.

Interestingly, Brandes and Fleischer [94] prove that current-flow closeness centrality is equal to information centrality. Stephenson and Zelen [533] introduced information centrality to account for information that flows along all paths in a network, rather than just along shortest paths. Information centrality also takes into account that certain paths carry a larger amount of information than others. Mathematically, information centrality $c_I : V \rightarrow \mathbb{R}$ is defined by

$$c_I(v)^{-1} = nM_{vv} + \text{trace}(M) - \frac{2}{n},$$

where the matrix M is defined as $(L+U)^{-1}$, with L being the Laplacian matrix, and U being a matrix of the same size with all entries equal to one.

3.8   Random Processes

Sometimes, it may not be possible for a vertex to compute shortest paths because of a lack of global knowledge. In such a case, shortest paths based centralities make no sense, and a random-walk model provides an alternative way of traversing the network. In a random walk something walks from vertex to vertex, following the edges of the network. Reaching some vertex v, it chooses one of the edges of v randomly to follow it to the next vertex.

The 'travel' of a bank note is a typical example for such a random walk. Somebody gets a brand new bill from her bank and gives it to someone else she encounters later on. Normally, nobody has any intention to give the bank note to someone special and the same bill may get to the same person more than once. For a marketing study, it could be of interest to find out the person or company who mediates most of these transactions. In the next section, we will have a closer look at the so-called random walk betweenness centrality that calculates the hot spots of mediation in such transactions.

3.8.1   Random Walks and Degree Centrality

In the case of undirected graphs, an observation can be made that relates the random-walk centrality with its complex definition to the most basic of all centralities, degree.

In the following theorem we prove that the stationary probabilities in the canonical random walk on a graph are proportional to the degree of the vertex.

Theorem 3.8.1. $p_{ij} = \frac{a_{ij}}{d(i)} \implies \pi_i = \frac{d(i)}{\sum_{v \in V} d(v)}$

Proof.

$$(\pi P)_j = \sum_{i \in V} \pi_i p_{ij} = \frac{\sum_{i \in V} d(i) p_{ij}}{\sum_{v \in V} d(v)} = \frac{\sum_{i \in V} a_{ij}}{\sum_{v \in V} d(v)} = \frac{d(j)}{\sum_{v \in V} d(v)} = \pi_j$$

$\square$

3.8.2   Random-Walk Betweenness Centrality

The random-walk betweenness centrality introduced in [443] is based on the following idea. Suppose that vertex s has a message for vertex t but neither s nor any other vertex knows how to send it to t on a shortest path. Each vertex that gets the message for vertex t will just send it to any of its adjacent vertices at random. We assume that the graph is unweighted, undirected and connected.

This so-called random walk is modeled by a discrete-time stochastic process. At time 0, vertex s sends a message to one of its neighbors. If the message reaches vertex t at any time it will not be forwarded any further and such be absorbed by t. More formally, let m_{ij} describe the probability that vertex j sends the message to vertex i in time $k+1$ if it had it at time k:

$$m_{ij} = \begin{cases} \frac{a_{ij}}{d(j)} & \text{if } j \neq t \\ 0 & \text{else} \end{cases} \qquad (3.24)$$

where a_{ij} denotes the ij-th element of the adjacency matrix A (see Section 2.3) and $d(j)$ is the degree of vertex j. The resulting matrix is denoted by M. Let D be the degree matrix of the graph:

$$d_{ij} = \begin{cases} d(i) & \text{if } i = j \\ 0 & \text{else} \end{cases} \qquad (3.25)$$

The inverse D^{-1} of this matrix has the inverted vertex degrees on its diagonal, and is zero elsewhere. Because of the special behavior of vertex t the matrix notation $M = A \cdot D^{-1}$ is not correct. Removing the t-th row and column of all matrices yields a correct relation between the three matrices:

$$M_t = A_t \cdot D_t^{-1}, \qquad (3.26)$$

where the index denotes the missing row and column, respectively.

Random-walk betweenness centrality considers all paths that a random walk can use, as well as the probabilities that such paths are used. Thus, the question arises how to compute the set of used paths, and how to compute the probability of using a single one of these paths. To guide the reader on his way, we first discuss how many different $i - j$ paths of length r exist in a given graph, where i and j are arbitrarily chosen vertices. It can easily be seen that the answer is $(A^r)_{ij}$, where A^r denotes the rth power of A. However, we are not interested in the number of random walks, but in the probability that a random walk of r steps, that starts at s, ends in vertex j. This is given by the r-th power of M_t at row j, column s, denoted by $(M_t^r)_{js}$. With this, the probability that the message is sent to vertex i in step $r + 1$ is given by:

$$\left(M_t^{r+1}\right)_{js} = m_{ij}^{-1} \left(M_t^r\right)_{js} \qquad (3.27)$$

Now, we are interested in the probability that vertex j is sending a message that is starting at s to vertex i, summing over all paths, beginning at length 0 to ∞.

Note that all entries in any matrix M_t^r are values between 0 and 1, and thus the sum over all paths is convergent (see Theorem 3.9.2):

$$\sum_{r=0}^{\infty} m_{ij}^{-1} \left(M_t^r\right)_{js} = m_{ij}^{-1}[(I_{n-1} - M_t)^{-1}]_{js} \qquad (3.28)$$

where I_{n-1} is the identity matrix of dimension $n - 1$.

Let s be a vector with dimension $n - 1$ that is 1 at vertex s and 0 else. Writing equation 3.28 in matrix notation we get:

$$v^{st} = D_t^{-1} \cdot (I - M_t)^{-1} \cdot s \qquad (3.29)$$
$$= (D_t - A_t)^{-1} \cdot s \qquad (3.30)$$

The vector $\boldsymbol{v^{st}}$ describes the probability to find the message at vertex i while it is on its random walk from vertex s to vertex t. Of course, some of the random walks will have redundant parts, going from vertex a to vertex b and back again to vertex a. It does not seem reasonable to give a vertex a high centrality if most of the random walks containing it follow this pattern. Since the network is undirected every cycle will be accounted for in both directions, thus extinguishing each other. It is important to note that $\boldsymbol{v^{st}}$ contains only the net probability that disregards these cycles.

At this point, it becomes clear that random walks are closely related to current flows in electrical networks, see Section 3.7. Indeed, consider an electrical network $N = (G, c)$ with unit edge weights $c(e) = 1$ for all $e \in E$. The unit edge weights yield a Laplacian matrix $L(N) = D - A$, where D is the degree matrix and A the adjacency matrix of the graph G. So, a potential p_{st} in N for a unit s-t-supply b_{st} is a solution to the system $Lp_{st} = b_{st}$. The matrix L is not of full rank, but this problem can be circumvented by fixing one potential, say for vertex v, since potentials are unique up to an additive factor. Removing the rows and columns corresponding to the fixed vertex v yields the matrices L_v, D_v, and A_v, where L_v has full rank and is thus invertible. We conclude that a potential p_{st} for the unit s-t-supply b_{st} is given by $p_{st} = L_v^{-1} b_{st} = (D_v - A_v)^{-1} b_{st}$. The latter is equivalent to Equation (3.29) above, which shows the relation between electrical currents and potentials and random walks. For a more in-depth discussion of this relation, we refer to [67].

Thus, the random-walk betweenness centrality $c_{RWB} : V \to \mathbb{R}$ that we are looking for is equivalent to current-flow betweenness, that is, $c_{RWB}(v) = c_{CB}(v)$ for all $v \in V$. Newman [443] and Brandes and Fleischer [94] describe this betweenness equivalence in more detail.

3.8.3 Random-Walk Closeness Centrality

The same approach gives a kind of random-walk closeness centrality, where we look for the mean first passage time (MFPT). A centrality based on MFPT is introduced as Markov centrality in [580]. The mean first passage time m_{st} is defined as the expected number of nodes a particle or message starting at vertex s has encountered until it encounters vertex t for the first time. It is given by the following series:

$$m_{st} = \sum_{n=1}^{\infty} n \cdot f_{st}^{(n)} \tag{3.31}$$

where $f_{st}^{(n)}$ denotes the probability that t is arrived for the first time after exactly n steps. Let M denote the MFPT matrix in which m_{st} is given for all pairs s, t. M can be computed by the following equation:

$$M = (I - EZ_{dg})\, D \tag{3.32}$$

where I denotes the identity matrix, E is a matrix containing all ones, and S is a diagonal matrix with:

$$s_{ij} = \begin{cases} \frac{1}{\pi(v)} & \text{if } i = j \\ 0 & \text{else} \end{cases} \tag{3.33}$$

π denotes the stationary distribution of the random walk in the given graph (see Section 2.4), i.e., the expected relative time a particle will be on vertex v during the random walk. (This model assumes that the transport of the message or particle to another nodes takes virtually no time.) The matrix Z_{dg} agrees with the so called fundamental matrix Z on the diagonal but is 0 everywhere else. Matrix Z itself is given by:

$$Z = \left(I - A - 1_n \pi^T\right)^{-1} \tag{3.34}$$

where 1_n is a column vector of all ones. The Markov centrality $c_M(v)$ is now defined as the inverse of the average MFPT for all random walks starting in any node s with target v (or vice versa):

$$c_M(v) = \frac{n}{\sum_{s \in V} m_{sv}} \tag{3.35}$$

This centrality can be defined for both directed and undirected networks. In directed networks the centrality is meaningfully defined for both, the average MFPT for random walks ending in v or leaving v. The expected number of steps from v to all other vertices or from all other vertices to v might be interpreted as a distance from v to all other vertices if a particle or information uses a random walk. Thus, the Markov centrality of a vertex is a kind of a (averaged) random-walk closeness centrality.

3.9 Feedback

This section presents centralities in which a node is the more central the more central its neighbors are. Some of these measures like Katzs status index belong to the oldest centralities presented in this chapter, others have their roots in the analysis of social networks. A third group belongs to the big class of analysis methods for the Web graph that is defined as the set of pages in the WWW connected by Web links.

Note, that in the following subsections centrality indices will be denoted as vectors. All feedback centralities are calculated by solving linear systems, such that the notation as a vector is much more convenient than using a function expressing the same. We just want to state here that all centrality indices presented here are fulfilling the definition of a structural index in Definition 3.2.1 if $c_X(i)$ is defined as $(c_X)_i$.

3.9.1 Counting All Paths – The Status Index of Katz

One of the first ideas with respect to feedback centralities was presented by Leo Katz [352] in 1953. It is based on the following observation: To determine the

importance or *status* of an individual in a social network where directed edges
(i, j) can, for example, be interpreted as "*i votes for j*", it is not enough to
count direct votes. If, e.g., only two individuals k and l vote for i but all other
persons in the network vote either for k or for l, then it may be that i is the
most important person in the network – even if she got only two direct votes.
All other individuals voted for her indirectly.

The idea of Katz is therefore to count additionally all indirect votes where
the number of intermediate individuals may be arbitrarily large.

To take the number of intermediate individuals into account, a damping
factor $\alpha > 0$ is introduced: the longer the path between two vertices i and j is,
the smaller should its impact on the status of j be.

The associated mathematical model is hence an unweighted (i.e. all weights
are 1) directed simple graph $G = (V, E)$ without loops and associated adjacency
matrix A. Using the fact that $(A^k)_{ji}$ holds the number of paths from j to i with
length k we hence have as status of vertex i

$$c_K(i) = \sum_{k=1}^{\infty} \sum_{j=1}^{n} \alpha^k (A^k)_{ji}$$

if the infinite sum converges.

In matrix notation we have

$$c_K = \sum_{k=1}^{\infty} \alpha^k (A^T)^k \mathbf{1}_n. \tag{3.36}$$

(Note that $\mathbf{1}_n$ is the n-dimensional vector where every entry is 1, cf. also Chapter
2.)

To guarantee convergence we have to restrict α.

Theorem 3.9.1. *If A is the adjacency matrix of a graph G, $\alpha > 0$, and λ_1 the
largest eigenvalue of A, then*

$$\lambda_1 < \frac{1}{\alpha} \iff \sum_{k=1}^{\infty} \alpha^k A^k \ \text{converges.}$$

For the proof see, e.g., [208].

Assuming convergence we find a closed form expression for the status index
of Katz:

$$c_K = \sum_{k=1}^{\infty} \alpha^k (A^T)^k \mathbf{1}_n = \left((I - \alpha A^T)^{-1} \right) \mathbf{1}_n$$

or, in another form

$$(I - \alpha A^T) c_K = \mathbf{1}_n,$$

an inhomogeneous system of linear equations emphasizing the feedback nature
of the centrality: the value of $c_K(i)$ depends on the other centrality values $c_K(j)$,
$j \neq i$.

3.9.2   General Feedback Centralities

In this subsection three centralities that are well known in the area of social network analysis are described.

Bonacich's Eigenvector Centrality. In 1972 Phillip Bonacich introduced a centrality measure based on the eigenvectors of adjacency matrices [71]. He presented three different approaches for the calculation and all three of them result in the same valuation of the vertices, the vectors differ only in a constant factor. In the following we assume that the graph G to be analyzed is undirected, connected, loop-free, simple, and unweighted. As the graph is undirected and loop-free the adjacency matrix $A(G)$ is symmetric and all diagonal entries are 0.

The three methods of calculation are:

a. the factor analysis approach,
b. the convergence of an infinite sequence, and
c. the solving of a linear equation system

In the following we describe all three approaches and call the results s^a, s^b, and s^c.

First, we explain the factor analysis approach. For a better understanding think of the graph as a friendship network, where an edge denotes friendship between the persons that are modeled as vertices. We want to define a centrality that measures the ability to 'find friends'. Thus, we are interested in a vector $s^a \in \mathbb{R}^n$, such that the i-th entry s_i^a should hold the interaction or 'friendship' potential of the vertex i. We declare that $s_i^a S_j^a$ should be close to a_{ij} and interprete the problem as the minimization of the least squared difference. We are therefore interested in the vector s^a that minimizes the following expression:

$$\sum_{i=1}^{n}\sum_{j=1}^{n}(s_i^a s_j^a - a_{ij})^2 \tag{3.37}$$

A second approach presented by Bonacich is based on an infinite sequence. For a given $\lambda_1 \neq 0$ we define

$$s^{b_0} = \mathbf{1}_n \quad \text{and} \quad s^{b_k} = A\frac{s^{b_{k-1}}}{\lambda_1} = A^k\frac{s^{b_0}}{\lambda_1^k} \; .$$

According to Theorem 3.9.2, the sequence

$$s^b = \lim_{k\to\infty} s^{b_k} = \lim_{k\to\infty} A^k\frac{s^{b_0}}{\lambda_1^k}$$

converges towards an eigenvector s^b of the adjacency matrix A if λ_1 equals the largest eigenvalue.

Theorem 3.9.2. *Let $A \in \mathbb{R}^{n \times n}$ be a symmetric matrix and λ_1 the largest eigenvalue of A, then*

$$\lim_{k \to \infty} A^k \frac{s^{b_0}}{\lambda_1^k}$$

converges towards an eigenvector of A with eigenvalue λ_1.

The third approach follows the idea of calculating an eigenvector of a linear equation system. If we define the centrality of a vertex to be equal to the sum of the centralities of its adjacent vertices, we get the following equation system:

$$s_i^c = \sum_{j=1}^{n} a_{ij} s_j^c \quad \text{resp.} \quad s^c = A * s^c \qquad (3.38)$$

This equation system has a solution only if $\det(A - I) = 0$. We solve $\lambda s = As$, the eigenvalue problem for A, instead. According to Theorem 3.9.3, under the given conditions for the graph defined above, exactly one eigenvector contains entries that are either all positive or all negative. Therefore, we use the absolute value of the entries of this eigenvector as the solution.

Theorem 3.9.3. *Let $A \in \mathbb{R}^{n \times n}$ be the adjacency matrix of an undirected and connected graph. Then:*

– *The largest eigenvalue λ_1 of A is simple.*
– *All entries of the eigenvector for λ_1 are of the same sign and not equal to zero.*

We have seen three methods for the calculation of the solution vectors s^a, s^b, s^c. These vectors differ only by a constant factor. The eigenvector centrality is therefore (independently from the solution method) defined by:

$$c_{\text{EV}} = \frac{|s^c|}{||s^c||} \qquad (3.39)$$

In general, whenever one has a graph with multiple, poorly spanned dense clusters, no single eigenvector will do a satisfactory job of characterizing walk-based centrality. This is because each eigenvector will tend to correspond to loadings on a given cluster: Everett and Borgatti [194] explain this behavior via their core-periphery model, where in the idealized case the core corresponds to a complete subgraph and the nodes in the periphery do not interact with each other. To measure how close a graph is to the ideal core-periphery structure (or, in other words, how *concentrated* the graph is) they define the ρ-measure

$$\rho = \sum_{i,j} a_{ij} \delta_{ij}$$

with $\delta_{ij} = c_i c_j$, where a_{ij} are the components of the adjacency matrix and c_i measures the *coreness* of a node, $c_i \in [0, 1]$.

To determine the coreness of the nodes, the authors propose to minimize the sum of squared distances of a_{ij} and the product $c_i c_j$, which is nothing else than one approach to compute Bonacich's Standard Centrality, see 3.37, hence nothing else then computing the principal eigenvector of the adjacency matrix. Thus, only the core-vertices get high c-values, nodes in smaller clusters not belonging to the core will get values near zero.

According to [71], the eigenvector centrality can be applied to disconnected graphs. In this case several eigenvectors have to be taken into account, one for every component of the graph.

Hubbell Index. Even earlier than Bonacich, Charles Hubbell [319] suggested in 1965 a centrality measure based on the solution of a system of linear equations. His approach uses directed weighted graphs where the weights of the edges may be real numbers. A graph may contain loops but has to be simple, too. Please note that the adjacency matrix $W(G)$ of a graph G is asymmetric and contains real numbers instead of zeros and ones.

The general assumption of Hubbell's centrality measure is similar to the idea of Bonacich: the value of a vertex v depends on the sum of the values of each adjacent vertex w multiplied with the weight of the incident edge $e = (v, w)$. Therefore, the following equation should hold: $e = We$. To make the equation system solvable an additional parameter called the exogenous input or the boundary condition E has to be added. This is a column vector containing external information for every vertex. Hubbell suggested that if this boundary condition is unknown $E = 1$ may be used.

The final equation is

$$s = E + Ws \tag{3.40}$$

Through a simple transformation this equation can be rewritten into $s = (I - W)^{-1} E$. This system has a solution if the matrix $(I - W)$ is invertible. Since $\frac{I}{(I-W)} = \sum_{k=1}^{\infty} W^k$ holds, this is identical to the problem of the convergence of the geometric series. According to Theorem 3.9.1, the series converges against $\frac{I}{(I-W)}$ if and only if the largest eigenvalue λ_1 of W is less than one.

The solution S of the equation system 3.40 is called Hubbell centrality c_{HBL} or Hubbell Index.

Bonacich's Bargaining Centrality. Both feedback centralities presented so far follow the idea of positive feedback: the centrality of a vertex is higher if it is connected to other high-valued vertices. In 1987 Phillip Bonacich [72] suggested a centrality which is not restricted to this concept. His idea supports both, the positive influence as seen for example in communication networks, and the negative influence as seen in bargaining situations. In bargaining situations a participant is strong if he is connected to individuals having no other options and are therefore weak.

Bonacich's bargaining centrality is defined for unweighted and directed graphs $G = (V, E)$ without loops. Therefore the adjacency matrix is not necessarily symmetric and contains only zeros and ones. The definition is

$$c_{\alpha,\beta}(i) = \sum_{j=1}^{n} (\alpha + \beta * c_{\alpha,\beta}(j))a_{ij}$$

or, in matrix notation,

$$c_{\alpha,\beta} = \alpha(I - \beta A)^{-1} A \mathbf{1} \tag{3.41}$$

As can easily be seen from the matrix notation, the parameter α is just a scaling factor. Bonacich suggests a value such that $\sum_{i=1}^{n} c_{\alpha,\beta}(i)^2 = n$ holds. Therefore only the second parameter β is of interest. This parameter may be chosen either positive or negative, covering positive or negative influence, respectively. The choice $\beta = 0$ leads to a trivial solution where the centrality correlates with the degree of the vertices. A negative value for β may lead to negative values for the centralities of the vertices. Additionally it follows from the equation that the larger $|\beta|$ the higher the impact of the structure of the network on the centrality index is.

Equation 3.41 is solvable if the inverse of $(I - \beta A)$ exists. According to Theorem 3.9.4, this inverse exists if no eigenvalue of A is equal to 1.

Theorem 3.9.4. *Let $M \in \mathbb{R}^{n \times x}$ be a matrix and $\lambda_1, \ldots, \lambda_n$ the eigenvalues of M.*

$$(I - M) \text{ is invertible} \iff \forall i \in \{1 \ldots n\} \ \lambda_i \neq 1 \,.$$

We call $c_{\alpha,\beta}$ the bargaining centrality c_{BRG}.

In this subsection three different approaches to measure feedback centrality values where presented. They seem very similar but differences are for example the coverage of weighted versus unweighted edges or positive versus negative influence networks.

3.9.3 Web Centralities

Many people use the World Wide Web to search for information about interesting topics. Due to the immense size of the network consisting of Web pages that are connected by hyperlinks powerful search engines are required. But how does a search engine decide which Web pages are appropriate for a certain search query? For this, it is necessary to score the Web pages according to their relevance or importance. This is partly done by a pure text search within the content of the pages. Additionally, search engines use the structure of the network to rank pages and this is where centrality indices come into play.[6]

In this section we discuss three main representatives of Web-scoring algorithms:

---

[6] Many concepts used for the 'Web centralities' are not new, especially the idea of eigenvectors as a centrality was known long before the Web was established. We decided to use this headline due to the interest of the last years into this topic.

- PageRank
- Hubs & Authorities
- SALSA

Whereas PageRank only takes the topological structure into account, the latter two algorithms combine the 'textual importance' of the Web page with its 'topological importance'. Moreover, Hubs & Authorities (sometimes also called HITS algorithm) assigns two score values to each Web page, called hub and authority. The third approach, SALSA, discussed at the end of this section, is in some sense a combination of the others.

In the following we assume that the Web is represented by a digraph $G = (V, E)$ with a one-to-one-correspondence between the Web pages and the vertices $v \in V$ as well as between the links and the directed edges $(v, w) \in E$.

The Model of a Random-Surfer. Before defining centrality indices suitable for the analysis of the Web graph it might be useful to model the behavior of a Web surfer. The most common model simulates the navigation of a user through the Web as as a random walk within the Web graph.

In Section 2.4 the concept of random walks in graphs was introduced. The Web graph $G = (V, E)$ is formally defined as V the set of all Web pages p_i where an edge $e = (p_i, p_j) \in E$ is drawn between two pages if and only if page p_i displays a link to page p_j. As the Web graph is usually not strongly connected the underlying transition matrix T is not irreducible and may not even be stochastic as 'sinks' (vertices without outgoing links) may exist. Therefore, the transition matrix T of the Web graph has to be modified such that the corresponding Markov chain converges to a stationary distribution.

To make the matrix T stochastic we assume that the surfer jumps to a random page after he arrived at a sink, and therefore we set all entries of all rows for sinks to $\frac{1}{n}$. The definition of the modified transition matrix T' is

$$t'_{ij} = \begin{cases} \frac{1}{d^+(i)}, & \text{if } (i, j) \in E \\ \frac{1}{n}, & \text{if } d^+(i) = 0 \end{cases}$$

This matrix is stochastic but not necessarily irreducible and the computation of the stationary distribution π' may not be possible. We therefore modify the matrix again to get an irreducible version T''. Let $E = \frac{1}{n}\mathbf{1}_n^T\mathbf{1}_n$ be the matrix with all entries $\frac{1}{n}$. This matrix can be interpreted as a 'random jump' matrix. Every page is directly reachable from every page by the same probability. To make the transition matrix irreducible we simply add this new matrix E to the existing matrix T':

$$T'' = \alpha T' + (1 - \alpha)E$$

Factor α is chosen from the range 0 to 1 and can be interpreted as the probability of either following a link on the page by using T' or performing a jump to a random page by using E. The matrix T'' is by construction stochastic and irreducible and the stationary distribution π'' may be computed for example with the power method (see Section 4.1.5).

By modifying E, the concept of a random jump may be adjusted for example more towards a biased surfer. Such modifications leads directly to a personalized version of the Web centrality indices presented here. For more details on this topic, see Section 5.2.

PageRank. PageRank is one of the main ingredients of the search engine Google [101] and was presented by Page et al. in 1998 [458]. The main idea is to score a Web page with respect to its topological properties, i.e., its location in the network, but independent of its content. PageRank is a feedback centrality since the score or centrality of a Web page depends on the number and centrality of Web pages linking to it

$$c_{\mathrm{PR}}(p) = d \sum_{q \in \Gamma_p^-} \frac{c_{\mathrm{PR}}(q)}{d^+(q)} + (1 - d), \tag{3.42}$$

where $c_{\mathrm{PR}}(q)$ is the PageRank of page q and d is a damping factor.

The corresponding matrix notation is

$$c_{\mathrm{PR}} = dP c_{\mathrm{PR}} + (1 - d)\mathbf{1}_n, \tag{3.43}$$

where the *transition matrix* P is defined by

$$p_{ij} = \begin{cases} \frac{1}{d^+(j)}, & \text{if } (j, i) \in E \\ 0, & \text{otherwise} \end{cases}$$

This is equivalent to $p_{ij} = \frac{1}{d^+(j)} a_{ji}$ or $P = D^+ A$ in matrix notation, where D^+ denotes the diagonal matrix where the i-th diagonal entry contains the out degree $d^+(i)$ of vertex i.

Mostly, the linear system 3.43 is solved by a simple power (or Jacobi) iteration:

$$c_{\mathrm{PR}}^k = dP c_{\mathrm{PR}}^{k-1} + (1 - d)\mathbf{1}_n. \tag{3.44}$$

The following theorem guarantees the convergence and a unique solution of this iteration if $d < 1$.

Theorem 3.9.5. *If $0 \leq d < 1$ then Equ. 3.43 has a unique solution $c_{PR}^* = (1 - d)(I_n - dP)^{-1}\mathbf{1}_n$ and the solutions of the dynamic system 3.44 satisfy $\lim_{k \to \infty} c_{PR}^k = c_{PR}^*$ for any initial state-vector c_{PR}^0.*

A slightly different approach is to solve the following dynamic system

$$c_{\mathrm{PR}}^k = dP c_{\mathrm{PR}}^{k-1} + \frac{\alpha^{k-1}}{n}\mathbf{1}_n, \tag{3.45}$$

where $\alpha^{k-1} = \|c_{\mathrm{PR}}^{k-1}\| - \|dP c_{\mathrm{PR}}^{k-1}\|$. The solutions of this system converge to $\frac{c_{\mathrm{PR}}^*}{\|c_{\mathrm{PR}}^*\|}$, the normalized solution of 3.44.

Hubs & Authorities. Shortly after the presentation of PageRank, Kleinberg introduced the idea of scoring Web pages with respect to two different 'scales' [359], called hub and authority, where

"A good hub is a page that points to many good authorities"

and

"A good authority is a page that is pointed to by many good hubs".

In contrast to PageRank, Kleinberg proposed to include also the content of a Web page into the scoring process. The corresponding algorithm for determining the hub and authority values of a Web page consists of two phases, where the first phase depends on the search query and the second phase deals only with the link structure of the associated network.

Given the search query σ, in the first phase of the algorithm an appropriate subgraph $G[V_\sigma]$ induced by a set of Web pages $V_\sigma \subseteq V$ is extracted, where

- V_σ should be comparably small,
- V_σ should contain many pages relevant for the search query σ, and
- V_σ should contain many important authorities.

This goal is achieved by using algorithm 1 to calculate V_σ, the set of relevant Web pages.

Algorithm 1: Hubs & Authorities, 1st Phase

Output: V_σ, the set of relevant pages

Use a text based search engine for search query σ
Let W_σ be the list of results
Choose $t \in \mathbb{N}$
Let $W_\sigma^t \subset W_\sigma$ contain the t pages ranked highest
$V_\sigma := W_\sigma^t$
forall $i \in W_\sigma^t$ **do**
    $V_\sigma := V_\sigma \cup \Gamma^+(i)$
    if $|\Gamma^-(i)| \leq r$ *(r is a user-specified bound)* **then**
        $V_\sigma := V_\sigma \cup \Gamma^-(i)$
    else
        choose $\Gamma_r^-(i) \subseteq \Gamma^-(i)$ such that $|\Gamma_r^-(i)| = r$
        $V_\sigma := V_\sigma \cup \Gamma_r^-(i)$
return V_σ

The second phase of the Hubs & Authorities algorithm consists of computing the hub and authority scores for the Web pages in $G[V_\sigma]$ which is done by taking into account the mutual dependence between hubs and authorities. This mutual dependence can be expressed by

$$c_{\text{HA-H}} = A_\sigma c_{\text{HA-A}} \text{ assuming } c_{\text{HA-A}} \text{ is known and} \tag{3.46}$$

$$c_{\text{HA-A}} = A_\sigma^T c_{\text{HA-H}} \text{ assuming } c_{\text{HA-H}} \text{ is known,} \tag{3.47}$$

where A_σ is the adjacency matrix of $G[V_\sigma]$.

Algorithm 2: Hubs & Authorities Iteration

Output: Approximations for $c_{\text{HA-H}}$ and $c_{\text{HA-A}}$

$c^0_{\text{HA-A}} := \mathbf{1}_n$
for $k = 1 \ldots$ **do**

$\quad c^k_{\text{HA-H}} := A_\sigma c^{k-1}_{\text{HA-A}}$

$\quad c^k_{\text{HA-A}} := A^T_\sigma c^k_{\text{HA-H}}$

$\quad c^k_{\text{HA-H}} := \dfrac{c^k_{\text{HA-H}}}{\|c^k_{\text{HA-H}}\|}$

$\quad c^k_{\text{HA-A}} := \dfrac{c^k_{\text{HA-A}}}{\|c^k_{\text{HA-A}}\|}$

Since neither $c_{\text{HA-H}}$ nor $c_{\text{HA-A}}$ are known, Kleinberg proposes an iterative procedure including a normalization step shown in algorithm 2. He shows

Theorem 3.9.6. *If A_σ is the adjacency matrix of $G[V_\sigma]$ then* $\lim\limits_{k \to \infty} c^k_{HA\text{-}A} = c_{\text{HA-A}}$ *and* $\lim\limits_{k \to \infty} c^k_{HA\text{-}H} = c_{\text{HA-H}}$, *where* $c_{\text{HA-A}}$ ($c_{\text{HA-H}}$) *is the first eigenvector of* $A^T_\sigma A_\sigma$ ($A_\sigma A^T_\sigma$)

Therefore, the given iterative procedure is nothing but solving the eigenvector-equations

$$\lambda c_{\text{HA-A}} = (A^T_\sigma A_\sigma) c_{\text{HA-A}}$$
$$\lambda c_{\text{HA-H}} = (A_\sigma A^T_\sigma) c_{\text{HA-H}}$$

for the largest eigenvalue by a power iteration, see Section 4.1.5. The vector $c_{\text{HA-A}}$ then contains the scores for the vertices with respect to their authority, whereas $c_{\text{HA-H}}$ is the vector of hub scores.

SALSA. In 2000, Lempel and Moran developed the SALSA (Stochastic Approach for Link Structure Analysis) algorithm [387]. The authors introduced this new Web-scoring approach to retain on the one hand the intuitive and appealing idea of hubs and authorities and to provide the index on the other hand with a higher robustness against the so called 'TKC effect'. TKC stands for *Tightly-Knit Community*, a small set of highly connected Web pages that in some cases may cause the Hubs & Authorities algorithm to rank the corresponding Web pages high even if they cover only a small (or no) aspect of the query. To this end Lempel and Moran combined the ideas of PageRank with those of Hubs & Authorities.

SALSA is a 3-phase algorithm where the first phase is identical to the first phase of the Hubs & Authorities algorithm: it constructs the graph $G[V_\sigma]$ for a certain search query σ (see algorithm 1). In the second phase an artificial bipartite undirected graph $\bar{G}_\sigma = (V^h_\sigma \dot\cup V^a_\sigma, \bar{E})$ according to the algorithm 3 is built. For the third phase of SALSA recall that the PageRank algorithm works with the transition matrix P which is the transposed adjacency matrix of the underlying graph with the non-zero columns weighted by their column sums. The Hubs & Authorities algorithm uses the product of the adjacency matrix A_σ of $G[V_\sigma]$ with its transpose. For SALSA the following matrices are defined:

P_σ: A_σ with each non-zero column weighted by its column sum
R_σ: A_σ with each non-zero row weighted by its row sum

Algorithm 3: SALSA, 2nd phase

Output: The bipartite undirected graph $\bar{G}_\sigma$

forall $i \in V_\sigma$ **do**
  if $d^+(i) > 0$ **then**
    $\lfloor$ create a copy i^h of i in V_σ^h
  if $d^-(i) > 0$ **then**
    $\lfloor$ create a copy i^a of i in V_σ^a

forall $e = (i,j) \in E(G[V_\sigma])$ **do**
  $\lfloor$ create an undirected edge $\bar{e} = \{i^h, j^a\}$ in $\bar{E}$

Then the indices of the non-zero columns (rows) of $R_\sigma P_\sigma^T$ correspond to the elements in V_σ^h and those of $P_\sigma^T R_\sigma$ to V_σ^a. Define

A_σ^h: non-zero rows and columns of $R_\sigma P_\sigma^T$
A_σ^a: non-zero rows and columns of $P_\sigma^T R_\sigma$

and use power iteration (see Section 4.1.5) to compute the SALSA authority scores $c_{\text{S-A}}$ and the SALSA hub scores $c_{\text{S-H}}$.

3.10   Dealing with Insufficient Connectivity

Most of the centrality-measures presented so far assume that the underlying network is connected. If this is not the case, computing these centralities might be a problem. For local centrality indices, such as degree centrality, this connectivity assumption has no implications. However, this is not the case in general. In this section, we investigate how to deal with disconnected undirected graphs and weakly connected digraphs.

Consider, for example, the centralities based on shortest paths, such as the measures based on eccentricity or closeness. Both centralities depend on the knowledge of the shortest paths length $d(u,v)$ between all pairs of vertices u and v. For a disconnected undirected graph or a weakly connected digraph there are pairs of vertices for which this length is not defined, and it is not clear how to deal with them. A very naive approach would be to restrict the computation of centrality values to subgraphs where the measure is well defined, i.e., to compute the centrality measure for a vertex with respect to its component or strong components in the case of digraphs. This approach is not very reasonable in most applications. Consider, for example, a (directed) network consisting of two (strong) components, where one is the complete graph of two vertices, and the other one is the complete graph with $n - 2$ vertices, where n is large. Then the above approach yields a closeness value of 1 for all vertices, but it seems obvious that the vertices in the large component are much more central than the two other vertices.

3.10.1 Intuitive Approaches

A common way to deal with this problem is to simply multiply the centrality values with the size of the component, following the intuition that the vertices in large components are more important. This seems to be reasonable, but it is not proper unless the centrality measure behaves proportional to the size of the network. Computational experiments of Poulin, Boily and Mâsse [481] indicate that this is not the case for closeness and eccentricity.

Two other repair mechanisms use inverse path lengths, and arbitrary fixed values for the distance between unconnected vertices. The latter possibility yields an approximation of the desired centrality values. However, Botafogo et al. [88] have shown that the result strongly depends on the fixed value k for the unconnected vertex pairs. They defined a closeness-based measure for digraphs

$$c_{C'}(u) = \frac{\sum_{v \in V} \sum_{w \in V} d(v, w)}{\sum_{v \in V} d(u, v)} \tag{3.48}$$

where the distance $d(u, v)$ between any unconnected vertex pair u and v is set to k. Clearly, an appropriate value for k is the number of vertices n, since the maximum distance between any two vertices is at most $n - 1$. In the digraph of Fig. 3.10 the vertex reaching all other vertices is w. For $k = 2n$ w becomes the vertex with highest centrality value but for $k = n$ the vertex v which does not reach w has highest value. This example shows that the choice of k will crucially influence the order of centrality index values assigned to the vertices.

Moreover, the centrality based on the eccentricity does not make sense anymore in non-connected graphs or in non-strongly connected digraphs. If the fixed value is large enough, then it dominates all other distances in the graph and yields centrality values that differ only in a very small range.

The usage of inverse path lengths makes it more difficult to interpret and compare centrality values. By substituting the path lengths in the closeness centrality by their inverses, and multiplying the sum of the inverse path length by $(n - 1)$, we do not obtain the closeness centrality but an entirely different centrality measure.

3.10.2 Cumulative Nominations

A more sophisticated approach was presented by Poulin, Boily and Mâsse [481]. Their starting point is a measure that is very similar to Bonacich's eigenvector

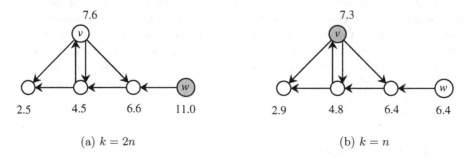

(a) $k = 2n$ (b) $k = n$

Fig. 3.10. The centralities with respect to the measure due to Botafogo et al. are shown. In each subfigure the vertex with the maximum value is colored grey

centrality. The *cumulative number of nominations* centrality $c_{\text{CNN}}(i)$ of vertex i is defined to be the ith component of the l_1-normalized eigenvector corresponding to the largest eigenvalue of $A + I$, where A is the adjacency matrix. In other words, c_{CNN} is the solution of $(A + I - \lambda_1 I)p = 0$ under the restriction $\sum_i p_i = 1$. Therefore, Bonacich's centrality and the cumulative number of nominations only differ by a constant. Poulin, Boily and Mâsse claim that their measure when computed by an iterative algorithm converges faster and is more stable. Moreover, their centrality may be applied to bipartite graphs as the graph corresponding to $(A + I)$ is not bipartite, even if the graph for A is.

Due to the normalization, c_{CNN} is not independent of the size of the connected component. The more vertices the component contains, the smaller the absolute centrality values become. But, using the approach of iteratively solving

$$c_{\text{CNN}}^{k+1} = (A + I)c_{\text{CNN}}^k,$$

the authors obtain the *cumulative nominations index of centrality*

$$c_{\text{CN}}(i) = c_{\text{CS}}(i) \lim_{k \to \infty} c_{\text{CNN}}^k(i),$$

where $c_{\text{CS}}(i)$ is the size of the component containing vertex i. This cumulative nominations index assigns a value of 1 to a vertex having an average structural position in a connected component.

In addition, the *cumulated nominations growth rate centrality index* of a vertex is defined as

$$c_{\text{CNG}}(i) = \lim_{k \to \infty} \left[\left(\sum_j a_{ij} c_{\text{CNN}}^k(j) + c_{\text{CNN}}^k(i) \right) \frac{1}{c_{\text{CNN}}^k(i)} \right],$$

and is the same for each vertex in a connected component.

This growth rate allows a comparison between different connected components. To this end, the *multi-component cumulated nominations centrality index* c_{MCN} is defined by

$$c_{\mathrm{MCN}}(i) = c_{\mathrm{CN}}(i)c_{\mathrm{CNG}}(i),$$

and, to take into account the (relative) size of the components (vertices in larger components should get a larger centrality score), we get the *corrected multi-component cumulated nominations centrality index*

$$c_{\mathrm{CMCN}}(i) = c_{\mathrm{MCN}}(i)c_{\mathrm{CS}}(i).$$

The authors report on computational experiments which indicate that neither c_{MCN} nor c_{CMCN} depends on n, hence both are centrality measures well suited for networks consisting of more than one component.

3.11   Graph- vs. Vertex-Level Indices

This section makes a connection between the analysis of a network on the level of vertices and on the level of the whole graph: Intuitively, it is clear that some graphs are more centralized than others, i.e., some graphs are more depending on the most central nodes than others. The star topology in which only one vertex v is connected to all others but all other vertices are only connected to v is a very centralized graph. A clique where every vertex is connected to every other vertex is not centralized.

Freeman [226] has proposed a very general approach with which the centralization $c_X(G)$ of a graph G can be calculated in relation to the values of any vertex centrality index c_X :

$$c_X(G) = \frac{\sum_{i \, \in \, V} c_X(j)^* - c_X(i)}{n - 1} \tag{3.49}$$

where $c_X(j)^*$ denotes the largest centrality value associated with any vertex in the graph under investigation. This approach measures the average difference in centrality between the most central point and all others. If normalized centralities in the range of $[0, 1]$ are used, the centralization value will also be in the range $[0, 1]$ (for further details to the normalization of centrality indices see Section 5.1). Other obvious possibilities to generate a graph index from the distribution of centrality indices are to compute the variance of the values or the maximal difference between centrality values or any other statistics on these values.

On the other hand, also a structural index for graphs like the Wiener Index (see Section 3.6.2) can be transformed into a structural index for vertices. We want to formalize this idea by first defining a structural index for graphs.

Definition 3.11.1 (Structural Index for Graphs). *Let* $G = (V, E)$ *be a weighted, directed or undirected multigraph. A function* $C \colon G \to \mathbb{R}$ *is called a structural index for graphs if and only if the following condition is satisfied:* $\forall G' \simeq G : \Longrightarrow C(G') = C(G))$.

Let $f : V \to \mathbb{R}$ be any structural index on the vertices of a graph and let $\odot$ be an operator on the set of all vertices V, like the summation over

$f(v)$, the average of all terms $f(v)$, the calculation of the variance of all $f(v)$ or the maximum/minimum operator. Then $\odot V =: f(G)$ defines a graph measure because all structural indices on vertices are stable under isomorphism. On the other hand, let $f : G \leftarrow \mathbb{R}$ be a structural index on the whole graph. Let $G(v,d)$ be the induced subgraph in which all vertices are contained with a hopping distance to v of no more than d. I.e. $G(v,d) = (V', E')$ is a subset of $G = (V,E)$ with $V' = \{w \in V | d(w,v) \leq d\}$ and $E' = \{(x,y) \in V' \times V' | (x,y) \in E\}$. Then $f(G(d,v))$ defines at least a structural index on the vertices of this graph, and in most cases also a reasonable vertex centrality index.

With this we can for example derive a centrality index from the Wiener Index by constraining the calculation of it to subgraphs with a small diameter. Such an approach might be useful in networks, where a message will not be transported more than k steps before it dies, as it is the case in some peer-to-peer network protocols. The new centrality index would then measure how well connected a node is within the subgraph of diameter k. It should be noted, however, that these subgraphs will be of different sizes in most cases. How centrality index values can be compared with each other in this case is discussed in the section about applying centrality indices to disconnected graphs (see Section 3.10).

3.12   Chapter Notes

Many interesting facts and a good overview of centrality indices used in social network analysis are given in [569]. Hage and Harary carried some of these ideas to a graph theoretic notation [269].

The notion of 'centrality' is very graphic and can be supported by adequate visualization. An approach to visualizing centrality measures in an intuitive way is [96] (see also Figure 1.2).

Closeness Centrality. Closeness centrality is often cited in the version of Sabidussi [500]. Nonetheless, it was also mentioned by Shimbel [519] but not as a centrality index. He defined the *dispersion* as the sum of all distances in a graph. Thus, it is a synonym for the Wiener Index [583] (see also Section 3.6.2). For directed graphs he defined the accessibility $A(i,G)$ of G from vertex i as $A(i,G) = \sum_{j \in V} d(i,j)$ and the accessibility $A^{-1}(i,G)$ of vertex i from G as $A^{-1}(i,G) = \sum_{j \in V} d(j,i)$. These two values are easily recognized as directed version of the closeness centrality.

Betweenness Centrality. Betweenness centrality was introduced by Freeman [226] and, independently, Anthonisse [32]. He was inspired by ideas of Bavelas [50]. Bavelas was the first who tried to map psychological situations to graphs. His main interest was the notion of centers (called 'innermost regions'), but he additionally discussed the following example: A group of Italian speaking women is employed in a large garment factory. Only one of them speaks English. Bavelas states: "It is difficult to imagine that the English speaking member would be other than central with respect to communication which had of necessity to pass through her (...) It is interesting in passing to point out the importance of the English speaking member with respect to the group's perception of the 'outside'. (...)To the extent that policy decisions are based upon information, as to the state of affairs 'outside', withholding information, coloring or distroting it in transmission, or in other ways misrepresenting the state of the outside will fundamentally affect these decisions."

Both edge and vertex betweenness have found many applications in the analysis of social networks (for example [457]), sexual intercourse networks (see [81]), or terrorist networks (for example [111]). Another interesting application is a graph clustering algorithm based on edge betweenness centrality [445]. Modern techniques try to approximate the expected congestion in a communication network using vertex betweenness [522]. According to this, the probability for congestion can be decreased by scaling the bandwidth proportional to betweenness centrality of a vertex. Nonetheless, betweenness centrality does not always scale with the expected congestion, as indicated in [304] (see also the introduction to Chapter 4).

The algorithmic complexity of this index is $\mathcal{O}(nm)$ for unweighted networks and $\mathcal{O}(nm + n^2 \log n)$ for weighted networks (for details see Section 4.2. Since this runtime makes it very hard to compute the betweenness centrality for graphs bigger than approximately 10,000 vertices, one should consider alternatives. In Section 4.3.1 we will discuss a way to approximate betweenness centrality. In Section 5.2.1 a personalized variant of the betweenness centrality is presented. A directed version of shortest-path betweenness centrality was first discussed in [32] and reinvented in [578].

Feedback Centralities. As far as we know, the first paper that defined a feedback centrality (without actually naming it in this way) was published by Seeley [510]. The status index of Katz was presented shortly afterwards in 1953 [352]. The index defined by Hubbell [319] and the approach presented by Bonacich [71] focus on the idea of propagating strength, where a high value vertex influences all vertices in his vicinity. All of these approaches solely focus on positive feedback relations. The first centrality index that covered negative feedback relation was presented by Bonacich [72].

Web Centralities. We covered three Web centralities: PageRank ([101, 458]), Hubs & Authorities ([359]) and SALSA ([387]). Especially for PageRank a whole bunch of papers is available and therefore we just give three references ([61, 378, 379]) which are a good starting point for further investigations of the topic.

4 Algorithms for Centrality Indices

Riko Jacob,* Dirk Koschützki, Katharina Anna Lehmann, Leon Peeters,*
and Dagmar Tenfelde-Podehl

The usefulness of centrality indices stands or falls with the ability to compute them quickly. This is a problem at the heart of computer science, and much research is devoted to the design and analysis of efficient algorithms. For example, shortest-path computations are well understood, and these insights are easily applicable to all distance based centrality measures. This chapter is concerned with algorithms that efficiently compute the centrality indices of the previous chapters.

Most of the distance based centralities can be computed by directly evaluating their definition. Usually, this naïve approach is reasonably efficient once all shortest path distances are known. For example, the closeness centrality requires to sum over all distances from a certain vertex to all other vertices. Given a matrix containing all distances, this corresponds to summing the entries of one row or column. Computing all closeness values thus traverses the matrix once completely, taking n^2 steps. Computing the distance matrix using the fastest known algorithms will take between n^2 and n^3 steps, depending on the algorithm, and on the possibility to exploit the special structure of the network. Thus, computing the closeness centrality for all vertices can be done efficiently in polynomial time. Nevertheless, for large networks this can lead to significant computation times, in which case a specialized algorithm can be the crucial ingredient for analyzing the network at hand. However, even a specialized exact algorithm might still be too time consuming for really large networks, such as the Web graph. So, for such huge networks it is reasonable to approximate the outcome with very fast, preferably linear time, algorithms.

Another important aspect of real life networks is that they frequently change over time. The most prominent example of this behavior is the Web graph. Rather than recomputing all centrality values from scratch after some changes, we prefer to somehow reuse the previous computations. Such dynamic algorithms are not only valuable in a changing environment. They can also increase performance for vitality based centrality indices, where the definition requires to repeatedly remove an element from the network. For example, dynamic all-pairs shortest paths algorithms can be used in this setting.

This chapter not only lists the known results, but also provides the ideas that make such algorithms work. To that end, Section 4.1 recapitulates some basic shortest paths algorithms, to provide the background for the more special-

---

* Lead authors

U. Brandes and T. Erlebach (Eds.): Network Analysis, LNCS 3418, pp. 62–82, 2005.
© Springer-Verlag Berlin Heidelberg 2005

ized centrality algorithms presented in Section 4.2. Next, Section 4.3 describes fast approximation algorithms for closeness centrality as well as for web centralities. Finally, algorithms for dynamically changing networks are considered in Section 4.4.

4.1 Basic Algorithms

Several good text books on basic graph algorithms are available, such as Ahuja, Magnanti, and Orlin [6], and Cormen, Leiserson, Rivest, and Stein [133]. This section recapitulates some basic and important algorithmic ideas, to provide a basis for the specific centrality algorithms in Section 4.2. Further, we briefly review the running times of some of the algorithms to indicate how computationally expensive different centrality measures are, especially for large networks.

4.1.1 Shortest Paths

The computation of the shortest-path distances between one specific vertex, called the source, and all other vertices is a classical algorithmic problem, known as the Single Source Shortest Path (SSSP) problem.

Dijkstra [146] provided the first polynomial-time algorithm for the SSSP for graphs with non-negative edge weights. The algorithm maintains a set of shortest-path labels $d(s, v)$ denoting the length of the shortest path found so-far between s and v. These labels are initialized to infinity, since no shortest paths are known when the algorithm starts. The algorithm further maintains a list P of permanently labeled vertices, and a list T of temporarily labeled vertices. For a vertex $v \in P$, the label $d(s, v)$ equals the shortest-path distance between s and v, whereas for vertices $v \in T$ the labels $d(s, v)$ are upper bounds (or estimates) on the shortest-path distances.

The algorithm starts by marking the source vertex s as permanent and inserting it into P, scanning all its neighbors $N(s)$, and setting the labels for the neighbors $v \in N(s)$ to the edge lengths: $d(s, v) = \omega(s, v)$. Next, the algorithm repeatedly removes a non-permanent vertex v with minimum label $d(s, v)$ from T, marks v as permanent, and scans all its neighbors $w \in N(v)$. If this scan discovers a new shortest path to w using the edge (v, w), then the label $d(s, w)$ is updated accordingly. The algorithm relies upon a priority queue for finding the next node to be marked as permanent. Implementing this priority queue as a Fibonacci heap, Dijkstra's algorithm runs in time $\mathcal{O}(m + n \log n)$. For unit edge weights, the priority queue can be replaced by a regular queue. Then, the algorithm boils down to Breadth-First Search (BFS), taking $\mathcal{O}(m + n)$ time. Algorithm 4 describes Dijkstra's algorithm more precisely.

Often, one is not only interested in the shortest-path distances, but also in the shortest paths themselves. These can be retraced using a function $pred(v) \in V$, which stores the predecessor of the vertex v on its shortest path from s. Starting at a vertex v, the shortest path from s is obtained by recursively applying $pred(v), pred(pred(v)), \ldots$, until one of the $pred()$ functions returns s. Since

---

Algorithm 4: Dijkstra's SSSP algorithm

Input: Graph $G = (V, E)$, edge weights $\omega : E \to \mathbb{R}$, source vertex $s \in V$
Output: Shortest path distances $d(s, v)$ to all $v \in V$

$P = \emptyset, T = V$
$d(s, v) = \infty$ for all $v \in V, d(s, s) = 0, pred(s) = 0$
while $P \neq V$ **do**
$\quad$ $v = \text{argmin}\{d(s, v) | v \in T\}$
$\quad$ $P := P \cup v, T := T \setminus v$
$\quad$ **for** $w \in N(v)$ **do**
$\quad\quad$ **if** $d(s, w) > d(s, v) + \omega(v, w)$ **then**
$\quad\quad\quad$ $d(s, w) := d(s, v) + \omega(v, w)$
$\quad\quad\quad$ $pred(w) = v$

---

the algorithm computes exactly one shortest path to each vertex, and no such shortest path can contain a cycle, the set of edges $\{(pred(v), v) \mid v \in V\}$, defines a spanning tree of G. Such a tree, which need not be unique, is called a shortest-paths tree.

Since Dijkstra's original work in 1954 [146], many improved algorithms for the SSSP have been developed. For an overview, we refer to Ahuja, Magnanti, and Orlin [6], and Cormen, Leiserson, Rivest, and Stein [133].

4.1.2 Shortest Paths Between All Vertex Pairs

The problem of computing the shortest path distances between all vertex pairs is called the All-Pairs Shortest Paths problem (APSP). All-pairs shortest paths can be straightforwardly computed by computing n shortest paths trees, one for each vertex $v \in V$, with v as the source vertex s. For sparse graphs, this approach may very well yield the best running time. In particular, it yields a running time of $\mathcal{O}(nm + n^2)$ for unweighted graphs.

For non-sparse graphs, however, this may induce more work than necessary. The following shortest path label optimality conditions form a crucial observation for improving the above straightforward APSP algorithm.

Lemma 4.1.1. *Let the distance labels $d(u, v), u, v \in V$, represent the length of some path from u to v. Then the labels d represent shortest path distances if and only if*

$$d(u, w) \leq d(u, v) + d(v, w) \text{ for all } u, v, w, \in V.$$

Thus, given some set of distance labels, it takes n^3 operations to check if these optimality conditions hold. Based on this observation and a theorem of Warshall [568], Floyd [217] developed an APSP algorithm that achieves an $\mathcal{O}(n^3)$ time bound, see Algorithm 5. The algorithm first initializes all distance labels to infinity, and then sets the distance labels $d(u, v)$, for $\{u, v\} \in E$, to the edge lengths $\omega(u, v)$. After this initialization, the algorithm basically checks whether there exists a vertex triple u, v, w for which the distance labels violate the condition in Lemma 4.1.1. If so, it decreases the involved distance label $d(u, w)$. This

check is performed in a triple for-loop over the vertices. Since we are looking
for all-pairs shortest paths, the algorithm maintains a set of predecessor indices
$pred(u, v)$ that contain the predecessor vertex of v on some shortest path from
u to v.

Algorithm 5: Floyd-Warshall's APSP algorithm

Input: Graph $G = (V, E)$, edge weights $\omega : E \to R$
Output: Shortest path distances $d(u, v)$ between all $u, v \in V$

$d(u, v) = \infty, pred(u, v) = 0$ for all $u, v \in V$
$d(v, v) = 0$ for all $v \in V$
$d(u, v) = \omega(u, v), pred(u, v) = u$ for all $\{u, v\} \in E$
for $v \in V$ **do**
$\quad$ **for** $\{u, w\} \in V \times V$ **do**
$\quad\quad$ **if** $d(u, w) > d(u, v) + d(v, w)$ **then**
$\quad\quad\quad$ $d(u, w) := d(u, v) + d(v, w)$
$\quad\quad\quad$ $pred(u, w) := pred(v, w)$

4.1.3 Dynamic All-Pairs Shortest Paths

The dynamic variant of the APSP problem is particularly interesting in the context of network analysis. The dynamic APSP problem consists of maintaining
an optimal set of shortest path distance labels $d(u, v), u, v \in V$, in a graph that
changes by edge insertions and deletions. Typically, one also wants to simultaneously maintain the corresponding shortest paths themselves, rather than only
the distances.

Thus, dynamic APSP's are of importance for vitality related questions, such
as how shortest path distances change upon removing an edge. Since removing
a vertex from a graph results in the removal of its incident edges, vertex vitality
corresponds to sequences of edge removals in a dynamic APSP setting. Further,
the dynamic APSP is clearly applicable in the setting of the changing Web graph.

The challenge for the dynamic APSP problem is to do better than recomputing a set of optimal distance labels from scratch after an update. Recently, Demetrescu and Italiano [142] described an algorithm for the dynamic APSP problem
on directed graphs with non-negative real-valued edge weights. Per edge insertion, edge deletion, or edge weight change, their algorithm takes $\mathcal{O}(n^2 \log^3 n)$
amortized time to maintain the all-pairs shortest path distance labels. As the
algorithm and its analysis are quite involved, their discussion falls outside the
scope of this book. Instead, we refer to Demetrescu and Italiano [142] for details
on the dynamic APSP.

Further, Thorup [549] provides an alternative description of the algorithm,
as well as an improved amortized update time of $\mathcal{O}(n^2(\log n + \log^2(m + n/n)))$.
Moreover, the improved algorithm allows for negative weights. Roditty and

Zwick [496] argue that the dynamic SSSP problem on weighted graphs is as difficult as the static APSP problem. Further, they present a randomized algorithm for the dynamic APSP, returning correct results with very high probability, with improved amortized update time for sparse graphs.

4.1.4   Maximum Flows and Minimum-Cost Flows

For flow betweenness (see Section 3.6.1), the maximum flow between a designated source node s and a designated sink node t needs to be computed. The maximum-flow problem has been studied extensively in the literature, and several algorithms are available. Some are generally applicable, some focus on restricted cases of the problem, such as unit edge capacities, and others provide improvements that may have more theoretical than practical impact. The same applies to minimum-cost flows, with the remark that minimum-cost flow algorithms are even more complex.

Again, we refer to the textbooks by Ahuja, Magnanti, and Orlin [6], and Cormen, Leiserson, Rivest, and Stein [133] for good in-depth descriptions of the algorithms. To give an idea of flow algorithms' worst-case running times, and of the resulting impact on centrality computations in large networks, we briefly mention the following algorithms. The preflow-push algorithm by Goldberg and Tarjan [252] runs in $\mathcal{O}(nm \log(n^2/m))$, and the capacity scaling algorithm by Ahuja and Orlin [8] runs in $\mathcal{O}(nm \log U)$, where U is the largest edge capacity. For minimum cost flows, the capacity scaling algorithm by Edmonds and Karp [172] runs in $\mathcal{O}((m \log U)(m + n \log n))$.

Alternatively, both maximum flow and minimum-cost flow problems can be solved using linear programming. The linear program for flow problems has a special structure which guarantees an integer optimal solution for any integer inputs (costs, capacities, and net inflows). Moreover, specialized network simplex algorithms for flow-based linear programs with polynomial running times are available.

4.1.5   Computing the Largest Eigenvector

Several centrality measures described in this part of the book are based on the computation of eigenvectors of a given matrix. This section provides a short introduction to the computation of eigenvectors and eigenvalues. In general, the problem of computing eigenvalues and eigenvectors is non-trivial, and complete books are dedicated to this topic. We focus on a single algorithm and sketch the main idea. All further information, such as optimized algorithms, or algorithms for special matrices, are available in textbooks like [256, 482]. Furthermore, Section 14.2 (chapter on spectral analysis) considers the computation of all eigenvalues of the matrix representing a graph.

The eigenvalue with largest absolute value and the corresponding eigenvector can be computed by the power method, which is described by Algorithm 6. As input the algorithm takes the matrix A and a start vector $q^{(0)} \in \mathbb{R}^n$ with $||q^{(0)}||_2 = 1$. After the k-th iteration, the current approximation of the largest eigenvalue in absolute value and the corresponding eigenvector are stored in the variables $\lambda^{(k)}$ and $q^{(k)}$, respectively.

Algorithm 6: Power method for computating the largest eigenvalue

---

Input: Matrix $A \in \mathbb{R}^{n \times n}$ and vector $||q^{(0)}||_2 = 1$
Output: Largest eigenvalue $\lambda^{(k)}$ in absolute value
and corresponding eigenvector $q^{(k)}$
$k := 1$
repeat
$\quad z^{(k)} := Aq^{(k-1)}$
$\quad q^{(k)} := z^{(k)}/||z^{(k)}||_2$
$\quad \lambda^{(k)} := (q^{(k)})^T Aq^{(k)}$
$\quad k := k + 1$
until $\lambda^{(k)}$ and $q^{(k)}$ are acceptable approximations

---

The power method is guaranteed to converge if the matrix $A \in \mathbb{C}^{n \times n}$ has a dominant eigenvalue, i.e., $|\lambda_1| > |\lambda_i|$ for $i \in \{2 \ldots n\}$, or, alternatively, if the matrix $A \in \mathbb{R}^{n \times n}$ is symmetric. The ratio $\frac{|\lambda_2|}{|\lambda_1|}$ of the second largest and the largest eigenvalues determines the rate of convergence, as the approximation error decreases with $\mathcal{O}((\frac{|\lambda_2|}{|\lambda_1|})^k)$. Further details on the power method can be found in many textbooks on linear algebra, e.g., Wilkinson [587].

As the power method only requires matrix-vector multiplication, it is particularly suited for large matrices. For one iteration, it suffices to scan over the matrix once. So, the power method can be reasonably efficient, even without storing the complete matrix in main memory.

4.2 Centrality-Specific Algorithms

As already mentioned, most centrality indices can be computed reasonably fast by directly following their definition. Nevertheless, improvements over this straightforward approach are possible. This section elaborates on two algorithmic ideas for such an improvement.

4.2.1 Betweenness Centrality

Recall the definition of the betweenness centrality of a vertex $v \in V$:

$$c_B(v) = \sum_{s \neq v \neq t \in V} \frac{\sigma_{st}(v)}{\sigma_{st}},$$

with σ_{st} being the number of shortest paths between vertices s and t, and $\sigma_{st}(v)$ the number of those paths passing through vertex v. A straightforward idea for computing $c_B(v)$ for all $v \in V$ is the following. First compute tables with the length and number of shortest paths between all vertex pairs. Then, for each vertex v, consider all possible pairs s and t, use the tables to identify the fraction of shortest s-t-paths through v, and sum these fractions to obtain the betweenness centrality of v.

For computing the number of shortest paths in the first step, one can adjust Dijkstra's algorithm as follows. From Lemma 4.1.1, observe that a vertex v is on a shortest path between two vertices s and t if and only if $d(s,t) = d(s,v) + d(v,t)$. We replace the predecessor vertices by predecessor sets $pred(s,v)$, and each time a vertex $w \in N(v)$ is scanned for which $d(s,t) = d(s,v) + d(v,t)$, that vertex is added to the predecessor set $pred(s,v)$. Then, the following relation holds:

$$\sigma_{sv} = \sum_{u \in pred(s,v)} \sigma_{su}.$$

Setting $pred(s,v) = s$ for all $v \in N(s)$, we can thus compute the number of shortest paths between a source vertex s and all other vertices. This adjustment can easily be incorporated into Dijkstra's algorithm, as well as in the BFS for unweighted graphs.

As for the second step, vertex v is on a shortest s-t-path if $d(s,t) = d(s,v) + d(v,t)$. If this is the case, the number of shortest s-t-paths using v is computed as $\sigma_{st}(v) = \sigma_{sv} \cdot \sigma_{vt}$. Thus, computing $c_B(v)$ requires $\mathcal{O}(n^2)$ time per vertex v because of the summation over all vertices $s \neq v \neq t$, yielding $\mathcal{O}(n^3)$ time in total. This second step dominates the computation of the length and the number of shortest paths. Thus, the straightforward idea for computing betweenness centrality has an overall running time of $\mathcal{O}(n^3)$.

Brandes [92] describes a specific algorithm that computes the betweenness centrality of all vertices in a graph in $\mathcal{O}(nm + n^2 \log n)$ time for weighted graphs, and $\mathcal{O}(nm)$ time for unweighted graphs. Note that this basically corresponds to the time complexity for the n SSSP computations in the first step of the straightforward idea. We describe this betweenness algorithm below.

The pair-dependency of a vertex pair $s, t \in V$ on an intermediate vertex v is defined as $\delta_{st}(v) = \sigma_{st}(v)/\sigma_{st}$, and the dependency of a source vertex $s \in V$ on a vertex $v \in V$ as

$$\delta_{s\bullet}(v) = \sum_{t \in V} \delta_{st}(v).$$

So, the betweenness centrality of a vertex v can be computed as $c_B(v) = \sum_{s \neq v \in V} \delta_{s\bullet}(v)$.

The betweenness centrality algorithm exploits the following recursive relations for the dependencies $\delta_{s\bullet}(v)$.

Theorem 4.2.1 (Brandes [92]). *The dependency $\delta_{s\bullet}(v)$ of a source vertex $s \in V$ on any other vertex $v \in V$ satisfies*

$$\delta_{s\bullet}(v) = \sum_{w : v \in pred(s,w)} \frac{\sigma_{sv}}{\sigma_{sw}} (1 + \delta_{s\bullet}(w)).$$

Proof. First, extend the variables for the number of shortest paths and for the dependency as follows. Define $\sigma_{st}(v, e)$ as the number of shortest paths from s to t that contain both the vertex $v \in V$ and the edge $e \in E$. Further, define the pair-dependency of a vertex pair s, t on both a vertex v and an edge e as $\delta_{st}(v, e) = \sigma_{st}(v, e)/\sigma_{st}$. Using these, we write

$$\delta_{s\bullet}(v) = \sum_{t \in V} \delta_{st}(v) = \sum_{t \in V} \sum_{w:v \in pred(s,w)} \delta_{st}(v, \{v, w\}).$$

Consider a vertex w for which $v \in pred(s, w)$. There are σ_{sw} shortest paths from s to w, of which σ_{sv} go from s to v and then use the edge $\{v, w\}$. Thus, given a vertex t, a fraction σ_{sv}/σ_{sw} of the number of shortest paths $\sigma_{st}(w)$ from s to $t \neq w$ using w also uses the edge $\{v, w\}$. For the pair-dependency of s and t on v and $\{v, w\}$, this yields

$$\delta_{st}(v, \{v, w\}) = \begin{cases} \dfrac{\sigma_{sv}}{\sigma_{sw}} & \text{if } t = w, \\ \dfrac{\sigma_{sv}}{\sigma_{sw}} \cdot \dfrac{\sigma_{st}(w)}{\sigma_{st}} & \text{if } t \neq w. \end{cases}$$

Exchanging the sums in the above summation, and substituting this relation for $\delta_{st}(v, \{v, w\})$ gives

$$\sum_{w:v \in pred(s,w)} \sum_{t \in V} \delta_{st}(v, \{v, w\}) = \sum_{w:v \in pred(s,w)} \left(\frac{\sigma_{sv}}{\sigma_{sw}} + \sum_{t \in V \setminus w} \frac{\sigma_{sv}}{\sigma_{sw}} \cdot \frac{\sigma_{st}(w)}{\sigma_{st}} \right)$$

$$= \sum_{w:v \in pred(s,w)} \frac{\sigma_{sv}}{\sigma_{sw}} (1 + \delta_{s\bullet}(w)).$$

$\square$

The betweenness centrality algorithm is now stated as follows. First, compute n shortest-paths trees, one for each $s \in V$. During these computations, also maintain the predecessor sets $pred(s, v)$. Second, take some $s \in V$, its shortest-paths tree, and its predecessor sets, and compute the dependencies $\delta_{s\bullet}(v)$ for all other $v \in V$ using the dependency relations in Theorem 4.2.1. For vertex s, the dependencies can be computed by traversing the vertices in non-increasing order of their distance from s. In other words, start at the leaves of the shortest-paths tree, work backwardly towards s, and afterwards proceed with the next vertex s. To finally compute the centrality value of vertex v, we merely have to add all dependencies values computed during the n different SSSP computations. The resulting $\mathcal{O}(n^2)$ space usage can be avoided by immediately adding the dependency values to a 'running centrality score' for each vertex.

This algorithm computes the betweenness centrality for each vertex $v \in V$, and requires the computation of one shortest-paths tree for each $v \in V$. Moreover, it requires a storage linear in the number of vertices and edges.

Theorem 4.2.2 (Brandes [92]). *The betweenness centrality $c_B(v)$ for all $v \in V$ can be computed in $\mathcal{O}(nm + n^2 \log n)$ time for weighted graphs, and in $\mathcal{O}(nm)$ time for unweighted graphs. The required storage space is $\mathcal{O}(n + m)$.*

Other shortest-path based centrality indices, such as closeness centrality, graph centrality, and stress centrality can be computed with similar shortest-paths tree computations followed by iterative dependency computations. For further details on this, we refer to Brandes [92].

4.2.2 Shortcut Values

Another algorithmic task is to compute the shortcut value for all edges of a directed graph $G = (V, E)$, as introduced in Section 3.6.3. More precisely, the task is to compute the shortest path distance from vertex u to vertex v in $G_e = (V, E \setminus \{e\})$ for every directed edge $e = (u, v) \in E$. The shortcut value for edge e is a vitality based centrality measure for edges, defined as the maximum increase in shortest path length (absolute, or relative for non-negative distances) if e is removed from the graph.

The shortcut values for all edges can be naïvely computed by $m = |E|$ calls to a SSSP routine. This section describes an algorithm that computes the shortcut values for all edges with only $n = |V|$ calls to a routine that is asymptotically as efficient as a SSSP computation. To the best of our knowledge this is the first detailed exposition of this algorithm, which is based on an idea of Brandes.

We assume that the directed graph G contains no negative cycles, such that $d(i, j)$ is well defined for all vertices i and j. To simplify the description we assume that the graph contains no parallel edges, such that an edge is identified by its endpoints.

The main idea is to consider some vertex u, and to execute one computation to determine the shortcut values for all edges starting at u. These shortcut values are defined by shortest paths that start at vertex u and reach an adjacent vertex v, without using the edge (u, v). To compute this, define $\alpha_i = d(u, i)$ to be the length of a shortest path from u to i, the well known shortest path distance. Further, let the variable $\tau_i \in V$ denote the second vertex (identifying the first edge of the path) of all paths from u to i with length α_i, if this is unique, otherwise it is undefined, $\tau_i = \perp$. Thus, $\tau_i = \perp$ implies that there are at least two paths of length α_i from u to i that start with different edges. Finally, the value β_i is the length of the shortest path from u to i that does not have τ_i as the second vertex, ∞ if no such path exists, or $\beta_i = \alpha_i$ if $\tau_i = \perp$.

Assume that the values α_v, τ_v, and β_v are computed for a neighbor v of u. Then, the shortcut value for the edge (u, v) is α_v if $\tau_v \neq v$, i.e., the edge (u, v) is not the unique shortest path from u to v. Otherwise, if $\tau_v = v$, the value β_v is the shortcut value for (u, v). Hence, it remains to compute the values $\alpha_i, \tau_i, \beta_i$ for $i \in V$. The algorithm exploits that the values $\alpha_i, \tau_i, \beta_i$ obey some recursions. At the base of these recursions we have:

$$\alpha_u = 0, \quad \tau_u = \emptyset, \quad \beta_u = \infty$$

The values α_j obey the shortest paths recursion:

$$\alpha_j = \min_{i:(i,j)\in E} \left(\alpha_i + \omega(i,j) \right)$$

To define the recursion for τ_j, it is convenient to consider the set of incoming neighbors I_j of vertices from which a shortest path can reach j,

$$I_j = \{i \mid (i,j) \in E \text{ and } \alpha_j = \alpha_i + \omega(i,j)\}.$$

It holds that

$$\tau_j = \begin{cases} j & \text{if } I_j = \{u\}, \\ a & \text{if } a = \tau_i \text{ for all } i \in I_j \text{(all predecessors have first edge } (u,a)), \\ \bot & \text{otherwise.} \end{cases}$$

The value τ_j is only defined if all shortest paths to vertex j start with the same edge, which is the case only if all τ_i values agree on the vertices in I_j. For the case $\tau_j = \bot$ it holds that $\beta_j = \alpha_j$, otherwise

$$\beta_j = \min \left\{ \min_{\substack{i:(i,j)\in E, \\ \tau_i = \tau_j}} \beta_i + \omega(i,j) \quad , \quad \min_{\substack{i:(i,j)\in E, \\ \tau_i \neq \tau_j}} \alpha_i + \omega(i,j) \right\}.$$

To see this, consider the path p that achieves β_j, i.e., a shortest path p from u to j that does not start with τ_j. If the last vertex i of p before j has $\tau_i = \tau_j$, the path p up to i does not start with τ_j, and this path is considered in β_i and hence in β_j. If instead the path p has as the next to last vertex i, and $\tau_i \neq \tau_j$, then one of the shortest paths from u to i does not start with τ_j, and the length of p is $\alpha_i + \omega(i,j)$.

With the above recursions, we can efficiently compute the values $\alpha_i, \tau_i, \beta_i$. For the case of positive weights, any value α_i depends only on values α_j that are smaller than α_i, so these values can be computed in non-decreasing order (just as Dijkstra's algorithm does). If all edge weights are positive, the directed graph containing all shortest paths (another view on the sets I_j) is acyclic, and the values τ_i can be in topological order. Otherwise, we have to identify the strongly connected components of G, and contract them for the computation of τ. Observe that β_i only depends upon β_j if $\beta_j \leq \beta_i$. Hence, these values can be computed in non-decreasing order in a Dijkstra-like algorithm. In the unweighted case, this algorithm does not need a priority queue and its running time is only that of BFS.

If there are negative edge weights, but no negative cycles, the Dijkstra-like algorithm is replaced by a Bellman-Ford type algorithm to compute the α values. The computation of τ remains unchanged. Instead of computing β_i, we compute $\beta'_i = \beta_i - \alpha_i$, i.e., we apply the shortest-paths potential to avoid negative edge weights. This replaces all $\omega(i,j)$ terms with terms of the form $\omega(i,j) - \alpha_j + \alpha_i \geq 0$, and hence the β'_i values can be set in increasing order, and this computes the β_i values as well.

Note that the above method can be modified to also work in networks with parallel edges. There, the first edge of a path is no longer identified by the second vertex of the path, such that this edge should be used instead. We can even modify the method to compute the shortcut value of the vertex v, i.e., the two neighbors of v whose distance increases most if v is deleted from the network. To achieve this, negate the length and direction of the incoming edges, run the above algorithm, and subtract the length of the outgoing edges from the resulting β_i values on the neighbors of v. In this way, for all pairs of neighbors that can reach each other through v the difference between the direct connection and the shortest alternative are computed.

Summarizing, we showed that in the above mentioned types of graphs all shortcut values can be computed in the time of computing n times a SSSP.

4.3   Fast Approximation

Most of the centralities introduced in Chapter 3 can be computed in polynomial time. Although this is a general indication that such computations are feasible, it might still be practically impossible to analyze huge networks in reasonable time. As an example, it may be impossible to compute betweenness centrality for large networks, even when using the improved betweenness algorithm of Section 4.2.1. This phenomenon is particularly prominent when investigating the web graph. For such a huge graph, we typically do not want to invest more than a small number of scans over the complete input.

With this limited computational investment, it might not be possible to determine exact centrality values. Instead, the focus should be on approximate solutions and their quality. In this setting, approximation algorithms provide a guaranteed compromise between running time and accuracy.

Below, we describe an approximation algorithm for the calculation of closeness centrality, and then adapt this algorithm to an approximative calculation for betweenness centrality. Next, Section 4.3.2 discusses approximation methods for the computation of web centralities.

4.3.1   Approximation of Centralities Based on All Pairs Shortest Paths Computations

We have argued above that the calculation of centrality indices can require a lot of computing time. This also applies to the computation of all-pairs shortest paths, even when using the algorithms discussed in Section 4.1.2. In many applications, it is valuable to instead compute a good approximate value for the centrality index, if this is faster. With the random sampling technique introduced by Eppstein and Wang [179], the closeness centrality of all vertices in a weighted, undirected graph can be approximated in $\mathcal{O}(\frac{\log n}{\epsilon^2}(n \log n + m))$ time. The approximated value has an additive error of at most $\epsilon \Delta_G$ with high probability, where ϵ is any fixed constant, and Δ_G is the diameter of the graph. We adapt this technique for the approximative calculation of betweenness centrality, yielding an approximation of the betweenness centrality of all vertices in a weighted, directed graph with an additive error of $(n-2)\epsilon$, and with the same time bound as above.

The following randomized approximative algorithm estimates the closeness centrality of all vertices in a weighted graph by picking K sample vertices and computing single source shortest paths (SSSP) from each sample vertex to all other vertices. Recall the definition of closeness centrality of a vertex $v \in V$:

$$c_C(v) = \frac{\sum\limits_{x \in V} d(v, x)}{n - 1}. \tag{4.1}$$

The centrality $c_C(v)$ can be estimated by the calculation of the distance of v to K other vertices $v_1, \ldots, v_K$ as follows

$$\hat{c}_C(v) = \frac{n}{K \cdot (n - 1)} \sum_{i=1}^{K} d(v, v_i). \tag{4.2}$$

For undirected graphs, this calculates the average distance from v to K other vertices, then scales this to the sum of distances to/from all other n vertices, and divides by $n - 1$. As both c_C and $\hat{c}_C$ consider average distances in the graph, the expected value of $\hat{c}_C(v)$ is equal to $c_C(v)$ for any K and v. This leads to the following algorithm:

1. Pick a set of K vertices $\{v_1, v_2, \ldots, v_K\}$ uniformly at random from V.
2. For each vertex $v \in \{v_1, v_2, \ldots, v_K\}$, solve the SSSP problem with that vertex as source.
3. For each vertex $v \in V$, compute $\hat{c}_C(v) = \dfrac{n}{K \cdot (n - 1)} \sum\limits_{i=1}^{K} d(v, v_i)$

We now recapitulate the result from [179] to compute the required number of sample vertices K that suffices to achieve the desired approximation. The result uses Hoeffding's Bound [299]:

Lemma 4.3.1. *If $x_1, x_2, \ldots, x_K$ are independent with $a_i \leq x_i \leq b_i$, and $\mu = E[\sum x_i / K]$ is the expected mean, then for $\xi > 0$*

$$Pr\left\{ \left| \frac{\sum_{i=1}^{K} x_i}{K} - \mu \right| \geq \xi \right\} \leq 2 \cdot e^{-2K^2 \xi^2 / \sum_{i=1}^{K} (b_i - a_i)^2}. \tag{4.3}$$

By setting x_i to $\frac{n \cdot d(v_i, u)}{n - 1}$, μ to $c_C(v)$, a_i to 0, and b_i to $\frac{n\Delta}{n-1}$, we can bound the probability that the error of estimating $c_C(v)$ by $\hat{c}_C(v)$, for any vertex, is more than ξ:

$$Pr\left\{ \left| \frac{\sum_{i=1}^{K} x_i}{K} - \mu \right| \geq \xi \right\} \leq 2 \cdot e^{-2K^2 \xi^2 / \sum_{i=1}^{K} (b_i - a_i)^2} \tag{4.4}$$

$$= 2 \cdot e^{-2K^2 \xi^2 / K (\frac{n\Delta}{n-1})^2} \tag{4.5}$$

$$= 2 \cdot e^{-\Omega(K \xi^2 / \Delta^2)} \tag{4.6}$$

If we set ξ to $\epsilon \cdot \Delta$ and use $\Theta(\frac{\log n}{\epsilon^2})$ samples, the probability of having an error greater than $\epsilon \cdot \Delta$ is at most $1/n$ for every estimated value.

The running time of an SSSP algorithm is $\mathcal{O}(n + m)$ in unweighted graphs, and $\mathcal{O}(m + n \log n)$ in weighted graphs, yielding a total running time of $\mathcal{O}(K \cdot (n + m))$ and $\mathcal{O}(K(m + n \log n))$ for this approach, respectively. With K set to $\Theta(\frac{\log n}{\epsilon^2})$, this results in running times of $\mathcal{O}(\frac{\log n}{\epsilon^2}(n + m))$ and $\mathcal{O}(\frac{\log n}{\epsilon^2}(m + n \log n))$.

We now adapt this technique to the estimation of betweenness centrality in weighted and directed graphs. As before, a set of K sample vertices is randomly picked from V. For every source vertex v_i, we calculate the total dependency $\delta_{v_i \bullet}(v)$ (see Section 3.4.2) for all other vertices v, and sum them up. The estimated betweenness centrality $\hat{c}_B(v)$ is then defined as

$$\hat{c}_B(v) = \sum_{i=1}^{K} \frac{n}{K} \delta_{v_i \bullet}(v). \tag{4.7}$$

Again, the expected value of $\hat{c}_B(v)$ is equal to $c_B(v)$ for all K and v. For this new problem, we set x_i to $n \cdot \delta_{v_i \bullet}$, μ to $c_B(v)$, and a_i to 0. The total dependency $\delta_{v_i \bullet}(v)$ can be at most $n - 2$ if and only if v is the only responsible vertex for all shortest paths leaving v_i. Thus, we set b_i to $n(n - 2)$. Using the bound (4.3.1), it follows that the probability that the difference between the estimated betweenness centrality $\hat{c}_B(v)$ and the betweenness centrality $c_B(v)$ is more than ξ is

$$\Pr\{|\hat{c}_B(v) - c_B(v)| \geq \xi\} \leq 2e^{-2K^2\xi^2/K \cdot (n(n-2))^2} \tag{4.8}$$

$$= 2 \cdot e^{-2K\xi^2/(n(n-2))^2} \tag{4.9}$$

Setting ξ to $\epsilon(n(n-2))$, and the number of sample vertices K to $\Theta(\log n/\epsilon^2)$, the difference between the estimated centrality value and the correct value is at most $\epsilon n(n - 1)$ with probability $1/n$. As stated above, the total dependency $\delta_{v_i \bullet}(v)$ of a vertex v_i can be calculated in $\mathcal{O}(n + m)$ in unweighted graphs and in $\mathcal{O}(m + n \log n)$ in weighted graph. With K set as above, this yields running times of $\mathcal{O}(\frac{\log n}{\epsilon^2}(n + m))$ and $\mathcal{O}(\frac{\log n}{\epsilon^2}(m + n \log n))$, respectively. Hence, the improvement over the exact betweenness algorithm in Section 4.2.1 is the factor K which replaces a factor n, for the number of SSSP-like computations.

Note that this approach can be applied to many centrality indices, namely those that are based on summations over some primitive term defined for each vertex. As such, those indices can be understood as taking a normalized average, which makes them susceptible to random vertex sampling.

4.3.2   Approximation of Web Centralities

Most of the approximation and acceleration techniques for computing Web-centralities are designed for the PageRank method. Therefore, in the following we concentrate on this method. A good short overview of existing acceleration PageRank techniques can be found in [378]. We distinguish the following acceleration approaches:

- approximation by cheaper computations, usually by avoiding matrix multiplications,
- acceleration of convergence,
- solving a linear system of equations instead of solving an eigenvector problem,
- using decomposition of the Web-graph, and
- updating instead of recomputations.

We discuss these approaches separately below.

Approximation by Cheaper Computations. In [148] and [149], Ding et al. report on experimental results indicating that the rankings obtained by both PageRank and Hubs & Authorities are strongly correlated to the in-degree of the vertices. This especially applies if only the top-20 query results are taken into consideration. Within the unifying framework the authors propose, the ranking by in-degree can be viewed as an intermediate between the rankings produced by PageRank and Hubs & Authorities. This result is claimed to also theoretically show that the in-degree is a good approximation of both PageRank and Hubs & Authorities. This seems to be true for graphs in which the rankings of PageRank and Hubs & Authorities are strongly related. However, other authors performed computational experiments with parts of the Web graph, and detected only little correlation between in-degree and PageRank, see, e.g., [463]. A larger scale study confirming the latter result can be found in [380].

Acceleration of Convergence. The basis for this acceleration technique is the power method for determining the eigenvector corresponding to the largest eigenvalue, see Section 4.1.5.

Since each iteration of the power-method consists of matrix multiplication, and is hence very expensive for the Web graph, the goal is to reduce the number of iterations. One possibility was proposed by Kamvar et al. [340] and extended by Haveliwala et al. [292]. In the first paper the authors propose a quadratic extrapolation that is based on the so-called Aitken Δ^2 method. The Aitken extrapolation assumes that an iterate $x^{(k-2)}$ can be written as a linear combination of the first two eigenvectors u and v. With this assumption, the next two iterates are linear combinations of the first two eigenvectors as well:

$$\begin{aligned} x^{(k-2)} &= u + \alpha v \\ x^{(k-1)} = Ax^{(k-2)} &= u + \alpha\lambda_2 v \\ x^{(k)} = Ax^{(k-1)} &= u + \alpha\lambda_2^2 v. \end{aligned}$$

By defining

$$y_i = \frac{\left(x_i^{(k-1)} - x_i^{(k-2)}\right)^2}{x_i^{(k)} - 2x_i^{(k-1)} + x_i^{(k-2)}}$$

and some algebraic reductions (see [340]) we get $y = \alpha v$ and hence

$$u = x^{(k-2)} - y. \tag{4.10}$$

Note that the assumption that $x^{(k-2)}$ can be written as a linear combination of u and v is only an approximation, hence (4.10) is also only an approximation of the first eigenvector, which is then periodically computed during the ordinary power method.

For the quadratic extrapolation the authors assume that an iterate $x^{(k-2)}$ is a linear combination of the first three eigenvectors u, v and w. Using the characteristic polynomial they arrive at an approximation of u only depending on the iterates:

$$u = \beta_2 x^{(k-2)} + \beta_1 x^{(k-1)} + \beta_0 x^{(k)}.$$

As in the Aitken extrapolation, this approximation is periodically computed during the ordinary power method. The authors report on computational experiments indicating that the accelerated power method is much faster than the ordinary power method, especially for large values of the damping factor d, for which the power method converges very slowly. As we discuss in Section 5.5.2, this is due to the fact that d equals the second largest eigenvalue (see [290]), hence a large value for d implies a small eigengap.

The second paper [292] is based on the ideas described above. Instead of having a linear combination of only two or three eigenvector approximations, the authors assume that $x^{(k-h)}$ is a linear combination of the first $h+1$ eigenvector approximations. Since the corresponding eigenvalues are assumed to be the h-th roots of unity, scaled by d, it is possible to find a simple closed form for the first eigenvector. This acceleration step is used as above.

Kamvar et al. [338] presented a further idea to accelerate the convergence, based on the observation that the speed of convergence in general varies considerably from vertex to vertex. As soon as a certain convergence criteria is reached for a certain vertex, this vertex is taken out of the computation. This reduces the size of the matrix from step to step and therefore accelerates the power method.

The Linear System Approach. Each eigenvalue problem

$$Ax = \lambda x$$

can be written as homogeneous linear system of equations

$$(A - \lambda I)\, x = 0_n.$$

Arasu et al. [33] applied this idea to the PageRank algorithm and conducted some experiments with the largest strongly connected component of a snapshot of the Web graph from 1998. The most simple linear system approach for the PageRank system

$$(I - dP)\, c_{\mathrm{PR}} = (1 - d)\, 1_n$$

is probably the Jacobi iteration. But, as was mentioned in the description of the PageRank algorithm, the Jacobi iteration is very similar to the power method, and hence does not yield any acceleration.

Arasu et al. applied the Gauss-Seidel iteration defined by

$$c_{\mathrm{PR}}^{(k+1)}(i) = (1-d) + d\sum_{j<i} p_{ij}c_{\mathrm{PR}}^{(k+1)}(j) + d\sum_{j>i} p_{ij}c_{\mathrm{PR}}^{(k)}(j).$$

For $d = 0.9$, their experiments on the above described graph are very promising: the Gauss-Seidel iteration converges much faster than the power iteration. Arasu et al. then combine this result with the fact that the Web graph has a so-called *bow tie* structure. The next paragraph describes how this structure and other decomposition approaches may be used to accelerate the computations.

Decomposition Techniques. Since the Web graph is very large, and grows larger every day, some researchers propose to decompose the graph. So, it is possible to determine centrality values in smaller components of the Web in a first step, and to adjust them to the complete Web graph in the second step, if necessary. As noted above, Arasu et al. [33] exploit the observation of Broder et al. [102] that the Web graph has a so-called *bow tie* structure, see Figure 4.1 and Section 15.3.2. Note that the Web crawl of Broder et al. was carried out in 1999, and it is not clear whether the web structure has changed since.

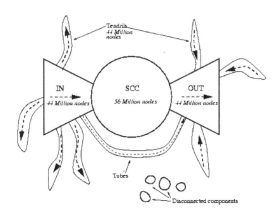

Fig. 4.1. Bow tie structure of the Web graph (from `http://www9.org/w9cdrom/160/160.html`)

This structure may be used for the power method, but the authors claim that it is especially well suited for the linear system approach, since the corresponding link-matrix has the *block upper triangular* form:

$$P = \begin{pmatrix} P_{11} & P_{12} & P_{13} & \cdots & P_{1K} \\ 0 & P_{22} & P_{23} & \cdots & P_{2K} \\ \vdots & \ddots & P_{33} & \cdots & P_{3K} \\ \vdots & & \ddots & \ddots & \vdots \\ 0 & \cdots & \cdots & 0 & P_{KK} \end{pmatrix}.$$

By partitioning c_{PR} in the same way, the large problem may be solved by the following sequence of smaller problems

$$(I - dP_{KK})\,c_{\text{PR},K} = (1 - d)\mathbf{1}_{n_K}$$

$$(I - dP_{ii})\,c_{\text{PR},i} = (1 - d)\mathbf{1}_{n_i} + d \sum_{j=i+1}^{K} P_{ij} c_{\text{PR},j}$$

A second approach was proposed by Kamvar et al. [339]. They investigated, besides a smaller partial Web graph, a Web crawl of 2001, and found the following interesting structure:

1. There is a block structure of the Web.
2. The individual blocks are much smaller than the entire Web.
3. There are nested blocks corresponding to domains, hosts and sub-directories within the path.

Algorithm 7: PageRank exploiting the block structure: BlockRank

---

1. For each block I,
   compute the local PageRank scores $c_{\text{PR(I)}}(i)$ for each vertex $i \in I$
2. Weight the local PageRank scores
   according to the importance of the block the vertices belongs to
3. Apply the standard PageRank algorithm
   using the vector obtained in the first two steps

Based on this observation, the authors suggest the three-step-algorithm 7. In the first and third step the ordinary PageRank algorithm can be applied. The question is how to formalize the second step. This is done via a *block graph* B where each block I is represented by a vertex, and an edge (I, J) is part of the block graph if there exists an edge (i, j) in the original graph satisfying $i \in I$ and $j \in J$, where (i, j) may be a loop. The weight ω_{IJ} associated with an edge (I, J) is computed as the sum of edge weights from vertices $i \in I$ to $j \in J$ in the original graph, weighted by the local PageRank scores computed from Step 1:

$$\omega_{IJ} = \sum_{i \in I, j \in J} a_{ij} c_{\text{PR(I)}}(i).$$

If the local PageRank vectors are normalized using the 1-norm, then the weight matrix $\Omega = (\omega_{IJ})$ is a stochastic matrix, and the ordinary PageRank algorithm can be applied to the block graph B to obtain the block weights b_I.
The starting vector for Step 3 is then determined by

$$c_{\mathrm{PR}}^{(0)}(i) = c_{\mathrm{PR(I)}}(i) b_I \; \forall \, I, \forall \, i \in I.$$

Another decomposition idea was proposed by Avrachenkov and Litvak [367] who showed that if a graph consists of several connected components (which is obviously true for the Web graph), then the final PageRank vector may be computed by determining the PageRank vectors in the connected components and combining them appropriately using the following theorem.

Theorem 4.3.2.

$$\mathbf{c}_{PR} = \left(\frac{|V_1|}{|V|} \mathbf{c}_{PR(1)}, \frac{|V_2|}{|V|} \mathbf{c}_{PR(2)}, \ldots, \frac{|V_K|}{|V|} \mathbf{c}_{PR(K)}, \right),$$

where $G_k = (V_k, E_k)$ are the connected components, $k = 1, \ldots, K$ and $\mathbf{c}_{PR(k)}$ is the PageRank vector computed for the kth connected component.

Finally, we briefly mention the 2-step-algorithm of Lee et al. [383] that is based on the observation that the Markov chain associated with the PageRank matrix is *lumpable*.

Definition 4.3.3. *If $\mathcal{L} = \{L_1, L_2, \ldots, L_K\}$ is a partition of the states of a Markov chain P then P is* lumpable *with respect to $\mathcal{L}$ if and only if for any pair of sets $L, L' \in \mathcal{L}$ and any state i in L the probability of going from i to L' doesn't depend on i, i.e. for all $i, i' \in L$*

$$\Pr[X_{t+1} \in L' | X_t = i] = \sum_{j \in L'} p_{ij} = \Pr[X_{t+1} \in L' | X_t = i'] = \sum_{j \in L'} p_{i'j} \, .$$

The common probabilities define a new Markov chain, the lumped chain P_L *with state space $\mathcal{L}$ and transition probabilities $p_{LL'} = \Pr[X_{t+1} \in L' | X_t \in L]$.*

The partition the authors use is to combine the dangling vertices (i.e., vertices without outgoing edges) into one block and to take all dangling vertices as singleton-blocks. This is useful since the number of dangling vertices is often much larger than the number of non-dangling vertices (a Web crawl from 2001 contained 290 million pages in total, but only 70 million non-dangling vertices, see [339]). In a second step, the Markov chain is transformed into a chain with all non-dangling vertices combined into one block using a state aggregation technique.

For the lumped chain of the first step, the PageRank algorithm is used for computing the corresponding centrality values. For the second Markov chain, having all non-dangling vertices combined, the authors prove that the algorithm to compute the limiting distribution consists of only three iterations (and one Aitken extrapolation step, if necessary, see Section 4.3.2). The vectors obtained in the two steps are finally concatenated to form the PageRank score vector of the original problem.

4.4   Dynamic Computation

In Section 4.3.2, several approaches for accelerating the calculation of page importance were described. In this section, we focus on the 'on the fly' computation of the same information, and on the problem of keeping the centrality values up-to-date in the dynamically changing Web.

4.4.1   Continuously Updated Approximations of PageRank

For the computation of page importance, e.g. via PageRank, the link matrix has to be known in advance. Usually, this matrix is created by a crawling process. As this process takes a considerable amount of time, approaches for the 'on the fly' computation of page importance are of interest. Abiteboul et al. [1] describe the 'On-line Page Importance Computation' (OPIC) algorithm, which computes an approximation of PageRank, and does not require to store the possibly huge link matrix.

The idea is based on the distribution of 'cash.' At initialization, every page receives an amount of cash and distributes this cash during the iterative computation. The estimated PageRank can then be computed directly from the current cash distribution, even while the approximation algorithm is still running.

Algorithm 8 describes the OPIC algorithm. The array c holds the actual distribution of cash for every page, and the array h holds the history of the cash for every page. The scalar g is just a shortcut for $\sum_{i=1}^n h[i]$.

An estimate of the PageRank of page i is given by $c_{\text{PRapprox}}(i) = \frac{h[i]+c[i]}{g+1}$. To guarantee that the algorithm calculates a correct approximation of PageRank, the selection of the vertices is crucial. Abiteboul et al. discuss three strategies: random, greedy, and circular. The strategies 'randomly select a page' and 'circularly select all pages' are obvious. Greedy selects the page with the highest cash. For the convergence of the computation, the selection of the vertices has to be fair, and this has to be guaranteed in all selection strategies.

After several iterations the algorithm converges towards the page importance information defined by the eigenvector for the largest eigenvalue of the adjacency matrix of the graph. To guarantee the convergence of the calculation similar concepts as for the random surfer (see Section 3.9.3) have to be applied. These are, for example, the inclusion of a 'virtual page' that every page links upon. The original work contains an adaptive version that covers link additions and removals, and in some parts vertex additions and removals. This modified adaptive OPIC algorithm is not discussed here, and can be found in [1].

4.4.2   Dynamically Updating PageRank

An interesting approach to accelerate the calculation of page importance lies in the recomputation of the PageRank for the 'changed' part of the network only. In case of the Web these changes are page additions and removals and link additions and removals. For this idea, Chien et al. [124] described an approach for link additions.

Algorithm 8: OPIC: On-line Page Importance Computation

Input: The graph G
Output: c and h: arrays for cash and history, g: sum of the history values

Initialization
for $i \leftarrow 1$ *to* n **do**
 $\quad c[i] \leftarrow 1/n$
 $\quad h[i] \leftarrow 0$
$g \leftarrow 0$

repeat
 $\quad$ choose a vertex i from G
 $\quad$ *See text for vertex selection strategies*

 $\quad$ *Update the history of i*
 $\quad h[i] \leftarrow h[i] + c[i]$

 $\quad$ *Distribute the cash from i to children*
 $\quad$ **for** *each child j of i* **do**
 $\quad\quad c[j] \leftarrow c[j] + c[i]/d^{+}[i]$

 $\quad$ *Update the global history value*
 $\quad g \leftarrow g + c[i]$

 $\quad$ *Reset cash for i*
 $\quad c[i] \leftarrow 0$
until *hell freezes over*

The idea is founded on an observation regarding the perturbation of the probability matrix P of the PageRank Markov chain for the Web graph W. This perturbation, stemming from link additions, can be modeled by the relation $P = \tilde{P} + E$, where E is an error matrix and $\tilde{P}$ is the perturbed matrix. For a single edge addition[1], E contains only changes in some row i. Therefore, the matrix $\tilde{P}$ differs from the original matrix P only in this row. Chien et al. observed that the recomputation of PageRank is required for a small area around the perturbation to achieve a good approximation for the modified Web graph W'. This small area is defined by the graph structure and can be extracted from the original Web graph W. The extraction yields a graph G that contains the new edge between i and j, and further every vertex and edge which are 'near' to the new edge. Additionally, the graph G contains a 'supervertex' that models all vertices from the graph W that are not in G. A transition matrix T for the graph G is constructed, and its stationary distribution τ is calculated.

For all vertices of the graph G (except for the supervertex), the stationary distribution $\tilde{\pi}$ of the perturbed matrix $\tilde{P}$ can, therefore, be approximated by the stationary distribution τ of the matrix T. For the vertices in W that are not covered by G, the stationary distribution $\tilde{\pi}$ of $\tilde{P}$ is simply approximated by the stationary distribution π of the matrix P. Several experiments showed that

---

[1] In the original work a description for the single edge case is given and extended towards multiple edge changes. We only cover the single edge case here.

this approach gives a good approximation for the modified Web graph W', and that the computation time decreases due to the computation of the stationary distribution of the smaller matrix T instead of $\tilde{P}$.

5 Advanced Centrality Concepts

Dirk Koschützki, *Katharina Anna Lehmann,* *Dagmar Tenfelde-Podehl, and Oliver Zlotowski*

The sheer number of different centrality indices introduced in the literature, or even only the ones in Chapter 3, is daunting. Frequently, a new definition is motivated by the previous ones failing to capture the notion of centrality of a vertex in a new application. In this chapter we will discuss the connections, similarities and differences of centralities. The goal of this chapter is to present an overview of such connections, thus providing some kind of map of the existing centrality indices. For that we focus on formal descriptions that hold for all networks. However, this approach has its limits.

Usually such approaches do not consider the special structure of the network that might be known for a concrete application, and it might not be able to convey the intuitive appeal of certain definitions in a concrete application. Nevertheless we consider such an approach appropriate to investigate the abstract definitions of different centrality indices. This is in a certain contrast to some of the literature, that only intuitively justifies a new definition of a centrality index on small example graphs.

Such connection between different definitions have been studied before, though usually not in a mathematical setting. One typical example is the work by Holme [304]. He considers a connection of betweenness centrality and congestion of a simulated particle hopping network. The particles are routed along shortest paths, but two particles are not allowed to occupy the same vertex. He investigates two policies of dealing with this requirement, namely that a particle waits if the next scheduled vertex is occupied, thus creating the possibility of deadlocks. Alternatively the particles can be allowed to continue their journey on a detour. He finds that such a prediction is only possible if the total number of particles in the network is small. Thus shortest-path betweenness for the application of the particle hopping model is the wrong choice, as it fails to predict congestion. In retrospect this is not really surprising because the definition of betweenness does not account for one path being blocked by another path, thus assuming that the particles do not interfere with each other. In particular the possibility of spill-backs as a result of overcrowded vertices is well known for car traffic flow on road networks, as for example addressed by the traffic-simulation presented by Gawron in [242]. Nagel [437] gives a more general overview of traffic considerations.

---

* Lead authors

U. Brandes and T. Erlebach (Eds.): Network Analysis, LNCS 3418, pp. 83–111, 2005.
© Springer-Verlag Berlin Heidelberg 2005

Unfortunately, the only general lesson to be learned from this is that it does matter which precise definition of centrality one uses in a concrete application. This sheds another light on our attempts to classify centrality indices, namely to help identify the 'right' centrality index for a particular application. This is perhaps not possible in general, just because we have no idea what kind of applications might be of interest, and how the network is constructed. However, for a concrete application the considerations here might give valuable ideas on how to model the situation precisely or as a reasonable approximation.

In Section 5.1 we start with some general approaches to normalize centrality indices. Many of these techniques are so general that they can be applied to all indices presented in Chapter 3. We will differentiate between approaches that facilitate the comparison of centrality values within the same graph and between different graphs.

We then consider the possibility to modify a centrality index by letting it focus on a certain subset of vertices. This set can, e.g., be a subset of Web pages that a Web surfer is most interested in. With such a subset a ranking can be personalized to the interests of an user. This idea of personalization is explained in more detail in Section 5.2. As in the case of normalization some of the techniques are virtually applicable to all centrality indices presented in Chapter 3, whereas others are designed especially for only one centrality index.

An informal approach to structure the wide field of centrality indices presented in this book is given in Section 5.3. For that we dissect these indices into different components, namely a basic term, a term operator, personalization, and normalization and thereby we define four categories of centrality indices. This approach finally leads to a flow chart that may be used to 'design' a new centrality index.

Section 5.4 elaborates on fundamental properties that any general or application specific centrality index should respect. Several such properties are proposed and discussed, resulting in different sets of axioms for centrality indices.

Finally, in Section 5.5 we discuss how centrality indices react on changes on the structure of the network. Typical examples are experimentally attained networks, where a new experiments or a new threshold changes the valuation or even existence of elements, or the Web graph, where the addition of pages and links happens at all times. For this kind of modifications the question of stability of ranking results is of interest and we will provide several examples of centrality indices and their reactions on such modifications.

5.1 Normalization

In Chapter 3 we saw different centrality concepts for vertices and edges in a graph. Many of them were restricted to the nonnegative reals, and some to the interval $[0, 1]$, such as the Hub- & Authority-scores which are obtained using normalization with respect to the Euclidean norm.

The question that arises is what it means to have a centrality of, say, 0.8 for an edge or vertex? Among other things, this strongly depends on the maximum centrality that occurs in the graph, on the topology of the graph, and on the number of vertices in the graph. In this section we discuss whether there are general concepts of normalization that allow a comparison of centrality scores between the elements of a graph, or between the elements of different graphs. Most of the material presented here stems from Ruhnau [499] and Möller [430].

In the following, we restrict our investigations to the centrality concepts of vertices, but the ideas can be carried over to those for edges.

5.1.1 Comparing Elements of a Graph

We start by investigating the question how centrality scores, possibly produced by different centrality concepts, may be compared in a given graph $G = (V, E)$ with n vertices. To simplify the notation of the normalization approaches we will use here a centrality vector instead of a function. For any centrality c_X, where X is a wildcard for the different acronyms, we will define the vector $\boldsymbol{c_X}$ where $\boldsymbol{c_X}_i = c_X(i)$ for all vertices $i \in V$. Each centrality vector $\boldsymbol{c_X}$ may then be normalized by dividing the centrality by the p-norm of the centrality vector

$$\|\boldsymbol{c_X}\|_p = \begin{cases} \left(\sum_{i=1}^{n} |\boldsymbol{c_X}_i|^p\right)^{1/p}, & 1 \leq p < \infty \\ \max_{i=1,\dots,n}\{|\boldsymbol{c_X}_i|\}, & p = \infty \end{cases}$$

to produce centrality scores $\boldsymbol{c_X}_i \leq 1$.

The main difference between the p-norm for $p < \infty$ and $p = \infty$ (the maximum norm) is that, when normalizing using $p = \infty$, the maximum centrality score in the graph is 1, and this value is attained for at least one vertex. Therefore, the normalization using the maximum norm yields a 'relative' centrality for each vertex in a graph. Note that this normalization is not appropriate for comparing vertices in different graphs, since the value of 1 (or -1, if negative values are allowed) is attained in each graph, independent of its topology.

For $p < \infty$, the centrality concepts that may produce negative centrality scores (e.g. Bonacich's bargaining centrality, see Section 3.9.2) have to be treated in a special way. Möller [430] proposes to separate the negative and positive components:

$$c'_{\boldsymbol{X}_i} = \begin{cases} c_{\boldsymbol{X}_i} / \left(\sum_{j:\boldsymbol{c_X}_j>0} |\boldsymbol{c_X}_j|^p\right)^{1/p}, & \boldsymbol{c_X}_i > 0, \\ 0, & \boldsymbol{c_X}_i = 0, \\ c_{\boldsymbol{X}_i} / \left(\sum_{j:\boldsymbol{c_X}_j<0} |\boldsymbol{c_X}_j|^p\right)^{1/p}, & \boldsymbol{c_X}_i < 0. \end{cases}$$

Taking $p = 1$, this means (for non-negative centralities) that each of the vertices is assigned their associated percentage of centrality within a graph. It might be worth discussing whether a similar approach is reasonable when using the maximum norm – or whether one should normalize using the maximum value instead of the maximum absolute value. The latter would have the advantage that in each graph we would obtain a 1 as the maximal normalized centrality value.

A normalization with the p-norm is in general not appropriate for comparing vertices of different graphs. We will see that the Euclidean norm ($p = 2$) forms an exception for eigenvector centralities in that the maximal value that can be attained is independent of the number of vertices, see the end of Section 5.4.2.

5.1.2   Comparing Elements of Different Graphs

When vertices in different graphs have to be compared, the varying size of the graphs can be problematic. Let $\mathcal{G}_n$ be the set of connected graphs $G = (V, E)$ with n vertices. Freeman [227] proposed to define the *point-centrality*

$$c''_{Xi} = \frac{c_{Xi}}{c^*_X}, \tag{5.1}$$

where $c^*_X = \max_{G \in \mathcal{G}_n} \max_{i \in V(G)} c_{Xi}$ is the maximum centrality value that a vertex can obtain taken over all graphs with n vertices.

Using the point-centrality c''_{Xi}, the maximum value 1 is always attained by at least one vertex in at least one graph of size n. Thus, a comparison of centrality values in different graphs is possible. Unfortunately, this is often only possible in theory, since the determination of c^*_X is not trivial in general, and even impossible for some centrality concepts. Consider, for example, the status-index of Katz (see Section 3.9.1), where the centrality scores are related to the chosen damping factor. Theorem 3.9.1 states that the damping factor α is itself strongly related to the maximum eigenvalue λ_1 of the adjacency matrix. Hence, it is not clear that a feasible damping factor for the graph under investigation is also feasible for all other graphs of the same size.

Möller provides a nice example with the following two adjacency matrices:

$$A_1 = \begin{pmatrix} 0 & 1 \\ 0 & 0 \end{pmatrix}, \quad A_2 = \begin{pmatrix} 0 & 1 \\ 1 & 0 \end{pmatrix}.$$

Since A_1^k is the zero matrix for $k \geq 2$, convergence is guaranteed for any $\alpha \in \,]0, 1]$. If we choose the maximum possible value $\alpha = 1$, then the infinite sum $\sum_{k=1}^{\infty} \alpha^k A_2^k$ does not converge, since it is equal to $\lim_{K \to \infty} \sum_{k=1}^{K} 1_2 1_2^T$. This example shows that it is not clear which damping factor to choose in order to determine the value c^*_K (especially if we have to do that for different n).

Nevertheless, there are centrality concepts that allow the computation of c^*_X. A very simple example is the degree centrality. It is obvious that in a simple, undirected graph with n vertices the maximum centrality value a vertex can obtain (with respect to degree centrality) is $n-1$. Another example is the shortest paths betweenness centrality (s. Section 3.4.2): The maximum value any vertex can obtain is given in a star with a value of $\frac{n^2 - 3n + 2}{2}$ [227].

Further, the minimum total distance from a vertex i to all other vertices is attained when i is incident to all other vertices, that is, when i has maximum degree. So, it is clear that for the closeness centrality (see Section 3.2) we have $c^*_C = (n - 1)^{-1}$.

Möller shows that, in addition, the eccentricity centrality as well as the Hubs & Authorities centrality allow the calculation of the value c_X^*. For the eccentricity centrality we just note that a vertex with maximum degree has an eccentricity value of 1 and all other vertices have smaller eccentricity values, hence $c_E^* = 1$. Similarly, the maximum values for hub- and authority centrality values (centrality vectors are assumed to be normalized using the Euclidean norm) are 1 and they are attained by the center of a star (either all edges directed to the center of the star or all edges directed away from the center).

Shortest-path betweenness centrality and the Euclidean normalized eigenvector centrality provide other, more sophisticated, examples, see, e.g., Ruhnau [499]: These two centralities have the additional property that the maximum centrality score of 1 is attained exactly for the central vertex of a star. This property is useful when comparing vertices of different graphs, and is explained in more detail in the Section 5.4.2.

Finally we note that Everett, Sinclair and Dankelmann found an expression for the maximum betweenness in bipartite graphs, see [195].

5.2 Personalization

The motivation for a personalization of centrality analysis of networks is easily given: Imagine that you could configure your favorite Web search engine to order the WWW according to your interests and liking. In this way every user would always get the most relevant pages for every search, in an individualized way.

There are two major approaches to this task: The first is to change weights on the vertices (pages) or edges (links) of the Web graph with a personalization vector v. The weights on vertices can describe something like the time spent each day on the relevant page and a weight on the edge could describe the probability that the represented link will be used. With this, variants of Web-centrality algorithms can be run that take these personal settings into account. The other approach is to choose a 'rootset' $R \subseteq V$ of vertices and to measure the importance of other vertices and edges relative to this rootset.

We will see in Section 5.3 that these two approaches can be used as two operators. The first approach changes the description of the graph itself and the corresponding operator is denoted by P_v. Then the corresponding term for each vertex (or edge) is evaluated on the resulting graph. The second personalization approach chooses a subset of all terms that is given by the rootset R. This operator is denoted by P_R.

We will first discuss personalization approaches for distance and shortest paths based centralities and then discuss approaches for Web centralities.

5.2.1 Personalization for Distance and Shortest Paths Based Centralities

All centralities that were presented in Chapter 3 rank every vertex relative to all other vertices in the graph. In this subsection we will be concerned with variants of these centralities that determine the relative importance of vertices with respect to a set R of root vertices. R is chosen such that the vertices in R are assumed to be important and the question is how all other vertices should be ranked in importance with respect to R. The approach presented by White and Smith in [580] is very general and deserves some attention.

Let $c(v)$ be some centrality index on vertices. Then, $c(v|R)$ denotes the relative importance of vertex v with respect to the given rootset R. Let $P(s,t)$ denote any well defined set of paths between vertex s and t. The authors suggest different kinds of path sets:

- a set of shortest paths
- a set of k-shortest paths, defined as the set of all paths with length smaller than a given k
- a set of k-shortest vertex-disjoint paths[1]

The set of shortest paths is used e.g. in the shortest-path betweenness centrality (see Section 3.4.2). The *relative betweenness centrality* $c_{RBC}(v)$ can be defined in three ways. In the first variant we define a vertex v as important if the fraction of shortest paths leaving a vertex r from R contains v. We will denote this *source relative betweenness centrality* by

$$c_{sRBC}(v) = \sum_{r \in R} \sum_{t \in V} \delta_{rt}(v) \; . \tag{5.2}$$

If an element v is important if it is contained in a large fraction of shortest paths ending in a vertex r of R we denote the *target relative betweenness centrality* as

$$c_{tRBC}(v) = \sum_{s \in V} \sum_{r \in R} \delta_{sr}(v) \; . \tag{5.3}$$

In the last case, an element is supposed to be important if it is contained in a large fraction of shortest paths leading from R to R, denoted by

$$c_{RBC}(v) = \sum_{r_s \in R} \sum_{r_t \in R} \delta_{r_s r_t}(v) \; . \tag{5.4}$$

If any other set of paths $P(s,t)$, e.g. the set of k-shortest paths, is chosen, then the definition of $\delta_{st}(v)$ has to be changed, denoted by

$$\delta_{st|P}(v) = \frac{\sigma_{st}(v)}{|P(s,t)|} \tag{5.5}$$

where $\sigma_{st}(v)$ denotes the number of paths $p \in P(s,t)$ that contain vertex v.

This example demonstrates the general idea behind this kind of personalization. It can be easily expanded to all centralities that are based on distance.

---

[1] We just want to note that this set of paths is not unique in most graphs. For a deterministic centrality it is of course important to determine a unique path set, so this last path set should only be used on graphs where there is only one set for each vertex pair.

5.2.2 Personalization for Web Centralities

Consider again the random surfer model (see Section 3.9.3) for Web centralities and assume the random surfer arrived at a page where there is no outlink or where the existing out links are not relevant. The original assumption in this case is a jump to a random page where each page has equal probability. It is obvious that the assumption of equal probability is not very realistic: some surfers prefer Web pages about sports if they get stuck in a sink, others continue with a news-page etc. The question at hand is hence how to model the many different types of Web users.

A very intuitive approach is to replace $\mathbf{1}_n$ (cf. Equation 3.44) by a *personal-ization vector* $\mathbf{v}$ satisfying $v_i > 0 \; \forall \; i$ and $\sum_i v_i = 1$. White and Smyth [580], for example, proposed to score the vertices relative to a kernel set R using

$$v_i = \begin{cases} \frac{1-\varepsilon}{|R|}, & i \in R \\ \frac{\varepsilon}{n-|R|}, & i \notin R \end{cases},$$

where $0 < \varepsilon \ll 1$.

They also proposed a very similar approach for the Hubs & Authorities algorithm. Instead of applying the iterative procedure given in Algorithm 2 on page 55 they added in each step a portion of the personalization vector and obtained the following modified equations:

$$\begin{aligned} c_{\text{HA-H}}^k &= dA_\sigma c_{\text{HA-A}}^{k-1} + (1-d)\mathbf{v} \\ c_{\text{HA-A}}^k &= dA_\sigma^T c_{\text{HA-H}}^k + (1-d)\mathbf{v} \\ c_{\text{HA-H}}^k &= \frac{c_{\text{HA-H}}^k}{\|c_{\text{HA-H}}^k\|} \\ c_{\text{HA-A}}^k &= \frac{c_{\text{HA-A}}^k}{\|c_{\text{HA-A}}^k\|}, \end{aligned}$$

where $d \in [0,1]$ is chosen to control the influence of $\mathbf{v}$.

Going back to the PageRank algorithm it is clear that as long as all elements of $\mathbf{v}$ are positive and $\mathbf{v}$ is a stochastic vector, the associated Markov chain is still irreducible hence the convergence of the PageRank algorithm is not touched. Thus, at a first glance, this approach seems to be appealing. But there is one big disadvantage: As already known the computations of PageRank vectors for the non-personalized version is very time consuming, there is, at least at the moment, no chance to compute PageRank centralities for many different types of Web users. Nevertheless there are some promising approaches to obtain personalized PageRank vectors in an adequate amount of time.

To this end we give a general approach of personalization for PageRank, taken from Haveliwala et al. [291]. As noted above the personalized PageRank vector is given as the solution of the following equation

$$c_{\text{PR}} = dP^T c_{\text{PR}} + (1-d)\mathbf{v}.$$

Since $(I - dP^T)$ is a strictly diagonally dominant matrix, it is invertible and hence

$$c_{\mathrm{PR}}^v := c_{\mathrm{PR}} = (1 - d)\left(I - dP^T\right)^{-1} v =: Qv. \qquad (5.6)$$

(We write c_{PR}^v to emphasize the dependence of c_{PR} on v.)

If we choose v to be the ith unit vector $v = e^i$, then $c_{\mathrm{PR}}^{e^i} = Q_{.j}$, hence the set of columns of Q may be seen as a basis for the personalized PageRanks.

The Problem that occurs is that the determination of Q needs to invert a matrix which is very time consuming if the matrices are large. To reduce the computational complexity Q is approximated by $\hat{Q} \in \mathbb{R}^{n \times K}$ and hence we consider only a subset of K basis vectors (independent columns of Q) taking a convex combination to obtain an estimate for

$$c_{\mathrm{PR}}^w = \hat{Q}w$$

where $w \in \mathbb{R}^K$ is a stochastic vector, $w_i > 0 \ \forall \ i$.

Haveliwala et al. show that the following three personalization approaches can be subsumed under the general approach described above:

- Topic sensitive PageRank [289],
- Modular PageRank [326],
- BlockRank [339].

They only differ in how the approximation is conducted. We describe these approaches briefly in the following subsections.

Topic Sensitive PageRank. Haveliwala [289] proposes to proceed in a combined offline-online algorithm where the first phase (offline) consists of the following two steps

1. Choose the K most important topics $t_1, \ldots, t_K$ and define v_i^k to be the (normalized) degree of membership of page i to topic t_k, $i = 1, \ldots, n$, $k = 1, \ldots, K$.
2. Compute $\hat{Q}_{.k} = c_{\mathrm{PR}}^{v^k}$, $k = 1, \ldots, K$

The second phase that is run online is as follows

1. For query σ compute (normalized) topic-weights $w_1^\sigma, \ldots, w_K^\sigma$
2. Combine the columns of $\hat{Q}$ with respect to the weights to get

$$c_{\mathrm{PR}}^\sigma = \sum_{k=1}^K w_k^\sigma \hat{Q}_{.k}.$$

Note that to apply this approach it is important that

- K is small enough (e.g. $K = 16$) and
- the range of topics is broad enough.

Modular PageRank. A second approach was proposed by Jeh and Widom [326]. Their algorithm consists of an offline and an online step. In the offline step K pages $i_1, \ldots, i_K$ with high rank are chosen. These high-ranked pages form the set of hubs.

Using personalization vectors e^{i_k}, the associated PageRank vectors called *basis vectors* or *hub vectors* $c_{PR}^{e^{i_k}}$ are computed. By linearity for each personalization vector v that is a convex combination of $e^{i_1}, \ldots, e^{i_K}$ the corresponding personalized PageRank vector can be computed as a convex combination of the hub vectors. But if K gets larger, it is neither possible to compute all hub vectors in advance nor to store them efficiently. To overcome this deficiency, Jeh and Widom propose a procedure using *partial vectors* and a *hubs skeleton*. They are able to show that in contrast to the hub vectors it is possible to compute and store the partial vectors efficiently. These partial vectors together with the hubs skeleton are enough to compute all hub vectors and hence (by transitivity) the final personalized PageRank. Essentially the idea is to reduce the computations to the set of hubs, which is much smaller than the Web graph (but $K \geq 10^4$ is possible). Note that the larger K may be chosen, the better the Q-matrix is represented.

The online step then consists of determining a personalization vector $v^\sigma = \sum_{k=1}^{K} \alpha_k^\sigma e^{i_k}$ and the corresponding PageRank vector

$$c_{PR}^\sigma = \sum_{k=1}^{K} c_{PR}^{\alpha_k^\sigma e^{i_k}}$$

(again by using partial vectors and the hubs skeleton).

BlockRank. This approach of Kamvar et al. [339] was originally invented for accelerating the computation of PageRank, see Section 4.3.2. It consists of a 3-phase-algorithm where the main idea is to decompose the Web graph according to hosts. But, as already proposed by the authors, this approach may also be applied to find personalized PageRank scores: In the second step of the algorithm the host-weights have to be introduced, hence the algorithm is the following:

1. (offline) Choose K blocks (hosts) and let v_i^k be the local PageRank of page i in block k, $i = 1, \ldots, n$, $k = 1, \ldots, K$. Compute $\hat{Q}_{\cdot k} = c_{PR}^{v^k}$ (the authors claim that $K \geq 10^3$ is possible if the Web structure is exploited).
2. (online) For query σ find appropriate host-weights to combine the hosts.
3. (online) Apply the (standard) PageRank algorithm to compute the associated centralities. Use as input the local PageRank scores computed in the first step, weighted by the host-weights of step 2.

Both, the concept of personalization from this section and normalization from the previous section will be rediscussed in the following two sections to introduce the four dimensions of centrality indices.

5.3   Four Dimensions of a Centrality Index

In this section we present a four dimension approach which is an attempt to structure the wide field of different centrality measures and related personalization and normalization methods presented so far. The idea to this model emerged from the observation that there is currently no consistent axiomatic schema that captures all the centrality measures considered in Chapter 3, for more details see Section 5.4. But it is important to note, that the following contribution does not constitute a formal approach or even claims completeness. Nevertheless, we believe that it may be a helpful tool in praxis.

The analysis of the centrality measures in Chapter 3 has led to the idea of dividing the centralities into four categories according to their fundamental computation model. Each computation model is represented by a so-called basic term. Given a basic term, a term operator (e.g. the sum or the maximum), and several personalization and normalization methods may be applied to it. In the following we want to discuss the idea in more detail. At the end of this section we provide a scheme based on our perception that helps to classify new centrality measures, or helps to customize existing ones.

Basic Term. The classification of the centrality measures into four categories and the representation of each category by a basic term constitutes the first dimension of our approach. Once again, we want to mention that this classification is only a proposal which emerged form the analysis of existing measures described so far.

Reachability. The first category of centrality measures is based on the notion of 'reachability'. A vertex is supposed to be central if it reaches many other vertices. Centrality measures of this category are the degree centrality (cf. Section 3.3.1), the centrality based on eccentricity and closeness (cf. Section 3.3.2), and the random walk closeness centrality (cf. Section 3.8.3). All of these centralities rely on the distance concept $d(u, v)$ of two vertices u and v. In the degree centrality, for example, we count the number of vertices that can be reached within distance 1. The closeness of a vertex u is measured by the reciprocal of the sum over the distances to all other vertices v. The same is true for the centrality based on eccentricity, where the maximum is taken instead of the total distance. In the case of the random walk closeness centrality the notion of distance is equivalently given as the mean first passage time from vertex u to all other vertices v in a random walk.

Amount of flow. The second category of centrality measures is based on the amount of flow $f_{st}(x)$ from a vertex s to a vertex t that goes through a vertex or an edge x. This can be easily seen at centrality measures based on current flow processes (cf. Section 3.7) and random walks as described in Section 3.8.1 and 3.8.2. But also measures based on the enumeration of shortest paths belong to this category. The stress centrality presented in Section 3.4.1 may also be interpreted as measuring the amount of flow going through an element x if every vertex s sends to every other vertex t one unit flow along each shortest path connecting them. In the same context, the shortest-path betweenness centrality introduced in Section 3.4.2 measures the expected fraction of times a unit flow goes through the element if every vertex s sends one unit flow consecutively to every other vertex t, and each time choosing one of all shortest paths connecting them uniformly, independently at random. The basic term covering these measures is $f_{st}(x)$.

Vitality. A third category of centrality measures is based on the vitality as defined in Section 3.6. Here, the centrality value of an element x is defined as the difference of a real-valued function f on G with and without the element. Recall, a general vitality measure was denoted by $\mathcal{V}(G, x) = f(G) - f(G \backslash \{x\})$. The maximum flow betweenness vitality presented in Sect. 3.6.1 belongs to this category.

Feedback. A fourth category of centrality measures is based on a implicit definition of a centrality (cf. Section 3.9). These measures might be subsumed by the abstract formula $c(v_i) = f(c(v_1), \ldots, c(v_n))$, where the centrality value of a certain vertex v_i depends on the centrality values of all vertices $v_1, \ldots, v_n$.

Term Operator. The second dimension is represented by the term operator. Consider the first three categories: here we observed that often a set of suitable operators can be applied to a basic term to obtain meaningful centrality measures. We want to illustrate this idea on some centrality measures: If we have carefully defined the distance for a given application, we can choose whether the centrality index is given by the maximum of all distances from u to any other vertex v (as in the eccentricity), or the sum over all distances (as in the closeness centrality), or the average distance to all other vertices (as a normalized closeness centrality). In some cases even a special operator as the variance of all the distance might led a meaningful centrality index. Thus, for all centrality indices of the first three categories, it makes sense to separate the choice of a term operator from the basic term.

Personalization. The third dimension is given by the methods that help to personalize centrality measures. In Section 5.2 we differentiate two variants of personalization. The first approach, denoted by P_v, is applicable to all centrality measure that can deal with vertex or edge weights. This personalization applies a weight vector v to V, E, or to the transition matrix of the random surfer model in the case of the Web centralities. The second personalization method, denoted by P_R, considers a subset of vertices, the so called rootset R. The centrality of a vertex is measured with respect to this rootset. This method is applicable to all distance based centrality indices. Both personalization methods and all other approaches to personalization build the third dimension.

Normalization. All of the centrality measures presented in this book can be normalized. Thus, the normalization forms a fourth dimension. Recall, a common normalization applicable to most centrality measures is to divide every value by the maximum centrality value. In Section 5.1 several normalization methods were considered.

Independence of the Dimensions. All of these four dimensions: basic term, term operator, personalization, and normalization are independent of each other and we have outlined that the centrality measures presented in this book can be meaningfully dissected into them. Of course, we cannot claim that all centrality indices ever published will fall into one of these categories or can be dissected as demonstrated. Moreover, since we lack any strict definition of centrality indices, we cannot ensure that every possible combinations will result in meaningful centrality index. Our aim is to provide a model that helps to structure the design of a suitable centrality index according to our four-dimensional approach.

Designing a Centrality Index. The diagram in Figure 5.1 shows an approach that demonstrates how an appropriate centrality can be found or adapted for a given application. The first step in choosing an appropriate centrality index is to find the question that should be answered by the centrality measure. That determines the category and the corresponding basic term. In general, however, the basic term refers only to an abstract concept. The distance between two vertices, for example, could be measured by the mean first passage time in a random walk or by the classic definition of distance on shortest paths. Thus a concrete computational model must be developed for the chosen basic term. After this step, a first personalization might be applied. This personalization leads to a personalized graph with modified or added weights on the vertices or edges, respectively. Afterwards, a second personalization might be applicable by choosing a 'rootset' if the basic term corresponds to one of the categories *reachability, amount of flow or vitality*. The centrality of a vertex is then measured with respect to this rootset. If the resulting term belongs to the first three categories, 'reachability', 'amount of flow', or 'vitality', we have to chose a term operator which will be applied to the term with respect to the personalized graph. We want to mention here as examples the maximum-operator or the summation over all terms.

If the chosen centrality index is a feedback centrality a personalization with a rootset is not always applicable. Thus, the route through the diagram follows a special path for these indices. The next step here is to determine the appropriate linear equation system and to solve it.

In all four categories the resulting centrality values might be normalized, as discussed in Section 5.1. Usually this normalization is performed by a multiplication with a scalar.

As a tool for describing, structuring, and developing centrality measures our four dimension approach provides a flexible alternative to classical approaches even though more formalization and refinement is needed. In the next section

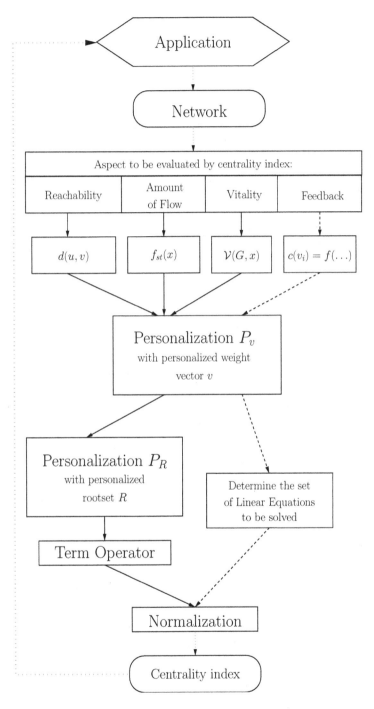

Fig. 5.1. A flow chart for choosing, adapting or designing an appropriate centrality measure for a given application

we consider several classical approaches which may also be used to characterize centrality measures.

5.4   Axiomatization

In Chapter 3, we saw that there are many different centrality indices fitting for many different applications. This section discusses the question whether there exist general properties a centrality should have.

We will first cover two axiomatizations of distance-based approaches of centrality indices and in a second subsection discuss two aximatisations for feedback centralities.

5.4.1   Axiomatization for Distance-Based Vertex Centralities

In the fundamental paper of Sabidussi [500], several axioms are defined for a vertex centrality of an undirected connected graph $G = (V, E)$. In the following we restate these in a slightly modified way. Sabidussi studied two operations on graphs:

Adding an edge (u, v): Let u and v be distinct vertices of G where $(u, v) \notin E(G)$. The graph $H = (V, E \cup \{(u, v)\})$ is obtained from G by adding the edge (u, v).

Moving an edge (u, v): Let u, v, w be three distinct vertices of G such that $(u, v) \in E(G)$ and $(u, w) \notin E(G)$. The graph $H = (V, (E \setminus \{(u, v)\}) \cup \{(u, w)\})$ is obtained by removing (u, v) and inserting (u, w). Moving an edge must be admissible, i.e., the resulting graph must still be connected.

Let $\mathcal{G}_n$ be the class of connected undirected graphs with n vertices. Furthermore, let $c \colon V \to \mathbb{R}_0^+$ be a function on the vertex set V of a graph $G = (V, E) \in \mathcal{G}_n$ which assigns a non-negative real value to each vertex $v \in V$. Recall, in Section 3.3.3 we denoted by $\mathcal{S}_c(G) = \{u \in V \colon \forall v \in V\ c(u) \geq c(v)\}$ the set of vertices of G of maximum centrality with respect to a given vertex centrality c.

Definition 5.4.1 (Vertex Centrality (Sabidussi [500])). *A function c is called a* vertex centrality *on $G \in \mathcal{G}'_n \subseteq \mathcal{G}_n$, and $\mathcal{G}'_n$ is called c-admissible, if and only if the following conditions are satisfied:*

1. *$\mathcal{G}'_n$ is closed under isomorphism, i.e., if $G \in \mathcal{G}'_n$ and H is isomorphic to G then also $H \in \mathcal{G}'_n$.*
2. *If $G = (V, E) \in \mathcal{G}'_n$, $u \in V(G)$, and H is obtained from G by moving an edge to u or by adding an edge to u, then $H \in \mathcal{G}'_n$, i.e., $\mathcal{G}'_n$ is closed under moving and adding an edge.*
3. *Let $G \simeq_\phi H$, then $c_G(u) = c_H(\phi(u))$ for each $u \in V(G)$.[2]*

---

[2] By $c_G(u)$ and $c_H(u)$ we denote the centrality value of vertex u in G and H, respectively.

4. *Let* $u \in V(G)$, *and* H *be obtained from* G *by adding an edge to* u, *then* $c_G(u) < c_H(u)$ *and* $c_G(v) \leq c_H(v)$ *for each* $v \in V$.
5. *Let* $u \in \mathcal{S}_c(G)$, *and* H *be obtained from* G *either by moving an edge to* u *or by adding an edge to* u, *then* $c_G(u) < c_H(u)$ *and* $u \in \mathcal{S}_c(H)$.

The first two conditions provide a foundation for Condition 3 and 5. Note that certain classes of graphs fail to satisfy Condition 2, e.g., the class of all trees is closed under moving of edges, but not under addition of edges. Condition 3 describes the invariance under isomorphisms, also claimed in Definition 3.2.1. The idea behind Condition 4 is that adding an edge increases the degree of centralization of a network. Condition 5 is the most significant one. If an edge is moved or added to a vertex $u \in \mathcal{S}_c(G)$, then the centrality of u should be increased and it should contained in $\mathcal{S}_c(H)$, i.e., u must have maximal centrality in the new graph H.

For the degree centrality introduced in Section 3.3.1, it is easy to verify that the axioms are satisfied. Thus, the degree centrality is a vertex centrality in terms of Sabidussi's definition.

We shall now see that the vertex centrality $c_E(u)$ based on the eccentricity $e(u)$ introduced in Section 3.1 is not a vertex centrality according to Sabidussi's definition. In Figure 5.2 two graphs are shown where the eccentricity value for each vertex is indicated. The first graph is a simple path with one central vertex u_5. After adding the edge (u_5, u_9) the new central vertex is u_4. Thus, adding an edge according to Condition 5 does not preserve the center of a graph. Note, also Condition 4 is violated.

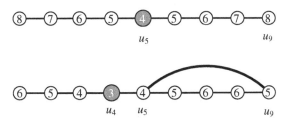

Fig. 5.2. The eccentricity $e(u)$ for each vertex $u \in V$ is shown. The example illustrates that the eccentricity centrality given by $c_E(u) = e(u)^{-1}$ is not a vertex centrality according to Sabidussi's definition (see Definition 5.4.1)

In Section 3.2 the closeness centrality of a vertex was defined by $c_C(u) = s(u)^{-1}$. Kishi [357] showed that this centrality is not a vertex centrality respecting Sabidussi's definition. An example is given in Figure 5.3, where the value of the total distance for each vertex is indicated. The median $\mathcal{M}(G) = \{u \in V : s(G) = s(u)\}$ of the left graph G consists of the vertices u, u', and u''. The insertion of edge (u, v) yields a graph H with $\mathcal{M}(H) \cap \mathcal{M}(G) = \emptyset$.

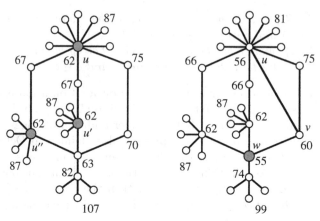

Fig. 5.3. The total distance $s(u)$ for each vertex $u \in V$ is shown. The example depicts that the closeness centrality defined by $c_C(u) = s(u)^{-1}$ is not a vertex centrality according to Sabidussi's definition (see Definition 5.4.1)

Kishi [357] provides a definition for distance-based vertex centralities relying on Sabidussi's definition. Let c be a real valued function on the vertices of a connected undirected graph $G = (V, E)$, and let u and v be two distinct non-adjacent vertices of G. The insertion of (u, v) leads to a graph $H = (V, E \cup \{(u, v)\})$ where the difference of the centrality values is measured by $\Delta_{uv}(w) = c_H(w) - c_G(w)$ for each vertex $w \in G$.

Definition 5.4.2 (Vertex Centrality (Kishi [357])). *The function* c *is called a* vertex centrality *if and only if the following conditions are satisfied*

1. *$\Delta_{uv}(u) > 0$, i.e., $c_G(u) < c_H(u)$.*
2. *For each $w \in V$ with $d(u, w) \leq d(v, w)$ it holds that $\Delta_{uv}(u) \geq \Delta_{uv}(w)$ for any pair of non-adjacent vertices u and v.*

The conditions of Definition 5.4.2 are quite similar to Condition 4 and 5 of Sabidussi's definition 5.4.1. Therefore, it is not surprising that the eccentricity is not a vertex centrality according to Kishi's definition. To see that, reconsider Figure 5.2 where vertex u_5 violates the Condition 2 of Kishi's definition. However, Kishi [357] showed that the closeness centrality is a vertex centrality with respect to Definition 5.4.2.

As these two examples show, it will still be a challenge to find minimal requirements which can be satisfied by a distance-based centrality index. In Section 3.2 we claimed that the centrality index only depends on the structure of the graph (cf. Def. 3.2.1). But as mentioned already, not every structural index will be accepted as a centrality index.

Finally, we want to note that there are also attempts to define requirements for a vertex centrality of an weakly connected directed graphs, see e.g. Nieminen [451].

5.4.2   Axiomatization for Feedback-Centralities

Up to now, we mainly discussed sets of axioms that defined and admitted centralities that are based either on shortest path distances or on the degree of the vertex. This section reviews axiomatizations that lead to feedback centralities or feedback-like centralities.

Far from being complete we want to give two examples of how an axiomatization could work. To our knowledge there are several approaches concerning axiomatization, but up to now there is a lack of structure and generality: Many properties a centrality should have are proposed in the literature, but those sets of properties in most cases depend very much on the application the authors have in mind and exclude known and well-established centralities.

We start with a paper by van den Brink and Gilles [563], which may serve as a bridge between degree-based and feedback-like centralities. This is continued by presenting results of Volij and his co-workers that axiomatically characterize special feedback-centralities.

From Degree to Feedback. In [563], van den Brink and Gilles consider directed graphs. In the main part of their paper the graphs are unweighted, but the axiomatic results are generalized to the weighted case. We only review the results for the unweighted case - the weighted case is strongly related but much more complicated with respect to notation.

The goal is to find an axiomatic characterization of centralities, or, to be more specific, of what they call *relational power measures* which assign to each directed network with n vertices an n-dimensional vector of reals such that the ith component of the vector is a measure of the relational power (or dominance) of vertex i.

The first measure is the β-*measure*, that was developed by the same authors [562] for hierarchical economic organizations. It measures the potential influence of agents on trade processes.

Let $\mathcal{G}_n$ be the set of unweighted directed graphs having n vertices. For a directed edge $(i, j) \in E$, vertex i is said to *dominate* vertex j.

Definition 5.4.3. *Given a set of vertices V with $|V| = n$, the β-measure on V is the function $\beta : \mathcal{G}_n \to \mathbb{R}^n$ given by*

$$\beta_G(i) = \sum_{j \in N_G^+(i)} \frac{1}{d_G^-(j)} \qquad \forall\, i \in V,\ G \in \mathcal{G}_n$$

(Remember that $d^-(j)$ is the in-degree of vertex j and $N^+(i)$ is the set of vertices j for which a directed edge (i, j) exists.)

This β-measure may be viewed as a feedback centrality, since the score for vertex i depends on properties of the vertices in its forward neighborhood.

A set of four axioms uniquely determines the β-measure. To state the four axioms, let $f : \mathcal{G}_n \to \mathbb{R}^n$ be a relational power measure on V. Moreover, we need the following definition:

Definition 5.4.4. *A partition of $G \in \mathcal{G}_n$ is a subset $\{G_1, \ldots, G_K\} \subseteq \mathcal{G}_n$ such that*

- $\bigcup_{k=1}^{K} E_k = E$ *and*
- $E_k \cap E_l = \emptyset \ \forall \ 1 \leq k, l \leq K, \ k \neq l$.

The partition is called independent *if in addition*

$$\left| \left\{ k \in \{1, \ldots, K\} : d_{G_k}^-(i) > 0 \right\} \right| \leq 1 \ \forall \ i \in V,$$

i.e., if each vertex is dominated in at most one directed graph of the partition.

Which properties should a centrality have in order to measure the relational power or dominance of a vertex?

First of all it would be good to normalize the measure in order to compare dominance values of different vertices - possibly in different networks. Due to the domination structure of their approach van den Brink and Gilles propose to take the number of dominated vertices as the total value that is distributed over the vertices according to their relational power:

| Axiom 1: *Dominance normalization* | For every $G \in \mathcal{G}_n$ it holds that

$$\sum_{i \in V_G} f_G(i) = \left| \left\{ j \in V_G : d_G^-(j) > 0 \right\} \right|.$$

The second axiom simply says that a vertex that does not dominate any other vertex has no relational power and hence gets the value zero:

| Axiom 2: *Dummy vertex property* | For every $G \in \mathcal{G}_n$ and $i \in V$ satisfying $N_G^+(i) = \emptyset$ it holds that $f_G(i) = 0$.

In the third axiom the authors formalize the fact that if two vertices have the same dominance structure, i.e. the same number of dominated vertices and the same number of dominating vertices, then they should get the same dominance-value:

| Axiom 3: *Symmetry* | For every $G \in \mathcal{G}_n$ and $i, j \in V$ satisfying $d_G^+(i) = d_G^+(j)$ and $d_G^-(i) = d_G^-(j)$ it holds that $f_G(i) = f_G(j)$.

Finally, the fourth axiom addresses the case of putting together directed graphs. It says that if several directed graphs are combined in such a way that a vertex is dominated in at most one directed graph (i.e. if the result of the combination may be viewed as an independent partition), then the total dominance value of a vertex should simply be the sum of its dominance values in the directed graphs.

| Axiom 4: *Additivity over independent partitions* | For every $G \in \mathcal{G}_n$ and every independent partition $\{G_1, \ldots, G_K\}$ of G it holds

$$f_G = \sum_{k=1}^{K} f_{G_k}.$$

Interestingly, these axioms are linked to the preceding sections on degree-based centralities: If the normalization axiom is changed in a specific way, then the unique centrality score that satisfies the set of axioms is the out-degree centrality. The authors call this *score-measure*. Note that an analogous result also holds for the weighted case.

In more detail, after substituting the dominance normalization by the score normalization (see Axiom 1b below), the following function is the unique relational power measure that satisfies Axioms 2 – 4 and 1b:

$$\sigma_G(i) = d_G^+(i) \ \forall \ i \in V, \ G \in \mathcal{G}_n$$

Instead of taking the number of dominated vertices as the total value that is distributed over the vertices according to their dominance, the total number of relations is now taken as a basis for normalization:

Axiom 1b: *Score normalization*   For every $G \in \mathcal{G}_n$ it holds that

$$\sum_{i \in V} f_G(i) = |E|.$$

Above, we presented a set of axioms that describe a certain measure that has some aspects of feedback centralities but also links to the preceding section via its strong relation to the score measure. We now pass over to feedback centralities in the narrower sense.

Feedback Centralities. In terms of citation networks, Palacios-Huerta and Volij [460] proposed a set of axioms for which a centrality with normalized influence proposed by Pinski and Narin [479] is the unique centrality that satisfies all of them. This Pinski-Narin-centrality is strongly related to the PageRank score in that it may be seen as the basis (of PageRank) that is augmented by the addition of a stochastic vector that allows for leaving the sinks.

To state the axioms properly we need some definitions. We are given a directed graph $G = (V, E)$ with weights ω on the edges and weights α on the vertices. In terms of citation networks V corresponds to the set of journals and $(i, j) \in E$ iff journal i is cited by journal j. The weight $\omega(i, j)$ is defined to be the number of citations to journal i by journal j if $(i, j) \in E$ and 0 otherwise, while the vertex weight $\alpha(i)$ corresponds to the number of articles published in journal i. The authors consider strongly connected subgraphs with the additional property that there is no path from a vertex outside the subgraph to a vertex contained in it. (Note that they allow loops and loop weights.) Palacios-Huerta and Volij call such subgraphs a *discipline*, where a discipline is a special *communication class* (a strongly connected subgraph) which itself is defined to be an equivalence class with respect to the equivalence relation of communication. Two journals i and j *communicate*, if either $i = j$ or if i and j *impact* each other, where i impacts j if there is a sequence of journals $i = i_0, i_1, \ldots, i_{K-1}, i_K = j$ such that i_{l-1} is cited by i_l, that is, if there is a path from i to j.

Define the $(|V| \times |V|)$-matrices

$$W = (\omega(i,j)), \; D_\omega = \mathrm{diag}(\omega(\cdot j)) \text{ with } \omega(\cdot j) = \sum_{i \in V} \omega(i,j),$$

and set WD_ω^{-1} to be the normalized weight matrix, and $D_\alpha = \mathrm{diag}\,(\alpha(i))$. Then the *ranking problem* $\langle V, \alpha, W \rangle$ is defined for the vertex set V of a discipline, the associated vertices weights α and the corresponding citation matrix W, and considers the ranking (a centrality vector) $c_{\mathrm{PHV}} \geq 0$ that is normalized with respect to the l_1-norm: $\|c_{\mathrm{PHV}}\|_1 = 1$.

The authors consider two special classes of ranking problems:

1. ranking problems with all vertex weights equal, $\alpha(i) = \alpha(j) \; \forall \; i,j \in V$ (*isoarticle problems*) and
2. ranking problems with all *reference intensities* equal, $\frac{\omega(\cdot i)}{\alpha(i)} = \frac{\omega(\cdot j)}{\alpha(j)} \; \forall \; i,j \in V$ (*homogeneous problems*).

To relate small and large problems, the *reduced ranking problem* R^k for a ranking problem $R = \langle V, \alpha, W \rangle$ with respect to a given vertex k is defined as $R^k = \langle V \setminus \{k\}, (\alpha(i))_{i \in V \setminus \{k\}}, (\omega_k(i,j))_{(i,j) \in V \setminus \{k\} \times V \setminus \{k\}} \rangle$, with

$$\omega_k(i,j) = \omega(i,j) + \omega(k,j) \frac{\omega(i,k)}{\sum_{l \in V \setminus \{k\}} \omega(l,k)} \; \forall \; i,j \in V \setminus \{k\}.$$

Finally, consider the problem of splitting a vertex j of a ranking problem $R = \langle V, \alpha, W \rangle$ into $|T_j|$ sets of identical vertices (j, t_j) for $t_j \in T_j$. For $V' = \{(j, t_j) : j \in V, t_j \in T_j\}$, the *ranking problem resulting from splitting j* is denoted by

$$R' = \langle V', (\alpha'((j,t_j)))_{j \in J, t_j \in T_j}, (\omega'((i,t_i)(j,t_j)))_{((i,t_i)(j,t_j)) \in V' \times V'} \rangle,$$

with

$$\alpha'((j,t_j)) = \frac{\alpha(j)}{|T_j|}, \; \omega'((i,t_i)(j,t_j)) = \frac{\omega(i,j)}{|T_i||T_j|}.$$

Note that the latter two definitions of special ranking problems are needed to formulate the following axioms.

A *ranking method* Φ assigning to each ranking problem a centrality vector should then satisfy the following four axioms (at least the weak formulations):

Axiom 1: *invariance with respect to reference intensity*

Φ is *invariant with respect to reference intensity* if

$$\Phi(\langle V, \alpha, W\Gamma \rangle) = \Phi(\langle V\alpha, W \rangle)$$

for all ranking problems $\langle V, \alpha, W \rangle$ and every Matrix $\Gamma = \mathrm{diag}(\gamma_j)_{j \in V}$ with $\gamma_j > 0 \; \forall \; j \in V$.

Axiom 2: *(weak) homogeneity*

a) Φ satisfies *weak homogeneity* if for all two-journal problems $R = \langle \{i, j\},$ $\alpha, W \rangle$ that are homogeneous and isoarticle, it holds that

$$\frac{\Phi_i(R)}{\Phi_j(R)} = \frac{\omega(i, j)}{\omega(j, i)}. \tag{5.7}$$

b) Φ satisfies *homogeneity* if (Equation 5.7) holds for all homogeneous problems.

Axiom 3: *(weak) consistency*

a) Φ satisfies *weak consistency* if for all homogeneous, isoarticle problems $R = \langle V, \alpha, W \rangle$ with $|V| > 2$ and for all $k \in V$

$$\frac{\Phi_i(R)}{\Phi_j(R)} = \frac{\Phi_i(R^k)}{\Phi_j(R^k)} \ \forall \ i, j \in V \setminus \{k\}. \tag{5.8}$$

b) Φ satisfies *consistency* if (Equation 5.8) holds for all homogeneous problems.

Axiom 4: *invariance with respect to the splitting of journals*

Φ is *invariant to splitting of journals*, i.e. for all ranking problems R and for all splittings R' of R holds

$$\frac{\Phi_i(R)}{\Phi_j(R)} = \frac{\Phi_{(i,t_i)}(R')}{\Phi_{(j,t_j)}(R')} \ \forall i, j \in V, \ \forall \ i \in T_i, \ \forall \ j \in T_j.$$

Palacios-Huerta and Volij show that the ranking method assigning the Pinski-Narin centrality c_{PN} given as the unique solution of

$$D_\alpha^{-1} W D_W^{-1} D_\alpha c = c$$

is the only ranking method that satisfies

- invariance to reference intensity (Axiom 1),
- weak homogeneity (Axiom 2a),
- weak consistency (Axiom 3a), and
- invariance to splitting of journals (Axiom 4).

Slutzki and Volij [526] also consider the axiomatization of ranking problems, which they call *scoring problems*. Although their main field of application is shifted from citation networks to (generalized) tournaments, it essentially considers the same definitions as above, excluding the vertex weights α. Further, they consider strongly connected subgraphs (not necessarily disciplines), and set $\omega(i, i) = 0$ for all $i \in V$, meaning that there are no self-references, i.e. no loops in the corresponding graph. For this case, the Pinski-Narin centrality may be characterized by an alternative set of axioms, and again it is the only centrality satisfying this set.

The Link to Normalization. Above, we saw that normalization is a question when dealing with axiomatizations. Either it is explicitly stated as an axiom (see the centralities of van den Brink and Gilles) or the normalization is implicitly assumed when talking about centralities (see the papers of Volij and his coworkers). The topic of normalization was already investigated in Section 5.1. Here, we report on investigations of Ruhnau [499] about normalizing centralities.

Her idea is based on an intuitive understanding of centrality, already formulated by Freeman in 1979 [227]:

"A person located in the center of a star is universally assumed to be structurally more central than any other person in any other position in any other network of similar size."

She formalizes this in the definition of a vertex-centrality for undirected connected graphs $G = (V, E)$.

Definition 5.4.5 (Ruhnau's vertex centrality axioms). *Let $G = (V, E)$ be an undirected and connected graph with $|V| = n$ and let $c_V : V \to \mathbb{R}$. c_V is called a* vertex-centrality *if*

1. *$c_V(i) \in [0, 1]$ for all $i \in V$ and*
2. *$c_V(i) = 1$ if and only if G is a star with n vertices and i the central vertex of it.*

(Note: Ruhnau calls this a node *centrality. For consistency with the rest of the chapter we used the equivalent term vertex centrality here.)*

The property of being a vertex-centrality may be very useful when comparing vertices of different graphs. To see this, compare the central vertex of a star of order n with any vertex in a complete graph of order n. Both have a degree of $n - 1$, but intuitively the central vertex of a star has a much more prominent role in the graph than any of the vertices in a complete graph.

Freeman [226] showed that the betweenness centrality satisfies the conditions of the above definition. Due to the fact that the eigenvector centrality normalized by the Euclidean norm has the property that the maximal attainable value is $1/\sqrt{2}$ (independent of n), and that it is attained exactly at the center of a star (see [465]), it is also a vertex-centrality (multiplied by $\sqrt{2}$). For more information about normalization, see Section 5.1.

5.5   Stability and Sensitivity

Assume that a network is modified slightly for example due to the addition of a new link or the inclusion of a new page in case of the Web graph. In this situation the 'stability' of the results are of interest: does the modification invalidate the computed centralities completely?

In the following subsection we will discuss the topic of stability for distance based centralities, i.e., eccentricity and closeness, introduce the concept of stable, quasi-stable and unstable graphs and give some conditions for the existence of stable, quasi-stable and unstable graphs.

A second subsection will cover Web centralities and present results for the numerical stability and rank stability of the centralities discussed in Section 3.9.3.

5.5.1 Stability of Distance-Based Centralities

In Section 5.4.1 we considered the axiomatization of connected undirected graphs $G = (V, E)$ and presented two definitions for distance-based vertex centralities. Moreover, we denoted by $\mathcal{S}_c(G) = \{u \in V : \forall v \in V \; c(u) \geq c(v)\}$ the set of maximum centrality vertices of G with respect to a centrality c and we studied the change of the centrality values if we add an edge (u, v) between two distinct non-adjacent vertices in G. In this section we focus on the stability of the center $\mathcal{S}_c(G)$ with respect to this graph operation (cf. Condition 5 of Definition 5.4.1).

Let $u \in \mathcal{S}_c(G)$ be a central vertex with respect to c, and $(u, v) \notin G$. Then the insertion of an edge (u, v) to G yields a graph $H = (V, E \cup (u, v))$. Regarding $\mathcal{S}_c(H)$ two cases can occur, either

$$\mathcal{S}_c(H) \subseteq \mathcal{S}_c(G) \cup \{v\} \tag{5.9}$$

or

$$\mathcal{S}_c(H) \nsubseteq \mathcal{S}_c(G) \cup \{v\} \tag{5.10}$$

for every vertex $v \in V$. Kishi [357] calls a graph for which the second case (Equation 5.10) occurs an *unstable graph* with respect to c. Figures 5.2 and 5.3 in Section 5.4.1 show unstable graphs with respect to the eccentricity and the closeness centrality. The first case (Equation 5.9) can be further classified into

$$\mathcal{S}_c(H) \subseteq \mathcal{S}_c(G) \quad \text{and} \quad u \in \mathcal{S}_c(H) \tag{5.11}$$

and

$$\mathcal{S}_c(H) \nsubseteq \mathcal{S}_c(G) \quad \text{or} \quad u \notin \mathcal{S}_c(H) \tag{5.12}$$

A graph G is called a *stable graph* if the first case (Equation 5.11) occurs, otherwise G is called a *quasi-stable graph*. The definition of stable graphs with respect to c encourages Sabidussi's claim [500] that an edge added to a central vertex $u \in \mathcal{S}_c(G)$ should strengthen its position.

In Figure 5.4 an example for a quasi-stable graph with respect to closeness centrality is shown. For each vertex the status value $s(u) = \sum_{v \in V} d(u, v)$ is indicated. Adding the edge (u, v) leads to a graph with a new central vertex v.

In [357] a more generalized form of closeness centrality is presented by Kishi: The centrality value $c_{GenC}(u)$ of a vertex $u \in V$ is

$$c_{GenC}(u) = \frac{1}{\sum_{k=1}^{\infty} a_k n_k(u)} \tag{5.13}$$

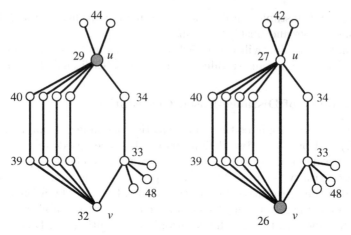

Fig. 5.4. A quasi-stable graph with respect to the closeness centrality. The values indicate the total distances $s(u)$. After inserting the edge (u, v) the new median is vertex v

where $n_k(u)$ is the number of vertices whose distance from u is k and each a_k is a real constant. With $a_k = k$ it is easy to see that

$$\frac{1}{\sum_{k=1}^{\infty} a_k n_k(u)} = \frac{1}{\sum_{v \in V} d(u, v)} = c_C(u).$$

Kishi and Takeuchi [358] have shown under which conditions there exists a stable, quasi-stable, and unstable graph for generalized centrality functions c_{GenC} of the form in Equation 5.13:

Theorem 5.5.1. *For any generalized vertex centrality c_{GenC} of the form in Equation 5.13 holds:*

1. *if $a_2 < a_3$, then there exists a quasi-stable graph, and*
2. *if $a_3 < a_4$, then there exists an unstable graph.*

Theorem 5.5.2. *Any connected undirected graph G is stable if and only if the generalized vertex centrality c_{GenC} given in Equation 5.13 satisfies $a_2 = a_3$. Moreover, G is not unstable if and only if c_{GenC} satisfies $a_3 = a_4$.*

Sabidussi has shown in [500] that the class of undirected trees are stable graphs with respect to the closeness centrality.

Theorem 5.5.3. *If an undirected graph G forms a tree, then G is stable with respect to the closeness centrality.*

5.5.2 Stability and Sensitivity of Web-Centralities

First, we consider stability with respect to the centrality *values*, later on we report on investigations on the centrality *rank*. We call the former *numerical stability* and the latter rank stability.

Numerical Stability. Langville and Meyer [378] remark that it is not reasonable to consider the linear system formulation of, e.g., the PageRank approach and the associated condition number[3], since it may be that the solution vector of the linear system changes considerable but the normalized solution vector stays almost the same. Hence, what we are looking for is to consider the stability of the eigenvector problem which is the basis for different Web centralities mentioned in Section 3.9.3.

Ng et al. [449] give a nice example showing that an eigenvector may vary considerably even if the underlying network changes only slightly. They considered a set of Web pages where 100 of them are linked to *algore.com* and the other 103 pages link to *georgewbush.com*. The first two eigenvectors (or, in more detail, the projection onto their nonzero components) are drawn in Figure 5.5(a). How the scene changes if five new Web pages linking to both *algore.com* and *georgewbush.com* enter the collection is then depicted in Figure 5.5(b).

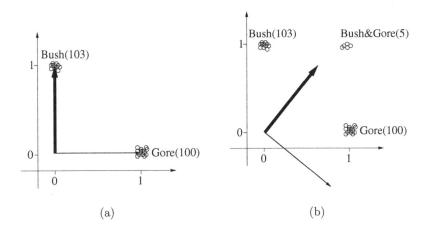

(a)   (b)

Fig. 5.5. A small example showing instability resulting from perturbations of the graph. The projection of the eigenvector is shown and the perturbation is visible as a strong shift of the eigenvector

Regarding the Hubs & Authorities approach Ng et al. the authors give a second example, cf. Figs 5.6(a) and 5.6(b). In the Hubs & Authorities algorithm the largest eigenvector for $S = A^T A$ is computed. The solid lines in the figures represent the contours of the quadratic form $x^T S_i x$ for two matrices S_1, S_2 as well as the contours of the slightly (but equally) perturbed matrices. In both figures the associated eigenvectors are depicted. The difference (strong shift in the eigenvectors in the first case, almost no change in the eigenvectors in the

---

[3] $cond(A) = \|A\|\|A^{-1}\|$ (for A regular)

second case) between the two figures consists of the fact that S_1 has an eigengap[4] $\delta_1 \sim 0$ whereas S_2 has eigengap $\delta_2 = 2$. Hence in the case that the eigengap is almost zero, the algorithm may be very sensitive about small changes in the matrix whereas in case the eigengap is large the sensitivity is small.

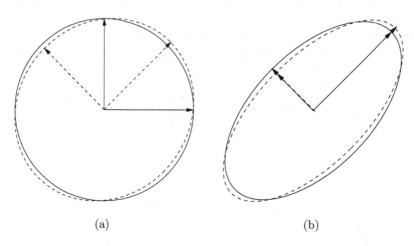

(a)   (b)

Fig. 5.6. A simple example showing the instability resulting from different eigengaps. The position of the eigenvectors changes dramatically in the case of a small eigengap (a)

Ng et al. also show this behavior theoretically

Theorem 5.5.4. *Given* $S = A^T A$, *let* $c_{HA\text{-}A}$ *be the principal eigenvector and* δ *the eigengap of* S. *Assume* $d^+(i) \le d$ *for every* $i \in V$ *and let* $\varepsilon > 0$. *If the Web graph is perturbed by adding or deleting at most* k *links from one page,* $k < \left(\sqrt{d+\alpha} - \sqrt{d}\right)^2$, $\alpha = \frac{\varepsilon\delta}{4+\sqrt{2}\varepsilon}$ *then the perturbed principal eigenvector* $\tilde{c}_{HA\text{-}A}$ *of the perturbed matrix* $\tilde{S}$ *satisfies* $\|c_{HA\text{-}A} - \tilde{c}_{HA\text{-}A}\|_2 \le \varepsilon$.

Theorem 5.5.5. *If* S *is a symmetric matrix with eigengap* δ, *then there exists a perturbed version* $\tilde{S}$ *of* S *with* $\|S - \tilde{S}\|_F = \mathcal{O}(\delta)$ *that causes a large* $(\Omega(1))$ *change in the principal eigenvector.*

(Note that $\|X\|_F = \left(\sum_i \sum_j (x_{ij}^2)\right)^{1/2}$ *denotes the Frobenius norm.)*

If we consider the PageRank algorithm, then the first fact that we have to note is that for a Markov chain having transition matrix P the sensitivity of the principal eigenvector is determined by the difference of the second eigenvalue to 1. As shown by Haveliwala and Kamvar [290] the second eigenvalue for the PageRank-matrix with P having at least two irreducible closed subsets satisfies

---

[4] Difference between the first and the second largest eigenvalue.

$\lambda_2 = d$. This is true even in the case that in Formula 3.43 the vector $\mathbf{1}_n$ is substituted by any stochastic vector $\boldsymbol{v}$, the so-called *personalization vector*, cf. Section 5.2 for more information about the personalization vector.

Therefore a damping factor of $d = 0.85$ (this is the value proposed by the founders of Google) yields in general much more stable results than $d = 0.99$ which would be desirable if the similarity of the original Web graph with its perturbed graph should be as large as possible.

Ng et al. [449] proved

Theorem 5.5.6. *Let $U \subseteq V$ be the set of pages where the outlinks are changed, $\boldsymbol{c}_{PR}$ be the old PageRank score and $\boldsymbol{c}^U_{PR}$ be the new PageRank score corresponding to the perturbed situation. Then*

$$\|\boldsymbol{c}_{PR} - \boldsymbol{c}^U_{PR}\|_1 \leq \frac{2}{1-d} \sum_{i \in U} c_{PR}(i).$$

Bianchini, Gori and Scarselli [61] were able to strengthen this bound. They showed

Theorem 5.5.7. *Under the same conditions as given in Theorem 5.5.6 it holds*

$$\|\boldsymbol{c}_{PR} - \boldsymbol{c}^U_{PR}\|_1 \leq \frac{2d}{1-d} \sum_{i \in U} c_{PR}(i).$$

(Note that $d < 1$.)

Rank Stability. When considering Web centralities, the results are in general returned as a list of Web pages matching the search-query. The scores attained by the Web pages are in most cases not displayed and hence the questions that occurs is whether numeric stability also implies stability with respect to the rank in the list (called *rank-stability*). Lempel and Moran [388] investigated the three main representatives of Web centrality approaches with respect to rank-stability.

To show that numeric stability does not necessarily imply rank-stability they used the graph $G = (V, E)$ depicted in Figure 5.7. Note that in the graph any undirected edge $[u, v]$ represents two directed edges (u, v) and (v, u). From G two different graphs $G_a = (V, E \cup \{(y, h_a)\})$ and $G_b = (V, E \cup \{(y, h_b)\})$ are derived (they are not displayed). It is clear that the PageRank vector $\boldsymbol{c}^a_{\mathrm{PR}}$ corresponding to G_a satisfies

$$0 < c^a_{\mathrm{PR}}(x_a) = c^a_{\mathrm{PR}}(y) = c^a_{\mathrm{PR}}(x_b),$$

and therefore $c^a_{\mathrm{PR}}(h_a) > c^a_{\mathrm{PR}}(h_b)$.

Analogously in G_b we have

$$0 < c^b_{\mathrm{PR}}(x_a) = c^b_{\mathrm{PR}}(y) = c^b_{\mathrm{PR}}(x_b),$$

hence $c^b_{\mathrm{PR}}(h_a) < c^b_{\mathrm{PR}}(h_b)$.

Concluding we see that by shifting one single outlink from a very low-ranking vertex y induces a complete change in the ranking:

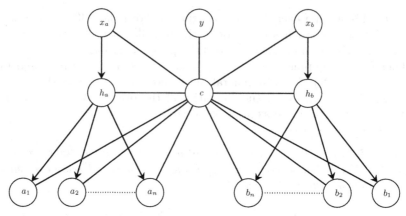

Fig. 5.7. The graph G used for the explanation of the rank stability effect of PageRank. Please note that for G_a a directed edge from y to h_a is added and in the case of G_b from y to h_b

$$c_{\mathrm{PR}}^a(a_i) > c_{\mathrm{PR}}^a(b_i) \text{ and } c_{\mathrm{PR}}^b(a_i) < c_{\mathrm{PR}}^b(b_i) \; \forall \; i.$$

To decide whether an algorithm is rank-stable or not we have to define the term *rank-stability* precisely. Here we follow the lines of Borodin et al. [87] and [388]. Let $\mathcal{G}$ be a set of directed graphs and $\mathcal{G}_n$ the subset of $\mathcal{G}$ where all directed graphs have n vertices.

Definition 5.5.8. *1. Given two ranking vectors r^1 and r^2, associated to a vertex-set of order n, the ranking-distance between them is defined by*

$$d_r(r^1, r^2) = \frac{1}{n^2} \sum_{i,j=1}^n l_{ij}^{r^1, r^2}$$

$$\text{where } l_{ij}^{r^1, r^2} = \begin{cases} 1, r_i^1 < r_j^1 \text{ and } r_i^2 > r_j^2 \\ 0, \qquad\qquad\qquad \text{otherwise} \end{cases}$$

2. An algorithm $\mathcal{A}$ is called **rank-stable** *on $\mathcal{G}$ if for each k fixed we have*

$$\lim_{n \to \infty} \max_{\substack{G_1, G_2 \in \mathcal{G}_n \\ d_e(G_1, G_2) \leq k}} d_r(\mathcal{A}(G_1), \mathcal{A}(G_2)) \longrightarrow 0,$$

where $d_e(G_1, G_2) = |(E_1 \cup E_2) \setminus (E_1 \cap E_2)|$.

Hence an algorithm $\mathcal{A}$ is rank-stable on $\mathcal{G}$ if for each k the effect on the ranking of the nodes of changing k edges vanishes if the size of the node-set of a graph tends to infinity.

Borodin et al. were able to show that neither the Hubs & Authorities algorithm nor the SALSA method are rank-stable on the set of all directed graphs $\bar{\mathcal{G}}$.

However, they obtained a positive result by considering a special subset of $\bar{\mathcal{G}}$, the set of authority connected directed graphs $\mathcal{G}^{ac}$:

Definition 5.5.9. *1. Two vertices* $p, q \in V$ *are called* co-cited, *if there is a vertex* $r \in V$ *satisfying* $(r, p), (r, q) \in E$.

2. p, q *are connected by a* co-citation path *if there exist vertices* $p = v_0, v_1, \ldots,$ $v_{k-1}, v_k = q$ *such that* (v_{i-1}, v_i) *are co-cited for all* $i = 1, \ldots, k$.

3. A directed graph $G = (V, E)$ *is* authority connected *if for all* p, q *satisfying* $d^-(p), d^-(q) > 0$ *there is a co-citation path.*

Lempel and Moran argue that it is reasonable to restrict the stability investigations to this subset of directed graphs due to the following observation:

− if p, q are co-cited then they cover the same subject,
− the relevance of p and q should be measured with respect to the same bar, and
− there is no interest in answering questions like "is p a better geography resource that q is an authority on sports?"

For authority connected subgraphs it holds that

− SALSA is rank-stable on $\mathcal{G}^{ac}$,
− The Hubs & Authorities algorithm is not rank-stable on $\mathcal{G}^{ac}$, and
− PageRank is not rank-stable on $\mathcal{G}^{ac}$.

Note that the latter two results were obtained by Lempel and Moran [388].

With this result we finish the discussion of sensitivity and stability of Web centralities. Interested readers are directed to the original papers shortly mentioned in this section.

6 Local Density

Sven Kosub

Actors in networks usually do not act alone. By a selective process of establishing relationships with other actors, they form groups. The groups are typically founded by common goals, interests, preferences or other similarities. Standard examples include personal acquaintance relations, collaborative relations in several social domains, and coalitions or contractual relationships in markets. The cohesion inside these groups enables them to influence the functionality of the whole network.

Discovering cohesive groups is a fundamental aspect in network analysis. For a computational treatment, we need formal concepts reflecting some intuitive meaning of cohesiveness. At a general level, the following characteristics have been attributed to cohesive groups [569]:

- *Mutuality:* Group members choose each other to be included in the group. In a graph-theoretical sense, this means that they are adjacent.
- *Compactness:* Group members are well reachable for each other, though not necessarily adjacent. Graph-theoretically, this comes in two flavors: being well reachable can be interpreted as having short distances or high connectivity.
- *Density:* Group members have many contacts to each other. In terms of graph theory, that is group members have a large neighborhood inside the group.
- *Separation:* Group members have more contacts inside the group than outside.

Depending on the network in question, diverse concepts can be employed, incorporating cohesiveness characteristics with different emphases. Notions where density is a dominant aspect are of particular importance.

Density has an outstanding relevance in social networks. On the one hand, recent studies have found that social networks show assortative mixing on vertices [441, 444, 446], i.e, they tend to have the property that neighbors of vertices with high degree have also high degree. Assortative mixing is an expression of the typical observation that social networks are structured by groups of high density.[1] On the other hand, there are several mathematical results demonstrating that high density implies the other characteristics of cohesiveness. For instance, one classical result [431] says that if each member of a group shares ties with at least

---

[1] Assortativity is now considered as one statistical criterion separating social networks and non-social networks [446]. For instance, some experimental analyses have shown that in the Internet at the level of autonomous systems, the mean degree of the neighbors of an autonomous system with k neighbors is approximately $k^{-1/2}$ [468]. At this level, the Internet is disassortatively mixed.

U. Brandes and T. Erlebach (Eds.): Network Analysis, LNCS 3418, pp. 112–142, 2005.
© Springer-Verlag Berlin Heidelberg 2005

a $\frac{1}{k}$-fraction of the other members of the group, then the tie distance within the group is at most k. Results comparable to that can be proven for connectivity as well. Here, however, the dependency from density is not as strong as in the case of distances (see Chapter 7).

In this chapter, we survey computational approaches and solutions for discovering *locally* dense groups. A graph-theoretical group property is local if it is definable over subgraphs induced by the groups only. Locality does not correspond to the above-mentioned separation characteristic of cohesiveness, since it neglects the network outside the group. In fact, most notions that have been defined to cover cohesiveness have a maximality condition. That is, they require for a group to be cohesive with respect to some property Π, in addition to fulfilling Π, that it is not contained in any larger group of the network that satisfies Π as well. Maximality is non-local. We present the notions on the basis of their underlying graph-theoretical properties and without the additional maximality requirements. Instead, maximality appears in connection with several computational problems derived from these notions. This is not a conceptual loss. Actually, it emphasizes that locality reflects an important hidden aspect of cohesive groups: being invariant under network changes outside the group. Interior robustness and stability is an inherent quality of groups. Non-local density notions and the corresponding algorithmic problems and solutions are presented in Chapter 8. A short list of frequently used non-local notions is also discussed in Section 6.4.

The prototype of a cohesive group is the clique. Since its introduction into sociology in 1949 [401], numerous efforts in combinatorial optimization and algorithms have been dedicated to solving computational problems for cliques. Therefore, the treatment of algorithms and hardness results for clique problems deserves a large part of this chapter. We present some landmark results in detail in Section 6.1. All other notions that we discuss are relaxations of the clique concept. We make a distinction between structural and statistical relaxations. A characteristic of structural densities is that all members of a group have to satisfy the same requirement for group membership. These notions (plexes, cores) admit strong statements about the structure within the group. Structurally dense groups are discussed in Section 6.2. In contrast, statistical densities average over members of a group. That is, the property that defines group membership needs only be satisfied in average (or expectation) over all group members. In general, statistically dense groups reveal only few insights into the group structure. However, statistical densities can be applied under information uncertainty. They are discussed in Section 6.3.

All algorithms are presented for the case of unweighted, undirected simple graphs exclusively. Mostly, they can be readily translated for directed or weighted graphs. In some exceptional cases where new ideas are needed, we mention these explicitly.

6.1   Perfectly Dense Groups: Cliques

The graph with perfect cohesion is the clique [401].

Definition 6.1.1. *Let $G = (V, E)$ be an undirected graph. A subset $U \subseteq V$ is said to be a* clique *if and only if $G[U]$ is a complete graph.*

In a clique, each member has ties with each other member. A clique U is a maximal clique in a graph $G = (V, E)$ if and only if there is no clique U' in G with $U \subset U'$. A clique is a maximum clique in graph G if and only if it has maximum cardinality among all cliques in G.

The striking reasons for perfectness of cliques as cohesive structures are obvious:

1. Cliques are perfectly dense, i.e., if U is a clique of size k, then $\delta(G[U]) = \bar{d}(G[U]) = \Delta(G[U]) = k - 1$. A higher degree is not possible.
2. Cliques are perfectly compact, i.e., $\mathrm{diam}(G[U]) = 1$. A shorter distance between any two vertices is not possible.
3. Cliques are perfectly connected, i.e., if U is a clique of size k, then U is $(k-1)$-vertex-connected and $(k-1)$-edge-connected. A higher connectivity is not possible.

The structural properties of a clique are very strong. In real-world settings, large cliques thus should be rarely observable. The famous theorem of Turán [554] gives precise sufficient conditions for the guaranteed existence of cliques of certain sizes with respect to the size of the entire network.

Theorem 6.1.2 (Turán, 1941). *Let $G = (V, E)$ be an undirected graph. If $m > \frac{n^2}{2} \cdot \frac{k-2}{k-1}$, then there exists a clique of size k within G.*

An immediate consequence of this theorem is that a network itself needs to be dense in order to surely possess a large clique. However, as social networks are usually sparse, we have no *a priori* evidence for the existence of a clique. Identifying cliques becomes an algorithmic task. Note that, as we will see below, even if we knew that there is a clique of a certain size in a network, we would not be able to locate it in reasonable time.

Maximal cliques do always exist in a graph. In fact, there are many of them and they tend to overlap, i.e., in general it can be the case that maximal cliques U_1 and U_2 exist satisfying $U_1 \neq U_2$ and $U_1 \cap U_2$ is non-empty. Another classical result due to Moon and Moser [432] gives a tight estimation of the number of maximal cliques:

Theorem 6.1.3 (Moon and Moser, 1965). *Every undirected graph G with n vertices has at most $3^{\lceil \frac{n}{3} \rceil}$ maximal cliques.*

In reality, the expected enormous number of maximal cliques leads to the serious problem of how to identify the more important ones among them. There are only few algorithmic techniques available providing helpful interpretation of the maximal-clique collection. Prominent examples for methods are based on the co-membership matrix or the clique overlap centrality [192].

The family of all cliques of a certain graph shows some structure:

1. Cliques are closed under exclusion, i.e., if U is a clique in G and $v \in U$, then $U - \{v\}$ is also a clique.[2]
2. Cliques are nested, i.e., each clique of size n contains a clique of size $n - 1$ (even n cliques of size $n - 1$). Though this is an immediate consequence of the closure under exclusion, it is a property to be proved for related notions that are not closed under exclusion.

Distance-based cliques. There is a number of approaches to generalize the notion of a clique that are relevant in several settings of social-network theory. We list some of them [400, 14, 429]. Let $G = (V, E)$ be an undirected graph, let U be a vertex subset of V, and let $N > 0$ be any natural number.

1. U is said to be an N-clique if and only if for all $u, v \in U$, $d_G(u, v) \leq N$.
2. U is said to be an N-club if and only if $\operatorname{diam}(G[U]) \leq N$.
3. U is said to be an N-clan if and only if U is a maximal N-clique and $\operatorname{diam}(G[U]) \leq N$.

N-cliques are based on non-local properties, as the distance between vertices u and v is measured with respect to graph G, and not with respect to $G[U]$. An immediate consequence is that N-cliques need not be connected for $N > 1$. Though clubs and clans are local structures (except the maximality condition), they are of minor interest in our context, since they emphasize distances rather than density. Moreover, there has been some criticism of distance-based cliques, which was sparked off by at least two facts (cf., e.g., [514, 189]). First, in many cases real-world networks have globally a small diameter, thus, the distance is a rather coarse measure to identify meaningful network substructures. Second, distance-based cliques are in general neither closed under exclusion nor nested.

6.1.1 Computational Primitives

In many respects, cliques are simple objects, easily manageable from an algorithmic point of view. We have fast algorithms with run-time $\mathcal{O}(n + m)$ at hand for several computational primitives:

1. *Determine if a given set $U \subseteq V$ of vertices is a clique in G.* We simply test whether each pair of vertices of U is an edge in G. Note that these are up to $\binom{n}{2}$ pairs, but even if we have much fewer edges, after testing m pairs we are done in any case.
2. *Determine if a given clique $U \subseteq V$ is maximal in G.* We simply test whether there exists a vertex in $V - U$ which is adjacent to all vertices in U. Again, after testing m edges we are done in the worst case.

---

[2] In graph theory, a property Π is called *hereditary* if and only if, whenever a graph satisfies Π, so does every induced subgraph. Being a clique is a hereditary property of graphs.

Another efficiently computable primitive is finding some maximal clique. For later use, we state this in a more general form. Suppose that the vertex set V of a graph $G = (V, E)$ is ordered. We say that a set $U \subseteq V$ is lexicographically smaller than a set $U' \subseteq V$ if and only if the first vertex that is not in both U and U' belongs to U. Our third primitive is the following:

3. *Compute the lexicographically smallest maximal clique containing some clique* U'. We start with setting $U := U'$, iterate over all $v \in V - U$ in increasing order, and test for each v whether $U \subseteq N(v)$; if this is the case then add vertex v to U. After completing the iteration, U is a maximal clique containing U'. This works in time $\mathcal{O}(n + m)$.

Algorithmic difficulties appear only when we are interested in finding cliques of certain sizes or maximum cliques. For these problems, no algorithms with running times comparable to the one above are known (and, probably, no such algorithms exist).

6.1.2   Finding Maximum Cliques

We discuss several aspects of the maximum clique problem. Of course, it is easy to compute a clique of maximum size, if we do not care about time. The obvious approach is exhaustive search. In an exhaustive search algorithm, we simply enumerate all possible candidate sets $U \subseteq V$ and examine if U is a clique. We output the largest clique found. A simple estimation gives a worst-case upper bound $\mathcal{O}(n^2 \cdot 2^n)$ on the time complexity of the algorithm.

Computational hardness. The problem arises whether we can improve the exhaustive search algorithm significantly with respect to the amount of time. Unfortunately, this will probably not be the case. Computationally, finding a maximum clique is an inherently hard problem. We consider the corresponding decision problem:

Problem:	CLIQUE
Input:	Graph G, Parameter $k \in \mathbb{N}$
Question:	Does there exist a clique of size at least k within G?

Let $\omega(G)$ denote the size of a maximum clique of a graph G. Note that if we have an algorithm that decides CLIQUE in time $T(n)$ then we are able to compute $\omega(G)$ in time $\mathcal{O}(T(n) \cdot \log n)$ using binary search. The other way around, any $T(n)$ algorithm for computing $\omega(G)$, gives a $T(n)$ algorithm for deciding CLIQUE. Thus, if we had a polynomial algorithm for CLIQUE, we would have a polynomial algorithm for maximum-clique sizes, and vice versa. However, CLIQUE was among the first problems for which $\mathcal{NP}$-completeness was established [345].

Theorem 6.1.4. CLIQUE *is* $\mathcal{NP}$-*complete.*

Proof. Note that testing whether some guessed set is a clique is possible in polynomial time. This shows the containment in $\mathcal{NP}$. In order to prove the $\mathcal{NP}$-hardness, we describe a polynomial-time transformation of SATISFIABILITY into CLIQUE. Suppose we are given a Boolean formula H in conjunctive normal form consisting of m clauses $C_1, \ldots, C_k$. For H we construct a k-partite graph G_H where vertices are the literals of H labeled by their clause, and where edges connect literals that are not negations of each other. More precisely, define $G_H = (V_H, E_H)$ to be the following graph:

$$V_H =_{\text{def}} \left\{ (L, i) \mid i \in \{1, \ldots, k\} \text{ and } L \text{ is a literal in clause } C_i \right\}$$
$$E_H =_{\text{def}} \left\{ \{(L, i), (L', j)\} \mid i \neq j \text{ and } L \neq \neg L' \right\}$$

Clearly, the graph G_H can be computed in time polynomial in the size of the formula H. We show that H is satisfiable if and only if the graph G_H contains a clique of size k.

Suppose that H is satisfiable. Then there exists a truth assignment to variables $x_1, \ldots, x_n$ such that in each clause at least one literal is true. Let $L_1, \ldots, L_k$ be such literals. Then, of course, it must hold that $L_i \neq \neg L_j$ for $i \neq j$. We thus obtain that the set $\{(L_1, 1), \ldots, (L_k, k)\}$ is a clique of size k in G_H.

Suppose now that $U \subseteq V_H$ is a clique of size k in graph G_H. Since G_H is k-partite, U contains exactly one vertex from each part of V_H. By definition of set V_H, we have that for all vertices (L, i) and (L', j) of U, $L \neq \neg L'$ whenever $i \neq j$. Hence, we can assign truth values to variables in such a way that all literals contained in U are satisfied. This gives a satisfying truth assignment to formula H. $\square$

So, unless $\mathcal{P} = \mathcal{NP}$, there are no algorithms with a running time polynomial in n for solving CLIQUE with arbitrary clique size or computing the maximum clique. On the other hand, even if we have a guarantee that there is a clique of size k in graph G, then we are not able to find it in polynomial time.

Corollary 6.1.5. *Unless $\mathcal{P} = \mathcal{NP}$, there is no algorithm running in polynomial time to find a clique of size k in a graph which is guaranteed to have a clique of size k.*

Proof. Suppose we have an algorithm A that runs in polynomial time on each input (G, k) and outputs a clique of size k, if it exists, and behaves in an arbitrary way in the other cases. A can be easily modified into an algorithm A' that decides CLIQUE in polynomial time. On input (G, k), run algorithm A, if A produces no output, then reject the instance. If A outputs a set U, then test whether U is a clique. If so, accept, otherwise reject. This procedure is certainly polynomial time. $\square$

Note that the hardness of finding the hidden clique does not depend on the size of the clique. Even very large hidden cliques (of size $(1 - \varepsilon)n$ for $\varepsilon > 0$) cannot be found unless $\mathcal{P} = \mathcal{NP}$ (see, e.g., [308, 37]). The situation becomes slightly better if we consider randomly chosen graphs, i.e., graphs where each edge appears with probability $\frac{1}{2}$. Suppose we additionally place at random a clique of size k in such a random graph of size n. How fast can we find this clique? It has been observed that, if $k = \Omega(\sqrt{n \log n})$, then almost surely the k vertices with highest degree form the clique [374]. This gives a trivial $\mathcal{O}((n + m) \log n)$ algorithm (which can be improved to an $\mathcal{O}(n+m)$ algorithm with a technique discussed in Theorem 6.2.7). For $k = \Omega(\sqrt{n})$, algorithms based on spectral techniques have been proven to find hidden cliques of size k in polynomial time [22] (even in some weaker random graph models [202]). However, many natural algorithmic techniques do not achieve the goal of finding hidden cliques of size $k = o(\sqrt{n})$ [328].

Better exponential algorithms. Even though we will probably never have a polynomial-time algorithm for finding maximum cliques, we can try to design fast, super-polynomial algorithms. Exhaustive search gives the upper bound $\mathcal{O}(n^2 \cdot 2^n)$, or $\mathcal{O}^*(2^n)$ when omitting polynomial factors. Our goal is to design algorithms having running times $\mathcal{O}^*(\beta^n)$ with β as small as possible. The following theorem that can be found in [590] shows that we can do better than exhaustive search.

Theorem 6.1.6. *There exists an algorithm for computing a maximum clique in time $\mathcal{O}^*(1.3803^n)$.*

Sketch of Proof. We use a backtracking scheme with pruning of the recursion tree. Let G be a graph having n vertices and m edges. Let $v \in V$ be any vertex of minimum degree. If $\delta(G) \geq n - 3$ then the graph misses collections of pairwise disjoint cycles and paths, for being a complete graph. In this case, it is fairly easy to compute a maximum clique in $\mathcal{O}(n + m)$.[3] Assume that there is a vertex v with degree $d_G(v) \leq n - 4$. Every maximum clique either contains v or not. Corresponding to these two cases, a maximum clique of G is either $\{v\}$ combined with a maximum clique of the induced subgraph $G[N(v)]$ or a maximum clique of the induced subgraph $G[V - \{v\}]$. We recursively compute maximum cliques in both subgraphs and derive from them a solution for G (breaking ties arbitrarily). The worst-case time $T(n)$ essentially depends on the following recursive inequality:

$$T(n) \leq T(n - 4) + T(n - 1) + c \cdot (n + m) \quad \text{for some } c > 0$$

Using standard techniques based on generating functions, we calculate that $T(n)$ is within a polynomial factor of β^n where $\beta \approx 1.3803$ is the largest real zero of the polynomial $\beta^4 - \beta^3 - 1$. $\square$

---

[3] It might be easier to think of independent sets rather than cliques. An independent set in a graph G is a set U of vertices such that $G[U]$ has no edges. A clique in graph G corresponds to an independent set in graph $\overline{G}$, where in $\overline{G}$ exactly those vertices are adjacent that are not adjacent in G. Independent sets are a little bit easier to handle, since we do not have to reason about edges that are not in the graph. In fact, many algorithms in the literature are described for independent sets.

The intuitive algorithm in the theorem captures the essence of a series of fast exponential algorithms for the maximum clique problem. It started with an $\mathcal{O}^*(1.286^n)$ algorithm [543] that follows essentially the ideas of the algorithm above. This algorithm has been subsequently improved to $\mathcal{O}^*(1.2599^n)$ [545], by using a smart and tedious case analysis of the neighborhood around a low-degree vertex. The running time of the algorithm has been further improved to $\mathcal{O}^*(1.2346^n)$ [330], and, using combinatorial arguments on connected regular graphs, to $\mathcal{O}^*(1.2108^n)$ [495]. Unfortunately, the latter algorithm needs exponential space. This drawback can be avoided: there is a polynomial-space algorithm with a slightly weaker $\mathcal{O}^*(1.2227^n)$ time complexity [54]. A non-trivial lower bound on the basis of the exponential is still unknown (even under some complexity-theoretic assumptions).

6.1.3 Approximating Maximum Cliques

Since we are apparently not able to compute a maximum clique in moderate time, we could ask up to what size we can recognize cliques in justifiable time. Recall that $\omega(G)$ denotes the size of the largest clique in G. We say that an algorithm approximates $\omega(G)$ within factor $f(n)$ if and only if the algorithm produces, on input G, a clique U in G such that $\omega(G) \leq f(n) \cdot |U|$. Note that, since a maximum clique consists of at most n vertices, we can trivially approximate maximum clique within factor $\mathcal{O}(n)$, simply by outputting some edge, if there is one in the graph. With a lot of work and combinatorial arguments, we arrive at the next theorem [79], which is unfortunateley not very much better than the trivial ratio.

Theorem 6.1.7. *There exists a polynomial-time algorithm whose output, for graph G with n vertices, is a clique of size within factor $\mathcal{O}\left(\frac{n}{(\log n)^2}\right)$ of $\omega(G)$.*

The approximation ratio stated in the theorem is the best known. The following celebrated result [287] indicates that in fact, there is not much space for improving over that ratio.

Theorem 6.1.8. *Unless $\mathcal{NP} = \mathcal{ZPP}$,[4] there exists no polynomial-time algorithm whose output for a graph G with n vertices is a clique of size within factor $n^{1-\varepsilon}$ of $\omega(G)$ for any $\varepsilon > 0$.*

The complexity-theoretic assumption used in the theorem is almost as strong as $\mathcal{P} = \mathcal{NP}$. The inapproximability result has been strengthened to subconstant values of ε, first to $\mathcal{O}\left(\frac{1}{\sqrt{\log \log n}}\right)$ [177] and further to $\mathcal{O}\left(\frac{1}{(\log n)^\gamma}\right)$ [353] for some $\gamma > 0$. These results are based on much stronger complexity assumptions – essentially, that no $\mathcal{NP}$-complete problem can be solved by randomized algorithms with quasi-polynomial running time, i.e., in time $2^{(\log n)^{\mathcal{O}(1)}}$. Note that the ratio

---

[4] $\mathcal{ZPP}$ is the class of all problems that can be solved with randomized algorithms running in expected polynomial time while making no errors. Such algorithms are also known as (polynomial-time) Las Vegas algorithms.

$\frac{n}{(\log n)^2}$ is expressible as $\Omega\left(\frac{\log\log n}{\log n}\right)$ in terms of ε. The gap between the lower bound and the upper bound for approximability is thus pretty close.

Also many heuristic techniques for finding maximum cliques have been proposed. They often show reasonable behavior, but of course, they cannot improve over the theoretical inapproximability ratio. An extensive discussion of heuristics for finding maximum cliques can be found in [70].

In the random graph model, we known that, with high probability, $\omega(G)$ is either $(2 + o(1))\log n$ rounded up or rounded down, for a random graph of size n (see, e.g., [24]). There are several polynomial-time algorithms producing cliques of size $(1+o(1))\log n$, i.e., they achieve an approximation ratio of roughly two [263]. However, it is conjectured that there is no polynomial-time algorithm outputting a clique of size at least $(1 + \varepsilon)\log n$ for any $\varepsilon > 0$ [328, 347].

6.1.4   Finding Fixed-Size Cliques

In many cases, it might be appropriate to search only for cliques of bounded sizes. Technically that is, we consider the clique size not as part of the input. For instance, exhaustive search has running time $\Theta(n^k)$ when the clique size k is fixed. A nice trick helps us to obtain an algorithm for detecting cliques of size three (triangles) faster than $\mathcal{O}(n^3)$. The idea to the algorithm in the following theorem can be found in [321].

Theorem 6.1.9. *There exists an algorithm for testing a graph for triangles that runs in time $\mathcal{O}(n^{2.376})$.*

Proof. Let G be any graph with n vertices. Let $A(G)$ denote the adjacency matrix of G, i.e., entry a_{ij} of $A(G)$ is one if vertices v_i and v_i are adjacent, and zero otherwise. Consider the matrix $A(G)^2 = A(G) \cdot A(G)$ where $\cdot$ is the usual matrix multiplication. The entry b_{ij} of the matrix $A(G)^2$ is exactly the number of walks of length two between v_i and v_j. Suppose there exists an entry $b_{ij} \geq 1$. That is, there is at least one vertex $u \in V$ different to v_i and v_j which is adjacent to both v_i and v_j. If the graph G has an edge $\{v_i, v_j\}$, then we know that G contains the triangle $\{v_i, v_j, u\}$. Thus, an algorithm for triangle-testing simply computes $A(G)^2$ and checks whether there exists an edge $\{v_i, v_j\}$ for some non-zero entry b_{ij} in $A(G)^2$. Since fast square matrix multiplication can be done in time $\mathcal{O}(n^\alpha)$ where $\alpha < 2.376$ [132], the algorithm runs in time $\mathcal{O}(n^{2.376})$. □

Note that for sparse graphs there is an even faster algorithm running in time $\mathcal{O}(m^{\frac{2\alpha}{\alpha+1}}) = \mathcal{O}(m^{1.41})$ for finding triangles which makes use of the same technique [26] (see also Section 11.5).

Once we have reached this point, we would like to apply the matrix-multiplication technique to come up with algorithms for clique size larger than three as well. However, the direct argument does not work for some reasons. For instance, there exists always a walk of length three between adjacent vertices. This makes the matrix $A(G)^3$ and all higher powers ambiguous. We need a more sophisticated approach [174, 440].

Theorem 6.1.10. *For every $k \geq 3$ there exists an algorithm for finding a clique of size k in a graph with n vertices that runs in time $\mathcal{O}(n^{\beta(k)})$ where $\beta(k) = \alpha(\lfloor k/3 \rfloor, \lceil (k-1)/3 \rceil, \lceil k/3 \rceil)$ and multiplying an $n^r \times n^s$-matrix with an $n^s \times n^t$-matrix can be done in time $\mathcal{O}(n^{\alpha(r,s,t)})$.*

Proof. Let k_1 denote $\lfloor k/3 \rfloor$, let k_2 denote $\lceil (k-1)/3 \rceil$, and let k_3 denote the value $\lceil k/3 \rceil$. Note that $k = k_1 + k_2 + k_3$. Let G be any graph with n vertices and m edges. We first construct a tripartite auxiliary graph $\tilde{G}$ as follows: the vertex set $\tilde{V}$ is divided into three sets $\tilde{V}_1$, $\tilde{V}_2$, and $\tilde{V}_3$ where $\tilde{V}_i$ consists of all cliques of size k_i in G. Define two vertices $U \in \tilde{V}_i$ and $U' \in \tilde{V}_j$ to be adjacent in $\tilde{G}$ if and only if $i \neq j$ and $U \cup U'$ is a clique of size $k_i + k_j$ in G. The algorithm now tests the auxiliary graph $\tilde{G}$ for triangles. If there is such a triangle $\{U_1, U_2, U_3\}$, then the construction of $\tilde{G}$ implies that $U_1 \cup U_2 \cup U_3$ is a clique of size k in G. Testing the graph $\tilde{G}$ for triangles can be done by matrix multiplication as described in Theorem 6.1.9. However, we now have to multiply an $n^{k_1} \times n^{k_2}$ adjacency matrix, representing edges between $\tilde{V}_1$ and $\tilde{V}_2$, with an $n^{k_2} \times n^{k_3}$ adjacency matrix, representing edges between $\tilde{V}_2$ and $\tilde{V}_3$. This step can be done in time $\mathcal{O}(n^{\beta(k)})$. Computing the three matrices needs in the worst case $\mathcal{O}(n^{\max\{k_1+k_2, k_1+k_3, k_2+k_3\}}) = \mathcal{O}(n^{\lceil \frac{2k}{3} \rceil})$, which is asymptotically dominated by the time for the fast rectangular matrix multiplication [318]. $\square$

We give an impression of the algorithmic gain of using matrix multiplication (see, e.g., [260]).

Clique size	Exhaustive search	Matrix multiplication
3	$\mathcal{O}(n^3)$	$\mathcal{O}(n^{2.376})$
4	$\mathcal{O}(n^4)$	$\mathcal{O}(n^{3.376})$
5	$\mathcal{O}(n^5)$	$\mathcal{O}(n^{4.220})$
6	$\mathcal{O}(n^6)$	$\mathcal{O}(n^{4.751})$
7	$\mathcal{O}(n^7)$	$\mathcal{O}(n^{5.751})$
8	$\mathcal{O}(n^8)$	$\mathcal{O}(n^{6.595})$

The theorem has a nice application to the membership counting problem for cliques of fixed size. The following result is due to [260].

Theorem 6.1.11. *For every $k \geq 3$, there exists an algorithm that counts the number of cliques of size k to which each vertex of a graph on n vertices belongs, in time $\mathcal{O}(n^{\beta(k)})$ where $\beta(k)$ is the same function as in Theorem 6.1.10.*

Proof. The theorem is based on the observation that for the case $k = 3$ (see Theorem 6.1.9), it is not only easy to check whether two vertices v_i and v_j belong to some triangle in G, but also to compute in how many triangles they lie: if the edge $\{v_i, v_j\}$ exists in G, then the number is just the entry b_{ij} in the square of the adjacency matrix $A(G)$. In general, we apply this observation to the auxiliary graph $\tilde{G}$. For any vertex $v \in V$, let $C_k(v)$ denote the number of different cliques of size k in G in which v is contained. Similarly, let $\tilde{C}_3(U)$ denote the number of triangles to which node U of $\tilde{G}$ belongs. Notice that U is a clique in G of size smaller than k. Clearly, cliques of G of size k may have many representations in graph $\tilde{G}$. The exact number is the number of partitionings of a set of cardinality k into sets of cardinalities k_1, k_2, and k_3, i.e., $\binom{k}{k_1,k_2,k_3}$ where k_1, k_2, and k_3 are defined as in the proof of Theorem 6.1.10. Without loss of generality, let k_1 be the minimum of these three parameters. Let $\mathcal{U}(v)$ be the set of all cliques U of size k_1 in G such that $v \in U$. We then obtain the following equation:

$$\sum_{U \in \mathcal{U}(v)} \tilde{C}_3(U) = \binom{(k-1)}{(k_1-1), k_2, k_3} \cdot C_k(v) \tag{6.1}$$

Clearly, using Theorem 6.1.10, the left-hand side of this equation can be computed in time $\mathcal{O}(n^{\beta(k)})$ (first, compute the matrices and second, search entries for all U containing v). We now easily calculate $C_k(v)$ from Equation 6.1.   □

A recent study of the corresponding decremental problem [260], i.e., the scenario where starting from a given graph vertices and edges can be removed, has shown that we can save roughly $n^{0.8}$ time compared to computing the number of size-k cliques to which the vertices belong each time from the scratch. For example, the problem of counting triangles in a decremental setting now takes $\mathcal{O}(n^{1.575})$.

Fixed-parameter tractability. A way to study which time bounds we might expect for fixed-parameter clique problems is parameterized complexity [168]. The goal here is to figure out which input parameter makes a problem computationally hard. We say that a parameterized problem (with parameter k) is fixed-parameter tractable if and only if there is an algorithm for the problem that needs time polynomial in input size n, if k is fixed, and which is asymptotically independent of k. More precisely, the time complexity of the algorithm has the form $\mathcal{O}(f(k) \cdot p(n))$ where p is some polynomial independent of k and f is an arbitrary function independent of n. Note that the algorithm above does not satisfy such a bound. A good bound would be, e.g., $\mathcal{O}(k^k \cdot n^2)$. However, we are far from proving such bounds, and in fact, we should not even expect to obtain such algorithms. Let $\mathcal{FPT}$ denote the class of fixed-parameter tractable problems. We know that parameterized CLIQUE is complete for the class $\mathcal{W}[1]$, a superclass of $\mathcal{FPT}$ [167]. However, it is widely believed that $\mathcal{FPT} \neq \mathcal{W}[1]$, which would imply both $\mathcal{P} \neq \mathcal{NP}$ and CLIQUE is not fixed parameter tractable.

6.1.5   Enumerating Maximal Cliques

Enumerative algorithms for the clique problem have some tradition (cf., e.g., [70]), with probably the first appearing already in 1957 [284]. Several other, now classical, algorithms were proposed (e.g., [473, 103]). Most recently, also algorithms for the dynamic graph setting have been investigated [534].

We are interested in having efficient algorithms for enumerating maximal cliques. There are some gradations in the meaning of 'efficient.' Most of the interesting combinatorial problems have an exponential number of configurations; in our case indicated by the $3^{\lceil \frac{n}{3} \rceil}$ matching upper bound for the number of maximal cliques. A typical requirement for an enumerative algorithm to be efficient is polynomial *total* time. That is, the algorithm outputs all C possible configurations in time bounded by a polynomial in C and the input size n. Exhaustive search is not polynomial total time. In contrast, one of the classical algorithms [473] first runs $\mathcal{O}(n^2 C)$ steps with no output and then outputs all C maximal cliques all at once. However, an algorithm for the enumeration of all *maximum* cliques that runs in polynomial total time does not exist, unless $\mathcal{P} = \mathcal{NP}$ [382].

We next review enumerative algorithms for maximal cliques with polynomial total time having some further desirable properties.

Polynomial delay. An algorithm fulfilling this condition generates the configurations, one after the other in some order, in such a way that the delay until the first output, the delay between any two consecutive configurations, and the delay until it stops after the last output is bounded by a polynomial in the input size. For maximal cliques we know such algorithms that in addition, require only linear space [553].

Theorem 6.1.12. *There is an algorithm enumerating all maximal cliques of a graph with polynomial delay $\mathcal{O}(n^3)$ using only $\mathcal{O}(n + m)$ space.*

Proof. We construct a binary tree with n levels and leaves only at level n. Each level is associated with a vertex, i.e., at level i we consider vertex v_i. The nodes of the tree at level i are all maximal cliques of $G[\{v_1, \ldots, v_i\}]$. It immediately follows that the leaves are exactly the maximal cliques of G. Fix level i and maximal clique U in $G[\{v_1, \ldots, v_i\}]$. We want to determine the children of U at level $i + 1$. We have two main cases:

1. Suppose all vertices of U are adjacent to v_{i+1} in G. Then $U \cup \{v_{i+1}\}$ is maximal clique in $G[\{v_1, \ldots, v_i, v_{i+1}\}]$. Note that this is the only way to obtain a maximal clique of $G[\{v_1, \ldots, v_i, v_{i+1}\}]$ that contains U. In this case U has only one child in the tree.
2. Suppose there is a vertex in U not adjacent to v_{i+1} in G. Here, we can obtain maximal cliques in $G[\{v_1, \ldots, v_i, v_{i+1}\}]$ in two different ways: U itself is certainly a maximal clique, and another clique is $(U - \overline{N}(v_{i+1})) \cup \{v_{i+1}\}$, where $\overline{N}(v_{i+1})$ are all vertices of G not adjacent to v_{i+1}. If the latter set is a maximal clique, U would have two children. However, as the set $(U - \overline{N}(v_{i+1})) \cup \{v_{i+1}\}$ is potentially a child of several sets, we define it to be the child of the lexicographically smallest set U, if it is maximal.

By this definition, we have a tree where all internal nodes have one or two children, thus a binary tree, and all leaves are at level n.

Our enumerative algorithm now simply traverses the tree using a depth-first search and outputs all leaves. All we need to be able to perform the computation, given a node U of the tree at level i, is the following:

– Parent(U, i): According to the definition of the tree, the parent node of U is the lexicographically smallest maximal clique in $G[\{v_1, \ldots, v_{i-1}\}]$ containing the clique $U - \{v_i\}$. This is one of our efficiently computable primitives: the set can be computed in time $\mathcal{O}(n + m)$.

- LeftChild(U, i): If $U \subseteq N(v_{i+1})$ (the first case above), then the left child is $U \cup \{v_{i+1}\}$. If $U \not\subseteq N(v_{i+1})$ (one part of the second case above), then the left child is U. Checking which case has to be applied needs $\mathcal{O}(n + m)$ time.
- RightChild(U, i): If $U \subseteq N(v_{i+1})$, then there is no right child defined. If $U \not\subseteq N(v_{i+1})$, then the right child of U is $(U - \overline{N}(v_{i+1})) \cup \{v_{i+1}\}$ if it is a maximal clique and $U = \mathsf{Parent}((U - \overline{N}(v_{i+1})) \cup \{v_{i+1}\}, i + 1)$, otherwise the right child is not defined. Note that we only need $\mathcal{O}(n + m)$ processing time.

The longest path between any two leaves in the tree is $2n - 2$ passing through $2n - 1$ nodes. For each node we need $\mathcal{O}(n + m)$ time. Since any subtree of our tree has a leaf at level n, this shows that the delay between outputs is $\mathcal{O}(n^3)$. Note that the algorithm only needs to store while processing a node, the set U, the level i, and a label indicating whether it is the left or the right child. Hence, the amount of space is $\mathcal{O}(n + m)$.   □

Specified order. A more difficult problem is generating maximal cliques in a specific order, such as lexicographic order. If we only insist in polynomial total time, this is obviously not a restriction, since we need only collect all outputs and sort them for outputting in lexicographic order. Considering orders is only interesting in the case of polynomial delay. Note that the DFS-based polynomial-delay algorithm in Theorem 6.1.12 does not produce its outputs in lexicographic order. Another DFS-based algorithm [395] has been proposed that produces the outputs in lexicographic order but is not polynomial delay. We first observe that it is not obvious how to break the tradeoff.

Theorem 6.1.13. *Deciding for any given graph G and any maximal clique U of G, whether there is a maximal clique U' lexicographically larger than U, is $\mathcal{NP}$-complete.*

The theorem is proven by a polynomial transformation from SATISFIABILITY [334]. It has some immediate consequences, e.g., it rules out polynomial-delay algorithms with respect to inverse lexicographic order.

Corollary 6.1.14. *1. Unless $\mathcal{P} = \mathcal{NP}$, there is no algorithm that generates for any given graph G and any maximal clique U in G the lexicographically next maximal clique in polynomial time.*
2. Unless $\mathcal{P} = \mathcal{NP}$, there is no algorithm that generates for any given graph all maximal cliques in inverse lexicographic order with polynomial delay.

It might seem surprising that algorithms exist generating all maximal cliques in lexicographic order, with polynomial delay. The idea of such an algorithm is simply that while producing the current output, we invest additional time in producing lexicographically larger maximal cliques. We store these cliques in a priority queue Q. Thus, Q contains a potentially exponential number of cliques and requires potentially exponential space. The following algorithm has been proposed in [334] and uses in a clever way the tree structure employed in Theorem 6.1.12.

Algorithm 9: Lexicographic enumeration of maximal cliques [334]

Input: Graph $G = (V, E)$
Output: Sequence of maximal cliques of G in lexicographic order

Let U_0 be the lexicographically first maximal clique;
Insert U_0 into priority queue Q;
while Q is not empty **do**
    U :=ExtractMin(Q);
    Output U;
    foreach vertex v_j of G not adjacent to some vertex $v_i \in U$ with $i < j$ **do**
        $U_j := U \cap \{v_1, \ldots, v_j\}$;
        if $(U_j - \overline{N}(v_j)) \cup \{v_j\}$ is a maximal clique in $G[\{v_1, \ldots, v_j\}]$ **then**
            Let T be the lexicographically smallest maximal clique which
            contains $(U_j - \overline{N}(v_j)) \cup \{v_j\}$;
            Insert T into Q

Theorem 6.1.15. *Algorithm 9 enumerates all maximal cliques of a graph with n vertices in lexicographic order, and with delay $\mathcal{O}(n^3)$.*

Proof. For the correctness of the algorithm, first observe that the set T being inserted into Q when considering U is lexicographically greater than U. Thus, we store only sets into the queue that have to be output after U. Hence, the sequence of maximal cliques we produce is indeed lexicographically ascending. We also have to show that all maximal cliques are in the sequence. We do this by proving inductively: if U is the lexicographically first maximal clique not yet output, then U is in Q.

Base of induction: Suppose $U = U_0$. Then the statement is obviously true.

Step of induction: Suppose U is lexicographically greater than U_0. Let j be the largest index such that $U_j = U \cap \{v_1, \ldots, v_j\}$ is *not* a maximal clique in the graph restricted to vertices $v_1, \ldots, v_j$. Such an index must exist, since otherwise we would have $U = U_0$. Moreover, we have that $j < n$, since U is a maximal clique in the whole graph G. By maximality of j, we must have $v_{j+1} \in U$. There exists a non-empty set S such that $U_j \cup S$ is a maximal clique in $G[\{v_1, \ldots, v_j\}]$. Again, by maximality of j, the vertex v_{j+1} is not adjacent to all vertices in S. We conclude that there is a maximal clique U' containing $U_j \cup S$ but not vertex v_{j+1}. Note that U' is lexicographically smaller than U, since they differ on set S. By induction hypothesis, U' has already been output. At the time when U' was output, the vertex v_{j+1} was found not to be adjacent to some vertex v_i in U' with index $i < j + 1$. Clearly, we have $(U'_{j+1} - \overline{N}(v_{j+1})) \cup \{v_{j+1}\} = U_{j+1}$ and U_{j+1} is a maximal clique in $G[\{v_1, \ldots, v_{j+1}\}]$. So the lexicographically first maximal clique T containing U_{j+1} was inserted into Q. Once more by maximality of j, U and T coincide on the first $j + 1$ vertices. Assume that $U \neq T$. Let k be the first index such that U and T disagree on v_k. It follows that $k > j + 1$. Since T is lexicographically less than U, we have $v_k \in T$ and $v_k \notin U$. Hence, U_k is not a maximal clique in $G[\{v_1, \ldots, v_k\}]$, a contradiction to maximality of j. Therefore, $U = T$ and so U is in Q. This proves the induction step.

For the time bound, the costly operations are the extraction of the lexicographically smallest maximal clique from Q (which needs $\mathcal{O}(n \log C)$), the n computations of maximal cliques containing a given set (which takes $\mathcal{O}(n+m)$ for each set), and attempting to insert a maximal clique into Q (at costs $\mathcal{O}(n \log C)$ per clique). Since $C \leq 3^{\lceil \frac{n}{3} \rceil}$, the total delay is $\mathcal{O}(n^3)$ in the worst case. □

Counting complexity. We conclude this section with some remarks on the complexity of counting the number of maximal cliques. An obvious way to count maximal cliques is to enumerate them with some of the above-mentioned algorithms and increment a counter each time a clique is output. This, however, would take exponential time. The question is whether it is possible to compute the number more directly and in time polynomial in the graph size. To study such issues the class $\#\mathcal{P}$ has been introduced [559], which can be considered as the class of all functions counting the number of solutions of instances of $\mathcal{NP}$-problems. It can be shown that counting the number of maximal cliques is $\#\mathcal{P}$-complete (with respect to an appropriate reducibility notion) [560]. An immediate consequence is that if there is a polynomial-time algorithm for computing the number of maximal cliques, then CLIQUE is in $\mathcal{P}$, and thus, $\mathcal{P} = \mathcal{NP}$. Note that in the case of planar, bipartite or bounded-degree graphs there are polynomial-time algorithms for counting maximal cliques [557].

6.2 Structurally Dense Groups

We review two relaxations of the clique concept based on minimal degrees [515, 514, 513]. Both relaxations are structural, as they impose universal constraints on individuals in a group.

6.2.1 Plexes

We generalize the clique concept by allowing members in a group to miss some ties with other group members, but only up to a certain number $N \geq 1$. This leads to the notion of an N-plex [514, 511].

Definition 6.2.1. *Let* $G = (V, E)$ *be any undirected graph and let* $N \in \{1, \ldots, n-1\}$ *be a natural number. A subset* $U \subseteq V$ *is said to be an N-plex if and only if* $\delta(G[U]) \geq |U| - N$.

Clearly, a clique is simply a 1-plex, and an N-plex is also an $(N+1)$-plex. We say that a subset $U \subseteq V$ is a maximal N-plex if and only if U is an N-plex and it is not strictly contained in any larger N-plex of G. A subset $U \subseteq V$ is a maximum N-plex if and only if U has a maximum number of vertices among all N-plexes of G.

It is easily seen that any subgraph of an N-plex is also an N-plex, that is, N-plexes are closed under exclusion. Moreover, we have the following relation between the size of an N-plex and its diameter [514, 189, 431].

Proposition 6.2.2. *Let $N \in \{1, \ldots, n-1\}$ be a natural number. Let $G = (V, E)$ be any undirected graph on n vertices.*

1. *If V is an N-plex with $N < \frac{n+2}{2}$, then $\mathrm{diam}(G) \leq 2$ and, if additionally $n \geq 4$, G is 2-edge-connected.*
2. *If V is an N-plex with $N \geq \frac{n+2}{2}$ and G is connected, then $\mathrm{diam}(G) \leq 2N - n + 2$.*

Proof. 1. Suppose $N < \frac{n+2}{2}$. Let $u, v \in V$ be vertices such that $u \neq v$. If u and v are adjacent, the distance between them is one. Now, suppose u and v are not adjacent. Assume that the distance between u and v is at least three, i.e., with respect to neighborhoods it holds $N(u) \cap N(v) = \emptyset$. We obtain

$$n - 2 \geq |N(u) \cup N(v)| \geq 2\delta(G) \geq 2(n - N) > 2 \left(n - \frac{n+2}{2} \right) = n - 2,$$

a contradiction. Thus, the distance between u and v is at most two. Hence, $\mathrm{diam}(G) \leq 2$. To verify that for $n \geq 4$, G is 2-edge-connected, assume to the contrary that there is a bridge, i.e., an edge e such that after removing it, $G - \{e\}$ consists of two connected components V_1 and V_2. Obviously, every shortest path from a vertex in V_1 to a vertex in V_2 must use that bridge. Since $\mathrm{diam}(G) \leq 2$, one component is a singleton. This implies that the vertex in this component has degree one. However, as V is an N-plex with $n \geq 4$ vertices, we obtain for the degree of this vertex $n - N > n - (n+2)/2 = (n-2)/2 \geq 1$, a contradiction. Thus, a bridge cannot exist in G.

2. Suppose $N \geq \frac{n+2}{2}$. Let $\{v_0, v_1, \ldots, v_r\}$ be the longest shortest path of G, i.e., a path that realizes the diameter r. We may suppose that $r \geq 4$. Since there is no shorter path between v_0 and v_r, we have that v_i is not adjacent to $v_0, \ldots, v_{i-2}, v_{i+2}, \ldots, v_r$ for all $i \in \{0, \ldots, r\}$ (where vertices with negative index do not exist). Furthermore, there cannot exist a vertex adjacent to both v_0 and v_3. Thus, the following inclusion is true:

$$\{v_0\} \ \cup \ \{v_2, v_3, \ldots, v_r\} \ \cup \ (N(v_3) - \{v_2, v_4\}) \subseteq \overline{N}(v_0)$$

Note that we have a disjoint union on the left-hand side. We thus obtain the inequality $1 + (r - 1) + d_G(v_3) - 2 \leq N$. It follows $r + n - N - 2 \leq N$. Hence, $\mathrm{diam}(G) = r \leq 2N - n + 2$. □

From a computational point of view, finding maximum plexes is not easier than finding maximum cliques. This is immediate when we consider the variable decision problem for plexes, where the problem instance consists of graph G, the size parameter k, and the plex parameter N. Since CLIQUE appears as instances of the form $(G, k, 1)$, the problem is $\mathcal{NP}$-complete. We discuss the complexity of finding N-plexes of certain sizes for fixed N. For any natural number $N > 0$, we define the following decision problem:

Problem:	N-PLEX
Input:	Graph G, Parameter $k \in \mathbb{N}$
Question:	Does there exist an N-plex of size at least k within G?

As 1-PLEX = CLIQUE, and thus 1-PLEX is $\mathcal{NP}$-complete, it is not surprising that finding maximum N-plexes is $\mathcal{NP}$-hard for all $N > 0$ as well.

Theorem 6.2.3. N-PLEX *is* $\mathcal{NP}$-*complete for all natural numbers* $N > 0$.

Proof. It suffices to consider the case $N > 1$. There is a generic proof of the theorem which is based on the fact that being an N-plex is a hereditary graph property (see, e.g., [240]). We give a direct proof in order to demonstrate the structural similarity between cliques and plexes. We describe a polynomial transformation of CLIQUE into N-PLEX. Let (G, k) be any instance of the clique problem. We construct a new graph G' in the following way: we take $N - 1$ copies of each vertex of G, connect them to each other by an edge, and all new vertices to the vertices of G except to the original one. More specifically, let $G' = (V', E')$ be the graph defined as follows:

$$
\begin{aligned}
V' =_{\text{def}} &\; V \times \{0, 1, \ldots, N - 1\} \\
E' =_{\text{def}} &\; \big\{ \{(u, 0), (v, 0)\} \mid \{u, v\} \in E \big\} \; \cup \\
&\cup \big\{ \{(u, i), (v, j)\} \mid u, v \in V \text{ and } i, j > 0 \big\} \; \cup \\
&\cup \big\{ \{(u, 0), (v, i)\} \mid u, v \in V \text{ with } u \neq v \text{ and } i > 0 \big\}
\end{aligned}
$$

The graph G' can certainly be computed in time polynomial in the size of G. A crucial observation is that copy vertices, i.e., vertices in $V \times \{1, \ldots, N - 1\}$ are adjacent to all vertices in V' except one. We will show that G contains a clique of size k if and only if G' contains an N-plex of size $k + (N - 1)n$.

Suppose there exists a clique $U \subseteq V$ of size exactly k in G. Let U' denote the vertex set in G' consisting of all original vertices of U and all copies of vertices of V, i.e., $U' = U \times \{0\} \cup V \times \{1, \ldots, N - 1\}$. Notice that U' has cardinality $k + (N - 1)n$. Each vertex with label $i \in \{1, \ldots, N - 1\}$ is directly connected to each other vertex in U' except one vertex with label zero, thus has degree $|U'| - 2 = k + (N - 1)n - 2$. Each vertex $(u, 0)$ is adjacent to all vertices in U' except (u, i) with $i > 0$. That is, $(u, 0)$ has degree $k + (N - 1)n - 1 - (N - 1)$. Hence, U' is an N-plex.

Suppose there is no clique of size k in G. Thus, any induced subgraph of G having $k' \geq k$ vertices has minimal degree at most $k' - 2$. Let $U \subseteq V'$ be any vertex set with $k + (N - 1)n$ vertices. Then there is another set $U' \subseteq V'$ on $k + (N - 1)n$ vertices such that $\delta(G'[U']) \geq \delta(G'[U])$ and U' contains all copy vertices of G', i.e., $U' \supseteq V \times \{1, \ldots, N-1\}$. This follows from the fact that there is always a vertex in $U_0 = U \cap (V \times \{0\})$ that is not adjacent to some other vertex in U_0 (otherwise U_0 would induce a clique of size $|U_0| \geq k$ in G). Remembering the observation above, we are now allowed to recursively exchange such vertices by vertices of $V \times \{1, \ldots, N-1\}$ as long as possible, without decreasing minimum degrees. We end up with a desired set $U' \subseteq V'$. Since we have no size-k clique in G, we may conclude $\delta(G'[U]) \leq \delta(G'[U']) \leq k + (N - 1)n - 2 - (N - 1)$. Hence, there is no N-plex in G'. $\qquad\square$

6.2.2   Cores

A concept dual to plexes is that of a core. Here, we do not ask how many edges are missing in the subgraph for being complete, but we simply fix a threshold in terms of a minimal degree for each member of the subgroup. One of the most important things to learn about cores is that there exist polynomial-time algorithms for finding maximum cores. Cores have been introduced in [513].

Definition 6.2.4. *Let $G = (V, E)$ be any undirected graph. A subset $U \subseteq V$ is said to be an N-core if and only if $\delta(G[U]) \geq N$.*

The parameter N of an N-core is the order of the N-core. A subset $U \subseteq V$ is a maximal N-core if and only if U is an N-core and it is not strictly contained in any larger N-core of G. A subset $U \subseteq V$ is a maximum N-core if and only if U has maximum number of vertices among all N-cores of G. Maximum cores are also known as main cores.

Any $(N+1)$-core is an N-core and any N-core is an $(n - N)$-plex. Moreover, if U and U' are N-cores, then $U \cup U'$ is an N-core as well. That means maximal N-cores are unique. However, N-cores are not closed under exclusion and are in general not nested. As an example, a cycle is certainly a 2-core but any proper subgraph has at least one vertex with degree less than two. N-cores need not be connected. The following proposition relates maximal connected N-cores to each other.

Proposition 6.2.5. *Let $G = (V, E)$ be any undirected graph and let $N > 0$ be any natural number. Let U and U' be maximal connected N-cores in G with $U \neq U'$. Then there exists no edge between U and U' in G.*

Proof. Assume there is an edge $\{u, v\}$ with $u \in U$ and $v \in U'$. It follows that $U \cup U'$ is an N-core containing both U and U'. Furthermore, it is connected, since U and U' are connected.   □

Some immediate consequences of the proposition are the following: the unique maximum N-core of a graph is the union of all its maximal connected N-cores, the maximum 2-core of a connected graph is connected (notice that the internal vertices of a path have degree two), and a graph is a forest if and only if it possesses no 2-cores. The next result is an important algorithmic property of N-cores, that was exhibited in [46].

Proposition 6.2.6. *Let $G = (V, E)$ be any undirected graph and let $N > 0$ be any natural number. If we recursively remove all vertices with degree strictly less than N, and all edges incident with them, then the remaining set U of vertices is the maximum N-core.*

Proof. Clearly, U is an N-core. We have to show that it is maximum. Assume to the contrary, the N-core U obtained is not maximum. Then there exists a non-empty set $T \subseteq V$ such that $U \cup T$ is the maximum N-core, but vertices of T have been removed. Let t be the first vertex of T that has been removed. At that time, the degree of t must have been strictly less than N. However, as t has at least N neighbors in $U \cup T$ and all other vertices have still been in the graph when t was removed, we have a contradiction.   □

The procedure described in the proposition suggests an algorithm for computing N-cores. We extend the procedure for obtaining auxiliary values which provide us with complete information on the core decomposition of a network. Define the core number of a vertex $v \in V$ to be the highest order N of a maximum N-core vertex v belongs to, i.e.,

$$\xi_G(v) =_{\text{def}} \max\{ N \mid \text{there is an } N\text{-core } U \text{ in } G \text{ such that } v \in U \}.$$

A method, according to [47], for computing all core numbers is shown in Algorithm 10. The algorithm is correct due to the following reasons: any graph G is certainly a $\delta(G)$-core, and each neighbor of vertex v having lower degree than v decrements the potential core number of v. A straightforward implementation of the algorithm yields a worst-case time bound of $\mathcal{O}(mn \log n)$ – the most costly operations being sorting vertices with respect to their degree. A more clever implementation guarantees linear time [47].

Algorithm 10: Computing core numbers [47]

Input: Graph $G = (V, E)$
Output: Array ξ_G containing the core numbers of all vertices in G

Compute the degrees of all vertices and store them into D;
Sort V in increasing degree-order D;
foreach $v \in V$ in sorted order **do**
  $\xi_G(v) := D[v]$;
  foreach vertex u adjacent to v **do**
    if $D[u] > D[v]$ **then**
      $D[u] := D[u] - 1$;
      Resort V in increasing degree-order of D

Theorem 6.2.7. *There is an implementation of Algorithm 10 that computes the core numbers of all vertices in a given graph $G = (V, E)$ having n vertices and m edges in time $\mathcal{O}(n + m)$.*

Proof. To reduce the running time of the algorithm, we have to speed up the sorting operations in the algorithm. This can be achieved by two techniques.

1. Since the degree of a vertex lies in the range $\{0, \ldots, n-1\}$, we do sorting using n buckets, one for each vertex degree. This gives us an $\mathcal{O}(n)$ time complexity for initially sorting the vertex-set array V.
2. We can save resorting entirely, by maintaining information about where in the array V, which contains the vertices in ascending order of their degree, a new region with higher degree starts. More specifically, we maintain an array

J where entry $J[i]$ is the minimum index j such that for all $r \geq j$, vertex $V[r]$ has degree at least i. We can now replace the 'resort'-line in Algorithm 10 by the following instructions:

> **if** $u \neq$ vertex w at position $J[D[u] + 1]$ **then** swap vertices u and w in V;
> Increment $J[D[u] + 1]$

Resorting the array V in order to maintain the increasing order of degrees now takes $\mathcal{O}(1)$ time. Notice that the array J can initially be computed in time $\mathcal{O}(n)$.

For the total running time of Algorithm 10, we now obtain $\mathcal{O}(n)$ for initializing and sorting and $\mathcal{O}(m)$ for the main part of the algorithm (since each edge is handled at most twice). This proves the $\mathcal{O}(n + m)$ implementation.   □

Corollary 6.2.8. *For all $N > 0$, the maximum N-core for a graph with n vertices and m edges can be computed in time $\mathcal{O}(n + m)$, which is independent of N.*

6.3 Statistically Dense Groups

In general, statistical measures over networks do not impose any universal structural requirements on individuals. This makes them more flexible than structural measures but usually harder to analyze. We turn to statistical measures for densities of graphs.

6.3.1 Dense Subgraphs

The natural notion of density of a graph is the following. Let $G = (V, E)$ be any undirected graph with n vertices and m edges. The density $\varrho(G)$ of G is the ratio defined as

$$\varrho(G) =_{\text{def}} \frac{m}{\binom{n}{2}}.$$

That is, the density of a graph is the percentage of the number of edges of a clique, observable in a graph. We are interested in subgraphs of certain densities.

Definition 6.3.1. *Let $G = (V, E)$ be an undirected graph and let $0 \leq \eta \leq 1$ be a real number. A subset $U \subseteq V$ is said to be an η-dense subgraph if and only if $\varrho(G[U]) \geq \eta$.*

In an η-dense subgraph, the interpretation is that any two members share with probability (or frequency) at least η a relationship with each other. It is, however, immediate that even graphs of fairly high density are allowed to have isolated vertices.

A clique, as the subgraph with highest density, is a 1-dense subgraph. An N-plex has density $1 - \frac{N-1}{n-1}$. Thus, for n approaching infinity, the density of an N-plex approaches one. A little bit more exactly, for all $N > 0$ and for all

$0 \leq \eta \leq 1$, an N-plex of size at least $\frac{N-\eta}{1-\eta}$ is an η-dense subgraph. But evidently, not every $(1 - \frac{N-1}{n-1})$-dense subgraph (when allowing non-constant densities) is an N-plex. An N-core is an $\frac{N}{n-1}$-dense subgraph, which can have a density arbitrarily close to zero for large n.

In general, η-dense subgraphs are not closed under exclusion. However, they are nested.

Proposition 6.3.2. *Let $0 \leq \eta \leq 1$ be real number. An η-dense subgraph of size k in a graph G contains an η-dense subgraph of size $k - 1$ in G.*

Proof. Let U be any η-dense subgraph of G, $|U| = k$. Let m_U denote the number of edges in $G[U]$. Let v be a vertex with minimal degree in $G[U]$. Note that $\delta(G[U]) \leq \bar{d}(G[U]) = \frac{2m_U}{k} = \varrho(G[U])(k - 1)$. Consider the subset U' obtained by excluding vertex v from U. Let $m_{U'}$ denote the number of edges of U'. We have

$$m_{U'} = m_U - \delta(G[U]) \geq \varrho(G[U])\binom{k}{2} - \varrho(G[U])(k - 1) = \varrho(G[U])\binom{k - 1}{2}$$

Hence, $\varrho(G[U']) \geq \varrho(G[U]) \geq \eta$. Thus, U' is an η-dense subgraph.   □

The proposition suggests a greedy approach for obtaining η-dense graphs: recursively deleting a vertex with minimal degree until an η-dense subgraph remains. However, this procedure can fail drastically. We will discuss this below.

Walks. The density averages over edges in subgraphs. An edge is a walk of length one. A generalization of density can involve walks of larger length. To make this more precise, we introduce some notations. Let $G = (V, E)$ be any undirected graph with n vertices. Let $\ell \in \mathbb{N}$ be any walk-length. For a vertex $v \in V$, we define its degree of order ℓ in G as the number of walks of length ℓ that start in v. Let $d_G^\ell(v)$ denote v's degree of order ℓ in G. We set $d_G^0(v) = 1$ for all $v \in V$. Clearly, $d_G^1(v)$ is the degree of v in G. The number of walks of length ℓ in a graph G is denoted by $W_\ell(G)$. We have the following relation between the degrees of higher order and the number of walks in a graph.

Proposition 6.3.3. *Let $G = (V, E)$ be any undirected graph. For all $\ell \in \mathbb{N}$ and for all $r \in \{0, \dots, \ell\}$, $W_\ell(G) = \sum_{v \in V} d_G^r(v) \cdot d_G^{\ell-r}(v)$.*

Proof. Any walk of length ℓ consists of vertices $v_0, v_1, \dots, v_\ell$. Fix an arbitrary $r \in \{0, \dots, \ell\}$. Consider the element v_r. Then the walk $v_0, v_1, \dots, v_r$ contributes to the degree of order r of v_r, and the walk $v_r, v_{r+1}, \dots, v_\ell$ contributes to the degree of order $\ell - r$ of v_r. Thus, there are $d_G^r(v_r) \cdot d_G^{\ell-r}(v_r)$ walks of length ℓ having vertex v_r at position r. Summing over all possible choices of a vertex at position r shows the statement.   □

It is clear that the maximum number of walks of length ℓ in a graph with n vertices is $n(n - 1)^\ell$. We thus define the density of order ℓ of a graph G as

$$\varrho_\ell(G) =_{\text{def}} \frac{W_\ell(G)}{n(n-1)^\ell}.$$

Note that $\varrho_1(G) = \varrho(G)$ as in $W_1(G)$ each edge counts twice. We easily conclude the following proposition.

Proposition 6.3.4. *It holds $\varrho_\ell(G) \leq \varrho_{\ell-1}(G)$ for all graphs G and all natural numbers $\ell \geq 2$.*

Proof. Let $G = (V, E)$ be any undirected graph with n vertices. By Proposition 6.3.3, $W_\ell(G) = \sum_{v \in V} d_G^1(v) \cdot d_G^{\ell-1}(v) \leq (n-1) \sum_{v \in V} d_G^{\ell-1}(v) = (n-1) \cdot W_{\ell-1}(G)$. Now, the inequality follows easily.   □

For a graph $G = (V, E)$ we can define a subset $U \subseteq V$ to be an η-*dense subgraph of order* ℓ if and only if $\varrho_\ell(G[U]) \geq \eta$. From the proposition above, any η-dense subgraph of order ℓ is an η-dense subgraph of order $\ell - 1$ as well. The η-dense subgraphs of order $\ell \geq 2$ inherit the property of being nested from the η-dense subgraphs. If we fix a density and consider dense subgraphs of increasing order, then we can observe that they become more and more similar to cliques. A formal argument goes as follows. Define the density of infinite order of a graph G as

$$\varrho_\infty(G) =_{\text{def}} \lim_{\ell \to \infty} \varrho_\ell(G).$$

The density of infinite order induces a discrete density function due to the following zero-one law [307].

Theorem 6.3.5. *Let $G = (V, E)$ be any undirected graph.*

1. *It holds that $\varrho_\infty(G)$ is either zero or one.*
2. *V is a clique if and only if $\varrho_\infty(G) = 1$.*

The theorem says that the only subgroup that is η-dense for some $\eta > 0$ and for all orders, is a clique. In a sense, the order of a density functions allows a scaling of how important compactness of groups is in relation to density.

Average degree. One can easily translate the density of a graph with n vertices into its average degree (as we did in the proof of Proposition 6.3.2): $\bar{d}(G) = \varrho(G)(n-1)$. Technically, density and average degree are interchangeable (with appropriate modifications). We thus can define dense subgraphs alternatively in terms of average degrees. Let $N > 0$ be any rational number. An N-*dense subgraph* of a graph $G = (V, E)$ is any subset $U \subseteq V$ such that $\bar{d}(G[U]) \geq N$. Clearly, an η-dense subgraph (with respect to percentage densities) of size k is an $\eta(k-1)$-dense subgraph (with respect to average degrees), and an N-dense subgraph (with respect to average degrees) of size k is an $\frac{N}{k-1}$-dense subgraph (with respect to percentage densities). Any N-core is an N-dense subgraph. N-dense subgraphs are neither closed under exclusion nor nested. This is easily seen by considering N-regular graphs (for $N \in \mathbb{N}$). Removing some vertices decreases the average degree strictly below N. However, average degrees allow a more fine-grained analysis of network structure. Since a number of edges quadratic in the number of vertices is required for a graph to be denser than some given percentage threshold, small graphs are favored. Average degrees avoid this pitfall.

Extremal graphs. Based upon Turán's theorem (see Theorem 6.1.2), a whole new area in graph theory has emerged which has been called extremal graph theory (see, e.g., [66]). It studies questions like the following: how many edges may a graph have such that some of a given set of subgraphs are not contained in the graph? Clearly, if we have more edges in the graph, then all these subgraphs must be contained in it. This has been applied to dense subgraphs as well. The following classical theorem due to Dirac [156] is a direct strengthening of Turán's theorem.

Theorem 6.3.6 (Dirac, 1963). *Let* $G = (V, E)$ *be any undirected graph. If* $m > \frac{n^2}{2} \cdot \frac{k-2}{k-1}$, *then* G *contains subgraphs of size* $k + r$ *having average degree at least* $k + r - 1 - \frac{r}{k+r}$ *for all* $r \in \{0, \ldots, k-2\}$ *and* $n \geq k + r$.

Notice that the case $r = 0$ corresponds to the existence of size-k cliques as expressed by Turán's theorem. In many cases, only asymptotic estimations are possible. For example, it can be shown that, for a graph $G = (V, E)$ on n vertices and m edges, if $m = \omega\left(n^{2-\sqrt{\frac{2}{d \cdot k}}}\right)$, then G has a subgraph with k vertices and average degree d [368, 262]. It follows that to be sure that there are reasonably dense subgraphs of sizes not very small, the graph itself has to be reasonably dense. Some more results are discussed in [262].

6.3.2 Densest Subgraphs

We review a solution for computing a densest subgraph with respect to average degrees. Let $\gamma^*(G)$ be the maximum average degree of any non-empty induced subgraph of G, i.e.,

$$\gamma^*(G) =_{\text{def}} \max\{ \bar{d}(G[U]) \mid U \subseteq V \text{ and } U \neq \emptyset \}.$$

As in the case of N-cores, the maximal subgraph realizing $\gamma^*(G)$ is uniquely determined. We consider the following problem:

> *Problem:*   DENSEST SUBGRAPH
> *Input:*   Graph G
> *Output:*   A vertex set of G that realizes $\gamma^*(G)$

This problem can be solved in polynomial time using flow techniques [477, 248, 239]; our proof is from [248].

Theorem 6.3.7. *There is an algorithm for solving* DENSEST SUBGRAPH *on graphs with* n *vertices and* m *edges in time* $\mathcal{O}(mn(\log n)(\log \frac{n^2}{m}))$.

Proof. We formulate DENSEST SUBGRAPH as a maximum flow problem depending on some parameter $\gamma \in \mathbb{Q}^+$. Let $G = (V, E)$ be any undirected graph with n vertices and m edges. Consider a flow network consisting of graph $G' = (V', E')$ and capacity function $u_\gamma : E' \to \mathbb{Q}^+$ given as follows. Add to V a source s and a sink t; replace each edge of G (which is undirected) by two directed edges of capacity one each; connect the source to all vertices of V by an edge of capacity m; and connect each vertex $v \in V$ to the sink by an edge of capacity $m + \gamma - d_G(v)$. More specifically, the network is defined as

$$V' =_{\text{def}} V \cup \{s, t\}$$
$$E' =_{\text{def}} \{(v, w) \mid \{v, w\} \in E\} \cup \{(s, v) \mid v \in V\} \cup \{(v, t) \mid v \in V\}$$

and for $v, w \in V'$ the capacity function u_γ is defined as

$$u_\gamma(v, w) =_{\text{def}} \begin{cases} 1 & \text{if } \{v, w\} \in E \\ m & \text{if } v = s \\ m + \gamma - d_G(v) & \text{if } w = t \\ 0 & \text{if } (v, w) \notin E' \end{cases}$$

We consider capacities of cuts in the network. Let S, T be any partitioning of V' into two disjoint vertex sets with $s \in S$ and $t \in T$, $S_+ = S - \{s\}$ and $T_+ = T - \{t\}$. Note that $S_+ \cup T_+ = V$. If $S_+ = \emptyset$, then the capacity of the cut is $c(S, \bar{S}) = m|V|$. Otherwise we obtain:

$$c(S, T) = \sum_{v \in S, w \in T} u_\gamma(v, w)$$

$$= \sum_{w \in T_+} u_\gamma(s, w) + \sum_{v \in S_+} u_\gamma(v, t) + \sum_{v \in S_+, w \in T_+} u_\gamma(v, w)$$

$$= m|T_+| + \left(m|S_+| + \gamma|S_+| - \sum_{v \in S_+} d_G(v) \right) + \sum_{\substack{v \in S_+, w \in T_+ \\ \{v, w\} \in E}} 1$$

$$= m|V| + |S_+| \left(\gamma - \frac{1}{|S_+|} \left(\sum_{v \in S_+} d_G(v) - \sum_{\substack{v \in S_+, w \in T_+ \\ \{v, w\} \in E}} 1 \right) \right)$$

$$= m|V| + |S_+|(\gamma - \bar{d}(G[S_+])) \tag{6.2}$$

It is clear from this equation that γ is our guess on the maximum average degree of G. We need to know how we can detect whether γ is too big or too small. We prove the following claim.

Claim. Let S and T be sets that realize the minimum capacity cut, with respect to γ. Then we have the following:

1. If $S_+ \neq \emptyset$, then $\gamma \leq \gamma^*(G)$.
2. If $S_+ = \emptyset$, then $\gamma \geq \gamma^*(G)$.

The claim is proven by the following arguments.

1. Suppose $S_+ \neq \emptyset$. Since $c(\{s\}, V' - \{s\}) = m|V| \geq c(S,T)$, we have $|S_+|(\gamma - \bar{d}(G[S_+])) \leq 0$. Hence, $\gamma \leq \bar{d}(G[S_+]) \leq \gamma^*(G)$.
2. Suppose $S_+ = \emptyset$. Assume further to the contrary, that $\gamma < \gamma^*(G)$. Let $U \subseteq V$ be any non-empty vertex subset satisfying $\bar{d}(G[U]) = \gamma^*(G)$. By Equation 6.2, we obtain

$$c(U \cup \{s\}, \overline{U} \cup \{t\}) = m|V| + |U|(\gamma - \gamma^*(G)) < m|V| = c(S,T),$$

a contradiction to the minimality of the cut capacity $c(S,T)$. Thus, $\gamma \geq \gamma^*(G)$.

The claim suggests an algorithm for finding the right guess for γ by binary search. Notice that $\gamma^*(G)$ can have only a finite number of values, i.e.,

$$\gamma^*(G) \in \left\{ \frac{2i}{j} \;\middle|\; i \in \{0, \ldots, m\} \text{ and } j \in \{1, \ldots, n\} \right\}.$$

It is easily seen that the smallest possible distance between two different points in the set is $\frac{1}{n(n-1)}$. A binary search procedure for finding a maximum average degree subgraph is given as Algorithm 11.

Algorithm 11: Densest subgraph by min-cut and binary search [248]

Input: Graph $G = (V, E))$
Output: A set of k vertices of G

Initialize $l := 0$, $r := m$, and $U := \emptyset$;
while $r - l \geq \frac{1}{n(n-1)}$ **do**

    $\gamma := \frac{l+r}{2}$;
    Construct flow network (V', E', u_γ);
    Find minimum cut S and T of the flow network;
    if $S = \{s\}$ **then**
        $r := \gamma$
    else
        $l := \gamma$;
        $U := S - \{s\}$

Return U

For a time bound, note that we execute the iteration $\lceil \log((m+1)n(n-1)) \rceil = \mathcal{O}(\log n)$ times. Inside each iteration we have to run an algorithm which finds a minimum capacity cut. If we use, e.g., the push-relabel algorithm [252] for max-flow computations, we can do this in time $\mathcal{O}(nm \log \frac{n^2}{m})$ in a network with n vertices and m edges. Out network has $n + 2$ vertices and $2m + 2n$ edges. This does not change the complexity of the max-flow algorithm asymptotically. We thus obtain the overall time bound $\mathcal{O}(nm(\log n)(\log \frac{n^2}{m}))$. $\qquad\square$

Parametric maximum flow algorithms [239, 6] have been employed to improve the time bound to $\mathcal{O}(nm \log \frac{n^2}{m})$ [239]. In [113], DENSEST SUBGRAPH has been solved by linear programming. This gives certainly a worse upper bound for the time complexity, but has some extensions to the case of directed graphs.

Directed graphs. There is no obvious way to define the notion of density in directed graphs. Since average in-degree and average out-degree in a directed graph are always equal, both measures are not sensitive to orientedness. One approach followed in the literature [342, 113] is based on considering two vertex sets S and T, which are not necessarily disjoint, to capture orientations. For any directed graph $G = (V, E)$ and non-empty sets $S, T \subseteq V$, let $E(S, T)$ denote the set of edges going from S to T, i.e., $E(S, T) = \{(u, v) \mid u \in S \text{ and } v \in T\}$. We define an average degree of the pair (S, T) in the graph as [342]:

$$\bar{d}_G(S, T) =_{\text{def}} \frac{|E(S, T)|}{\sqrt{|S| \cdot |T|}}.$$

This notion was introduced to measure the connectedness between hubs and authorities in web graphs. The set S is understood as the set of hubs, and the set T is understood as the set of authorities in the sense of [359], or fans and centers in the sense of [376]. If $S = T$ then $\bar{d}_G(S, T)$ is precisely the average degree of $G[S]$ (i.e., the sum of the average in-degree and the average out-degree of $G[S]$). The maximum average degree for a directed graph $G = (V, E)$ is defined as

$$\gamma^*(G) =_{\text{def}} \max\{\ \bar{d}_G(S, T) \mid S, T \subseteq V \text{ and } S \neq \emptyset, T \neq \emptyset\ \}.$$

DENSEST SUBGRAPH on directed graphs can be solved in polynomial time by linear programming [113]. To do so, we consider the following LP relaxations LP_γ, where γ ranges over all possible ratios $|S|/|T|$:

$$\begin{aligned}
\max\ & \textstyle\sum_{(u,v) \in E} x_{(u,v)} \\
\text{s.t.}\ & x_{(u,v)} \leq s_u \text{ for all } (u, v) \in E \\
& x_{(u,v)} \leq t_v \text{ for all } (u, v) \in E \\
& \textstyle\sum_{u \in V} s_u \leq \sqrt{\gamma} \\
& \textstyle\sum_{v \in V} t_v \leq \frac{1}{\sqrt{\gamma}} \\
& x_{(u,v)}, s_u, t_v \geq 0 \text{ for all } u, v \in V \text{ and } (u, v) \in E
\end{aligned}$$

It can be shown that the maximum average degree for G is the maximum of the optimal solutions for LP_γ over all γ. Each linear program can be solved in polynomial time. Since there are $\mathcal{O}(n^2)$ many ratios for $|S|/|T|$ and thus for γ, we can now compute the maximum average degree for G (and a corresponding subgraph as well) in polynomial time by binary search.

6.3.3 Densest Subgraphs of Given Sizes

The densest subgraph of a graph is highly fragile, as a graph with some average degree need not possess a subgraph with the same average degree. We are thus not able to deduce easily information on the existence of subgraphs with certain average degrees *and* certain sizes, from a solution of DENSEST SUBGRAPH. We discuss this problem independently. For an undirected graph $G = (V, E)$ and parameter $k \in \mathbb{N}$, let $\gamma^*(G, k)$ denote the maximum value of the average degrees of all induced subgraphs of G having k vertices, i.e.,

$$\gamma^*(G, k) =_{\text{def}} \max\{ \bar{d}(G[U]) \mid U \subseteq V \text{ and } |U| = k \}.$$

The following optimization problem has been introduced in [201]:

Problem: DENSE k-SUBGRAPH
Input: Graph G, Parameter $k \in \mathbb{N}$
Output: A vertex set of G that realizes $\gamma^*(G, k)$

In contrast to DENSEST SUBGRAPH, this problem is computationally difficult. It is clear that DENSE k-SUBGRAPH is $\mathcal{NP}$-hard (observe that the instance $(G, k, k-1)$ to the corresponding decision problem means searching for a clique of size k in G). The best we may hope for is a polynomial algorithm with moderate approximation ratio. A natural approach for approximating $\gamma^*(G, k)$ is based on greedy methods. An example of a greedy procedure due to [201] is given as Algorithm 12.

Algorithm 12: Greedy procedure

Input: Graph $G = (V, E)$ and even parameter $k \in \mathbb{N}$ (with $|V| \geq k$)
Output: A set of k vertices of G

Sort the vertices in decreasing order of their degrees;
Let H be the set of $\frac{k}{2}$ vertices of highest degree;
Compute $N_H(v) = |N(v) \cap H|$ for all vertices $v \in V - H$;
Sort the vertices in $V - H$ in decreasing order of the N_H-values;
Let R be the set of $\frac{k}{2}$ vertices of $V - H$ of highest N_H-values;
Return $H \cup R$

Theorem 6.3.8. *Let G be any graph with n vertices and let $k \in \mathbb{N}$ be an even natural number with $k \leq n$. Let $A(G, k)$ denote the average degree of the subgraph of G induced by the vertex set that is the output of Algorithm 12. We have*

$$\gamma^*(G, k) \leq \frac{2n}{k} \cdot A(G, k).$$

Proof. For subsets $U, U' \subseteq V$, let $E(U, U')$ denote the set of edges consisting of one vertex of U and one vertex of U'. Let m_U denote the cardinality of the edge set $E(G[U])$. Let d_H denote the average degree of the $\frac{k}{2}$ vertices of G with highest degree with respect to G. We certainly have, $d_H \geq \gamma^*(G, k)$. We obtain

$$|E(H, V - H)| = d_H \cdot |H| - 2m_H \geq \frac{d_H \cdot k}{2} - 2m_H \geq 0.$$

By the greedy rule, at least the fraction of

$$\frac{|R|}{|V - H|} = \frac{k}{2n - k} > \frac{k}{2n}$$

of these edges has been selected to be in $G[H \cup R]$. Hence, the total number of edges in $G[H \cup R]$ is at least

$$\left(\frac{d_H \cdot k}{2} - 2m_H\right) \cdot \frac{k}{2n} + m_H \geq \frac{d_H \cdot k^2}{4n}.$$

This proves the inequality for the average degree. □

The greedy procedure is the better the larger k is in relation to n. It is an appropriate choice if we want to find large dense regions in a graph. However, for very small parameters, e.g., for $k = \mathcal{O}(1)$, it is almost as bad as any trivial procedure. An approximation ratio $\mathcal{O}(\frac{n}{k})$ has been obtained by several other approximation methods, e.g., by greedy methods based on recursively deleting vertices of minimal degree [38] or by semidefinite programming [204, 531]. However, to overcome the connection between n and k, we need complementary algorithms that work well on smaller values of k. In the light of the following theorem [201], this seems possible for up to $k = \mathcal{O}(n^{\frac{2}{3}})$.

Theorem 6.3.9. DENSE k-SUBGRAPH *can be approximated in polynomial time within ratio* $\mathcal{O}(n^{\frac{1}{3} - \varepsilon})$ *for some* $\varepsilon > 0$.

No better bound for the general problem is known. In special graph classes, however, approximation can be done within better ratio. For instance, on families of dense graphs, i.e., graphs with $\Omega(n^2)$ edges, there exist polynomial-time approximation algorithms with ratio arbitrary close to one [35, 137]. A drawback here is that most of the social networks are sparse, not dense. As to lower bounds on the approximation ratio, it has recently been proven that an approximation ratio of $1 + \varepsilon$ for all $\varepsilon > 0$ cannot be achieved unless all $\mathcal{NP}$ problems can be simulated by randomized algorithms with double-sided error and sub-exponential running time (more specifically, in time $\mathcal{O}(2^{n^{\varepsilon}})$ for all $\varepsilon > 0$)[354]. Moreover, it is even conjectured that there is no polynomial-time algorithm with approximation ratio $\mathcal{O}(n^{\varepsilon})$ for all $\varepsilon > 0$ [201].

6.3.4 Parameterized Density

As we have argued, the decision version of DENSE k-SUBGRAPH is $\mathcal{NP}$-complete. In contrast to this variable decision problem (note that the density parameter is part of the input), we are now interested in studying the fixed-parameter version. A function $\gamma : \mathbb{N} \to \mathbb{Q}_+$ is a density threshold if and only if γ is computable in polynomial time and $\gamma(k) \leq k - 1$ for all $k \in \mathbb{N}$. For any density threshold γ, a γ-dense subgraph of a graph $G = (V, E)$ is any subset $U \subseteq V$ such that $\bar{d}(G[U]) \geq \gamma(|U|)$. We consider the following problem:

> *Problem:* γ-DENSE SUBGRAPH
> *Input:*   Graph G, Parameter $k \in \mathbb{N}$
> *Question:* Does there exist γ-dense subgraph of size k within G?

Clearly, on the one hand, if we choose $\gamma(k) = k - 1$ for all $k \in \mathbb{N}$, then we obtain γ-DENSE SUBGRAPH = CLIQUE, and thus an $\mathcal{NP}$-complete problem. On the other hand, if we choose $\gamma(k) = 0$, then any choice of k vertices induces a γ-dense subgraph and thus γ-DENSE SUBGRAPH is solvable in polynomial time. The question is: which choices of γ do still admit polynomial-time algorithms and for which γ does the problem become $\mathcal{NP}$-complete? This problem has been studied by several authors [204, 37, 308]. The following theorem due to [308] gives a sharp boundary, which also shows that a complexity jump appears very early.

Theorem 6.3.10. *Let γ be any density threshold.*

1. *If $\gamma = 2 + \mathcal{O}\left(\frac{1}{k}\right)$, then γ-DENSE SUBGRAPH is solvable in polynomial time.*
2. *If $\gamma = 2 + \Omega\left(\frac{1}{k^{1-\varepsilon}}\right)$ for some $\varepsilon > 0$, then γ-DENSE SUBGRAPH is $\mathcal{NP}$-complete.*

A direct application of the theorem gives the following result for the case of constant density functions.

Corollary 6.3.11. *Finding a k-vertex subgraph with average degree at least two can be done in polynomial time. However, there is no algorithm for finding a k-vertex subgraph with average degree at least $2 + \varepsilon$ for any $\varepsilon > 0$, unless $\mathcal{P} = \mathcal{NP}$.*

This result should be contrasted with the corresponding result for N-cores, where detecting N-cores of size k can be done in linear time in the graph size, even for all $N > 0$. This demonstrates a drastic computational difference between statistical and structural density.

Results similar to Theorem 6.3.10 have been proven for the case of special network classes with real-world characteristics, in particular, for power-law graphs and general sparse graphs [306].

6.4   Chapter Notes

In this chapter, we studied computational aspects of notions of local densities, i.e., density notions defined over induced subgraphs only, consequently suppressing network structure outside a subgroup. We considered structural (N-plexes, N-cores) and statistical relaxations (η-dense subgraphs) of the clique concept, which is the perfectly cohesive subgroup. Although many algorithmic problems for these notions are computationally hard, i.e., we do not know polynomial algorithms for solving them, there are several cases where fast algorithms exist producing desirable information on the density-based cohesive structure of a network, e.g., the number of small cliques in graphs, core numbers, or the maximum average degree reachable by a subgroup in a directed and undirected network.

An observation coming up from the presented results is that there is a seemingly hard tradeoff between mathematical soundness and meaningfulness of these notions and their algorithmic tractability. This is evident from the following table summarizing properties of our main notions:

subgroup	closed under exclusion	nested	tractable
clique	+	+	−
N-plex (for $N \in \mathbb{N}$)	+	+	−
N-core (for $N \in \mathbb{N}$)	−	−	+
η-dense subgraph (for $\eta \in [0,1]$)	−	+	−

Here, we see that nestedness, as a meaningful structure inside a group, excludes fast algorithms for computing subgroups of certain sizes. This exclusion is also inherited by some further relaxations. However, we have no rigorous proof for this observation in case of general locally definable subgroups. On the other hand, a similar relation is provably true for closure under exclusion and efficiently detecting subgroups of a given size: we cannot achieve both with an appropriate notion of density (see, e.g., [240, GT21,GT22]).

We conclude this chapter with a brief discussion of a selection of non-local concepts of cohesive subgroups that have attracted interest in social network analysis. Since non-locality emphasizes the importance for a cohesive subgroup to be separated from the remaining network, such notions play an important role in models for core/periphery structures [84, 193]. An extensive study of non-local density notions and their applications to network decomposition problems can be found in Chapter 8 and Chapter 10.

LS sets (Luccio-Sami sets). The notion of an LS set has been introduced in [399, 381]. An LS set can be seen as a network region where internal ties are more significant than external ties. More specifically, for a graph $G = (V, E)$ a vertex subset $U \subseteq V$ is said to be an *LS set* if and only if for all proper, non-empty subsets $U' \subset U$, we have

$$|E(U', V - U')| > |E(U, V - U)|.$$

Trivially, V is an LS set. Also the singleton sets $\{v\}$ are LS sets in G for each $v \in V$. LS sets have some nice structural properties. For instance, they do not non-trivially overlap [399, 381], i.e., if U_1 and U_2 are LS sets such that $U_1 \cap U_2 \neq \emptyset$, then either $U_1 \subseteq U_2$ or $U_2 \subseteq U_1$. Moreover, LS sets are rather dense: the minimum degree of a non-trivial LS set is at least half of the number of outgoing edges [512]. Note that the structural strength of LS sets depends heavily on the universal requirement that *all* proper subsets share more ties with the network outside than the set U does (see [512] for a discussion of this point). Some relaxations of LS sets can be found in [86].

Lambda sets. A notion closely related to LS sets is that of a lambda set. Let $G = (V, E)$ be any undirected graph. For vertices $u, v \in V$, let $\lambda(u, v)$ denote the number of edge-disjoint paths between u and v in G, i.e., $\lambda(u, v)$ measures the edge connectivity of u and v in G. A subset $U \subseteq V$ is said to be a *lambda set* if and only if

$$\min_{u, v \in U} \lambda(u, v) > \max_{u \in U, v \in V - U} \lambda(u, v).$$

In a lambda set, the members have more edge-disjoint paths connecting them to each other than to non-members. Each LS set is a lambda set [512, 86]. Lambda sets do not directly measure the density of a subset. However, they have some importance as they allow a polynomial-time algorithm for computing them [86]. The algorithm essentially consists of two parts, namely computing the edge-connectivity matrix for the vertex set V (which can be done by flow algorithms in time $\mathcal{O}(n^4)$ [258]) and based on this matrix, grouping vertices together in a level-wise manner, i.e., vertices u and v belong to the same lambda set (at level N) if and only if $\lambda(u, v) \geq N$. The algorithm can also be easily extended to compute LS sets.

Normal sets. In [285], a normality predicate for network subgroups has been defined in a statistical way over random walks on graphs. One of the most important reasons for considering random walks is that typically the resulting algorithms are simple, fast, and general. A random walk is a stochastic process by which we go over a graph by selecting the next vertex to visit at random among all neighbors of the current vertex. We can use random walks to capture a notion of cohesiveness quality of a subgroup. The intuition is that a group is the more cohesive the higher the probability is that a random walk originating at some group member does not leave the group. Let $G = (V, E)$ be any undirected graph. For $d \in \mathbb{N}$ and $\alpha \in \mathbb{R}_+$, a subset $U \subseteq V$ is said to be (d, α)-*normal* if and only if for all vertices $u, v \in U$ such that $d_G(u, v) \leq d$, the probability that a random walk starting at u will reach v before visiting any vertex $w \in V - U$, is at least α. Though this notion is rather intuitive, we do not know how to compute normal sets or decomposing a network into normal sets. Instead, some heuristic algorithms, running in linear time (at least on graphs with bounded degree), have been developed producing decompositions in the spirit of normality [285].

7 Connectivity

Frank Kammer and Hanjo Täubig

This chapter is mainly concerned with the strength of connections between vertices with respect to the number of vertex- or edge-disjoint paths. As we shall see, this is equivalent to the question of how many nodes or edges must be removed from a graph to destroy all paths between two (arbitrary or specified) vertices. For basic definitions of connectivity see Section 2.2.4.

We present algorithms which

- check k-vertex (k-edge) connectivity,
- compute the vertex (edge) connectivity, and
- compute the maximal k-connected components

of a given graph.

After a few definitions we present some important theorems which summarize fundamental properties of connectivity and which provide a basis for understanding the algorithms in the subsequent sections.

We denote the vertex-connectivity of a graph G by $\kappa(G)$ and the edge-connectivity by $\lambda(G)$; compare Section 2.2.4. Furthermore, we define the local (vertex-)connectivity $\kappa_G(s,t)$ for two distinct vertices s and t as the minimum number of vertices which must be removed to destroy all paths from s to t. In the case that an edge from s to t exists we set $\kappa_G(s,t) = n-1$ since κ_G cannot exceed $n-2$ in the other case[1]. Accordingly, we define $\lambda_G(s,t)$ to be the least number of edges to be removed such that no path from s to t remains. Note, that for undirected graphs $\kappa_G(s,t) = \kappa_G(t,s)$ and $\lambda_G(s,t) = \lambda_G(t,s)$, whereas for directed graphs these functions are, in general, not symmetric.

Some of the terms we use in this chapter occur under different names in the literature. In what follows, we mainly use (alternatives in parentheses): cut-vertex (articulation point, separation vertex), cut-edge (isthmus, bridge), component (connected component), biconnected component (non-separable component, block). A *cut-vertex* is a vertex which increases the number of connected components when it is removed from the graph; the term *cut-edge* is defined similarly. A *biconnected component* is a maximal 2-connected subgraph; see Chapter 2. A *block* of a graph G is a maximal connected subgraph of G containing no cut-vertex, that is, the set of all blocks of a graph consists of its isolated

---

[1] If s and t are connected by an edge, it is not possible to disconnect s from t by removing only vertices.

U. Brandes and T. Erlebach (Eds.): Network Analysis, LNCS 3418, pp. 143–177, 2005.

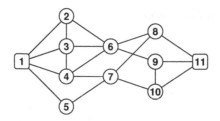

(a) A graph. We consider the connectivity between the vertices 1 and 11.

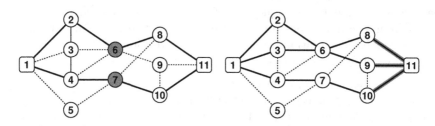

(b) 2 vertex-disjoint paths and a vertex-cutset of size 2.

(c) 3 edge-disjoint paths and an edge-cutset of size 3.

Fig. 7.1. Vertex-/edge-disjoint paths and vertex-/edge-cutsets

vertices, its cut-edges, and its maximal biconnected subgraphs. Hence, with our definition, a block is (slightly) different from a biconnected component.

The *block-graph* $B(G)$ of a graph G consists of one vertex for each block of G. Two vertices of the block-graph are adjacent if and only if the corresponding blocks share a common vertex (that is, a cut-vertex). The *cutpoint-graph* $C(G)$ of G consists of one vertex for each cut-vertex of G, where vertices are adjacent if and only if the corresponding cut-vertices reside in the same block of G. For the block- and the cutpoint-graph of G the equalities $B(B(G)) = C(G)$ and $B(C(G)) = C(B(G))$ hold [275]. The *block-cutpoint-graph* of a graph G is the bipartite graph which consists of the set of cut-vertices of G and a set of vertices which represent the blocks of G. A cut-vertex is adjacent to a block-vertex whenever the cut-vertex belongs to the corresponding block. The block-cutpoint-graph of a connected graph is a tree [283]. The maximal k-vertex-connected (k-edge-connected) subgraphs are called k-vertex-components (k-edge-components). A k-edge-component which does not contain any $(k+1)$-components is called a *cluster* [410, 470, 411, 412].

7.1 Fundamental Theorems

Theorem 7.1.1. *For all non-trivial graphs G it holds that:*

$$\kappa(G) \leq \lambda(G) \leq \delta(G)$$

Proof. The incident edges of a vertex having minimum degree $\delta(G)$ form an edge separator. Thus we conclude $\lambda(G) \leq \delta(G)$.

The vertex-connectivity of any graph on n vertices can be bounded from above by the connectivity of the complete graph $\kappa(K_n) = n - 1$.

Let $G = (V, E)$ be a graph with at least 2 vertices and consider a minimal edge separator that separates a vertex set S from all other vertices $\bar{S} = V \setminus S$. In the case that all edges between S and $\bar{S}$ are present in G we get $\lambda(G) = |S| \cdot |\bar{S}| \geq |V| - 1$. Otherwise there exist vertices $x \in S, y \in \bar{S}$ such that $\{x, y\} \notin E$, and the set of all neighbors of x in $\bar{S}$ as well as all vertices from $S \setminus \{x\}$ that have neighbors in $\bar{S}$ form a vertex separator; the size of that separator is at most the number of edges from S to $\bar{S}$, and it separates (at least) x and y. □

The following is the graph-theoretic equivalent of a theorem that was published by Karl Menger in his work on the general curve theory [419].

Theorem 7.1.2 (Menger, 1927). *If P and Q are subsets of vertices of an undirected graph, then the maximum number of vertex-disjoint paths connecting vertices from P and Q is equal to the minimum cardinality of any set of vertices intersecting every path from a vertex in P to a vertex in Q.*

This theorem is also known as the n-chain or n-arc theorem, and it yields as a consequence one of the most fundamental statements of graph theory:

Corollary 7.1.3 (Menger's Theorem). *Let s, t be two vertices of an undirected graph $G = (V, E)$. If s and t are not adjacent, the maximum number of vertex-disjoint s-t-paths is equal to the minimum cardinality of an s-t-vertex-separator.*

The analog for the case of edge-cuts is stated in the next theorem.

Theorem 7.1.4. *The maximum number of edge-disjoint s-t-paths is equal to the minimum cardinality of an s-t-edge-separator.*

This theorem is most often called the edge version of Menger's Theorem although it was first explicitly stated three decades after Menger's paper in publications due to Ford and Fulkerson [218], Dantzig and Fulkerson [141], as well as Elias, Feinstein, and Shannon [175].

A closely related result is the Max-Flow Min-Cut Theorem by Ford and Fulkerson (see Theorem 2.2.1, [218]). The edge variant of Menger's Theorem can be seen as a restricted version where all edge capacities have a unit value.

The following global version of Menger's Theorem was published by Hassler Whitney [581] and is sometimes referred to as 'Whitney's Theorem'.

Theorem 7.1.5 (Whitney, 1932). *Let $G = (V, E)$ be a non-trivial graph and k a positive integer. G is k-(vertex-)connected if and only if all pairs of distinct vertices can be connected by k vertex-disjoint paths.*

The difficulty in deriving this theorem is that Menger's Theorem requires the nodes to be not adjacent. Since this precondition is not present in the edge version of Menger's Theorem, the following follows immediately from Theorem 7.1.4:

Theorem 7.1.6. *Let $G = (V, E)$ be a non-trivial graph and k a positive integer. G is k-edge-connected if and only if all pairs of distinct vertices can be connected by k edge-disjoint paths.*

For a detailed review of the history of Menger's Theorem we refer to the survey by Schrijver [506].

Beineke and Harary discovered a similar theorem for a combined vertex-edge-connectivity (see [55]). They considered *connectivity pairs* (k, l) such that there is some set of k vertices and l edges whose removal disconnects the graph, whereas there is no set of $k - 1$ vertices and l edges or of k vertices and $l - 1$ edges forming a mixed vertex/edge cut set.

Theorem 7.1.7 (Beineke & Harary, 1967). *If (k, l) is a connectivity pair for vertices s and t in graph G, then there are $k + l$ edge-disjoint paths joining s and t, of which k are mutually non-intersecting.*

The following theorem gives bounds on vertex- and edge-connectivity (see [274]).

Theorem 7.1.8. *The maximum (vertex-/edge-) connectivity of some graph on n vertices and m edges is*
$$\lfloor \tfrac{2m}{n} \rfloor \quad , \quad \text{if } m \geq n - 1$$
$$0 \quad , \quad \text{otherwise.}$$
The minimum (vertex-/edge-) connectivity of some graph on n vertices and m edges is
$$m - \binom{n-1}{2} \quad , \quad \text{if } \binom{n-1}{2} < m \leq \binom{n}{2}$$
$$0 \quad , \quad \text{otherwise.}$$

A further proposition concerning the edge connectivity in a special case has been given by Chartrand [114]:

Theorem 7.1.9. *For all graphs $G = (V, E)$ having minimum degree $\delta(G) \geq \lfloor |V|/2 \rfloor$, the edge-connectivity equals the minimum degree of the graph: $\lambda(G) = \delta(G)$*

For more bounds on graph connectivity see [28, 62, 390, 63, 182, 523].

The following theorems deal with the k-vertex/edge-components of graphs. The rather obvious facts that two different components of a graph have no vertex in common, and two different blocks share at most one common vertex, have been generalized by Harary and Kodama [279]:

Theorem 7.1.10. *Two distinct k-(vertex-)components have at most $k - 1$ vertices in common.*

While k-vertex-components might overlap, k-edge-components do not.

Theorem 7.1.11 (Matula, 1968). *For any fixed natural number $k \geq 1$ the k-edge-components of a graph are vertex-disjoint.*

Proof. The proof is due to Matula (see [410]). Consider an (overlapping) decomposition $\tilde{G} = G_1 \cup G_2 \cup \ldots \cup G_t$ of a connected subgraph $\tilde{G}$ of G. Let $C = (A, \bar{A})$ be a minimum edge-cut of $\tilde{G}$ into the disconnected parts A and $\bar{A}$. To skip the trivial case, assume that $\tilde{G}$ has at least 2 vertices. For each subgraph G_i that contains a certain edge $e \in C$ of the min-cut, the cut also contains a cut for G_i (otherwise the two vertices would be connected in $G_i \setminus C$ and $\tilde{G} \setminus C$ which would contradict the assumption that C is a minimum cut). We conclude that there is a G_i such that $\lambda(\tilde{G}) = |C| \geq \lambda(G_i)$, which directly implies $\lambda(\tilde{G}) \geq \min_{1 \leq i \leq t}\{\lambda(G_i)\}$ and thereby proves the theorem. □

Although we can see from Theorem 7.1.1 that k-vertex/edge-connectivity implies a minimum degree of at least k, the converse is not true. But in the case of a large minimum degree, there must be a highly connected subgraph.

Theorem 7.1.12 (Mader, 1972). *Every graph of average degree at least $4k$ has a k-connected subgraph.*

For a proof see [404].

Several observations regarding the connectivity of *directed* graphs have been made. One of them considers directed spanning trees rooted at a node r, so called *r-branchings*:

Theorem 7.1.13 (Edmonds' Branching Theorem [171]). *In a directed multigraph $G = (V, E)$ containing a vertex r, the maximum number of pairwise edge-disjoint r-branchings is equal to $\kappa_G(r)$, where $\kappa_G(r)$ denotes the minimum, taken over all vertex sets $S \subset V$ that contain r, of the number of edges leaving S.*

The following theorem due to Lovász [396] states an interrelation of the maximum number of directed edge-disjoint paths and the in- and out-degree of a vertex.

Theorem 7.1.14 (Lovász, 1973). *Let $v \in V$ be a vertex of a graph $G = (V, E)$. If $\lambda_G(v, w) \leq \lambda_G(w, v)$ for all vertices $w \in V$, then $d^+(v) \leq d^-(v)$.*

As an immediate consequence, this theorem provided a proof for Kotzig's conjecture:

Theorem 7.1.15 (Kotzig's Theorem). *For a directed graph G, $\lambda_G(v, w)$ equals $\lambda_G(w, v)$ for all $v, w \in V$ if and only if the graph is pseudo-symmetric, i.e. the in-degree equals the out-degree for all vertices: $d^+(v) = d^-(v)$.*

7.2 Introduction to Minimum Cuts

For short, in an undirected weighted graph the sum of the weights of the edges with one endpoint in each of two disjoint vertex sets X and Y is denoted by $w(X, Y)$. For directed graphs, $w(X, Y)$ is defined in nearly the same way, but we only count the weight of edges with their origin in X and their destination in Y. A cut in a weighted graph $G = (V, E)$ is a set of vertices $\emptyset \subset S \subset V$ and its weight is $w(S, V \setminus S)$. In an unweighted graph, the weight of a cut is the number of edges from S to $V \setminus S$.

Definition 7.2.1. *A minimum cut is a cut S such that for all other cuts T,*

$$w(S, V \setminus S) \leq w(T, V \setminus T).$$

Observation 7.2.2. *A minimum cut in a connected graph G with edge weights greater than zero induces a connected subgraph of G.*

An algorithm that computes all *minimum cuts* has to represent these cuts. A problem is to store all minimum cuts without using too much space. A suggestion was made in 1976 by Dinitz et al. [153]. They presented a data structure called *cactus* that represents all minimum cuts of an undirected (weighted) graph. The size of a cactus is linear in the number of vertices of the input graph and a cactus allows us to compute a cut in a time linear in the size of the cut.

Karzanov and Timofeev outlined in [351] a first algorithm to construct a cactus for unweighted, undirected graphs. Their algorithm consists of two parts. Given an arbitrary input graph G, the first part finds a sequence of all minimum cuts in G and the second constructs the cactus C_G from this sequence. The algorithm also works on weighted graphs, as long as all weights are positive.

If negative weights are allowed, the problem of finding a minimum cut is $\mathcal{NP}$-hard [345]. Moreover, no generalization for directed graphs is known. An unweighted graph can be reduced to a weighted graph by assigning weight 1 to all edges. In the following, we will therefore consider the problem of finding minimum cuts only for undirected connected graphs with positive weights.

Consider a network N defined by the directed graph $G = (V, E)$, a capacity function u_N, a source s, a sink t and a flow f (Chapter 2). A *residual network* R_f consists of those edges that can carry additional flow, beyond what they already carry under f. Thus R_f is defined on the graph $G_{R_f} := \left(V, \left\{ (u,v) \,\middle|\, ((u,v) \in E \vee (v,u) \in E) \wedge u_{R_f}((u,v)) > 0 \right\} \right)$ with the same source s and sink t and the following capacity function

$$u_{R_f}((a,b)) := \begin{cases} c(a,b) - f(a,b) + f(b,a) & \text{if } (a,b) \in E \wedge (b,a) \in E \\ c(a,b) - f(a,b) & \text{if } (a,b) \in E \wedge (b,a) \notin E \\ f(b,a) & \text{if } (a,b) \notin E \wedge (b,a) \in E \end{cases}$$

Let $R_{f_{max}}$ be the residual network of N and f_{max}, where f_{max} is a maximum s-t-flow in N. As a consequence of Theorem 2.2.1 on page 11, the maximum flow saturates all minimum s-t-cuts and therefore each set $S \subseteq V \setminus t$ is a minimum s-t-cut iff $s \in S$ and no edges leave S in $R_{f_{max}}$.

7.3   All-Pairs Minimum Cuts

The problem of computing a minimum cut between all pairs of vertices can, of course, easily be done by solving $n(n-1)/2$ flow problems. As has been shown by Gomory and Hu [257], the computation of $n-1$ maximum flow problems is already sufficient to determine the value of a maximum flow / minimum cut for all pairs of vertices. The result can be represented in the *equivalent flow tree*, which is a weighted tree on n vertices, where the minimum weight of any edge on the (unique) path between two vertices s and t equals the maximum flow from s to t. They furthermore showed that there always exists an equivalent flow tree, where the components that result from removing the minimum weight edge of the s-t-path represent a minimum cut between s and t. This tree is called the *Gomory-Hu cut tree*.

Gusfield [265] demonstrated how to do the same computation without node contractions and without the overhead for avoiding the so called crossing cuts. See also [272, 344, 253].

If one is only interested in any edge cutset of minimum weight in an undirected weighted graph (without a specified vertex pair to be disconnected), this can be done using the algorithm of Stoer and Wagner, see Section 7.7.1.

7.4   Properties of Minimum Cuts in Undirected Graphs

There are $2^{|V|}$ sets and each of them is possibly a minimum cut, but the number of minimum cuts in a fixed undirected graph is polynomial in $|V|$. To see this, we need to discuss some well-known facts about minimum cuts. These facts also help us to define a data structure called *cactus*. A cactus can represent all minimum cuts, but needs only space linear in $|V|$.

For short, for a graph G, let in this chapter λ_G always denote the weight of a minimum cut. If the considered graph G is clear from the context, the index G of λ_G is omitted.

Lemma 7.4.1. *Let S be a minimum cut in $G = (V, E)$. Then, for all $\emptyset \neq T \subset S : w(T, S \setminus T) \geq \frac{\lambda}{2}$.*

Proof. Assume $w(T, S\setminus T) < \frac{\lambda}{2}$. Since $w(T, V \setminus S) + w(S \setminus T, V \setminus S) = \lambda$, w.l.o.g. $w(T, V \setminus S) \leq \frac{\lambda}{2}$ (if not, define T as $S \setminus T$). Then $w(T, V \setminus T) = w(T, S \setminus T) + w(T, V \setminus S) < \lambda$. Contradiction. $\square$

Lemma 7.4.2. *Let $A \neq B$ be two minimum cuts such that $T := A \cup B$ is also a minimum cut. Then*

$$w\left(A, \bar{T}\right) = w\left(B, \bar{T}\right) = w\left(A \setminus B, B\right) = w\left(A, B \setminus A\right) = \frac{\lambda}{2}.$$

Proof. As in the Figure 7.2, let $a = w\left(A, \bar{T}\right)$, $b = w\left(B, \bar{T}\right)$, $\alpha = w\left(A, B \setminus A\right)$ and $\beta = w\left(B, A \setminus B\right)$. Then $w\left(A, \bar{A}\right) = a + \alpha = \lambda$, $w\left(B, \bar{B}\right) = b + \beta = \lambda$ and $w\left(T, \bar{T}\right) = a + b = \lambda$. We also know that $w\left(A \setminus B, B \cup \bar{T}\right) = a + \beta \geq \lambda$ and $w\left(B \setminus A, A \cup \bar{T}\right) = b + \alpha \geq \lambda$. This system of equations and inequalities has only one unique solution: $a = \alpha = b = \beta = \frac{\lambda}{2}$. $\square$

Definition 7.4.3. *A pair $\langle S_1, S_2 \rangle$ is called crossing cut, if S_1, S_2 are two minimum cuts and neither $S_1 \cap S_2$, $S_1 \setminus S_2$, $S_2 \setminus S_1$ nor $\bar{S}_1 \cap \bar{S}_2$ is empty.*

Lemma 7.4.4. *Let $\langle S_1, S_2 \rangle$ be crossing cuts and let $A = S_1 \cap S_2$, $B = S_1 \setminus S_2$, $C = S_2 \setminus S_1$ and $D = \bar{S}_1 \cap \bar{S}_2$. Then*

a. A, B, C and D are minimum cuts

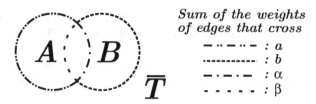

Sum of the weights of edges that cross

$- \cdot\cdot - \cdot\cdot -$: a
$\cdots\cdots\cdots$: b
$- \cdot - \cdot -$: α
$- - - - -$: β

Fig. 7.2. Intersection of two minimum cuts A and B

b. $w(A, D) = w(B, C) = 0$

c. $w(A, B) = w(B, D) = w(D, C) = w(C, A) = \frac{\lambda}{2}$.

Proof. Since we know that S_1 and S_2 are minimum cuts, we can conclude

$$w\left(S_1, \bar{S}_1\right) = w(A, C) + w(A, D) + w(B, C) + w(B, D) = \lambda$$
$$w\left(S_2, \bar{S}_2\right) = w(A, B) + w(A, D) + w(B, C) + w(C, D) = \lambda$$

and since there is no cut with weight smaller than λ, we know that

$$w\left(A, \bar{A}\right) = w(A, B) + w(A, C) + w(A, D) \geq \lambda$$
$$w\left(B, \bar{B}\right) = w(A, B) + w(B, C) + w(B, D) \geq \lambda$$
$$w\left(C, \bar{C}\right) = w(A, C) + w(B, C) + w(C, D) \geq \lambda$$
$$w\left(D, \bar{D}\right) = w(A, D) + w(B, D) + w(C, D) \geq \lambda$$

Summing up twice the middle and the right side of the first two equalities we obtain

$$2 \cdot w(A, B) + 2 \cdot w(A, C) + 4 \cdot w(A, D) + 4 \cdot w(B, C) + 2 \cdot w(B, D) + 2 \cdot w(C, D) = 4 \cdot \lambda$$

and summing up both side of the four inequalities we have

$$2 \cdot w(A, B) + 2 \cdot w(A, C) + 2 \cdot w(A, D) + 2 \cdot w(B, C) + 2 \cdot w(B, D) + 2 \cdot w(C, D) \geq 4 \cdot \lambda$$

Therefore $w(A, D) = w(B, C) = 0$. In other words, there are no diagonal edges in Figure 7.3.

For a better imagination, let us assume that the length of the four inner line segments in the figure separating A, B, C and D is proportional to the sum of the weights of all edges crossing this corresponding line segments. Thus the total length l of both horizontal or both vertical lines, respectively, is proportional to the weight λ.

Let us assume the four line segments have different length, in other words, the two lines separating the sets S_1 from $\bar{S}_1$ or S_2 from $\bar{S}_2$, respectively, do not cross each other exactly in the midpoint of the square, then the total length of the separating line segments of one vertex set $\Delta = A, B, C$ or D is shorter then l. Thus $w(\Delta, \bar{\Delta}) < \lambda$. Contradiction.

As a consequence, $w(A, B) = w(B, D) = w(D, C) = w(C, A) = \frac{\lambda}{2}$ and $w\left(A, \bar{A}\right) = w\left(B, \bar{B}\right) = w\left(C, \bar{C}\right) = w\left(D, \bar{D}\right) = \lambda$.   □

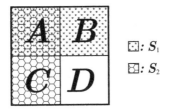

Fig. 7.3. Crossing cuts $\langle S_1, S_2 \rangle$ with $S_1 := A \cup B$ and $S_2 := A \cup C$

A crossing cut in $G = (V, E)$ partitions the vertex set V into exactly four parts. A more general definition is the following, where the vertex set can be divided in three or more parts.

Definition 7.4.5. *A* circular partition *is a partition of* V *into* $k \geq 3$ *disjoint sets* $V_1, V_2, \ldots, V_k$ *such that*

a. $w(V_i, V_j) = \begin{cases} \lambda/2 & : \quad |i - j| = 1 \bmod k \\ 0 & : \quad otherwise \end{cases}$

b. If S is a minimum cut, then

    1. S or $\bar{S}$ is a proper subset of some V_i or

    2. the circular partition is a refinement of the partition defined by the minimum cut S. In other words, the minimum cut is the union of some of the sets of the circular partition.

Let $V_1, V_2, \ldots, V_k$ be the disjoint sets of a *circular partition*, then for all $1 \leq a \leq b < k, S := \left(\cup_{i=a}^{b} V_i \right)$ is a minimum cut. Of course, the complement of S containing V_k is a minimum cut, too. Let us define these minimum cuts as *circular partition cuts*. Especially each $V_i, 1 \leq i \leq k$, is a minimum cut (property a. of the last definition).

Consider a minimum cut S such that neither S nor its complement is contained in a set of the circular partition. Since S is connected (Observation 7.2.2), S or its complement are equal to $\cup_{i=a}^{b} V_i$ for some $1 \leq a < b < k$.

Moreover, for all sets V_i of a circular partition, there exists no minimum cut S such that $\langle V_i, S \rangle$ is a crossing cut (property b. of the last definition).

Definition 7.4.6. *Two different circular partitions* $P := \{U_1, \ldots, U_k\}$ *and* $Q := \{V_1, \ldots, V_l\}$ *are* compatible *if there is a unique r and s, $1 \leq r, s \leq k$, such that for all $i \neq r : U_i \subseteq V_s$ and for all $j \neq s : V_j \subseteq U_r$.*

Lemma 7.4.7 ([216]). *All different circular partitions are pairwise compatible.*

Proof. Consider two circular partitions P and Q in a graph $G = (V, E)$. All sets of the partitions are minimum cuts. Assume a set $S \in P$ is equal to the union of more than one and less than all sets of Q. Exactly two sets $A, B \in Q$ contained in S are connected by at least an edge to the vertices $V \setminus S$. Obtain T from S by replacing $A \subset S$ by an element of Q connected to B and not contained in S. Then $\langle S, T \rangle$ is a crossing cut, contradiction.

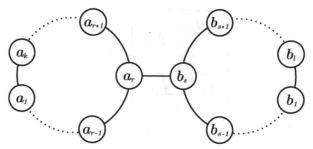

Fig. 7.4. Example graph $G = (\{a_1 \ldots a_r, b_1 \ldots b_s\}, E)$ shows two compatible partitions P, Q defined as follows:

$$P := \{\{a_1\}, \ldots, \{a_{r-1}\}, \{a_r, b_1, \ldots b_l\}, \{a_{r+1}\}, \ldots \{a_k\}\}$$

$$Q := \{\{b_1\}, \ldots, \{b_{s-1}\}, \{b_s, a_1, \ldots a_k\}, \{b_{s+1}\}, \ldots \{b_l\}\}$$

Therefore each set of P or its complement is contained in some set of Q.

Assume two sets of P are contained in two different sets of Q. Since each complement of the remaining sets of P cannot be contained in one set of Q, each remaining set of P must be contained in one subset of Q. Thus, $P = Q$. Contradiction.

Assume now all sets of P are contained in one set Y of Q. Then $Y = V$. Again a contradiction.

Since the union of two complements of sets in P is V and Q contains at least three sets, only one complement can be contained in one set of Q. Thus, there is exactly one set X of P that is not contained in Y of Q, but $\bar{X} \subset Y$.   □

Lemma 7.4.8. *If S_1, S_2 and S_3 are pairwise crossing cuts, then*

$$S_1 \cap S_2 \cap S_3 = \emptyset.$$

Proof. Assume that the lemma is not true. As shown in Figure 7.5, let

$$a = w\left(S_3 \setminus (S_1 \cup S_2), \overline{S_1 \cap S_2 \cap S_3}\right)$$

$$b = w\left((S_2 \cap S_3) \setminus S_1, S_2 \setminus (S_1 \cup S_3)\right)$$

$$c = w\left(S_1 \cap S_2 \cap S_3, (S_1 \cap S_2) \setminus S_3\right)$$

$$d = w\left((S_1 \cap S_3) \setminus S_2, S_1 \setminus (S_2 \cup S_3)\right)$$

On one hand $S_1 \cap S_2$ is a minimum cut (Lemma 7.4.4.a.) so that $c \geq \frac{\lambda}{2}$ (Lemma 7.4.1). On the other hand $c + b = c + d = \frac{\lambda}{2}$ (Lemma 7.4.4.c.). Therefore $b = d = 0$ and $(S_1 \cap S_3) \setminus S_2 = (S_2 \cap S_3) \setminus S_1 = \emptyset$.

If we apply Lemma 7.4.4.b. to S_1 and S_2, then $S_1 \cap S_2 \cap S_3$ and $S_3 \setminus (S_1 \cup S_2)$ are not connected. Contradiction.   □

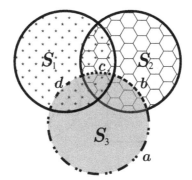

Fig. 7.5. Three pairwise crossing cuts S_1, S_2 and S_3

Lemma 7.4.9. *If S_1, S_2 and T are minimum cuts with $S_1 \subset S_2$, $T \not\subset S_2$ and $\langle S_1, T \rangle$ is a crossing cut, then $A := (S_2 \setminus S_1) \setminus T$, $B := S_1 \setminus T$, $C := S_1 \cap T$ and $D := (S_2 \setminus S_1) \cap T$ are minimum cuts, $w(A, B) = w(B, C) = w(C, D) = \frac{\lambda}{2}$ and $w(A, C) = w(A, D) = w(B, D) = 0$.*

Proof. Since $\langle S_1, T \rangle$ and therefore $\langle S_2, T \rangle$ is a crossing cut,

$$w(A \cup B, C \cup D) = \frac{\lambda}{2} \ (1), \ w(B, C) = \frac{\lambda}{2} \ (2),$$

$$w(A, B) + w\left(B, \overline{S_1 \cup S_2}\right) = w\left(B, A \cup \overline{S_1 \cup S_2}\right) = \frac{\lambda}{2} \ (3) \text{ and}$$

$$w\left(A, \overline{S_1 \cup S_2}\right) + w\left(B, \overline{S_1 \cup S_2}\right) = w\left(A \cup B, \overline{S_1 \cup S_2}\right) = \frac{\lambda}{2} \ (4).$$

All equalities follow from Lemma 7.4.4.c.. Moreover $w(A, T \setminus S_2) = 0$, $w\left(D, \overline{S_1 \cup S_2}\right) = 0$ (7.4.4.b.) and B, C are minimum cuts. Since (1), (2) and

$$w(A \cup B, C \cup D) = w(A, C) + w(A, D) + w(B, C) + w(B, D),$$

we can conclude that $w(A, C) = w(A, D) = w(B, D) = 0$.

A consequence of (3) and (4) is $w\left(A, \overline{S_1 \cup S_2}\right) = w(A, B)$. Moreover, $w(A, B) \geq \frac{\lambda}{2}$ (Lemma 7.4.1) and $w\left(A, \overline{S_1 \cup S_2}\right) \leq w\left(A, \overline{S_1} \cup S_2\right) = \frac{\lambda}{2}$. Therefore $w\left(A, \overline{S_1 \cup S_2}\right) = w(A, B) = \frac{\lambda}{2}$ and A is a minimum cut.

With a similar argument we can see, $w(C, D) = \frac{\lambda}{2}$ and D is a minimum cut. Therefore, the general case shown in Figure 7.6(a) can always be transformed into the Figure 7.6(b). $\qquad \square$

For short, given some sets $S_1, \ldots, S_k$, let

$$\mathcal{F}_{S_1, \ldots S_k}^{\alpha_1, \ldots, \alpha_k} = \bigcap_{i=1}^{k} \left\{ \begin{matrix} S_i \text{ if } \alpha_i = 1 \\ \overline{S_i} \text{ if } \alpha_i = 0 \end{matrix} \right\} \text{ and}$$

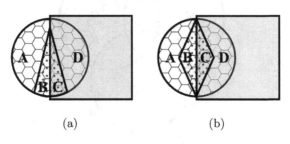

Fig. 7.6. Intersection of three minimum cuts

$$\mathcal{F}_{\{S_1,\ldots,S_k\}} = \left(\bigcup_{\alpha_1,\ldots,\alpha_k \in \{0,1\}^k} \mathcal{F}_{\{S_1,\ldots,S_k\}}^{\alpha_1,\ldots,\alpha_k} \right) \setminus \{\emptyset\}.$$

Lemma 7.4.10. *Let $\langle S_1, S_2 \rangle$ be a crossing cut and $A \in \mathcal{F}_{\{S_1,S_2\}}$. Choose $B \in \mathcal{F}_{\{S_1,S_2\}}$ such that $w(A,B) = \frac{\lambda}{2}$. For all crossing cuts $\langle B, T \rangle$:*

$$w(A, B \cap T) = \frac{\lambda}{2} \ or \ w(A, B \cap \bar{T}) = \frac{\lambda}{2}$$

Proof. W.l.o.g. $A = S_1 \cap S_2$ (if not, interchange S_1 and $\bar{S}_1$ or S_2 and $\bar{S}_2$), $B = S_1 \setminus S_2$ (if not, interchange S_1 and S_2). Let $C = S_2 \setminus S_1$ and $D = \bar{S}_1 \cap \bar{S}_2$. Then $(*) : w(B,C) = 0$ (Lemma 7.4.4.b.). Consider the following four cases:

$\mathbf{T} \subset (\mathbf{A} \cup \mathbf{B})$ (Figure 7.7(a)) $:$ $w(A, B \cap T) = \frac{\lambda}{2}$ (Lemma 7.4.9)

$\mathbf{T} \cap \mathbf{D} \neq \emptyset$ $:$ Because $\langle S_1, T \rangle$ is a crossing cut,

$$w(A \setminus T, A \cap T) + w(A \setminus T, B \cap T) + w(B \setminus T, A \cap T) + w(B \setminus T, B \cap T)$$

$$= w((A \setminus T) \cup (B \setminus T), (A \cap T) \cup (B \cap T))$$

$$= w(S_1 \setminus T, S_1 \cap T) = \frac{\lambda}{2}.$$

Together with $w(B \setminus T, B \cap T) \geq \frac{\lambda}{2}$ (Lemma 7.4.1), we can conclude
- $w(A \setminus T, A \cap T) = 0$ and therefore $A \cap T = \emptyset$ or $A \setminus T = \emptyset$,
- $w(A \setminus T, B \cap T) = 0$ (1) and
- $w(A \cap T, B \setminus T) = 0$ (2).

Note that $w(A, B) = \frac{\lambda}{2}$. If $A \cap T = \emptyset$, $w(A, B \cap T) \overset{(1)}{=} 0$ and $w(A, B \setminus T) = \frac{\lambda}{2}$. Otherwise $A \setminus T = \emptyset$, $w(A, B \setminus T) \overset{(2)}{=} 0$ and $w(A, B \cap T) = \frac{\lambda}{2}$.

$\mathbf{T} \not\subset (\mathbf{A} \cup \mathbf{B})$ and $\mathbf{T} \cap \mathbf{D} = \emptyset$ (3) and $(\mathbf{A} \cup \mathbf{C}) \subset \mathbf{T}$ (4) (Figure 7.7(b)) :

$$w\left(A, T \cap B\right) \overset{(*)}{=} w\left(A \cup C, T \cap B\right) \overset{(3),(4)}{=} w\left((A \cup C) \cap T, T \setminus (A \cup C)\right) \geq \frac{\lambda}{2},$$

since $(A \cup C)$ is a minimum cut (Lemma 7.4.1). Using the fact $w(A, B) = \frac{\lambda}{2}$, we get $w\left(A, T \cap B\right) = \frac{\lambda}{2}$.

$\mathbf{T} \not\subset (\mathbf{A} \cup \mathbf{B})$ and $\mathbf{T} \cap \mathbf{D} = \emptyset$ (5) and $(\mathbf{A} \cup \mathbf{C}) \not\subset \mathbf{T}$ (Figure 7.7(c)) :

$$w\left(A, T \cap B\right) \overset{(*)}{=} w\left(A \cup C, T \cap B\right) \overset{(5)}{=} w\left(A \cup C, T \setminus (A \cup C)\right) = \frac{\lambda}{2},$$

since $\langle A \cup C, T \rangle$ is a crossing cut.

This concludes the proof.   □

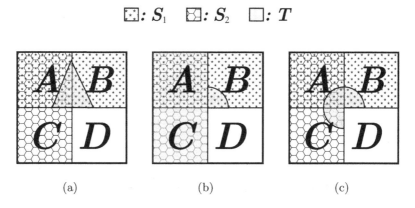

Fig. 7.7. A minimum cut T and a crossing cut $\langle S_1, S_2 \rangle$

Corollary 7.4.11. *The intersection of a crossing cut partitions the vertices of the input graph into four minimum cuts. Lemma 7.4.4.c. guarantees us that for each of the four minimum cuts A there exist two of the three remaining minimum cuts B, C such that $w(A, B) = w(A, C) = \frac{\lambda}{2}$. Although set B or C may be divided in smaller parts by further crossing cuts, there are always exactly two disjoint minimum cuts $X \subseteq B$ and $Y \subseteq C$ with $w(A, X) = w(A, Y) = \frac{\lambda}{2}$.*

Proof. Assume the corollary is not true. Let $\langle S, X_{1\&2} \rangle$ be the first crossing cut that divides the set $X_{1\&2}$ with $w(A, X_{1\&2}) = \frac{\lambda}{2}$ into the two disjoint sets X_1, X_2 with $w(A, X_1), w(A, X_2) \geq 0$. But then $\langle S, B \rangle$ or $\langle \bar{S}, B \rangle$ is also a crossing cut, which divides B into B_1 and B_2 with $X_1 \subseteq B_1$ and $X_2 \subseteq B_2$. Thus, $w(A, B_1), w(A, B_2) \geq 0$. This is a contradiction to Lemma 7.4.10.   □

Different crossing cuts interact in a very specific way, as shown in the next theorem.

Theorem 7.4.12 ([63, 153]). *In a graph $G = (V, E)$, for each partition P of V into 4 disjoint sets due to a crossing cut in G, there exists a circular partition in G that is a refinement of P.*

Proof. Given crossing cut $\langle S_1, S_2 \rangle$, choose the set

$$\Lambda := \left\{ S_1 \cap S_2, S_1 \setminus S_2, S_2 \setminus S_1, \overline{S_1 \cup S_2} \right\}$$

as a starting point.

As long as there is a crossing cut $\langle S, T \rangle$ for some $T \notin \Lambda$ and $S \in \Lambda$, add T to Λ. This process terminates since we can only add each set $T \in \mathcal{P}(V)$ into Λ once. All sets in Λ are minimum cuts. Definition 7.4.5.b. is satisfied for Λ.

The disjoint minimum cuts $\mathcal{F}(\Lambda)$ give us a partitioning of the graph. All sets in $\mathcal{F}(\Lambda)$ can be built by crossing cuts of minimum cuts in Λ. Therefore, each set in $\mathcal{F}(\Lambda)$ has exactly two neighbors, i.e., for each set $X \in \mathcal{F}(\Lambda)$, there exist exactly two different sets $Y, Z \in \mathcal{F}(\Lambda)$ such that $w(X, Y) = w(X, Z) = \frac{\lambda}{2}$ (Corollary 7.4.11). For all other sets $Z \in \mathcal{F}(\Lambda)$, $w(X, Z) = 0$. Since G is a connected graph, all sets in $\mathcal{F}(\Lambda)$ can be ordered, so that Definition 7.4.5.a. holds. Observe that Definition 7.4.5.b. is still true, since splitting the sets in Λ into smaller sets still allows a reconstruction of the sets in Λ. $\quad\square$

Lemma 7.4.13 ([63, 153]). *A graph $G = (V, E)$ has $\mathcal{O}\left(\binom{|V|}{2}\right)$ many minimum cuts and this bound is tight. This means that a graph can have $\Omega\left(\binom{|V|}{2}\right)$ many minimum cuts.*

Proof. The upper bound is a consequence of the last theorem. Given a graph $G = (V, E)$, the following recursive function Z describes the number of minimum cuts in G:

$$
Z(|V|) = \begin{cases}
\sum_{i=1}^{k} (Z(|V_i|)) + \binom{k}{2} & \begin{array}{l} \text{A circular partition} \\ V_1, \ldots, V_k \text{ exists in } G \end{array} \\
Z(|S|) + Z(|V - S|) + 1 & \begin{array}{l} \text{No circular partition, but a} \\ \text{minimum cut } S \text{ exists in } G \end{array} \\
0 & \text{otherwise}
\end{cases}
$$

It is easy to see that this function achieves the maximum in the case where a circular partition $W_1, \ldots, W_{|V|}$ exist. Therefore $Z(|V|) = \mathcal{O}\left(\binom{|V|}{2}\right)$.

The lower bound is achieved by a simple cycle of n vertices. There are $\Omega\left(\binom{n}{2}\right)$ pairs of edges. Each pair of edges defines another two minimum cuts S and $\bar{S}$. These two sets are separated by simply removing the pair of edges. $\quad\square$

7.5   Cactus Representation of All Minimum Cuts

In the following, a description of the *cactus* is given. First consider a graph $G = (V, E)$ without any circular partitions. Then due to the absence of all crossing cuts, all minimum cuts of G are laminar.

A set S of sets is called *laminar* if for every pair of sets $S_1, S_2 \in S$, either S_1 and S_2 are disjoint or S_1 is contained in S_2 or vice versa. Therefore each set $T \in S$ contained in some $S_1, S_2, \ldots \in S$ has a unique smallest superset. For clarity, we say that a tree has nodes and leaves, while a graph has vertices. Each laminar set S can be represented in a tree. Each node represents a set in S; the leaves represent the sets in S that contain no other sets of S. The parent of a node representing a set T represents the smallest superset of T. This construction ends with a set of trees called forest. Add an extra node r to the forest and connect all roots of the trees of the forest by an edge to this new node r, which is now the root of one big tree. Therefore, the nodes of one tree represent all sets of S, and the root of the tree represents the entire underlying set, i.e. the union of all elements of all $S \in S$. If this union has n elements, then such a tree can have at most n leaves and therefore at most $2n - 1$ nodes.

Since all minimum cuts G are laminar, these can be represented by a tree T_G defined as follows. Consider the smaller vertex set of every minimum cut. Denote this set of sets as Λ. If the vertex sets of a minimum cut are of same size, take one of these sets. Represent each set of Λ by a single node. Two nodes corresponding to minimum cuts A and B in G are connected by an edge if $A \subset B$ and there is no other minimum cut C such that $A \subset C \subset B$. The roots of the forest represent the minimum cuts in Λ that are contained in no other minimum cut in Λ. Again, connect all roots of the forest by an edge to a single extra node that we define as root of the tree.

Because removing one edge in the tree separates a subtree from the rest of the tree, let us define the following mapping: each vertex of the graph G is mapped to the node of the tree T_G that corresponds to the smallest cut containing this vertex. All vertices that are contained in no node of T_G are mapped to the root of T_G.

For each minimum cut S of G, the vertices of S are then mapped to some set of nodes X such that there is an edge and removing this edge separates the nodes X from the rest of the tree. Conversely, removing one edge from T_G separates the nodes of the tree into two parts such that the set of all vertices mapped into one part is a minimum cut.

If G has no circular partitions, the tree T_G is the *cactus* C_G for G. The number of nodes of a cactus is bounded by $2|V| - 1$.

Consider a graph $G = (V, E)$ that has only one circular partition $V_1, \ldots V_k$. The circular partition cuts can be represented by a circle of k nodes. For $1 \leq i \leq k$, the vertices of each part V_i are represented by one node N_i of the circle in such a way that two parts V_i and V_{i+1} are represented by two adjacent nodes.

Now we make use of the fact that for each minimum cut S that is no circular partition cut, either S or $\bar{S}$ is a proper subset of a V_i. Therefore, we can construct the tree $T_{(V_i, E)}$ for all minimum cuts that are a subset of V_i, but now with the restriction that only the vertices of V_i are mapped to this tree. The root of $T_{(V_i,E)}$ corresponds exactly to the set V_i. Thus we can merge node N_i of the circle and the root of $T_{(V_i,E)}$ for all $1 \le i \le k$. This circle connected with all the trees is the cactus C_G for G. The number of nodes is equal to the sum of all nodes in the trees $T_{(V_i,E)}$ with $1 \le i \le k$. Therefore, the number of nodes of the cactus is bounded by $2|V| - 1$ and again, there is a $1 - 1$ correspondence between minimum cuts in G and the separation of C_G into two parts.

Now consider a graph $G = (V, E)$ with the circular partitions $P_1, \ldots, P_z$. Take all circular partitions as a set of sets. Construct a cactus C_G representing the circular partition cuts of G in the following way.

The vertices of each set $F \in \mathcal{F}_{P_1 \cup \ldots \cup P_z}$ are mapped to one node and two nodes are connected, if for their corresponding sets F_1 and F_2, $w(F_1, F_2) > 0$. Then each circular partition creates one circle in C_G. Since all circular partitions are pairwise compatible, the circles are connected by edges that are not part of any circle. The cactus C_G is now a tree-like graph (Figure 7.8).

After representing the remaining minimum cuts that are not part of a circular partition, we get the cactus T_C for G. As before, the number of nodes of the cactus is bounded by $2|V| - 1$.

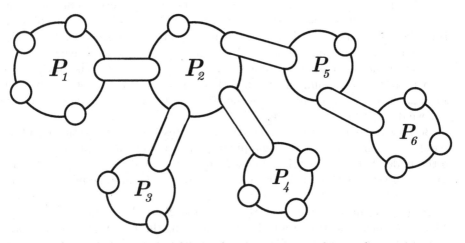

Fig. 7.8. A cactus representing the circular partition cuts of 6 circular partitions

7.6   Flow-Based Connectivity Algorithms

We distinguish algorithms that check k-vertex/edge-connectivity of a graph G for a given natural number k, and algorithms that compute the vertex/edge-connectivity $\kappa(G)$ or $\lambda(G)$ respectively. (A third kind of algorithms computes the maximal k-vertex/edge-connected subgraphs (k-components), which is the subject of discussion in Section 7.8.)

Most of the algorithms for computing vertex- or edge-connectivities are based on the computation of the maximum flow through a derived network. While the flow problem in undirected graphs can be reduced to a directed flow problem of comparable size [220], for the other direction only a reduction with increased capacities is known [478]. There were several algorithms published for the solution of (general) flow problems, see Table 7.1.

Table 7.1. The history of max-flow algorithms

1955 Dantzig & Fulkerson		[231, 141]
Network simplex method	$\mathcal{O}(n^2 m U)$	[140, 139]
1956 Ford & Fulkerson		[218, 219]
Augmenting path / Labeling	$\mathcal{O}(nmU)$	[220]
1969 Edmonds & Karp		[172]
Shortest augmenting path	$\mathcal{O}(nm^2)$	[593]
Capacity scaling	$\mathcal{O}(m^2 \log U)$	
1970 Dinitz		[150]
Layered network / blocking flow	$\mathcal{O}(n^2 m)$	
1973 Dinitz		[151, 234]
Capacity scaling	$\mathcal{O}(nm \log U)$	
1974 Karzanov		[350]
Preflow-push / layered network	$\mathcal{O}(n^3)$	
1977 Cherkassky	$\mathcal{O}(n^2 \sqrt{m})$	[122, 123]
1978 Malhotra, Kumar, Maheshwari	$\mathcal{O}(n^3)$	[406]
1978 Galil	$\mathcal{O}(n^{5/3} m^{2/3})$	[236]
1979 Galil & Naamad / Shiloach	$\mathcal{O}(nm(\log n)^2)$	[238, 518]
1980 Sleater & Tarjan		[525]
Dynamic trees	$\mathcal{O}(nm \log n)$	
1985 Goldberg		[249]
Push-relabel	$\mathcal{O}(n^3)$	
1986 Goldberg & Tarjan		[252]
Push-relabel	$\mathcal{O}(nm \log(n^2/m))$	
1987 Ahuja & Orlin		[7]
Excess scaling	$\mathcal{O}(nm + n^2 \log U)$	
1990 Cheriyan, Hagerup, Mehlhorn		[119]
Incremental algorithm	$\mathcal{O}(n^3 / \log n)$	
1990 Alon		[118, 20]
Derandomization	$\mathcal{O}(nm + n^{8/3} \log n)$	
1992 King, Rao, Tarjan		[118, 356]
Online game	$\mathcal{O}(nm + n^{2+\epsilon})$	
1993 Phillips & Westbrook		[476]
Online game	$\mathcal{O}(nm \log_{m/n} n + n^2 \log^{2+\varepsilon} n)$	
1998 Goldberg & Rao		[250]
Non-unit length function	$\mathcal{O}(\min(n^{2/3}, \sqrt{m}) m \log \frac{n^2}{m} \log U)$	

U denotes the largest possible capacity (integer capacities case only)

Better algorithms for the more restricted version of unit capacity networks exist.

Definition 7.6.1. *A network is said to be a* unit capacity network *(or 0-1 network) if the capacity is 1 for all edges. A unit capacity network is of* type 1 *if it has no parallel edges. It is called* type 2 *if for each vertex v ($v \neq s$, $v \neq t$) either the in-degree $d^-(v)$ or the out-degree $d^+(v)$ is only 1.*

Lemma 7.6.2. *1. For unit capacity networks, the computation of the maximum flow can be done (using Dinitz's algorithm) in $\mathcal{O}(m^{3/2})$.*

  2. For unit capacity networks of type 1, the time complexity of Dinitz's algorithm is $\mathcal{O}(n^{2/3}m)$.

  3. For unit capacity networks of type 2, the time complexity of Dinitz's algorithm is $\mathcal{O}(n^{1/2}m)$.

For a proof of the lemma see [188, 187, 349].

While the best bound for directed unit capacity flow problems differs only by logarithmic factors from the best known bound for integer capacities, even better bounds for the case of undirected unit capacity networks exist: $\mathcal{O}(\min(m, n^{3/2})\sqrt{m})$ by Goldberg and Rao [251], $\mathcal{O}(n^{7/6}m^{2/3})$ by Karger and Levine [343].

7.6.1 Vertex-Connectivity Algorithms

Table 7.2. The history of computing the vertex-connectivity κ

Year	Author(s)	MaxFlow calls	Compute κ	Ref.
1974	Even & Tarjan	$(\kappa+1)(n-\delta-1)$	$\mathcal{O}(\kappa n^{3/2}m)$ $\mathcal{O}(n^{1/2}m^2)$	[188]
1984	Esfahanian & Hakimi	$n-\delta-1+$ $\kappa(2\delta-\kappa-3)/2$	$\mathcal{O}((n-\delta+\kappa\delta-\kappa^2/2)\cdot$ $n^{2/3}m)$	[183]
1996	Henzinger, Rao, Gabow		$\mathcal{O}(\min\{\kappa^3+n, \kappa n\}\kappa n)$	[298]

Table 7.3. The history of checking vertex-connectivity

Year	Author(s)	MaxFlow calls	Check k-VC	Ref.
1969	Kleitman	$k(n-\delta)-\dbinom{k+1}{2}$	$\mathcal{O}(k^2n^3)$	[362]
1973	Even	$n-k+\dbinom{k}{2}$	$\mathcal{O}(k^3m+knm)$	[186]
1984	Esfahanian & Hakimi	$n-k+\dbinom{k-1}{2}$	$\mathcal{O}(k^3m+knm)$	[183]

The basis of all flow-based connectivity algorithms is a subroutine that computes the local connectivity between two distinct vertices s and t. Even [185, 186,

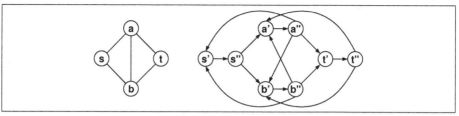

Fig. 7.9. Construction of the directed graph $\bar{G}$ that is derived from the undirected input graph G to compute the local vertex-connectivity $\kappa_G(s,t)$

187] presented a method for computing $\kappa_G(s,t)$ that is based on the following construction: For the given graph $G = (V, E)$ having n vertices and m edges we derive a directed graph $\bar{G} = (\bar{V}, \bar{E})$ with $|\bar{V}| = 2n$ and $|\bar{E}| = 2m+n$ by replacing each vertex $v \in V$ with two vertices $v', v'' \in \bar{V}$ connected by an (internal) edge $e_v = (v', v'') \in \bar{E}$. Every edge $e = (u, v) \in E$ is replaced by two (external) edges $e' = (u'', v'), e'' = (v'', u') \in \bar{E}$, see Figure 7.9.

$\kappa(s,t)$ is now computed as the maximum flow in $\bar{G}$ from source s'' to the target t' with unit capacities for all edges[2]. For a proof of correctness see [187]. For each pair $v', v'' \in \bar{V}$ representing a vertex $v \in V$ the internal edge (v', v'') is the only edge that emanates from v' and the only edge entering v'', thus the network $\bar{G}$ is of type 2. According to Lemma 7.6.2 the computation of the maximum flow resp. the local vertex-connectivity has time complexity $\mathcal{O}(\sqrt{n}m)$.

A trivial algorithm for computing $\kappa(G)$ could determine the minimum for the local connectivity of all pairs of vertices. Since $\kappa_G(s,t) = n - 1$ for all pairs (s,t) that are directly connected by an edge, this algorithm would make $\frac{n(n-1)}{2} - m$ calls to the flow-based subroutine. We will see that we can do much better.

If we consider a minimum vertex separator $S \subset V$ that separates a 'left' vertex subset $L \subset V$ from a 'right' subset $R \subset V$, we could compute $\kappa(G)$ by fixing one vertex s in either subset L or R and computing the local connectivities $\kappa_G(s,t)$ for all vertices $t \in V \setminus \{s\}$ one of which must lie on the other side of the vertex cut. The problem is: how to select a vertex s such that s does not belong to every minimum vertex separator? Since $\kappa(G) \leq \delta(G)$ (see Theorem 7.1.1), we could try $\delta(G) + 1$ vertices for s, one of which must not be part of all minimum vertex cuts. This would result in an algorithm of complexity $\mathcal{O}((\delta+1) \cdot n \cdot \sqrt{n}m)) = \mathcal{O}(\delta n^{3/2}m)$

Even and Tarjan [188] proposed Algorithm 13 that stops computing the local connectivities if the size of the current minimum cut falls below the number of examined vertices.

The resulting algorithm examines not more than $\kappa + 1$ vertices in the loop for variable i. Each vertex has at least $\delta(G)$ neighbors, thus at most $\mathcal{O}((n - \delta - 1)(\kappa + 1))$ calls to the maximum flow subroutine are carried out. Since $\kappa(G) \leq 2m/n$ (see Theorem 7.1.8), the minimum capacity is found not later than in call $2m/n + 1$. As a result, the overall time complexity is $\mathcal{O}(\sqrt{n}m^2)$.

---

[2] Firstly, Even used $c(e_v) = 1$, $c(e') = c(e'') = \infty$ which leads to the same results.

Algorithm 13: Vertex-connectivity computation by Even & Tarjan

---

Input   : An (undirected) graph $G = (V, E)$
Output: $\kappa(G)$

$\kappa_{\min} \leftarrow n - 1$
$i \leftarrow 1$
while $i \leq \kappa_{\min}$ **do**
    for $j \leftarrow i + 1$ **to** n **do**
        if $i > \kappa_{\min}$ **then**
        | **break**

        else if $\{v_i, v_j\} \notin E$ **then**
            compute $\kappa_G(v_i, v_j)$ using the MaxFlow algorithm
            $\kappa_{\min} \leftarrow \min\{\kappa_{\min}, \kappa_G(v_i, v_j)\}$

return $\kappa_{\min}$

Esfahanian and Hakimi [183] further improved the algorithm by the following observation:

Lemma 7.6.3. *If a vertex v belongs to all minimum vertex-separators then there are for each minimum vertex-cut S two vertices $l \in L_S$ and $r \in R_S$ that are adjacent to v.*

Proof. Assume v takes part in all minimum vertex-cuts of G. Consider the partition of the vertex set V induced by a minimum vertex-cut S with a component L (the 'left' side) of the remaining graph and the respective 'right' side R. Each side must contain at least one of v's neighbors, because otherwise v would not be necessary to break the graph into parts. Actually each side having more than one vertex must contain 2 neighbors since otherwise replacing v by the only neighbor would be a minimum cut without v, in contrast to the assumption.   □

These considerations suggest Algorithm 14. The first loop makes $n - \delta - 1$ calls to the MaxFlow procedure, the second requires $\kappa(2\delta - \kappa - 3)/2$ calls. The overall complexity is thus $n - \delta - 1 + \kappa(2\delta - \kappa - 3)/2$ calls of the maximum flow algorithm.

7.6.2   Edge-Connectivity Algorithms

Similar to the computation of the vertex-connectivity, the calculation of the edge-connectivity is based on a maximum-flow algorithm that solves the local edge-connectivity problem, i.e. the computation of $\lambda_G(s, t)$. Simply replace all undirected edges by pairs of antiparallel directed edges with capacity 1 and compute the maximum flow from the source s to the sink t. Since the resulting network is of type 1, the computation is, due to Lemma 7.6.2, of complexity $\mathcal{O}(\min\{m^{3/2}, n^{2/3}m\})$.

A trivial algorithm for computing $\lambda(G)$ could simply calculate the minimum of the local edge-connectivities for all vertex pairs. This algorithm would thus make $n(n - 1)/2$ calls to the MaxFlow subroutine. We can easily improve the

---

Algorithm 14: Vertex-connectivity computation by Esfahanian & Hakimi

---

Input : An (undirected) graph $G = (V, E)$
Output: $\kappa(G)$

$\kappa_{\min} \leftarrow n - 1$
Choose $v \in V$ having minimum degree, $d(v) = \delta(G)$
Denote the neighbors $N(v)$ by $v_1, v_2, \ldots, v_\delta$

foreach *non-neighbor* $w \in V \setminus (N(v) \cup \{v\})$ **do**
$\quad$ compute $\kappa_G(v, w)$ using the MaxFlow algorithm
$\quad$ $\kappa_{\min} \leftarrow \min\{\kappa_{\min}, \kappa_G(v, w)\}$

$i \leftarrow 1$
while $i \leq \kappa_{\min}$ **do**
$\quad$ **for** $j \leftarrow i + 1$ *to* $\delta - 1$ **do**
$\quad\quad$ **if** $i \geq \delta - 2$ *or* $i \geq \kappa_{\min}$ **then**
$\quad\quad\quad$ **return** $\kappa_{\min}$

$\quad\quad$ **else if** $\{v, w\} \notin E$ **then**
$\quad\quad\quad$ compute $\kappa_G(v_i, v_j)$ using the MaxFlow algorithm
$\quad\quad\quad$ $\kappa_{\min} \leftarrow \min\{\kappa_{\min}, \kappa_G(v_i, v_j)\}$

$\quad$ $i \leftarrow i + 1$
return $\kappa_{\min}$

---

complexity of the algorithm if we consider only the local connectivities $\lambda_G(s, t)$ for a single (fixed) vertex s and all other vertices t. Since one of the vertices $t \in V \setminus \{s\}$ must be separated from s by an arbitrary minimum edge-cut, $\lambda(G)$ equals the minimum of all these values. The number of MaxFlow calls is thereby reduced to $n - 1$. The overall time complexity is thus $\mathcal{O}(nm \cdot \min\{n^{2/3}, m^{1/2}\})$ (see also [188]). The aforementioned algorithm also works if the whole vertex set is replaced by a subset that contains two vertices that are separated by some minimum edge-cut. Consequently, the next algorithms try to reduce the size of this vertex set (which is called a λ-*covering*). They utilize the following lemma. Let S be a minimum edge-cut of a graph $G = (V, E)$ and let $L, R \subset V$ be a partition of the vertex set such that L and R are separated by S.

Lemma 7.6.4. *If $\lambda(G) < \delta(G)$ then each component of $G - S$ consists of more than $\delta(G)$ vertices, i.e. $|L| > \delta(G)$ and $|R| > \delta(G)$.*

Table 7.4. The history of edge-connectivity algorithms

Year	Author(s)	MaxFlow calls	Check k-EC Compute λ
1975	Even, Tarjan [188]		
		$n - 1$	$\mathcal{O}(nm \cdot \min\{n^{2/3}, m^{1/2}\})$
1984	Esfahanian, Hakimi [183]		
		$< n/2$	$\mathcal{O}(\lambda nm)$
1987	Matula [413]		$\mathcal{O}(kn^2)$
			$\mathcal{O}(\lambda n^2)$

Proof. Let the elements of L be denoted by $\{l_1, l_2, \ldots, l_k\}$ and denote the induced edges by $E[L] = E(G[L])$.

$$
\begin{aligned}
\delta(G) \cdot k &\leq \sum_{i=1}^{k} d_G(l_i) \\
&\leq 2 \cdot |E[L]| + |S| \\
&\leq 2 \cdot \frac{k(k-1)}{2} + |S| \\
&< k(k-1) + \delta(G)
\end{aligned}
$$

From $\delta(G) \cdot (k-1) < k(k-1)$ we conclude $|L| = k > 1$ and $|L| = k > \delta(G)$ (as well as $|R| > \delta(G)$). $\square$

Corollary 7.6.5. *If $\lambda(G) < \delta(G)$ then each component of $G - S$ contains a vertex that is not incident to any of the edges in S.*

Lemma 7.6.6. *Assume again that $\lambda(G) < \delta(G)$. If T is a spanning tree of G then all components of $G - S$ contain at least one vertex that is not a leaf of T (i.e. the non-leaf vertices of T form a λ-covering).*

Proof. Assume the converse, that is all vertices in L are leaves of T. Thus no edge of T has both ends in L, i.e. $|L| = |S|$. Lemma 7.6.4 immediately implies that $\lambda(G) = |S| = |L| > \delta(G)$, a contradiction to the assumption. $\square$

Lemma 7.6.6 suggests an algorithm that first computes a spanning tree of the given graph, then selects an arbitrary inner vertex v of the tree and computes the local connectivity $\lambda(v, w)$ to each other non-leaf vertex w. The minimum of these values together with $\delta(G)$ yields exactly the edge connectivity $\lambda(G)$. This algorithm would profit from a larger number of leaves in T but, unfortunately, finding a spanning tree with maximum number of leaves is $\mathcal{NP}$-hard. Esfahanian

---

Algorithm 15: Spanning tree computation by Esfahanian & Hakimi

---

Input : An (undirected) graph $G = (V, E)$
Output: Spanning Tree T with a leaf and an inner vertex in L and R, resp.

Choose $v \in V$
$T \leftarrow$ all edges incident at v
while $|E(T)| < n - 1$ **do**
  Select a leaf w in T such that for all leaves r in T:
  $|N(w) \cap (V - V(T))| \geq |N(r) \cap (V - V(T))|$
  $T \leftarrow T \cup G[w \cup \{N(w) \cap (V - V(T))\}]$
return T

---

and Hakimi [183] proposed an algorithm for computing a spanning tree T of G such that both, L and R of some minimum edge separator contain at least one leaf of T, and due to Lemma 7.6.6 at least one inner vertex (see Algorithm 15). The edge-connectivity of the graph is then computed by Algorithm 16. Since P is

---

Algorithm 16: Edge-connectivity computation by Esfahanian & Hakimi

---

Input : An (undirected) graph $G = (V, E)$
Output: $\lambda(G)$

Construct a spanning tree T using Algorithm 15
Let P denote the smaller of the two sets, either the leaves or the inner nodes of T
Select a vertex $u \in P$
$c \leftarrow \min\{\lambda_G(u, v) : v \in P \setminus \{u\}\}$
$\lambda \leftarrow \min(\delta(G), c)$
return λ

---

chosen to be the smaller of both sets, leaves and non-leaves, the algorithm requires at most $n/2$ calls to the computation of a local connectivity, which yields an overall complexity of $\mathcal{O}(\lambda mn)$.

This could be improved by Matula [413], who made use of the following lemma.

Lemma 7.6.7. *In case $\lambda(G) < \delta(G)$, each dominating set of G is also a λ-covering of G.*

Similar to the case of the spanning tree, the edge-connectivity can now be computed by choosing a dominating set D of G, selecting an arbitrary vertex $u \in D$, and calculating the local edge-connectivities between u and all other vertices in D. The minimum of all values together with the minimum degree $\delta(G)$ gives the result. While finding a dominating set of minimum cardinality is $\mathcal{NP}$-hard in general, the connectivity algorithm can be shown to run in time $\mathcal{O}(nm)$ if the dominating set is chosen according to Algorithm 17.

Algorithm 17: Dominating set computation by Matula

Input : An (undirected) graph $G = (V, E)$
Output: A dominating set D

Choose $v \in V$
$D \leftarrow \{v\}$
while $V \setminus (D \cup N(D)) \neq \emptyset$ **do**
$\quad$ Select a vertex $w \in V \setminus (D \cup N(D))$
$\quad$ $D \leftarrow D \cup \{w\}$
return D

7.7 Non-flow-based Algorithms

We consider now connectivity algorithms that are not based on network flow techniques.

7.7.1   The Minimum Cut Algorithm of Stoer and Wagner

In 1994 an algorithm for computing a minimum capacity cut of an edge-weighted graph was published by Stoer and Wagner [536]. It was unusual not only due to the fact that it did not use any maximum flow technique as a subroutine. Somewhat surprisingly, the algorithm is very simple in contrast to all other algorithms (flow-based and non-flow-based) that were published so far. In principle, each phase of the algorithm is very similar to Prim's minimum spanning tree algorithm and Dijkstra's shortest path computation, which leads to an equivalent running time of $\mathcal{O}(m + n \log n)$ per phase and overall time complexity of $\mathcal{O}(nm + n^2 \log n)$.

Algorithm 18: Minimum capacity cut computation by Stoer & Wagner

---

Input   : An undirected graph $G = (V, E)$
Output: A minimum cut $C_{\min}$ corresponding to $\lambda(G)$

Choose an arbitrary start vertex a
$C_{\min} \leftarrow$ undefined
$V' \leftarrow V$
while $|V'| > 1$ **do**
    $A \leftarrow \{a\}$
    while $A \neq V'$ **do**
        Add to A the most tightly connected vertex
        Adjust the capacities between A and the vertices in $V' \setminus A$
    $C :=$ cut of V' that separates the vertex added last to A from the rest of
    the graph
    if $C_{\min} =$ undefined *or* $w(C) < w(C_{\min})$ **then**
        $C_{\min} \leftarrow C$
    Merge the two vertices that were added last to A
return $C_{\min}$

After choosing an arbitrary start vertex a, the algorithm maintains a vertex subset A that is initialized with the start vertex and that grows by repeatedly adding a vertex $v \notin A$ that has a maximum sum of weights for its connections to vertices in A. If all vertices have been added to A, the last two vertices s and t are merged into one. While edges between s and t are simply deleted by the contraction, all edges from s and t to another vertex are replaced by an edge weighted with the sum of the old weights. The cut that separates the vertex added last from the rest of the graph is called the *cut-of-the-phase*.

Lemma 7.7.1. *The cut-of-the-phase is a minimum s-t-cut in the current (modified) graph, where s and t are the two vertices added last to A in the phase.*

Proof. Consider an arbitrary s-t-cut C for the last two vertices. A vertex $v \neq a$ is called *active* if v and its immediate predecessor with respect to the addition to A reside in different parts of C. Let A_v be the set of vertices that are in A just before v is added and let $w(S, v)$ for a vertex set S denote the capacity sum of all edges between v and the vertices in S.

The proof shows, by induction on the active vertices, that for each active vertex v the adjacency to the vertices added before (A_v) does not exceed the weight of the cut of $A_v \cup \{v\}$ induced by C (denoted by C_v). Thus it is to prove that

$$w(A_v, v) \leq w(C_v)$$

For the base case, the inequality is satisfied since both values are equal for the first active vertex. Assuming now that the proposition is true for all active vertices up to active vertex v, the value for the next active vertex u can be written as

$$
\begin{aligned}
w(A_u, u) &= w(A_v, u) + w(A_u \setminus A_v, u) \\
&\leq w(A_v, v) + w(A_u \setminus A_v, u) && (w(A_v, u) \leq w(A_v, v)) \\
&\leq w(C_v) + w(A_u \setminus A_v, u) && \text{(by induction assumption)} \\
&\leq w(C_u)
\end{aligned}
$$

The last line follows because all edges between $A_u \setminus A_v$ and u contribute their weight to $w(C_u)$ but not to $w(C_v)$.

Since t is separated by C from its immediate predecessor s, it is always an active vertex; thus the conclusion $w(A_t, t) \leq w(C_t)$ completes the proof.   □

Theorem 7.7.2. *A cut-of-the-phase having minimum weight among all cuts-of-the-phase is a minimum capacity cut of the original graph.*

Proof. For the case where the graph consists of only 2 vertices, the proof is trivial. Now assume $|V| > 2$. The following two cases can be distinguished:

1. Either the graph has a minimum capacity cut that is also a minimum s-t-cut (where s and t are the vertices added last in the first phase), then, according to Lemma 7.7.1, we conclude that this cut is a minimum capacity cut of the original graph.
2. Otherwise the graph has a minimum cut where s and t are on the same side. Therefore the minimum capacity cut is not affected by merging the vertices s and t.

Thus, by induction on the number of vertices, the minimum capacity cut of the graph is the cut-of-the-phase having minimum weight.   □

7.7.2 Randomized Algorithms

In 1982, Becker et al. [53] proposed a probabilistic variant of the Even/Tarjan vertex connectivity algorithm [188]. It computes the vertex connectivity of an undirected graph G with error probability at most ε in expected time $\mathcal{O}((-\log \varepsilon)n^{3/2}m)$ provided that $m \leq \frac{1}{2}dn^2$ for some constant $d < 1$. This improved the computation of κ for sparse graphs.

A few years later, Linial, Lovasz and Wigderson provided probabilistic algorithms [392, 393] that were based on a geometric, algebraic and physical interpretation of graph connectivity. As a generalization of the notion of s-t-numbering, they showed that a graph G is k-connected if and only if it has a certain non-degenerate convex embedding in $\mathbb{R}^{k-1}$, i.e., specifying any k vertices of G, the

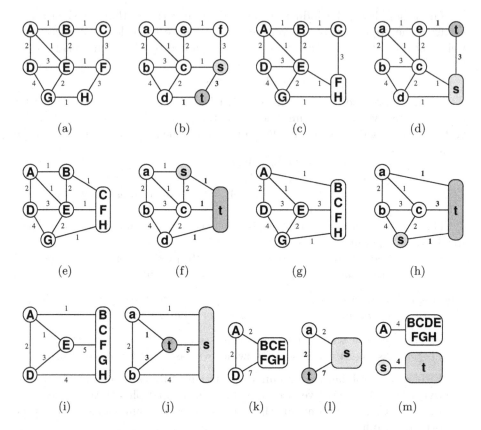

Fig. 7.10. Example for the Stoer/Wagner algorithm. Upper case letters are vertex names, lower case letters show the order of addition to the set S. The minimum cut $\{ABDEG\} \mid \{CFH\}$ has capacity 3 and is found in Part 7.10(f) (third phase)

vertices of G can be represented by points of $\mathbb{R}^{k-1}$ such that no k are in a hyperplane and each vertex is in the convex hull of its neighbors, except for the k specified vertices. As a result, they proposed a Monte-Carlo algorithm running in time $\mathcal{O}(n^{2.5} + n\kappa^{2.5})$ (that errs with probability less than $1/n$) and a Las Vegas algorithm with expected runtime of $\mathcal{O}(n^{2.5} + n\kappa^{3.5})$.

A subsequent work of Cheriyan and Reif [120] generalized this approach to directed graphs, which yielded a Monte Carlo algorithm with running time $\mathcal{O}((M(n) + nM(k)) \cdot \log n)$ and error probability $< 1/n$, and a Las Vegas algorithm with expected time $\mathcal{O}((M(n) + nM(k)) \cdot k)$, where $M(n)$ denotes the complexity for the multiplication of $n \times n$ matrices.

Henzinger, Rao and Gabow [298] further improved the complexities by giving an algorithm that computes the vertex connectivity with error probability at most $1/2$ in (worst-case) time $\mathcal{O}(nm)$ for digraphs and $\mathcal{O}(\kappa n^2)$ for undirected graphs. For weighted graphs they proposed a Monte Carlo algorithm that has error probability $1/2$ and expected running time $\mathcal{O}(nm \log(n^2/m))$.

7.8 Basic Algorithms for Components

Super-linear algorithms for the computation of the blocks and the cut-vertices as well as for the computation of the strongly connected components of a graph were proposed in [470] and [386, 484, 485, 435], respectively. Later on, linear time algorithms were published by Hopcroft and Tarjan [311, 542].

7.8.1 Biconnected Components

A problem that arises from the question which nodes of a network always remain connected in case one arbitrary node drops out is the computation of the *biconnected (or non-separable) components* of a graph, also called *blocks*.

Let us consider a depth-first search in an undirected and connected graph $G = (V, E)$ where we label the traversed vertices with consecutive numbers from 1 to $n = |V|$ using a pre-order numbering num. We observe that we inspect two kinds of edges: the ones that lead to unlabeled vertices become *tree edges*, and the ones that lead to vertices that were already discovered and labeled in a former step we call *backward edges*.

For each vertex v we keep the smallest label of any vertex that is reachable via arbitrary tree edges followed by not more than one backward edge, i.e. the smallest number of any vertex that lies on some cycle with v. Whenever a new vertex is discovered by the DFS, the low-entry of that vertex is initialized by its own number.

If we return from a descent to a child w – i.e. from a tree edge (v, w) –, we update low[v] by keeping the minimum of the child's entry low[w] and the current value low[v].

If we discover a backward edge (v, w), we update low[v] to be the minimum of its old value and the label of w.

To detect the cut-vertices of the graph we can now utilize the following lemma:

Lemma 7.8.1. *We follow the method described above for computing the values of* low *and* num *during a DFS traversal of the graph G. A vertex v is a cut-vertex if and only if one of the following conditions holds:*

1. *if v is the root of the DFS tree and is incident to at least 2 DFS tree edges,*
2. *if v is not the root, but there is a child w of v such that* low[w] ≥ num[v].

Proof. 1. Assume that v is the root of the DFS tree.
   → If v is incident to more than one tree edge, the children would be disconnected by removing vertex v from G.

← If v is a cut-vertex then there are vertices $x, y \in V$ that are disconnected by removing v, i.e. v is on every path connecting x and y. W.l.o.g. assume that the DFS discovers x before y. y can only be discovered after the descent to x returned to v, thus we conclude that v has at least two children in the DFS tree.

2. Assume now that v is not the root of the DFS tree.

→ If there is a child w of v such that $\texttt{low}[w] \geq \texttt{num}[v]$ this means that there is only one path connecting this successor w with all ancestors of v. Thus v is a cut-vertex.

← If v is a cut-vertex, there are vertices $x, y \in V$ such that v is on every path connecting x and y. If all children of v had an indirect connection (via arbitrary tree edges followed by one backward edge) to any ancestor of v the remaining graph would be connected. Therefore one of the children must have $\texttt{low}[w] \geq \texttt{num}[v]$.

This concludes the proof.   □

To find the biconnected components, i.e. the partition of the edges, we put every new edge on a stack. Whenever the condition $\texttt{low}[w] \geq \texttt{num}[v]$ holds after returning from a recursive call for a child w of v, the edges on top of stack including edge (v, w) form the next block (and are therefore removed from the stack).

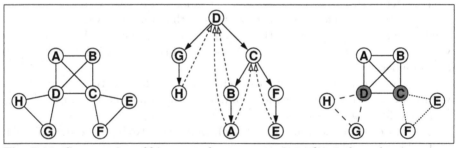

Fig. 7.11. Computation of biconnected components in undirected graphs. Left: the undirected input graph.   Middle: dfs tree with forward (straight) and backward (dashed) edges.   Right: the blocks and articulation nodes of the graph.

7.8.2 Strongly Connected Components

We now consider the computation of the strong components, i.e. the maximal strongly connected subgraphs in directed graphs (see Section 2.2.1). Analogously to the computation of biconnected components in undirected graphs, we use a modified depth-first search that labels the vertices by consecutive numbers from 1 to n. In case the traversal ends without having discovered all vertices we have to restart the DFS at a vertex that has not been labeled so far. The result is a spanning forest F.

The edges $e = (v, w)$ that are inspected during the DFS traversal are divided into the following categories:

1. All edges that lead to unlabeled vertices are called *tree edges* (they belong to the trees of the DFS forest).
2. The edges that point to a vertex w that was already labeled in a former step fall into the following classes:
   a) If $\texttt{num}[w] > \texttt{num}[v]$ we call e a *forward edge*.
   b) Otherwise, if w is an ancestor of v in the same DFS tree we call e a *backward edge*.
   c) Otherwise e is called a *cross edge* (because it points from one subtree to another).

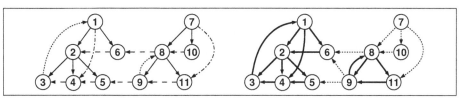

Fig. 7.12. DFS forest for computing strongly connected components in directed graphs: tree, forward, backward, and cross edges

An example is shown in Figure 7.12.

Two vertices v, w are in the same strong component if and only if there exist directed paths from v to w and from w to v. This induces an equivalence relation as well as a partition of the vertex set (in contrast to biconnected components where the edge set is partitioned while vertices may belong to more than one component).

During the DFS traversal we want to detect the roots of the strong components, i.e. in each component the vertex with smallest DFS label. As in the case of the biconnected components we must decide for each descendant w of a vertex v whether there is also a directed path that leads back from w to v. Now we define $\texttt{lowlink}[v]$ to be the smallest label of any vertex in the same strong component that can be reached via arbitrarily many tree arcs followed by at most one backward or cross edge.

Lemma 7.8.2. *A vertex v is the root of a strong component if and only if both of the following conditions are met:*

1. *There is no backward edge from v or one of its descendants to an ancestor of v.*
2. *There is no cross edge (v, w) from v or one of its descendants to a vertex w such that the root of w's strong component is an ancestor of v.*

This is equivalent with the decision whether $\texttt{lowlink}[v] = \texttt{num}[v]$.

Proof. → Assume conversely that the condition holds but u is the root of v's strong component with $u \neq v$. There must exist a directed path from v to u. The first edge of this path that points to a vertex w that is not a descendant of v in the DFS tree is a back or a cross edge. This implies $\text{lowlink}[v] \leq \text{num}[w] < \text{num}[v]$, since the highest numbered common ancestor of v and w is also in this strong component.

← If v is the root of some strong component in the actual spanning forest, we may conclude that $\text{lowlink}[v] = \text{num}[v]$. Assuming the opposite (i.e. $\text{lowlink}[v] < \text{num}[v]$), some proper ancestor of v would belong to the same strong component. Thus v would not be the root of the SCC.

This concludes the proof.   □

If we put all discovered vertices on a stack during the DFS traversal (similar to the stack of edges in the computation of the biconnected components) the lemma allows us to 'cut out' the strongly connected components of the graph.

It is apparent that the above algorithms share their similarity due to the fact that they are based on the detection of cycles in the graph. If arbitrary instead of simple cycles (for biconnected components) are considered, this approach yields a similar third algorithm that computes the bridge- (or 2-edge-) connected components (published by Tarjan [544]).

7.8.3   Triconnectivity

First results on graph triconnectivity were provided by Mac Lane [403] and Tutte [555, 556]. In the sixties, Hopcroft and Tarjan published a linear time algorithm for dividing a graph into its triconnected components that was based on depth-first search [309, 310, 312]. Miller and Ramachandran [422] provided another algorithm based on a method for finding open ear decompositions together with an efficient parallel implementation. It turned out that the early Hopcroft/Tarjan algorithm was incorrect, which was then modified by Gutwenger and Mutzel [267]. They modified the faulty parts to yield a correct linear time implementation of SPQR-trees. We now briefly review their algorithm.

Definition 7.8.3. *Let $G = (V, E)$ be a biconnected (multi-) graph. Two vertices $a, b \in V$ are called a* separation pair *of G if the induced subgraph on the vertices $V \setminus \{a, b\}$ is not connected.*

The pair (a, b) partitions the edges of G into equivalence classes $E_1, \ldots, E_k$ (*separation classes*), s.t. two edges belong to the same class exactly if both lie on some path p that contains neither a nor b as an inner vertex, i.e. if it contains a or b it is an end vertex of p. The pair (a, b) is a separation pair if there are at least two separation classes, except for the following special cases: there are exactly two separation classes, and one of them consists of a single edge, or if there are exactly three separation classes that all consist of a single edge. The graph G is triconnected if it contains no separation pair.

Definition 7.8.4. *Let (a, b) be a separation pair of a biconnected multigraph G and let the separation classes $E_{1..k}$ be divided into two groups $E' = \bigcup_{i=1}^{l} E_i$ and $E'' = \bigcup_{i=l+1}^{k} E_i$, s.t. each group contains at least two edges. The two graphs $G' = (V(E' \cup e), E' \cup e)$ and $G'' = (V(E'' \cup e), E'' \cup e)$ that result from dividing the graph according to the partition $[E', E'']$ and adding the new virtual edge $e = (a, b)$ to each part are called* split graphs *of G (and they are again biconnected). If the split operation is applied recursively to the split graphs, this yields the (not necessarily unique)* split components *of G.*

Every edge in E is contained in exactly one, and each virtual edge in exactly two split components.

Lemma 7.8.5. *Let $G = (V, E)$ be a biconnected multigraph with $|E| \geq 3$. Then the total number of edges contained in all split components is bounded by $3|E| - 6$.*

Proof. Induction on the number of edges of G: If $|E| = 3$, G cannot be split and the lemma is true. Assume now, the lemma is true for graphs having at most $m - 1$ edges. If the graph has m edges, the lemma is obviously true if G cannot be split. Otherwise G can be split into two graphs having $k + 1$ and $m - k + 1$ edges with $2 \leq k \leq m - 2$. By the assumption, the total number of edges is bounded by $3(k + 1) - 6 + 3(m - k + 1) - 6 = 3m - 6$. Thus, by induction on the number of edges, the proof is complete.   □

There are split components of three types: triple bonds (three edges between two vertices), triangles (cycles of length 3), and triconnected simple graphs. We now introduce the reverse of the split operation: the *merge graph* of two graphs $G_1 = (V_1, E_1)$ and $G_2 = (V_2, E_2)$, both containing the same virtual edge e, is defined as $G = (V_1 \cup V_2, (E_1 \cup E_2) \setminus \{e\})$. The *triconnected components* of a graph are obtained from its split components by merging the triple bonds as much as possible to multiple bonds and by merging the triangles as much as possible to form polygons. Mac Lane [403] showed that, regardless of the (possibly not unique) splitting and merging, we get the same triconnected components.

Lemma 7.8.6. *The triconnected components of a (multi)graph are unique.*

We now turn to the definition of SPQR-trees, which were initially defined for planar [143], later also for general graphs [144]. A *split pair* of a biconnected graph G is either a separation pair or a pair of adjacent vertices. A *split component* of a split pair $\{u, v\}$ is either an (u, v)-edge or an inclusion-maximal subgraph of G, were $\{u, v\}$ is not a split pair. A split pair $\{u, v\}$ of G is called a *maximal split pair* with respect to a split pair $\{s, t\}$ of G if for any other split pair $\{u', v'\}$, the vertices u, v, s, and t are in the same split component.

Definition 7.8.7. *Let $e = (s, t)$ be an edge of G. The* SPQR-tree $\mathcal{T}$ *of G with respect to this reference edge is a rooted ordered tree constructed from four different types of nodes (S,P,Q,R), each containing an associated biconnected multigraph (called the* skeleton*). $\mathcal{T}$ is recursively defined as follows:*

(Q) Trivial Case: If G consists of exactly two parallel s-t-edges, then $\mathcal{T}$ is a single Q-node with skeleton G.

(P) *Parallel Case: If the split pair $\{s,t\}$ has more than two split components $G_{1..k}$, the root of T is a P-node with a skeleton consisting of k parallel s-t-edges $e_{1..k}$ with $e_1 = e$.*

(S) *Series Case: If the split pair $\{s,t\}$ has exactly two split components, one of them is e; the other is denoted by G'. If G' has cut-vertices $c_{1..k-1}(k \geq 2)$ that partition G into blocks $G_{1..k}$ (ordered from s to t), the root of T is an S-node, whose skeleton is the cycle consisting of the edges $e_{0..k}$, where $e_0 = e$ and $e_i = (c_{i-1}, c_i)$ with $i = 1..k$, $c_0 = s$ and $c_k = t$.*

(R) *Rigid Case: In all other cases let $\{s_1, t_1\}, .., \{s_k, t_k\}$ be the maximal split pairs of G with respect to $\{s,t\}$. Further let G_i for $i = 1, .., k$ denote the union of all split components of $\{s_i, t_i\}$ except the one containing e. The root of T is an R-node, where the skeleton is created from G by replacing each subgraph G_i with the edge $e_i = (s_i, t_i)$.*

For the non-trivial cases, the children $\mu_{1..k}$ of the node are the roots of the SPQR-trees of $G_i \cup e_i$ with respect to e_i. The vertices incident with each edge e_i are the poles of the node μ_i, the virtual edge of node μ_i is the edge e_i of the node's skeleton. The SPQR-tree T is completed by adding a Q-node as the parent of the node, and thus the new root (that represents the reference edge e).

Each edge in G corresponds with a Q-node of T, and each edge e_i in the skeleton of a node corresponds with its child μ_i. T can be rooted at an arbitrary Q-node, which results in an SPQR-tree with respect to its corresponding edge.

Theorem 7.8.8. *Let G be a biconnected multigraph with SPQR-tree T.*

1. *The skeleton graphs of T are the triconnected components of G. P-nodes correspond to bonds, S-nodes to polygons, and R-nodes to triconnected simple graphs.*
2. *There is an edge between two nodes $\mu, \nu \in T$ if and only if the two corresponding triconnected components share a common virtual edge.*
3. *The size of T, including all skeleton graphs, is linear in the size of G.*

For a sketch of the proof, see [267].

We consider now the computation of SPQR-trees for a biconnected multigraph G (without self-loops) and a reference edge e_r. We assume a labeling of the vertices by unique indices from 1 to $|V|$. As a preprocessing step, all edges are reordered (using bucket sort), first according to the incident vertex with the lower index, and then according to the incident vertex with higher index, such that multiple edges between the same pair of vertices are arranged successively. In a second step, all such bundles of multiple edges are replaced by a new virtual edge. In this way a set of multiple bonds $C_1, .., C_k$ is created together with a simple graph G'.

In the second step, the split components $C_{k+1}, .., C_m$ of G' are computed using a dfs-based algorithm. In this context, we need the following definition:

Definition 7.8.9. *A palm tree P is a directed multigraph that consists of a set of tree arcs $v \rightarrow w$ and a set of fronds $v \hookrightarrow w$, such that the tree arcs form*

*a directed spanning tree of P (that is the root has no incoming edges, all other
vertices have exactly one parent), and if $v \hookrightarrow w$ is a frond, then there is a directed
path from w to v.*

Suppose now, P is a palm tree for the underlying simple biconnected graph
$G' = (V, E')$ (with vertices labeled $1, .., |V|$). The computation of the separation
pairs relies on the definition of the following variables:

$$\text{lowpt1}(v) = \min \left(\{v\} \cup \{w | v \xrightarrow{*} \hookrightarrow w\} \right)$$

$$\text{lowpt2}(v) = \min \left(\{v\} \cup \left(\{w | v \xrightarrow{*} \hookrightarrow w\} \setminus \{\text{lowpt1}(v)\} \right) \right)$$

These are the two vertices with minimum label, that are reachable from v by
traversing an arbitrary number (including zero) of tree arcs followed by exactly
one frond of P (or v itself, if no such option exists).

Let $\text{Adj}(v)$ denote the ordered adjacency list of vertex v, and let $D(v)$ be the
set of descendants of v (that is the set of vertices that are reachable via zero or
more directed tree arcs). Hopcroft and Tarjan [310] showed a simple method for
computing an *acceptable adjacency structure*, that is, an order of the adjacency
lists, which meets the following conditions:

1. The root of P is the vertex labeled with 1.
2. If $w_1, .., w_n$ are the children of vertex v in P according to the ordering in
   $\text{Adj}(v)$, then $w_i = v + |D(w_{i+1} \cup .. \cup D(w_n)| + 1$,
3. The edges in $\text{Adj}(v)$ are in ascending order according to $\text{lowpt1}(w)$ for tree
   edges $v \to w$, and w for fronds $v \hookrightarrow w$, respectively.
   Let $w_1, .., w_n$ be the children of v with $\text{lowpt1}(w_i)) = u$ ordered according
   to $\text{Adj}(v)$, and let i_0 be the index such that $\text{lowpt2}(w_i) < v$ for $1 \le i \le i_0$
   and $\text{lowpt2}(w_j) \ge v$ for $i_0 < j \le n$. Every frond $v \hookrightarrow w \in E'$ resides between
   $v \to w_{i_0}$ and $v \to w_{i_0+1}$ in $\text{Adj}(v)$.

An adequate rearrangement of the adjacency structure can be done in linear
time if a bucket sort with $3|V| + 2$ buckets is applied to the following sorting
function (confer [310, 267]), that maps the edges to numbers from 1 to $3|V| + 2$:

$$\phi(e) = \begin{cases} 3\text{lowpt1}(w) & \text{if } e = v \to w \text{ and } \text{lowpt2}(w) < v \\ 3w + 1 & \text{if } e = v \hookrightarrow w \\ 3\text{lowpt1}(w) + 2 & \text{if } e = v \to w \text{ and } \text{lowpt2}(w) \ge v \end{cases}$$

If we perform a depth-first search on G' according to the ordering of the edges
in the adjacency list, then this partitions G' into a set of paths, each consisting
of zero or more tree arcs followed by a frond, and each path ending at the vertex
with lowest possible label. We say that a vertex u_n is a *first descendant* of u_0 if
there is a directed path $u_0 \to \cdots \to u_n$ and each edge $u_i \to u_{i+1}$ is the first in
$\text{Adj}(u_i)$.

Lemma 7.8.10. *Let P be a palm tree of a biconnected graph $G = (V, E)$ that
satisfies the above conditions. Two vertices $a, b \in V$ with $a < b$ form a separation
pair $\{a, b\}$ if and only if one of the following conditions is true:*

Type-1 Case There are distinct vertices $r, s \in V \setminus \{a, b\}$ such that $b \to r$ is a tree edge, $lowpt1(r) = a$, $lowpt2(r) \geq b$, and s is not a descendant of r.

Type-2 Case There is a vertex $r \in V \setminus b$ such that $a \to r \overset{*}{\to} b$, b is a first descendant of r (i.e., a, r, b lie on a generated path), $a \neq 1$, every frond $x \hookrightarrow y$ with $r \leq x < b$ satisfies $a \leq y$, and every frond $x \hookrightarrow y$ with $a < y < b$ and $b \to w \overset{*}{\to} x$ has $lowpt1(w) \geq a$.

Multiple Edge Case (a, b) is a multiple edge of G and G contains at least four edges.

For a proof, see [310].

We omit the rather technical details for finding the split components $C_{k+1}, .., C_m$. The main loop of the algorithm computes the triconnected components from the split components $C_1, .., C_m$ by merging two bonds or two polygons that share a common virtual edge (as long as they exist). The resulting time complexity is $\mathcal{O}(|V| + |E|)$. For a detailed description of the algorithm we refer the interested reader to the original papers [309, 310, 312, 267].

7.9   Chapter Notes

In this section, we briefly discuss some further results related to the topic of this chapter.

Strong and biconnected components. For the computation of strongly connected components, there is another linear-time algorithm that was suggested by R. Kosaraju in 1978 (unpublished, see [5, p. 229]) and that was published by Sharir [517].

An algorithm for computing the strongly connected components using a non-dfs traversal (a mixture of dfs and bfs) of the graph was presented by Jiang [331]. This algorithm reduces the number of disk operations in the case where a large graph does not entirely fit into the main memory. Two space-saving versions of Tarjan's strong components algorithm (for the case of graphs that are sparse or have many single-node components) were given by Nuutila and Soisalon-Soininen [454].

One-pass algorithms for biconnected and strong components that do not compute auxiliary quantities based on the dfs tree (e.g., `low` values) were proposed by Gabow [235].

Average connectivity. Only recently, Beineke, Oellermann, and Pippert [56] considered the concept of average connectivity. This measure is defined as the average, over all pairs of vertices $a, b \in V$, of the maximum number of vertex-disjoint paths between a and b, that is, the average local vertex-connectivity. While the conventional notion of connectivity is rather a description of a worst case scenario, the average connectivity might be a better description of the global properties of a graph, with applications in network vulnerability and reliability. Sharp bounds for this measure in terms of the average degree were shown by

Dankelmann and Oellermann [138]. Later on, Henning and Oellermann considered the average connectivity of directed graphs and provided sharp bounds for orientations of graphs [294].

Dynamic Connectivity Problems. Quite a number of publications consider connectivity problems in a dynamical setting, that is, in graphs that are changed by vertex and/or edge insertions and deletions. The special case where only insertions are allowed is called semi-dynamic, partially-dynamic, or incremental. Since there is a vast number of different variants, we provide only the references for further reading: [490, 377, 237, 223, 341, 577, 144, 296, 297, 155, 295, 154, 303].

Directed graphs. As already mentioned, the local connectivity in directed graphs is not symmetric, which is the reason why many algorithms for undirected connectivity problems do not translate to the directed case. Algorithms that compute the edge-connectivity in digraphs were published by Schnorr [503] and by Mansour and Schieber [407]. Another problem of interest is the computation of edge-disjoint branchings, which is discussed in several publications [171, 232, 264, 551, 582].

Other measures. There are some further definitions that might be of interest. Matula [410] defines a *cohesiveness function* for each element of a graph (vertices and edges) to be the maximum edge-connectivity of any subgraph containing that element. Akiyama et al. [13] define the *connectivity contribution* or *cohesiveness* of a vertex v in a graph G as the difference $\kappa(G) - \kappa(G - v)$.

Connectivity problems that aim at dividing the graph into more than two components by removing vertices or edges are considered in conjunction with the following terms: A *shredder* of an undirected graph is a set of vertices whose removal results in at least three components, see for example [121]. The ℓ-*connectivity* of a graph is the minimum number of vertices that must be deleted to produce a graph with at least ℓ components or with fewer than ℓ vertices, see [456, 455]. A similar definition exists for the deletion of edges, namely the *i-th order edge connectivity*, confer [254, 255].

Acknowledgments. The authors thank the anonymous reviewer, the editors, and Frank Schilder for critical assessment of this chapter and valuable suggestions. We thank Professor Ortrud Oellermann for her support.

8 Clustering

Marco Gaertler

Clustering is a synonym for the decomposition of a set of entities into 'natural groups'. There are two major aspects to this task: the first involves algorithmic issues on how to find such decompositions, i.e., tractability, while the second concerns quality assignment, i.e., how good is the computed decomposition. Owing to the informal notion of natural groups, many different disciplines have developed their view of clustering independently. Originally, clustering was introduced to the data mining research as the unsupervised classification of patterns into groups [324]. Since that time a comprehensive framework has started to evolve. In the following, the simple yet fundamental paradigm of intra-cluster density versus inter-cluster sparsity will be discussed exclusively. This restriction is necessary in order to provide some insight into clustering theory, and to keep the scope of this chapter. However, many other interpretations of natural decomposition extend this framework, or are relatively similar. Another specialty is that the input data is represented as networks that are not complete in general. In the classic clustering theory, entities were embedded in metric spaces, and the distance between them was related to their similarity. Thus, all pairwise similarities were known at the beginning. In standard network analysis, the input networks are usually sparse. Even if they are not, it is very unlikely that they are complete. This will be the motivation for studying clustering methods that deal with network input data.

Clustering, based either on the simple paradigm of intra-cluster density versus inter-cluster sparsity or on other more sophisticated formulations, focuses on disjoint cliques as the ideal situation. In some instances the desired cases are totally different. Models where clusters and their connection between each other can have more complex structural properties are targeted in the following chapters about roles (Chapter 9) and blockmodels (Chapter 10).

The popularity of density-based clustering is due to its similarity to the human perception. Most things in our daily life are naturally grouped into categories. For example books are classified with respect to their content, e.g., scientific, fiction, guidebooks, law, etc. Each topic can be refined, e.g., scientific publications can be grouped according to their scientific discipline. The relationship of elements within the same group is strong, whereas elements in different groups typically have a weak relation. Most approaches that deal with information processing are based upon this fact. For example finding a book with a certain content. First, related topics are selected and only those books that

U. Brandes and T. Erlebach (Eds.): Network Analysis, LNCS 3418, pp. 178–215, 2005.

belong to these groups are examined closer. The recursive structure of topics and subtopics suggests a repeated application of this technique. Using the clustering information on the data set, one can design methods that explore and navigate within the data with a minimum of human interaction. Therefore, it is a fundamental aspect of automatic information processing.

Preliminaries

Let $G = (V, E)$ be a directed graph. A *clustering* $\mathcal{C} = \{C_1, \ldots, C_k\}$ of G is a partition of the node set V into non-empty subsets C_i. The set $E(C_i, C_j)$ is the set of all edges that have their origin in C_i and their destination in C_j; $E(C_i)$ is a short-hand for $E(C_i, C_i)$. Then $E(\mathcal{C}) := \bigcup_{i=1}^{k} E(C_i)$ is the set of *intra-cluster edges* and $\overline{E(\mathcal{C})} := E \setminus E(\mathcal{C})$ the set of *inter-cluster edges*. The number of intra-cluster edges is denoted by $m(\mathcal{C})$ and the number of inter-cluster edges by $\overline{m}(\mathcal{C})$. In the following, we often identify a cluster C_i with the induced subgraph of G, i.e., the graph $G[C_i] := (C_i, E(C_i))$. A clustering is called *trivial* if either $k = 1$ (*1-clustering*) or $k = n$ (*singletons*). A clustering with $k = 2$ is also called a cut (see also Section 2.2.3).

The set of all possible clusterings is denoted by $\mathcal{A}(G)$. The set $\mathcal{A}(G)$ is partially ordered with respect to inclusion. Given two clusterings $\mathcal{C}_1 := \{C_1, \ldots, C_k\}$ and $\mathcal{C}_2 := \{C_1', \ldots, C_\ell'\}$, Equation (8.1) shows the definition of the partial ordering.

$$\mathcal{C}_1 \leq \mathcal{C}_2 : \iff \forall\, 1 \leq i \leq k \colon \exists\, j \in \{1, \ldots, \ell\} \colon C_i \subseteq C_j' \tag{8.1}$$

Clustering $\mathcal{C}_1$ is called a *refinement* of $\mathcal{C}_2$, and $\mathcal{C}_2$ is called a *coarsening* of $\mathcal{C}_1$. A chain of clusterings, i.e., a subset of clusterings such that every pair is comparable, is also called a *hierarchy*. The hierarchy is called *total* if both trivial clusterings are contained. A hierarchy that contains exactly one clustering of k clusters for every $k \in \{1, \ldots, n\}$ is called *complete*. It is easy to see that such a hierarchy has n clusterings and that no two of these clusterings have the same number of clusters.

Besides viewing a clustering as a partition, it can also be seen as an equivalence relation $\sim_{\mathcal{C}}$ on $V \times V$, where $u \sim_{\mathcal{C}} v$ if u and v belong to the same cluster in $\mathcal{C}$. Note that the edge set E is also a relation over $V \times V$, and it is an equivalence relation if and only if the graph consists of the union of disjoint cliques.

The power set of a set X is the set of all possible subsets, and is denoted by $\mathcal{P}(X)$, see also Section 2.4. A *cut function* $S \colon \mathcal{P}(V) \to \mathcal{P}(V)$ maps a set of nodes to a subset of itself, i.e.,

$$\forall\, V' \subseteq V \colon S(V') \subseteq V' \ . \tag{8.2}$$

Cut functions formalize the idea of cutting a node-induced subgraph into two parts. For a given node subset V' of V the cut function S defines a cut by $(S(V'), V' \setminus S(V'))$. In order to exclude trivial functions, we require a cut function to assign a non-empty proper subset whenever possible. *Proper* cut functions in addition fulfill the condition (8.3).

$$\forall V' \subseteq V : |V'| > 1 \Longrightarrow \emptyset \neq S(V') \subset V' \; . \tag{8.3}$$

These functions are important for clustering techniques that are based on recursive cutting. These methods will be introduced in Section 8.2.1, and some examples are given in Section 8.2.2.

Graph model of this chapter. In this chapter, *graph* usually means simple and directed graphs with edge weights and without loops.

Content Organization

The following is organized into three parts. The first one introduces measurements for the quality of clusterings. They will provide a formal method to define 'good' clusterings. This will be important due to the informal foundation of clustering. More precisely, these structural indices rate partitions with respect to different interpretations of natural groups. Also, they provide the means to compare different clusterings with respect to their quality. In the second part, generic concepts and algorithms that calculate clusterings are presented. The focus is directed on the fundamental ideas, and not on concrete parameter choices. Finally, the last section discusses potential enhancements. These extensions are limited to alternative models for clusterings, some practical aspects, and Kleinberg's proposal of an axiomatic system for clustering.

8.1   Quality Measurements for Clusterings

As was pointed out in the introduction, clustering techniques are used to find groups that are internally dense and that are only sparsely connected with each other. Although this paradigm of intra-cluster density versus inter-cluster sparsity is more precise than the term 'natural groups', it is still based on our intuition and not on a formal quantification. One way to mathematically express it is by structural indices. These are mappings that assign a non-negative real number to each clustering. Often their range is normalized to the unit interval, where one means best possible structural behavior and zero means worst possible structural behavior. In the following a general framework for indices is presented. Most of the measurements can be expressed within it.

Let $G = (V, E, \omega)$ be a simple, weighted and directed graph, where $\omega : E \rightarrow \mathbb{R}_0^+$ represents the strength of the similarity relation modeled by the edges, and let $\mathcal{C} = \{C_1, \ldots, C_k\}$ be a clustering of G. It is indeed possible to have negative values as strength of similarity, which would express the dissimilarity of two nodes. However, it is rarely the case that similarity and dissimilarity are expressed within the same relation. A far more common case is to have two (weighted) relations, one for similarity and one for dissimilarity. In the following, we will focus on only one relation that expresses similarity. For the unweighted case, the weighting function ω is assumed to be constantly one. In many cases ω will be a mapping to $\mathbb{R}^+$, however, in some cases it is useful to distinguish between edges with weight zero and those node pairs that are not connected by an

edge. We will also use the following short-cut for summing up the weight of an
edge subset:

$$\omega(E') := \sum_{e \in E'} \omega(e) \qquad \text{for } E' \subseteq E \ .$$

For simplicity, we assume that $\omega(E) \neq 0$.

Before defining the actual indices, their framework is presented. The indices
will be composed of two independent functions $f, g \colon \mathcal{A}(G) \to \mathbb{R}_0^+$, where f
measures the density inside the clusters and g the sparsity between clusters.
The functions are combined in the following way:

$$\text{index}(\mathcal{C}) := \frac{f(\mathcal{C}) + g(\mathcal{C})}{\max\{f(\mathcal{C}') + g(\mathcal{C}') : \mathcal{C}' \in \mathcal{A}(G)\}} \qquad (8.4)$$

In order to guarantee the well-definition of Equation (8.4), we assume that there
is at least one clustering $\mathcal{C}'$ such that $f(\mathcal{C}') + g(\mathcal{C}')$ is not zero. For some indices
either f or g is constantly zero. These indices examine only the (internal) density
or the (external) sparsity.

Indices serve two different purposes simultaneously: first and foremost, they
rate a partition with respect to clustering paradigms, and, second, they compare
clusterings regarding quality. Before we explain both aspects in detail, please
note that, while our indices have these properties, there exist other measures
that realize only one aspect.

The quality of a partition as a clustering is expressed in quantitative terms,
i.e., an index (discretely) counts certain substructures like intra-cluster edges,
triangles, or cliques. These structural elements are related to clustering properties. Many intra-cluster edges, triangles, or cliques inside the clusters indicate
that the cluster is dense. In an analogous way the lack of these elements between
clusters imply the inter-cluster sparsity. Thus, the quality of clustering is reduced
to a quantitative aspect. Intuitively, there will always be a maximum number
of these indicators. The number of intra-cluster edges, triangles, or cliques is
limited by the size of clusters. In the ideal case the bounds are met. Thus, if
an index counts only half of them, then the clustering is only half as good as
possible. As we will see, these bounds are not tight for all indices and all input
graphs. Thus, most of them are 'absolute' in the sense that they do not depend
on the input graph, but rather on the clustering paradigms. In conclusion, the
actual range of indices provides useful information.

Because of this absolute, quantitative structure of our indices, they can
be used to compare clusterings regardless of whether the underlying graph
is the same or not. Different graphs can have different bounds for the number of substructures, but the indices rate the quality relative to the individual bounds. For example, in a graph with 10 nodes and 15 edges a clustering
could have 12 intra-cluster edges, and in another graph with 18 nodes and 30
edges another clustering could have 27 inter-cluster edges. Although we have
three inter-cluster edges in both cases, the second clustering would be better
since $27/30 = 9/10 > 4/5 = 12/15$. This property is required when algorithms

are evaluated with random instances, or the data of the input network is not re-liable, i.e., different data collections result into different networks. The clustering methods could be applied to all networks, and the clustering with the best score would be chosen. However, if the index uses characteristics of the input graph, like the number of edges, maximum weight, etc., then this comparison can only be done when the underlying graph is the same for all clusterings. Therefore, indices that depend on the input graph are not appropriate for all applications, such as benchmarks. Although this dependency will seldom occur, one has to consider these facts when designing new indices.

8.1.1 Coverage

The *coverage* $\gamma(\mathcal{C})$ measures the weight of intra-cluster edges, compared to the weight of all edges. Thus $f(\mathcal{C}) = \omega(E(\mathcal{C}))$ and $g \equiv 0$. The maximum value is achieved for $\mathcal{C} = \{V\}$. Equation (8.5) shows the complete formula.

$$\gamma(\mathcal{C}) := \frac{\omega(E(\mathcal{C}))}{\omega(E)} = \frac{\sum_{e \in E(\mathcal{C})} \omega(e)}{\sum_{e \in E} \omega(e)} \tag{8.5}$$

Coverage measures only the accumulated density within the clusters. Therefore, an individual cluster can be sparse or the number of inter-cluster edges can be large. This is illustrated in Figure 8.1. Coverage is also the probability of ran-

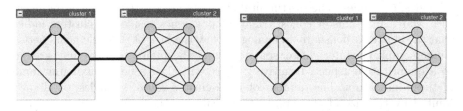

(a) intuitive clustering

(b) non-trivial clustering with best coverage

Fig. 8.1. A situation where coverage splits an intuitive cluster. The thickness of an edge corresponds to its weight. If normal edges have weight one and bold edges weight 100, then the intuitive clustering has $\gamma = 159/209 \approx 0.76$ while the optimal value for coverage is $413/418 \approx 0.99$

domly selecting an intra-cluster edge (where the probability of selection an edge is proportional to its weight, i.e., $\Pr[e] \sim \omega(e)$). The structure of clusterings with optimal coverage value is related to the connectivity structure of the graph. A clustering is *compatible* with the connectivity structure if clusters consist only of unions of connected components of the graph. Proposition 8.1.1 gives a char-acterization of clusterings with optimal coverage value.

Proposition 8.1.1. *A clustering has* $\gamma = 1$ *if and only if either the set of inter-cluster edges is empty or all inter-cluster edges have weight zero. Especially, clusterings that are compatible with the connectivity structure have coverage value 1.*

Sketch of Proof. The edge set is disjointly partitioned into intra-cluster edges and inter-cluster edges, therefore Equation (8.6) holds for any clustering $\mathcal{C}$.

$$\omega(E) = \omega(E(\mathcal{C})) + \omega(\overline{E(\mathcal{C})}) \tag{8.6}$$

Thus, coverage $\gamma(\mathcal{C}) = 1$ holds if and only if $\omega(\overline{E(\mathcal{C})}) = 0$.

Because clusterings that are compatible with the connectivity structure of the graph have no inter-cluster edge, one can use the equivalence to prove the proposition. $\square$

A conclusion of Proposition 8.1.1 is that the 1-clustering always has a coverage value 1. The case of disconnected input graphs is rather special and most techniques even assume that it is connected. Proposition 8.1.1 can be extended to characterize non-trivial clusterings that are optimal with respect to coverage.

A further characterization of non-trivial clusterings that are optimal when trivial clusterings are excluded is possible and given in Proposition 8.1.2.

Proposition 8.1.2. *Let $G = (V, E)$ be a connected graph where every cut has positive weight. Then the clusterings that have more than one cluster and have optimal coverage value are those that are induced by a minimum cut.*

Proof. First, it is obvious that every clustering $\mathcal{C}$ with $k > 1$ can be transformed into clustering $\mathcal{C}'$ with less than k clusters, with $\gamma(\mathcal{C}) \leq \gamma(\mathcal{C}')$. This is achieved by merging any two clusters. Therefore, a non-trivial clustering with optimal coverage has two clusters, and thus is a cut. Second, maximizing coverage is equivalent to minimizing the weight of the inter-cluster edges, and the edges that are contained in the cut have minimum weight. That completes the proof. $\square$

Because of the properties described in Propositions 8.1.1 and 8.1.2, coverage is rarely used as the only quality measurement of a clustering. Minimum cuts often cannot catch the intuition, and separate only a very small portion of the graph. However, there are a few exceptions [286], where both, the input graph and good clusterings, have a very special structure. An extended version will be used for the next index (Section 8.1.2).

8.1.2 Conductance

In contrast to coverage, which measures only the accumulated edge weight within clusters, one can consider further structural properties like connectivity. Intuitively, a cluster should be well connected, i.e., many edges need to be removed to bisect it. Two clusters should also have a small degree of connectivity between each other. In the ideal case, they are already disconnected. Cuts are a useful method to measure connectivity (see also Chapter 7). The standard minimum cut has certain disadvantages (Proposition 8.1.2), therefore an alternative cut measure will be considered: *conductance*. It compares the weight of the cut with the edge weight in either of the two induced subgraphs. Informally speaking, the conductance is a measure for bottlenecks. A cut is a bottleneck, if it separates two parts of roughly the same size with relatively few edges.

Definition 8.1.3. *Let* $C' = (C'_1, C'_2)$ *be a cut, i.e.,* $(C'_2 = V \setminus C'_1)$ *then the conductance-weight* $a(C'_1)$ *of a cut side and the conductance* $\varphi(C')$ *are defined in Equations (8.7) and (8.8).*

$$a(C'_1) := \sum_{(u,v)\in E(C'_1,V)} \omega((u,v)) \tag{8.7}$$

$$\varphi(C') := \begin{cases} 1, & \text{if } C'_1 \in \{\emptyset, V\} \\ 0, & \text{if } C'_1 \notin \{\emptyset, V\}, \omega(\overline{E(C)}) = 0 \\ \dfrac{\omega(\overline{E(C)})}{\min(a(C'_1), a(C'_2))}, & \text{otherwise} \end{cases} \tag{8.8}$$

The conductance of the graph G *is defined by*

$$\varphi(G) = \min_{C_1 \subseteq V} \varphi((C_1, V \setminus C_1)) . \tag{8.9}$$

Note that the case differentiation in Equation (8.8) is only necessary in order to prevent divisions by zero. Before presenting further general information about conductance, graphs with maximum conductance are characterized.

Lemma 8.1.4. *Let* $G = (V, E, \omega)$ *be an undirected and positively weighted graph. Then* G *has maximum conductance, i.e.,* $\varphi(G) = 1$ *if and only if* G *is connected and has at most three nodes, or is a star.*

Proof. Before the equivalence is shown, two short observations are stated:

1. All disconnected graphs have conductance 0 because there is a non-trivial cut that has zero weight and the second condition of the Formula (8.8) holds.
2. For a non-trivial cut $C' = (C'_1, V \setminus C'_1)$ the conductance-weight $a(C'_1)$ can be rewritten as

$$a(C'_1) = \sum_{e \in E(C'_1,V)} \omega(e) = \omega(E(C'_1)) + \omega(\overline{E(C)})$$

in undirected graphs. Thus, the third condition in Formula (8.8) can be simplified to

$$\frac{\omega(\overline{E(C)})}{\min\left(a(C'_1), a(V \setminus C'_1)\right)} = \frac{\omega(\overline{E(C)})}{\omega(\overline{E(C)}) + \min\left(\omega(E(C'_1)), \omega(E(V \setminus C'_1))\right)} . \tag{8.10}$$

'$\Longleftarrow$': If G has one node, then the first condition of Formula (8.8) holds and thus $\varphi(G) = 1$.

If G has two or three nodes or is a star, then every non-trivial cut $C' = (C_1', V \setminus C_1')$ isolates an independent set, i.e., $E(C_1') = \emptyset$. This is achieved by setting C_1' to the smaller cut set if G has at most three nodes and to the cut set that does not contain the center node if G is a star. Therefore $\omega(E(C_1')) = 0$ and Equation (8.10) implies $\varphi(C') = 1$. Because all non-trivial cuts have conductance 1, the graph G has conductance 1 as well.

'$\Longrightarrow$': If G has conductance one, then G is connected (observation 1) and for every non-trivial cut $C' = (C_1', V \setminus C_1')$ at least one edge set $E(C_1')$ or $E(V \setminus C_1')$ has 0 weight (observation 2). Because ω has only positive weight, at least one of these sets has to be empty.

It is obvious that connected graphs with at most three nodes fulfill these requirements, therefore assume that G has at least four nodes. The graph has a diameter of at most two because otherwise there is a path of length three with four pairwise distinct nodes $v_1, \ldots, v_4$, where $e_i := \{v_i, v_{i+1}\} \in E$ for $1 \leq i \leq 3$. Then the non-trivial cut $C' = (\{v_1, v_2\}, V \setminus \{v_1, v_2\})$ cannot have conductance 1, because first the inequality $\omega(E(C')) \geq \omega(e_2) \geq 0$ implies the third condition of Formula (8.8) and second both cut sides are non-empty ($e_1 \in E(\{v_1, v_2\})$ and $e_3 \in E(V \setminus \{v_1, v_2\})$). By the same argument, G cannot contain a simple cycle of length four or greater. It also cannot have a simple cycle of length three. Assume G has such a cycle v_1, v_2, v_3. Then there is another node v_4 that is not contained in the cycle but in the neighborhood of at least one v_i. Without loss of generality $i = 1$. Thus, the non-trivial cut $(\{v_1, v_4\}, V \setminus \{v_1, v_4\})$ is a counterexample. Thus G cannot contain any cycle and is therefore a tree. The only trees with at least four nodes, and a diameter of at most two, are stars.

$\square$

It is $\mathcal{NP}$-hard to calculate the conductance of a graph [39]. Fortunately, it can be approximated with a guarantee of $\mathcal{O}(\log n)$ [565] and $\mathcal{O}(\sqrt{\log n})$ [36]. For some special graph classes, these algorithms have constant approximation factors. Several of the involved ideas are found in the theory of Markov chains and random walks. There, conductance models the probability that a random walk gets 'stuck' inside a non-empty part. It is also used to estimate bounds on the rate of convergence. This notion of 'getting stuck' is an alternative description of bottlenecks. One of these approximation ideas is related to spectral properties. Lemma 8.1.5 indicates the use of eigenvalues as bounds.

Lemma 8.1.5 ([521, Lemma 2.6]). *For an ergodic[1] reversible Markov chain with underlying graph G, the second (largest) eigenvalue λ_2 of the transition matrix satisfies:*

---

[1] Aperiodic and every state can be reached an arbitrary number of times from all initial states.

$$\lambda_2 \geq 1 - 2 \cdot \varphi(G) \ . \tag{8.11}$$

A proof of that can be found in [521, p. 53]. Conductance is also related to isoperimetric problems as well as expanders, which are both related to similar spectral properties themselves. Section 14.4 states some of these characteristics.

For unweighted graphs the conductance of the complete graph is often a useful boundary. It is possible to calculate its exact conductance value. Proposition 8.1.6 states the result. Although the formula is different for even and odd number of nodes, it shows that the conductance of complete graphs is asymptotically $1/2$.

Proposition 8.1.6. *Let n be an integer, then equation (8.12) holds.*

$$\varphi(K_n) = \begin{cases} \frac{1}{2} \cdot \frac{n}{n-1} & , \text{if } n \text{ is even} \\ \frac{1}{2} + \frac{1}{n-1} & , \text{if } n \text{ is odd} \end{cases} \tag{8.12}$$

Proof. Evaluating Equation (8.8) in Definition 8.1.3 with $G = K_n$ leads to

$$\varphi(K_n) = \min_{C \subset V, 1 \leq |C| < n} \frac{|C| \cdot (n - |C|)}{\min(|C|(|C| - 1), (n - |C|)(n - |C| - 1))} \ . \tag{8.13}$$

Node subsets of size k of a complete graph are pairwise isomorphic, therefore only the size of the subset C matters. Thus, Equation (8.13) can be simplified to

$$\varphi(K_n) = \min_{1 \leq k < n} \frac{k(n - k)}{\min(k^2 - k, n^2 - 2nk - n - k)} \ . \tag{8.14}$$

The fraction in Equation (8.14) is symmetric, thus it is sufficient if k varies in the range from 1 to $\lceil n/2 \rceil$. Using the fact that the fraction is also monotonic decreasing with increasing k, the minimum is assumed for $k = \lfloor n/2 \rfloor$. A simple case differentiation for even and odd ks leads to the final Equation (8.12).   □

In the following, two clustering indices are derived with the help of conductance. These will be intra-cluster conductance and inter-cluster conductance, and both will focus on only one property. The first one measures internal density, while the second rates the connection between clusters.

The intra-cluster conductance α is defined as the minimum conductance occurring in the cluster-induced subgraphs $G[C_i]$, i.e.,

$$f(\mathcal{C}) = \min_{1 \leq i \leq k} \varphi(G[C_i]) \qquad \text{and} \qquad g \equiv 0 \ . \tag{8.15}$$

Note that $G[C_i]$ in $\varphi(G[C_i])$ denotes a subgraph, and therefore is independent of the rest of the original graph G. The conductance of a (sub)graph is small if it can naturally be bisected, and great otherwise. Thus, in a clustering with small intra-cluster conductance there is supposed to be at least one cluster containing a bottleneck, i.e., the clustering is possibly too coarse in this case. The minimum conductance cut itself can also be used as a guideline to split the cluster further (refer to Section 8.2.2 for further details).

The inter-cluster conductance δ considers the cuts induced by clusters, i.e.,

$$f \equiv 0 \quad \text{and} \quad g = \begin{cases} 1, & \text{if } \mathcal{C} = \{V\} \\ 1 - \max\limits_{1 \leq i \leq k} \varphi\left((C_i, V \setminus C_i)\right), & \text{otherwise} \end{cases} . \quad (8.16)$$

Note that $(C_i, V \setminus C_i)$ in $\varphi\left((C_i, V \setminus C_i)\right)$ denotes a cut within the graph G. A clustering with small inter-cluster conductance is supposed to contain at least one cluster that has relatively strong connections outside, i.e., the clustering is possibly too fine. In contrast to the intra-cluster conductance, one cannot directly use the induced cut information to merge two clusters.

For both indices the maximum of $f + g$ is one, which leads to the final formula:

$$\alpha\left(\mathcal{C}\right) := \min\limits_{1 \leq i \leq k} \varphi\left(G[C_i]\right) \quad (8.17)$$

$$\delta\left(\mathcal{C}\right) := \begin{cases} 1, & \text{if } \mathcal{C} = \{V\} \\ 1 - \max\limits_{1 \leq i \leq k} \varphi\left((C_i, V \setminus C_i)\right), & \text{otherwise} \end{cases} \quad (8.18)$$

Again a characterization of optimal clusterings is possible.

Proposition 8.1.7. *Only clusterings where the clusters consist of connected subgraphs that are stars or have size of at most three, have intra-cluster conductance 1.*

Proposition 8.1.8. *Only clusterings that have an inter-cluster edge weight of zero, including the 1-clustering, have inter-cluster conductance 1.*

Both Proposition 8.1.7 and 8.1.8 are immediate consequences of the Definition 8.1.3 and Lemma 8.1.4. Both measures, intra-cluster conductance and inter-cluster conductance, have certain disadvantages. Two examples are shown in Figures 8.2 and 8.3. Both examples explore the 'artificial' handling of small graphs considering their conductance. In practical instances, intra-cluster conductance values are usually below $1/2$. Small clusters with few connections have relatively small inter-cluster conductance values.

8.1.3 Performance

The next index combines two non-trivial functions for the density measure f and the sparsity measure g. It simply counts certain node pairs. According to the general intuition of intra-cluster density versus inter-cluster sparsity, we define for a given clustering a *'correct' classified* pair of nodes as two nodes either belonging to the same cluster and connected by an edge, or belonging to different clusters and not connected by an edge. The resulting index is called *performance*. Its density function f counts the number of edges within all clusters while its sparsity function g counts the number of nonexistent edges between clusters, i.e.,

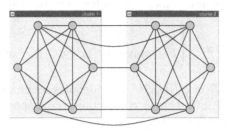

(a) intuitive clustering

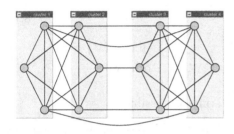

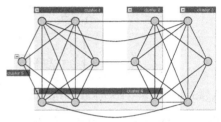

(b) non-trivial clustering with best intra-cluster conductance

(c) non-trivial clustering with best intra-cluster conductance

Fig. 8.2. A situation where intra-cluster conductance splits intuitive clusters. The intuitive clustering has $\alpha = 3/4$, while the other two clusterings have $\alpha = 1$. The split in Figure 8.2(b) is only a refinement of the intuitive clustering, while Figure 8.2(c) shows a clusterings with same intra-cluster conductance value that is skew to the intuitive clustering

$$
\begin{aligned}
f(\mathcal{C}) &:= \sum_{i=1}^{k} |E(C_i)| \quad \text{and} \\
g(\mathcal{C}) &:= \sum_{u,v \in V} [(u,v) \notin E] \cdot [u \in C_i, v \in C_j, i \neq j] \ .
\end{aligned}
\tag{8.19}
$$

The definition is given in Iverson Notation, first described in [322], and adapted by Knuth in [365]. The term inside the parentheses can be any logical statement. If the statement is true the term evaluates to 1, if it is false the term is 0. The maximum of $f+g$ has $n \cdot (n-1)$ as upper bound because there are $n(n-1)$ different node pairs. Please recall that loops are not present and each pair contributes with either zero or one. Calculating the maximum of $f + g$ is $\mathcal{NP}$-hard (see [516]), therefore this bound is used instead of the real maximum. By using some duality aspects, such as the number of intra-cluster edges and the number of inter-cluster edges sum up to the whole number of edges, the formula of performance can be simplified as shown in Equation (8.21).

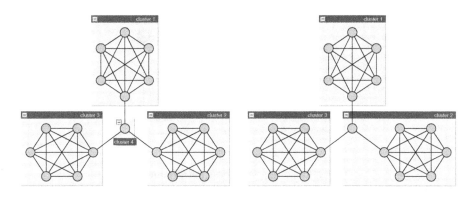

(a) intuitive clustering

(b) non-trivial clustering with best inter-cluster conductance

Fig. 8.3. A situation where two very similar clusterings have very different inter-cluster conductance values. The intuitive clustering has $\delta = 0$, while the other has the optimum value of $8/9$

$$
\begin{aligned}
\mathrm{perf}\,(\mathcal{C}) &= \frac{m\,(\mathcal{C}) + \left(n(n-1) - \sum_{i=1}^{k} |C_i|(|C_i| - 1) - \overline{m}\,(\mathcal{C})\right)}{n(n-1)} \\
&= \frac{n(n-1) - m + 2m\,(\mathcal{C}) - \sum_{i=1}^{k} |C_i|(|C_i| - 1)}{n(n-1)} \tag{8.20} \\
&= 1 - \frac{m(1 - 2\frac{m(\mathcal{C})}{m}) + \sum_{i=1}^{k} |C_i|(|C_i| - 1)}{n(n-1)} . \tag{8.21}
\end{aligned}
$$

Note that the derivation from Equation (8.20) to (8.21) applies the equality $m = m\,(\mathcal{C}) + \overline{m}\,(\mathcal{C})$, and that $m\,(\mathcal{C})\,/m$ is just the coverage $\gamma\,(\mathcal{C})$ in the unweighted case. Similarly to the other indices, performance has some disadvantages. Its main drawback is the handling of very sparse graphs. Graphs of this type do not contain subgraphs of arbitrary size and density. For such instances the gap between the number of feasible edges (with respect to the structure) and the maximum number of edges (regardless of the structure) is also huge. For example, a planar graph cannot contain any complete graph with five or more nodes, and the maximum number of edges such that the graph is planar is linear in the number of nodes, while in general it is quadratic. In conclusion, clusterings with good performance tend to have many small clusters. Such an example is given in Figure 8.4.

An alternative motivation for performance is given in the following. Therefore recall that the edge set induces a relation on $V \times V$ by $u \sim v$ if $(u, v) \in E$, and clusterings are just another notion for equivalence relations. The problem of finding a clustering can be formalized in this context as finding a transformation of the edge-induced relation into an equivalence relation with small cost. In

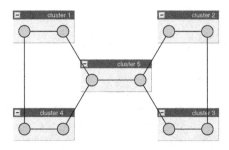

(a) clustering with best performance

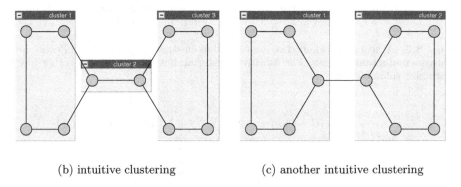

(b) intuitive clustering   (c) another intuitive clustering

Fig. 8.4. A situation where the clustering with optimal performance is a refinement (Figure 8.4(b)) of an intuitive clustering and is skew (Figure 8.4(c)) to another intuitive clustering

other words, add or delete edges such that the relation induced by the new edge set is an equivalence relation. As cost function, one simply counts the number of additional and deleted edges. Instead of minimizing the number of changes, one can consider the dual version: find a clustering such that the clustering-induced relation and the edge-set relation have the greatest 'intersection'. This is just the maximizing $f + g$. Thus performance is related to the 'distance' of the edge-set relation to the closed clustering. Because finding this maximum is $\mathcal{NP}$-hard, solving this variant of clustering is also $\mathcal{NP}$-hard. Although the problem is hard, it possesses a very simple integer linear program (ILP). ILPs are also $\mathcal{NP}$-hard, however there exist many techniques that lead to usable heuristics or approximations for the problems. The ILP is given by n^2 decision variables $X_{uv} \in \{0, 1\}$ with $u, v \in V$, and the following three groups of constraints:

reflexivity $\forall\, u\colon X_{uu} = 1$

symmetry $\forall\, u, v\colon X_{uv} = X_{vu}$

transitivity $\forall\, u, v, w\colon$ $\begin{cases} X_{uv} + X_{vw} - 2 \cdot X_{uw} \leq 1 \\ X_{uw} + X_{uv} - 2 \cdot X_{vw} \leq 1 \\ X_{vw} + X_{uw} - 2 \cdot X_{uv} \leq 1 \end{cases}$

and minimizing objective function:

$$\sum_{(u,v)\in V^2} (1 - E_{uv})X_{uv} + E_{uv}(1 - X_{uv})\ ,$$

$$\text{with}\ \ E_{uv} = \begin{cases} 1 & ,\ \text{if } (u, v) \in E \text{ or } u = v \\ 0 & ,\ \text{otherwise} \end{cases}.$$

The idea is that the X variables represent equivalence relations, i.e., two nodes $u, v \in V$ are equivalent if $X_{uv} = 1$ and the objective function counts the number of not 'correct' classified node pairs.

There exist miscellaneous variations of performance that use more complex models for classification. However, many modifications highly depend on their applicational background. Instead of presenting them, some variations to include edge weights are given. As pointed out in Section 8.1, indices serve two different tasks. In order to preserve the comparability aspect, we assume that all the considered edge weights have a meaningful maximum M. It is not sufficient to replace M with the maximum occurring edge weight because this value depends on the input graph. Also, choosing an extremely large value of M is not suitable because it disrupts the range aspects of the index. An example of such weightings with a meaningful maximum are probabilities where $M = 1$. The weighting represents the probability that an edge can be observed in a random draw. Using the same counting scheme of performance, one has to solve the problem of assigning a real value for node pairs that are not connected. This problem will be overcome with the help of the meaningful maximum M.

The first variation is straightforward and leads to the measure functions given in Equation (8.22):

$$f\left(\mathcal{C}\right) := \sum_{i=1}^{k} \omega\left(E(C_i)\right)\quad \text{and}$$

$$g\left(\mathcal{C}\right) := \sum_{u,v\in V} M \cdot [(u, v) \notin E] \cdot [u \in C_i, v \in C_j, i \neq j]\ . \tag{8.22}$$

Please note the similarity to the unweighted definition in Formula (8.19). However, the weight of the inter-cluster edges is neglected. This can be integrated by modifying g:

$$g'\left(\mathcal{C}\right) := g\left(\mathcal{C}\right) + \underbrace{M \cdot \left|\overline{E(\mathcal{C})}\right| - \omega\left(\overline{E(\mathcal{C})}\right)}_{=:g_w(\mathcal{C})}\ . \tag{8.23}$$

The additional term $g_w(\mathcal{C})$ corresponds to the difference of weight that would be counted if no inter-cluster edges were present and the weight that is assigned to the actual inter-cluster edges. In both cases the maximum is bounded by $M \cdot n(n-1)$, and the combined formula would be:

$$\mathsf{perf}_w(\mathcal{C}) = \frac{f(\mathcal{C}) + g(\mathcal{C}) + \vartheta \cdot g_w(\mathcal{C})}{n(n-1)M} \ , \tag{8.24}$$

where $\vartheta \in [0,1]$ is a scaling parameter that rates the importance of the weight of the inter-cluster edges (with respect to the weight of the intra-cluster edges). In this way there is a whole family of weighted performance indices.

An alternative variation is based on the duality. Instead of counting 'correct' classified node pairs the number/weight of the errors is measured. Equation (8.21) will be the foundation:

$$\tilde{f}(\mathcal{C}) = \sum_{i=1}^{k} \left(M|C_i|(|C_i| - 1) - \theta \cdot \omega(E(C_i)) \right) \quad \text{and}$$

$$\tilde{g}(\mathcal{C}) = \omega\left(\overline{E(\mathcal{C})} \right) \ , \tag{8.25}$$

where θ is a scaling parameter that rates the importance of the weight of the intra-cluster edges (with respect to the weight of the inter-cluster edges). The different symbols for density $\tilde{f}$ and sparsity $\tilde{g}$ functions are used to clarify that these functions perform inversely to the standard functions f and g with respect to their range: small values indicate better structural behavior instead of large values. Both can be combined to a standard index via

$$\mathsf{perf}_m(\mathcal{C}) = 1 - \frac{\tilde{f}(\mathcal{C}) + \tilde{g}(\mathcal{C})}{n(n-1)M} \ . \tag{8.26}$$

Note that the versions are the same for $\vartheta = \theta = 1$. In general, this is not true for other choices of ϑ and θ. Both families have their advantages and disadvantages. The first version (Equation (8.24)) should be used, if the clusters are expected to be heavy, while the other version (Equation (8.26)) handles clusters with inhomogeneous weights better.

8.1.4   Other Indices

Clearly these indices are only a small fraction of the whole spectrum used to formalize and evaluate clusterings. However, they clarify different concepts very well. This part covers some historical indices that have been used for clustering, as well as some measures that are based on certain assumptions.

Clustering was originally applied to entities embedded in metric spaces or vector spaces with a norm function. Many indices have been developed that use essential structural properties of these spaces, e.g., the possibility to calculate a barycenter or geometric coverings. Over time, some indices have been transferred to graph clustering as well. Because these spaces usually provide all pair information, the first problem is to estimate the similarity of nodes that are not connected with an edge. This is often solved using shortest path techniques that respect all information, like weight or direction, or only partial information, or none. The most common measures resulting are: diameter, edge weight variance, and average distance within the clusters. In contrast to the previous indices, these measures do not primarily focus on the intra-cluster density versus inter-cluster sparsity paradigm. Most of them even ignore the inter-cluster structure completely. Another difference is that these indices usually rate each cluster individually, regardless of its position within the graph. The resulting distribution is then rated with respect to the average or the worst case. Thus, a density measure[2] π can be transformed into an index by applying π on all cluster-induced subgraphs and rating the resulting distribution of values, e.g., via minimum, maximum, average, or mean:

$$\text{worst case:} \quad \min_i \{\pi(G[C_1]), \ldots, \pi(G[C_k])\}$$

$$\text{average case:} \quad \frac{1}{k} \sum_i \pi(G[C_i])$$

$$\text{best case:} \quad \max_i \{\pi(G[C_1]), \ldots, \pi(G[C_k])\}$$

Their popularity is partially based on the easy computation in metric or normed vector spaces, where the distance (inverse similarity) of (all) pairs is defined by their distance within the given space. The other reason is their use for greedy approaches that will be introduced in Section 8.2.1.

Another kind of measure compares the partition with an average case of the graph instead of an ideal case. This is especially preferred when the ideal situation is unknown, i.e., the network has very specific properties and thus general clustering paradigms could hardly be fulfilled. For example, Newman [442] postulated that a (uniform) random graph does not contain a clustering structure at all. Therefore, measuring the difference of numbers of edges within a cluster and the expected number of edges that would randomly fall within the cluster should be a good indicator whether the cluster is significant or not. The whole formula is shown in Equation (8.27).

$$\sum_{i=1}^{k} \left(|E(C_i)| - m \frac{|C_i| \cdot (|C_i| - 1)|}{n \cdot (n-1)} \right) . \tag{8.27}$$

One advantage is that one can now distinguish between good, 'random', and bad clusterings depending on the sign and magnitude of the index. However, the expected number of edges that fall at random in a subgraph may be a too global view. The local density may be very inhomogeneous and thus an average global view can be very inaccurate. Figure 8.5 sketches such an example. The graph consists of two groups, i.e., a cycle (on the left side) and a clique (on the right side), which are connected by a path. In this example there is no subgraph with average density, in fact the two groups have a significantly different density. One

---

[2] greater values imply larger density

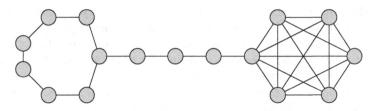

Fig. 8.5. Graph with inhomogeneous density

way to restrict the global view of average density is to fix the degree of each node and consider expected number of edges that fall at random in a subgraph. The modified formula is given in Equation (8.28).

$$\sum_{i=1}^{k} \left[\frac{\omega(E(C_i))}{\omega(E)} - \left(\frac{\sum_{e \in E(C_i, V)} \omega(e)}{\omega(E)} \right)^2 \right] . \qquad (8.28)$$

In general, these measures that compare the partition with an average case of the graph seem to introduce a new perspective that has not yet been fully explored.

8.1.5   Summary

The presented indices serve as quality measurements for clusterings. They provide a formal way to capture the intuitive notation of natural decompositions. Several of these indices can be expressed in a simple framework that models the paradigm of intra-cluster density versus inter-cluster sparsity. It has been introduced in Section 8.1. Table 8.1 summarizes these indices, the general form is given in Equation (8.29).

$$\text{index}(\mathcal{C}) := \frac{f(\mathcal{C}) + g(\mathcal{C})}{N} \qquad (8.29)$$

A commonality of these measures is that the associated optimization problem, i.e., finding the clustering with best score, usually is $\mathcal{NP}$-hard. If the optimal structure is known in advance, as is the case for coverage or conductance, the restriction to interesting or practically relevant clusterings leads to $\mathcal{NP}$-hard problems.

Bibliographic Notes. Further information about indices can be found in [324, 325]. In [569, Chapter 7] a collection of density concepts is presented that have nearly the same intuition. Benchmark tests that are based on indices can be found in [95].

There is also a lot of work done within the theory community, especially disciplines like machine learning or neural networks, which targets approximating indices or extracting structural properties to qualitatively rate partitions. Owing to our limited scope, these aspects could not be covered. [170, 565] may serve as an entry point to these issues.

Table 8.1. Summary of clustering indices

Name	$f(C)$	$g(C)$	N
coverage $\gamma(C)$	$\omega(E(C))$	0	$\omega(E)$
intra-cluster conductance $\alpha(C)$	$\min_{1 \leq i \leq k} \varphi(G[C_i])$	0	1
inter-cluster conductance $\delta(C)$	0	$\begin{cases} 1, & \text{if } C = \{V\} \\ 1 - \max\limits_{1 \leq i \leq k} \varphi((C_i, V \setminus C_i)), & \text{otherwise} \end{cases}$	1
performance $\text{perf}(C)$	$\sum_{i=1}^{k} |E(C_i)|$	$\sum_{u,v \in V} [(u,v) \notin E] \cdot [u \in C_i, v \in C_j, i \neq j]$	$n(n-1)$
performance $\text{perf}_w(C)$	$\sum_{i=1}^{k} \omega(E(C_i))$	$M \sum_{u,v \in V} [(u,v) \notin E] \cdot [u \in C_i, v \in C_j, i \neq j]$ $+ \vartheta \cdot \left(M \cdot \left| \overline{E(C)} \right| - \omega\left(\overline{E(C)} \right) \right)$	$M \cdot n(n-1)$
performance $1 - \text{perf}_m(C)$	$\sum_{i=1}^{k} \left(M|C_i|(|C_i| - 1) - \theta \cdot \omega(E(C_i)) \right)$	$\omega\left(\overline{E(C)} \right)$	$M \cdot n(n-1)$

8.2 Clustering Methods

This section covers the description of algorithmic paradigms and approaches used to calculate clusterings. It is split in three parts: First, some fundamental techniques are explained. Second, instances that embody these principles are described. Third, related problems are discussed.

8.2.1 Generic Paradigms

The description of generic methods is structured into three groups: Greedy and shifting approaches as well as general optimizing techniques utilized to find near-optimal clusterings. Mostly the concepts and algorithms apply a certain number of reductions to obtain an instance where a solution can easily be computed, and then extend it to the original input. This is very similar to the standard divide and conquer techniques, where instances are recursively divided until the problem is trivially solvable. In contrast to this, the reduction step is more general and can modify the instance completely. Thus, the reduced instance does not need to be a subinstance of the original problem. The reduction step is usually composed of two parts itself. First, significant substructures are identified. These can be bridges that are likely to separate clusters, or dense groups that indicate parts of clusters. Such an identification is often formulated as a hypothesis, and the recognized substructures are considered as evidence for its correctness. After the recognition phase a proper modification is applied to the current input graph. Such transformations can be arbitrary graph operations, however, the usual modifications are: addition and deletion of edges, as well as collapsing subgraphs, i.e., representing a subgraph by a new meta-node. A sketch of this idea is given in Figure 8.6. The 'shapes' of the (sub)instances indicate the knowledge of clustering, i.e., smooth shapes point to fuzzy information while the rectangles indicate exact knowledge. In the initial instance (left figure, upper row), no clustering information is present. During the reduction phases parts of the graph became more and more separated, which is indicated by the disjoint objects. The right instance in the upper row can be easily solved, and the fuzzy information is transformed into exact information with respect to the given instance. Based upon it, the expansion phase transfers this knowledge to the original instance (left figure, lower row). An additional difference to divide and conquer methods is that the size of the considered graphs can increase during the reduction phases.

Greedy Concepts. Most greedy methods fit into the following framework: start with a trivial and feasible solution and use update operations to lower its costs recursively until no further optimization is possible. This scheme for a greedy approach is shown in Algorithm 19, where $c(L)$ denotes the cost of solution L and $N_g(L)$ is the set of all solutions that can be obtained via an update operation starting with solution L. This iterative scheme can also be expressed for clusterings via hierarchies. A hierarchy represents an iterative refinement (or

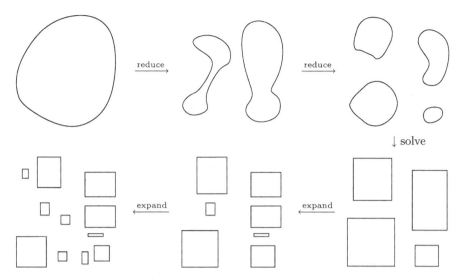

Fig. 8.6. Example of successive reductions. The 'shapes' of the (sub)instances indicate the knowledge of clustering, i.e., smooth shapes point to fuzzy information while the rectangles indicate exact knowledge. The first row shows the application of modifications, while the second indicates the expansion phases

---

Algorithm 19: Scheme for greedy methods

---

let L_0 be a feasible solution
$i \leftarrow 0$
while $\{L : L \in N_g(L_i), c(L) < c(L_i)\} \neq \emptyset$ **do**
  $\quad L_{i+1} \leftarrow \text{argmin}_{L \in N_g(L_i)} c(L)$
  $\quad i \leftarrow i + 1$
return L_i

---

coarsening) process. Greedy methods that use either merge or split operations as updates define a hierarchy in a natural way. The restriction to one of these operations guarantees the comparability of clusterings, and thus leads to a hierarchy. These two concepts will be formalized shortly, before that some facts of hierarchies are briefly mentioned.

Hierarchies provide an additional degree of freedom over clusterings: the number of clusters is not fixed. Thus, they represent the group structure independently of its granularity. However, this feature usually increases the space requirement, although a hierarchy can be implicitly represented by a tree, also called a dendrogram, that represents the merge operations. Algorithms tend to construct it explicitly. Therefore, their space consumption is often quadratic or larger. For a few special cases these costs can be reduced with the help of data structures [178].

The *Linkage process* (Agglomeration) iteratively coarses a given clustering by merging two clusters until the 1-clustering is reached. The formal description is shown in Definition 8.2.1.

Definition 8.2.1 (Linkage). *Given a graph* $G = (V, E, \omega)$, *an initial cluster-ing* $\mathcal{C}_1$ *and either a global cost function* $c_{global} \colon \mathcal{A}\,(G) \to \mathbb{R}_0^+$ *or a cost function* $c_{local} \colon \mathcal{P}(V) \times \mathcal{P}(V) \to \mathbb{R}_0^+$ *for merging operations, the* Linkage *process merges two clusters in the current clustering* $\mathcal{C}_i := \{C_1, \ldots, C_k\}$ *while possible in the following recursive way:*

global: *let* P *be the set of all possible clusterings resulting from* $\mathcal{C}_i$ *by merging two clusters, i.e.,*

$$P := \Big\{ \{C_1, \ldots, C_k\} \setminus \{C_\mu, C_\nu\} \cup \{C_\mu \cup C_\nu\} \mid \mu \neq \nu \Big\} \ ,$$

then the new clustering $\mathcal{C}_{i+1}$ *is defined as an element in* P *with minimum cost with respect to* c_{global}.

local: *let* μ, ν *be those two distinct indices such that* c_{local} *has one global min-imum in the pair* (C_μ, C_ν), *then the new clustering* $\mathcal{C}_{i+1}$ *is defined by merging* C_μ *and* C_ν, *i.e.,*

$$\mathcal{C}_{i+1} := \{C_1, \ldots, C_k\} \setminus \{C_\mu, C_\nu\} \cup \{C_\mu \cup C_\nu\} \ .$$

Although the definition is very formal, the basic idea is to perform a cheapest merge operation. The cost of such an operation can be evaluated using two different view points. A local version charges only merge itself, which depends only on the two involved clusters. The opposite view is a global version that considers the impact of the merge operation. These two concepts imply also the used cost functions, i.e., a global cost function has the set of clusterings as domain while the local cost function uses a pair of node subsets as arguments. An example of linkage is given in Figure 8.7. The process of linkage can be reversed, and, instead of merging two clusters, one cluster is split into two parts. This dual process is called *Splitting* (Diversion). The formal description is given in Definition 8.2.2.

Definition 8.2.2 (Splitting). *Let a graph* $G = (V, E, \omega)$, *an initial cluster-ing* $\mathcal{C}_1$, *and one of the following function sets be given:*

global: *a cost function* $c_{global} \colon \mathcal{A}\,(G) \to \mathbb{R}_0^+$
semi-global: *a cost function* $c_{global} \colon \mathcal{A}\,(G) \to \mathbb{R}_0^+$ *and a proper cut function* $S_{local} \colon$
    $\mathcal{P}(V) \to \mathcal{P}(V)$
semi-local: *a cost function* $c_{local} \colon \mathcal{P}(V) \times \mathcal{P}(V) \to \mathbb{R}_0^+$ *and a proper cut func-tion* $S_{local} \colon \mathcal{P}(V) \to \mathcal{P}(V)$
local: *a cost function* $c_{local} \colon \mathcal{P}(V) \times \mathcal{P}(V) \to \mathbb{R}_0^+$

The *Splitting process splits one cluster in the current clustering* $\mathcal{C}_i := \{C_1, \ldots C_k\}$ *into two parts. The process ends when no further splitting is possible. The cluster that is going to be split is chosen in the following way:*

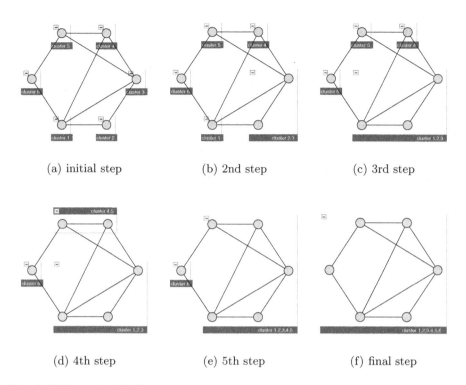

(a) initial step   (b) 2nd step   (c) 3rd step

(d) 4th step   (e) 5th step   (f) final step

Fig. 8.7. Example of linkage

global: let P be the set of all possible clusterings resulting from $\mathcal{C}_i$ by splitting one cluster into two non-empty parts, i.e.,

$$P := \left\{ \{C_1, \ldots, C_k\} \setminus \{C_\mu\} \cup \{C'_\mu, C_\mu \setminus C'_\mu\} \mid \emptyset \neq C'_\mu \subsetneq C_\mu \right\} ,$$

then the new clustering $\mathcal{C}_{i+1}$ is defined as an element in P with minimum cost with respect to c_{global}.

semi-global: let P be the set of all possible clusterings resulting from $\mathcal{C}_i$ by splitting one cluster into two non-empty parts according to S_{local}, i.e.,

$$P := \left\{ \{C_1, \ldots, C_k\} \setminus \{C_\mu\} \cup \{S_{local}(C_\mu), C_\mu \setminus S_{local}(C_\mu)\} \right\} ,$$

then the new clustering $\mathcal{C}_{i+1}$ is defined as an element in P with minimum cost with respect to c_{global}.

semi-local: let μ be an index such that c_{local} has one global minimum in the pair $(S_{local}(C_\mu), C_\mu \setminus S_{local}(C_\mu))$ then the new clustering $\mathcal{C}_{i+1}$ is defined by splitting cluster C_μ according to S_{local}, i.e.,

$$\mathcal{C}_{i+1} := \{C_1, \ldots, C_k\} \setminus \{C_\mu\} \cup \{S_{local}(C_\mu), C_\mu \setminus S_{local}(C_\mu)\} .$$

local: let μ be an index and C_ν be a proper subset of cluster C_μ such that c_{local} has one global minimum in the pair $(C_\nu, C_\mu \setminus C_\nu)$, then the new clustering $\mathcal{C}_{i+1}$ is defined by splitting cluster C_μ according to S_{local}, i.e.,

$$\mathcal{C}_{i+1} := \{C_1, \ldots, C_k\} \setminus \{C_\mu\} \cup \{C_\mu, C_\mu \setminus C_\nu\} \ .$$

Similar to the Linkage process, the definition is rather technical but the basic idea is to perform the cheapest split operation. In contrast to the Linkage method, the cost model has an additional degree of freedom because clusters can be cut in several ways. Again, there is a global and a local version that charge the impact of the split and the split itself, respectively. Both correspond to the views in the Linkage process. However, rating every possible non-trivial cut of the clusters is very time consuming and usually requires sophisticated knowledge of the involved cost functions. One way to reduce the set of possible splittings is to introduce an additional cut function S_{local}. It serves as an 'oracle' to produce useful candidates for splitting operations. The semi-global and semi-local versions have the same principles a the global and the local version, however, their candidate set is dramatically reduced. Therefore, they are often quite efficiently computable, and no sophisticated knowledge about the cost function is required. However, the choice of the cut function has usually a large impact on the quality.

Both, the Linkage and the Splitting process, are considered to be greedy for several reasons. One is the construction of the successive clusterings, i.e., an update operation chooses always the cheapest clustering. These can produce total or complete hierarchies quite easily. Total hierarchies can be achieved by simply adding the trivial clusterings to the resulting hierarchy. They are comparable to all other clusterings, therefore preserving the hierarchy property. Recall that complete hierarchies are hierarchies such that a clustering with k clusters is included for every integer $k \in [1, n]$. Both processes lead to a complete hierarchy when initialized with the trivial clusterings, i.e., singletons for Linkage and the 1-clustering for Splitting. Note that in the case of the Splitting process, it is essential that the cut functions are proper. Therefore, it is guaranteed that every cluster will be split until each cluster contains only one node. Although the cost can be measured with respect to the result or the operation itself, clairvoyance or projection of information into the future, i.e., accepting momentarily higher cost for a later benefit, is never possible.

Because of their simple structure, especially in the local versions, both concepts are frequently used and are also the foundation of clustering algorithms in general. The general local versions can be very efficiently implemented. For the Linkage process, a matrix containing all cluster pairs and their merging cost is stored. When an update operation takes place, only the cost of merging the new resulting cluster with another is recalculated. For certain cost functions this scheme can even be implemented with less than quadratic space and runtime consumption [178]. In the case of the Splitting process, only the cut information needs to be stored for each cluster. Whenever a cluster gets split, one has to recompute the cut information only for the two new parts. This is not true for any global version in general. However, a few very restricted cost and cut functions can be handled efficiently also in the global versions.

Shifting Concept. In contrast to the global action of the previous greedy strategies, the shifting approaches work more locally. They choose an initial clustering and iteratively modify it until a local optimum is found. Usually three operations are permitted: first, a node moves from one cluster to another existing cluster, second, a node moves from one cluster to create a new cluster, and, third, two nodes exchange their cluster assignments. Sometimes more complex operations, like instant removal of one cluster and reassigning nodes to already existing clusters, are allowed. However, these complex operations are usually only used for speed-up, or to avoid artifical situations, therefore they will not be discussed here. Note that one can typically simulate them by a sequence of simple operations. Regarding algorithmic aspects, shifting concepts are relatively close to the greedy ones. Algorithm 20 shows the general procedure where $N_s(L)$ denotes the set of all clusterings that can be obtained by applying the modifications to the clustering L. Step 1 can be based on cost or potential functions,

---

Algorithm 20: Scheme for shifting methods

---

let L_0 be an initial solution
$i \leftarrow 0$
while $N_s(L_i) \neq \emptyset$ **do**
1 | choose $L_{i+1} \in N_s(L_i)$
     | $i \leftarrow i + 1$
return L_i

---

random selecting, or on the applicational background. The technical definition of the shifting concept is given in Definition 8.2.3, and uses a potential function as selection criteria.

Definition 8.2.3. *Given a graph* $G = (V, E, \omega)$, *an initial clustering* $\mathcal{C}_1$ *and a potential function* $\Phi \colon \mathcal{A}(G) \times \mathcal{A}(G) \rightarrow \mathbb{R}$, *the* Shifting *process is defined as performing any operation on the current clustering* $\mathcal{C}_i$ *that results in a new clustering* $\mathcal{C}_{i+1}$ *such that* $\Phi(\mathcal{C}_i, \mathcal{C}_{i+1}) > 0$.

Shifting concepts have many degrees of freedom. The choice of potential functions is critical. There are two common subtypes of potentials: type-based and compressed. *Type-based potentials* are functions that heavily depend on the type of operations. They are often used to preserve certain properties. In the case that creating a new cluster via a node movement is very expensive in contrast to the other node movements, it is very likely that the number of clusters in the final clustering is roughly the same as in the initial clustering. *Compressed* or *sequenced shifts* collapse a series of operations into one meta-operation, and rate only the output of this operation with its argument. Thus, a certain number of operations are free of charge. These functions are often used in combination with a standard potential function that has many local optima. By ignoring a number of intermediate steps, it may be easier to reach a global optimum.

Owing to their contingent iterative nature, shifting approaches are rarely used on their own. There can be sequences of shifting operations where the initial and final clustering are the same, so-called loops. Also, bounds on the runtime are more difficult to establish than for greedy approaches. Nonetheless, they are a common postprocessing step for local improvements. An example of shifting is given in Figure 8.8.

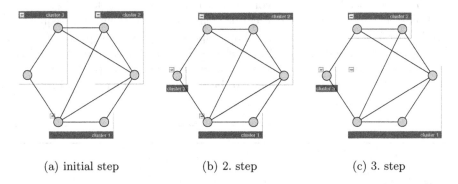

(a) initial step   (b) 2. step   (c) 3. step

Fig. 8.8. Example of shifting

General Optimization Concepts for Clustering. The two previous concepts, the greedy and the shifting framework, were fairly adapted. Both defined precisely the permitted operations and constraints, and the conditions of their application. The following concepts can be used for arbitrary optimization approaches. They are based on the idea that clusterings can be formulated as the result of a generic optimization process. The input data may be generated in a certain way with an implicit clustering structure. The optimization problem is to extract a clustering that is relatively close to the hidden one. Alternatively, the contained clustering is the result of an unknown optimization process. It is only known that this process respects certain paradigms, like intra-cluster density, inter-cluster sparsity, or both. The related problem is again to extract a clustering that is relatively close to the hidden one. The variety of techniques to solve optimization problems is gigantic, therefore only the following are considered here: parameter estimation, evolutionary and search-based approaches.

Parameter estimation is based on the assumption that the input data was created by a (random) sampling process: There is a hidden graph with a certain clustering, then a sample of the (whole) graph is drawn that will be the input graph. The approach then tries to estimate the parameters of this sampling process. These parameters will be used to reconstruct the clustering of the hidden graph. Originally, these methods were introduced to find clusterings for data embedded in metric spaces [325, Section 5.3]. There are clusterings that may be represented by a union of distributions, and the goal is to estimate the number of distributions and their parameters (mean, deviation, etc.). The *Expectation Maximization* (EM) is the most commonly used method. In general, it is only applied to data that is embedded in a metric space. Although graphs are typically not embedded, one can think of many processes that involve an implicit (metric) topology. An example is the following: let a finite space with a topology, a set of points, referred to as cluster centers, and a probability distribution for each cluster center be given; then n nodes/points are introduced by choosing one cluster center and a free spot around it that respects both the topology and its distribution. Two nodes will be connected by edges if their distance (with respect to the topology) is smaller than or equal to a given parameter. An example of this process is shown in Figure 8.9. Thus, the graph (Figure 8.9(c)) would be

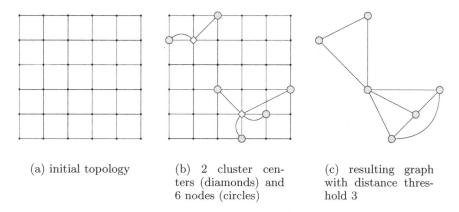

| (a) initial topology | (b) 2 cluster centers (diamonds) and 6 nodes (circles) | (c) resulting graph with distance threshold 3 |

Fig. 8.9. Example of generating a graph and its clustering using distributions.

the input graph, and the estimation approach would try to estimate the number of cluster centers as well as the assignment of each node to a cluster center. In the EM case, the resulting clustering should have the largest expectation to be the original hidden clustering, i.e., the same number of cluster points and the correct node cluster-point assignment.

Evolutionary approaches such as genetic algorithms (GA), evolution strategies (ES) and evolutionary programming (EP) iteratively modify a population of solution candidates by applying certain operations. 'Crossover' and 'mutation' are the most common ones. The first creates a new candidate by recombining two existing ones, whilst the second modifies one candidate. To each candidate a fitness value is associated, usually the optimization function evaluated on the candidate. After a number of basic operations, a new population is generated based on the existing one, where candidates are selected according to their fitness value. A common problem is to guarantee the feasibility of modified solutions. Usually this is accomplished by the model specification. In the context of clustering, the model can either use partitions or equivalence relations. As presented in Section 8.1.3, clusterings can be modeled as 0-1 vectors with certain constraints.

Search-based approaches use a given (implicit) topology of the candidate space and perform a random walk starting at an arbitrary candidate. Similar to evolutionary approaches, the neighborhood of a candidate can be defined by the result of simple operations like the mutations. The neighborhood of a clustering usually is the set of clusterings that result from node shifting, cluster merging, or cluster splitting. The selection of a neighborhood is also based on some fitness value, usually the optimization function evaluated on the candidate. The search usually stops after a certain number of iterations, after finding a local optimum, or a combination of both.

8.2.2   Algorithms

Clustering methods have been developed in many different fields. They were usually very adapted, either for specific tasks or under certain conditions. The reduction of algorithms to their fundamental ideas, and constructing a framework on top, started not that long ago. Thus, this part can only give a short synopsis about commonly used methods.

Instances of Linkage. The different instances of the Linkage framework were originally designed for distance edge weights. Distances are the 'dual' version of similarities. Historically, the input data for clusterings algorithms was metrically embedded and complete (the similarity/dissimilarity of every pair is known). In these scenarios, it is possible to find clusterings using distance functions instead of similarities, i.e., one has to search for spatially dense groups that are well-separated from each other. If the distance function is only partially known, it is no longer possible to derive information about the similarity of two objects from their distance to other objects.

However, the use of distance functions had certain advantages that can be carried over to similarity weights. One reason was that distances can be easily combined to estimate the distance of a path. The most common way is the summation of the edge weights along the path. The standard local cost functions are defined as:

$$c_{local}(C_i, C_j) := \bigodot \{d(u,v) \colon u \in C_i, v \in C_j\} , \qquad (8.30)$$

where $d(u,v)$ is the length of any shortest path connecting u and v, and $\bigodot$ is any set evaluation function, like minimum, average, or maximum. Indeed these three versions are called *Single Linkage*, *Average Linkage*, and *Complete Linkage*. A possible explanation of the name Single Linkage is that the cheapest, shortest path will be just an edge with minimum weight inside of $E(C_i, C_j)$. Note that the cost function can be asymmetric and have infinity as its value. Also, note that it is necessary to use the length of the shortest paths because not every node pair $(u,v) \in C_i \times C_j$ will be connected by an edge. In fact, the set $E(C_i, C_j)$ will be empty in the ideal case.

When dealing with similarities instead of distances one has to define a meaningful path 'length'. A simple way is to ignore it totally and define the cost function as

$$c_{local}\,(C_i, C_j) := \bigodot \{M - \omega(e)\colon e \in E(C_i, C_j)\}\ ,\qquad (8.31)$$

where M is the maximum edge weight in the graph. Alternatively, one can define the similarity of a path $P\colon v_1,\ldots,v_\ell$ by:

$$\omega(P) := \left(\sum_{i=1}^{\ell-1} \frac{1}{\omega(v_i, v_{i+1})}\right)^{-1}\,.\qquad (8.32)$$

Although this definition is compatible with the triangle inequality, the meaning of the original range can be lost along with other properties. Similar to cost definition in Equation (8.31), the distance value (in Equation(8.30)) would be replaced by $(n-1)M - \omega(P)$. These 'inversions' are necessary to be compatible with the range meaning of cost functions. Another definition that is often used in the context of probabilities is

$$\omega(P) := \prod_{i=1}^{\ell-1} \omega(v_i, v_{i+1})\,.\qquad (8.33)$$

If $\omega(v_i, v_{i+1})$ is the probability that the edge (v_i, v_{i+1}) is present and these probabilities are independent of each other, then $\omega(P)$ is the probability that the whole path exists.

Lemma 8.2.4. *The dendrogram of Single Linkage is defined by a Minimum Spanning Tree.*

Only a sketch of the proof will be given. A complete proof can be found in [324]. The idea is the following: consider the algorithm of Kruskal where edges are inserted in non-decreasing order, and only those that do not create a cycle. From the clustering perspective of Single Linkage, an edge that would create a cycle connects two nodes belonging to the same cluster, thus that edge cannot be an inter-cluster edge, and thus would have never been selected.

The Linkage framework is often applied in the context of sparse networks and networks where the expected number of inter-cluster edges is rather low. This is based on the observation that many Linkage versions tend to produce chains of clusters. In the case where either few total edges or few inter-cluster edges are present, these effects occur less often.

Instances of Splitting. Although arbitrary cut functions are permitted for Splitting, the idea of sparse cuts that separate different clusters from each other has been the most common one. Among these are: standard cuts (Equation (8.34)), Ratio Cuts (Equation (8.35)), balanced cuts (Equation (8.36)),

Table 8.2. Definition of various cut functions

$$S(V) := \min_{\emptyset \neq V' \subset V} \omega(E(V', V \setminus V')) \tag{8.34}$$

$$S_{\text{ratio}}(V) := \min_{\emptyset \neq V' \subset V} \frac{\omega(E(V', V \setminus V'))}{|V'| \cdot (|V| - |V'|)} \tag{8.35}$$

$$S_{\text{balanced}}(V) := \min_{\emptyset \neq V' \subset V} \frac{\omega(E(V', V \setminus V'))}{\min(|V'|, (|V| - |V'|))} \tag{8.36}$$

$$S_{\text{conductance}}(V) := \min_{\emptyset \neq V' \subset V} \delta(V') \tag{8.37}$$

$$S_{2\text{-sector}}(V) := \min_{\substack{V' \subset V, \\ \lfloor |V|/2 \rfloor \leq |V'| \leq \lceil |V|/2 \rceil}} \omega(E(V', V \setminus V')) \tag{8.38}$$

Conductance Cuts (Equation (8.37)), and Bisectors (Equation (8.38)). Table 8.2 contains all the requisite formulae.

Ratio Cuts, balanced cuts, and Bisectors (and their generalization, the k–Sectors) are usually applied when the uniformity of cluster size is an important constraint. Most of these measures are $\mathcal{NP}$-hard to compute. Therefore, approximation algorithms or heuristics are used as replacement. Note that balanced cuts and Conductance Cuts are based on the same fundamental ideas: rating the size/weight of the cut in relation to the size/weight of the smaller induced cut side. Both are related to node and edge expanders as well as isoperimetric problems. These problems focus on the intuitive notion of bottlenecks and their formalizations (see Section 8.1.2 for more information about bottlenecks). Some spectral aspects are also covered in Section 14.4, and [125] provides further insight. Beside these problems, the two cut measures have more in common. There are algorithms ([565]) that can be used to simultaneously approximate both cuts. However, the resulting approximation factor differs.

Splitting is often applied to dense networks or networks where the expected number of intra-cluster edges is extremely high. An example for dense graphs are networks that model gene expressions [286]. A common observation is that Splitting methods tend to produce small and very dense clusters.

Non-standard Instances of Linkage and Splitting. There are several algorithms that perform similar operations to 'linkage' or 'splitting', but do not fit into the above framework. In order to avoid application-specific details, only some general ideas will be given without the claim of completeness.

The *Identification of Bridge Elements* is a common Splitting variant, where cuts are replaced by edge or node subsets that should help to uncover individual clusters. One removal step can lead to further connected components, however it is not required. Figure 8.10 shows such an example. Most of the techniques that are used to identify bridge elements are based on structural indices, or properties

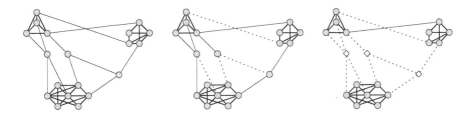

Fig. 8.10. Example for the removal of bridge elements. Removed elements are drawn differently: edges are dotted and nodes are reduced to their outline

derived from shortest path or flow computations. Also centralities can be utilized for the identification [445].

Multi-Level Approaches are generalizations of the Linkage framework, where groups of nodes are collapsed into a single element until the instance becomes solvable. Afterwards, the solution has to be transformed into a solution for the original input graph. During these steps the previously formed groups need not be preserved, i.e., a group can be split and each part can be assigned to individual clusters. In contrast to the original Linkage framework, here it is possible to tear an already formed cluster apart. Multi-level approaches are more often used in the context of equi-partitioning, where k groups of roughly the same size should be found that have very few edges connecting them. In this scenario, they have been successfully applied in combination with shiftings. Figure 8.11 shows an example.

Modularity – as presented at the beginning of Section 8.2.1, clustering algorithms have a very general structure: they mainly consist of 'invertible' transformations. Therefore, a very simple way to generate new clustering algorithms is the re-combination of these transformations, with modifications of the sequence where appropriate. A reduction does not need to reduce the size of an instance, on the contrary it also can increase it by adding new data. There are two different types of data that can be added. The first is information that is already present in the graph structure, but only implicitly. For example, an embedding such that the distance in the embedding is correlated to the edge weights. Spectral embeddings are quite common, i.e. nodes are positioned according to the entries of an eigenvector (to an associated matrix of the graph). More details about spectral properties of a graph can be found in Chapter 14. Such a step is usually placed at the beginning, or near the end, of the transformation sequence. The second kind is information that supports the current view of the data. Similar to the identification of bridge elements, one can identify cohesive groups. Bridges and these cohesive parts are dual to each other. Thus, while bridges would be removed, cohesive groups would be extended to cliques. These steps can occur during the whole transformation sequence.

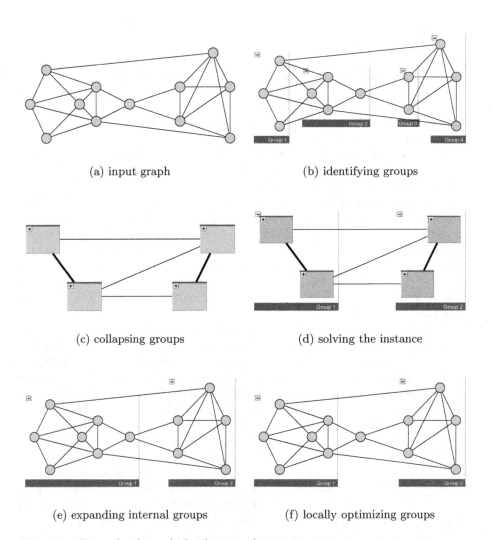

(a) input graph

(b) identifying groups

(c) collapsing groups

(d) solving the instance

(e) expanding internal groups

(f) locally optimizing groups

Fig. 8.11. Example of a multi-level approach

8.2.3   Bibliographic Notes

A good introduction to the variety of clustering techniques is given in [324, 325]. Although data mining is their primary motivation, other applicational aspects are covered as well. Equi-partitioning, i.e., finding a clustering where clusters have roughly the same size, is a slightly different problem from the general clustering problem. It is rooted in many divide and conquer methods. Many solving techniques involve splitting via cut functions and shiftings. Chip design (VLSI) is one major research field that has developed such algorithms. A survey that covers general methods as well as their relevance to VLSI is presented in [27]. More generally, finding good partitions, approximating sparse-cuts, and calculating (multi-commodity) flows has to be solved via (integer) linear programming. An introduction is given in [439, 507]. There are several other disciplines that deal with similar clustering and pattern matching problems, like Artifical Intelligence, neural networks, image processing [564], genetics [286], and facility location [439, 507]. Especially, facility location and its derivatives can be seen as extended clustering problems. There, a set of elements has to be covered by a subset of candidates, fulfilling certain constraints and optimizing a global target function. This is also closely related to the parameter estimation framework. An introduction to the general topic is provided in [169].

There is also a lot of work done within the theory community that develops sophisticated approximations and sampling techniques. These methods have the advantage that a provable guarantee of quality of resulting clustering can be given (see for example [36, 565, 170]). On the other hand, these techniques rarely operate on the clustering structure itself, but utilize other paradigms for finding a suitable decomposition. One of these tools is spectral decomposition. Unfortunately, these aspects go beyond the scope of this chapter. Some examples can be found in [27, 565, 170]. Spectral properties of networks are discussed in more detail in Chapter 14.

8.3   Other Approaches

The previous sections about quality measuring (Section 8.1) and clustering techniques (Section 8.2) provide an introduction both for application as well as for the topic as a whole. However, certain aspects have been neglected. Among them are extensions of clustering, alternative descriptions, axiomatics, dynamic updates, evaluation, pre- and post-processings. Discussing everything would definitely go beyond our scope, therefore only extensions of clusterings and axiomatics will be presented.

8.3.1   Extensions of Clustering

Clusterings were introduced as partitions of the node set. The extensions that are presented also group the node set.

Fuzzy clustering relaxes the disjoint constraint, thus clusters can overlap each other. The basic idea is that bridging elements belong to adjacent clusters rather than build their own one. In order to avoid redundancy, one usually requires that a cluster is not contained in the union of the remaining clusters. Figure 8.12 shows such an example where the two groups have the middle node in common. It is seldom used, due to its difficult interpretation. For example,

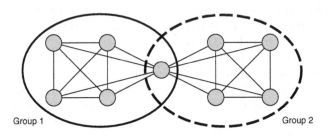

Fig. 8.12. Example of a fuzzy clustering

it is very difficult to judge single nodes or small node subsets that belong to a relatively large number of (fuzzy) clusters. When the number of clusters is also restricted, then artefacts occur more frequently. For example, if the number is restricted to a constant k, then a difficult situation is where more than k cliques of size at least k have a large number of nodes in common (see Figure 8.13(a)). If the number can scale according to a node's degree, then sparse clusters can be torn apart. For example, a star with k leaves may be decomposed into k fuzzy clusters, each containing one leaf and the central node (see Figure 8.13(b)).

Another extension is to enhance clusterings with *representatives*. Each cluster has a representative. It is very similar to the facility location problems (Section 8.2.3), where a candidate covers a subset of elements. This can be seen as 'representing' the group by one element. It is usually a node that is located 'in the center' of the cluster. This form of enhancement can be very effective when the graph is embedded in a metric or a vector space. In these cases the representative can also be an element of the space and not of the input. The concept is also used to perform speed-ups or approximate calculations. For example, if all the similarity/distance values between the nodes in two clusters are needed, then it can be sufficient to calculate the similarity/distance values between the representatives of the clusters.

Nested clustering represents a nested sequence of node subsets, i.e., a mapping $\eta\colon \mathbb{N} \to \mathcal{P}(V)$ such that:

1. the subsets are nested, i.e.,

$$\forall\, i \in \mathbb{N}\colon \eta(i+1) \subseteq \eta(i)$$

2. and the sequence is finite, i.e.,

$$\exists\, k \in \mathbb{N}\colon \forall \ell > k\colon \eta(\ell) = \emptyset \ .$$

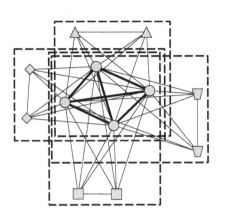
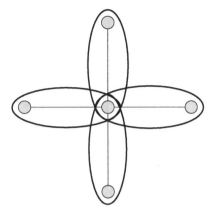

(a) Four cliques of size six that have a common K_4. Each maximum clique is a fuzzy cluster.

(b) A star with four leaves and each fuzzy cluster contains the center node and one leaf.

Fig. 8.13. Two examples of fuzzy clusterings where many clusters intersect each other

The smallest possible k is also called the *size* of the sequence, and $\eta(k)$ the top element. The intuition behind this structure is that the top element $\eta(k)$ consists of locally maximal dense groups. The density of the subsets $\eta(i)$ also decreases with decreasing argument i. Therefore the argument i can be seen as degree of density. One can distinguish two extreme types: The first one is called *hierarchies*, where each $\eta(i)$ induces a connected graph. The corresponding graphs can be seen as onions, i.e., having a unique core and multiple layers around it with different density. The second type is called *peaks*, and is complementary to hierarchies, i.e., at least one subset $\eta(i)$ induces a disconnected graph. An appropriate example may be boiling water, where several hotspots exist that are separated by cooler parts. If the graph has a skew density distribution then a hierarchy type can be expected. In this scenario, it can reveal structural information, unlike standard clustering. If the graph has a significant clustering, then peaks are more likely to occur. In that case, the top element consists of the core-parts of the original clusters. Figure 8.14 shows such examples. *Cores* are an often used realization of nested clusterings. They were introduced in [513, 47]. They can be efficiently computed and capture the intuition quite well.

8.3.2 Axiomatics

Axioms are the usual way to formulate generic properties and reduce concepts to their cores. They also provide a simple way to prove characteristics, once a property is proven for a system of axioms. Every structure fulfilling the axioms possesses this property. In the following, only Kleinberg's proposal introduced

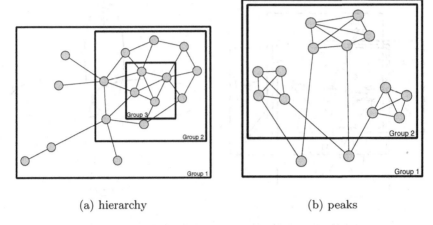

(a) hierarchy   (b) peaks

Fig. 8.14. Examples of a nested clustering

in [361] is presented. A graph clustering version of his result will be presented here. This version will be equivalent to the original.

Let $K_n = (V, E)$ be the complete graph on n nodes, and $\omega\colon E \to \mathbb{R}_0^+$ a distance function on the edges. The set of all possible distance functions is denoted by D.

Definition 8.3.1. *A clustering function f is a mapping $f\colon D \to \mathcal{A}(K_n)$ fulfilling the following axioms:*

Scale-Invariance:
$$\forall\, \alpha \in \mathbb{R}^+, \omega \in D\colon f(\omega) = f(\alpha \cdot \omega)\ ,$$

where $(\alpha \cdot \omega)(u, v) := \alpha \cdot (\omega(u, v))$
Richness: $f(D) = \mathcal{A}(K_n)$
Consistency: for all $\omega, \omega' \in D$ hold

$$\left(\forall\, u, v \in V\colon \omega'(u, v) \begin{cases} \leq \omega(u, v) & ,\ if\ u \sim_{f(\omega)} v \\ \geq \omega(u, v) & ,\ otherwise \end{cases}\right) \implies f(\omega) = f(\omega')\ .$$

A brief intuition of the axioms in Definition 8.3.1 is given before the consequences are presented. Scale-invariance ensures that a clustering does not depend on hard-coded values, but rather on ratios. The clustering should not change if the distances are homogeneously increased. Richness ensures that every possible clustering has at least one edge weighting as a preimage. It is self-evident that every clustering should be constructable by assigning suitable weights to the edges. Finally, consistency handles the relation between different clusterings. Assume that ω is fixed, so $f(\omega)$ represents a clustering. If the weights on the edges are changed respecting $f(\omega)$, then the clustering should be respected as well. The modifications on ω consist of non-increasing the distance inside of clusters and non-decreasing the distance between clusters. In this way clusters may become more compact (distances can decrease) and different clusters may become more separated (distances can increase).

Theorem 8.3.2 ([361, Theorem 2.1]). *For all $n \geq 2$ there exists no clustering function.*

Only a sketch of the proof is given here, but full details can be found in [361].

Sketch of Proof. Assume there exists a clustering function f for a fixed n. The basic idea is to show that $f(D)$, i.e., the image of f, cannot contain both a clustering and a refinement of it simultaneously. This is achieved by using the axiom consistency (and scale-invariance). There exists a clustering $\mathcal{C}$ in $\mathcal{A}(K_n)$ such that at least one cluster has more than one element, and an edge weighting ω with $f(\omega) = \mathcal{C}$. Informally speaking, it is possible to modify ω to isolate a proper subset of a cluster within $\mathcal{C}$. This can be done in such a way that consistency is fulfilled for the two weighting functions. However, one can show that the modified weighting also leads to a refinement of $\mathcal{C}$. Therefore, it is not possible that $f(D)$ contains a clustering and its refinement. Thus $f(D)$ is an antichain. This contradicts the fact that $\mathcal{A}(K_n)$ is not an antichain for $n \geq 2$.   □

Theorem 8.3.2 is a negative result for clustering functions and, as a consequence, for clustering algorithms. However, the set of axioms is very restrictive. In fact, there are many functions that are close to being clustering functions.

Lemma 8.3.3 ([361, Theorem 2.2]). *There are many functions that fulfill two of the three axioms of Definition 8.3.1.*

For every combination of two axioms, there is a selection criterion for Single Linkage such that it is a clustering function fulfilling these two axioms.

Further investigations reveal that relaxing the conditions of scale-invariance and consistency to refinements leads to clustering functions. This is not the only possibility. Another way is to explore the 'lack' of information (edges). Usually graphs are not complete, and cannot be complete for several reasons. The standard technique, completing the edge set and assigning only extremely large values to the additional edges, is not possible in this scenario. First of all, all values need to be finite, and, second, the two axioms, scale-invariance and consistency, allow general manipulations so there is no guarantee that an artificially introduced extreme value can keep its role. In order to include the lack of information, a clustering function could be defined as a mapping from the set of weighted relations to the set of partitions over the set. For a set of elements X, the set of all weighted (binary) relations over X is denoted by $\Omega(X)$ – or in short Ω, if the set X is clear. For every relation $\omega \in \Omega(X)$ its domain is given by $E(\omega)$.

Definition 8.3.4. *Given a set of elements X, a graph clustering function f is a mapping $f \colon \Omega(X) \to \mathcal{A}(X)$ fulfilling the following axioms:*

Scale-Invariance:

$$\forall \alpha \in \mathbb{R}^+, \omega \in \Omega(X) \colon f(\omega) = f(\alpha \cdot \omega)$$

Richness: $f(\Omega(X)) = \mathcal{A}(X)$

Consistency: for all $\omega, \omega' \in \Omega(X)$ with $E(\omega) \subseteq E(\omega')$ hold

$$\left(\forall\, u, v \in V : \omega'(u, v) \begin{cases} \leq \omega(u, v) & , \text{if } u \sim_{f(\omega)} v \\ \geq \omega(u, v) & , \text{otherwise} \end{cases} \right) \implies f(\omega) = f(\omega') \ ,$$

where $\omega(u, v) = \infty$ for $(u, v) \in E(\omega') \setminus E(\omega)$.

Note that the axioms in Definition 8.3.1 and Definition 8.3.4 are essentially the same. The complex rule for consistency is due to the fact that the relations in Ω can have different domains. This implies the additional constraints. Also the set D of all edge weightings (of a complete graph) is a subset of Ω. The name *graph clustering function* is inspired by the fact that X and $\omega \in \Omega$ can be seen as a graph: $G_\omega = (X, E(\omega), \omega)$. Thus, the graph clustering function f maps the set of (weighted and simple) graphs with n elements to the set of all partitions over n elements.

Lemma 8.3.5. *The function f_{comp} that maps $\omega \in \Omega$ to the clustering where the clusters are the connected components of G_ω is a graph clustering function.*

Proof. It is sufficient to check the axioms since f_{comp} has the correct domain and codomain. Scale-invariance and consistency are fulfilled. This is due to the fact that for every relation $\omega \in \Omega$, the clustering $f_{\text{comp}}(\omega)$ has no inter-cluster edges. Although additional edges can be introduced via the axiom consistency, these edges cannot be inter-cluster edges with respect to $f_{\text{comp}}(\omega)$. The axiom richness is also satisfied because, for every clustering $\mathcal{C}$, there exists a spanning forest F with the same connected components as the clusters. Its edge relation E_F induces a weighted relation ω via $\omega(u, v) = 1$ if $(u, v) \in E_F$. The relation ω will be mapped to $\mathcal{C}$ by f_{comp}. $\square$

Definition 8.3.4 and Lemma 8.3.5 clearly indicate that the lack of information indeed can be more explanatory than complete information. The presented graph clustering function does not provide further insight, but shows the potential of missing information. Alternatively to graph clustering functions, the axiom consistency may be redefined in various ways: restricted to a cluster or the connection between clusters, homogenous scalings, or predetermined breaking/merging points for controlled splits/linkages. These ideas are mostly unexplored, but have a certain potential.

A closing remark: it would also be interesting to know if there exist axiom systems that describe, or even characterize, already used frameworks. As mentioned previously, Single Linkage (with certain selection criteria) fulfills any two axioms, therefore it may be possible to extend current frameworks with the help of axioms. This might also be very valuable in identifying fundamental clustering concepts.

8.4   Chapter Notes

In this chapter, graph clustering has been introduced that was based on cohesion. The main objective of clustering is to find 'natural' decompositions of graphs. In our case, partitions of the node set modeled the clustering, and the paradigm of intra-cluster density versus inter-cluster sparsity had to be respected.

Both the term natural decomposition and the considered paradigm define clusterings only in an intuitive way, without mathematical parameters. To capture the informal notion more formally, structural indices were used that rate partitions with respect to clustering paradigms (see Section 8.1). Another benefit of these indices is that clusterings of different graphs can be compared, which is required for benchmark applications and to handle graphs with uncertainties. Most optimization problems related to indices, i.e., finding a non-trivial clustering with best index score, are $\mathcal{NP}$-hard.

In the first part of Section 8.2.1, generic frameworks were presented. They include simple methods like iteratively merging or splitting clusters, shifting concepts, and general optimization techniques. The frameworks are very flexible and can integrate a large number of problem-specific requirements. The next part dealt with the frameworks' parameters. It is still a high level view, and can be applied to most of the clustering problems. Some extensions with relatively similar methods concluded the section.

Alternative approaches for the representation of clusterings, as well as the methods to find suitable clusterings, were covered in the final section. First, the data structure partition was either enhanced with additional information or replaced by other decomposition types. Also, an axiom system was introduced to characterize clustering methods. In contrast to Section 8.2.2, where techniques were distinguished according to their different underlying ideas, the axiomatic framework describes the common foundation. Kleinberg's axioms were discussed in detail.

Owing to the large variety of applications for graph clustering, a complete and homogenous framework has not yet been established. However, several basic clustering techniques are fairly well understood, both from the informal and the mathematical points of view. The most difficult aspect is still the formalization of the 'natural decomposition', and, therefore, a mathematical framework for qualitatively good clusterings. Indices present a suitable method.

Alternative approaches to graph decompositions that are not based on density, but rather on structural properties, are presented in Chapters 9 and 10.

9 Role Assignments

Jürgen Lerner

Classification is the key to understand large and complex systems that are made up of many individual parts. For example in the study of food webs (networks that consist of living organisms and predator-prey relationships, flow of protein, etc.) it is, even for moderately small ecosystems, impossible to understand the relationship between each pair of individual organisms. Nevertheless, we can understand the system – to a certain extent – by classifying individuals and describing relationships on the class level. Classification in networks aims to describe regular patterns of interaction and to highlight essential structure, which remains stable over long periods of time.

In this chapter we formalize the classification of vertices in a graph, such that vertices in the same class can be considered to occupy the same *position*, or play the same *role* in the network. This idea of network position or role, see e.g., Nadel [436], has been formalized first by Lorrain and White [394] by a special type of vertex partition. They proposed that vertices play the same role if they have identical neighborhoods. Subsequent work like Sailer [501] and White and Reitz [579] generalized this early definition, weakening it sufficiently to make it more appropriate for modeling social roles. All these definitions have in common that vertices which are claimed to play the same role must have something in common w.r.t. the relations they have with other vertices, i.e., a generic problem definition for this chapter can be given by

> given a graph $G = (V, E)$,
> find a partition of V that is *compatible* with E.

The generic part here is the term 'compatible with E'. In this chapter, we present definitions for such compatibility requirements, and properties of the resulting classes of vertex-partitions.

Outline of this chapter. The remainder of this section treats preliminary notation. In Sections 9.1 through 9.3, different types of role assignments are introduced and investigated. In Section 9.4 definitions are adapted to graphs with multiple relations (see Definition 9.4.1) and in Section 9.5 composition of relations is introduced and its relationship to role assignments is investigated.

Sections 9.1 through 9.3 follow loosely a common pattern: After defining a compatibility requirement, some elementary properties of the so-defined set of role assignment are mentioned. Then, we investigate a partial ordering on this set, present an algorithm for computing specific elements, and treat the

U. Brandes and T. Erlebach (Eds.): Network Analysis, LNCS 3418, pp. 216–252, 2005.
© Springer-Verlag Berlin Heidelberg 2005

complexity of some decision problems. We provide a short conclusion for each type of vertex partition, where we dwell on the applicability for defining role assignments in empirical networks.

The most complete investigation is for regular equivalences in Section 9.2. Although there is some scepticism as to whether regular equivalences are a good formalization of role assignments in real social networks, we have chosen to treat them prominently in this chapter, since their investigation is exemplary for the investigation of types of role assignments. The results for regular equivalences are often translatable to other types of equivalences, often becoming easier or even trivial. We emphasize this generality when appropriate.

Graph model of this chapter. In this chapter, *graph* usually means directed graph, possibly with loops. Except for Sections 9.2.4 and 9.2.5, where graph means undirected graph, Section 9.3.1, where results are for undirected multigraphs, and Sections 9.4 and 9.5, where we consider graphs with multiple relations (see Definition 9.4.1).

9.0.1 Preliminaries

In the following, we will often switch between vertex partitions, equivalence relations on the vertex set, or role assignments, since, depending on the context, some point of view will be more intuitive than the other. Here we establish that these are just three different formulations for the same underlying concept.

Let V be a set. An *equivalence relation* $\sim$ is a binary relation on V that is *reflexive*, *symmetric*, and *transitive*, i.e., $v \sim v$, $u \sim v$ implies $v \sim u$, and $u \sim v \wedge v \sim w$ implies $u \sim w$, for all $u, v, w \in V$. If $v \in V$ then $[v] := \{u \,;\, u \sim v\}$ is its *equivalence class*.

A *partition* $\mathcal{P} = \{C_1, \ldots, C_k\}$ of V is a set of non-empty, disjoint subsets $C_i \subseteq V$, called *classes* or *blocks*, such that $V = \bigcup_{i=1}^{k} C_i$. That is, each vertex $v \in V$ is in exactly one class.

If $\sim$ is an equivalence relation on V, then the set of its equivalence classes is a partition of V. Conversely, a partition $\mathcal{P}$ induces an equivalence relation by defining that two vertices are equivalent iff they belong to the same class in $\mathcal{P}$. These two mappings are mutually inverse.

Definition 9.0.1. *A role assignment for V is a surjective mapping $r \colon V \to W$ onto some set W of roles.*

The requirement *surjective* is no big loss of generality since we can always restrict a mapping to its image set. One could also think of role assignments as vertex-colorings, but note that we do not require that adjacent vertices must have different colors. We use the terms role and position synonymously.

A role assignment defines a partition of V by taking the inverse-images $r^{-1}(w) := \{v \in V \,;\, r(v) = w\}$, $w \in W$ as classes. Conversely an equivalence relation induces a role assignment for V by the class mapping $v \mapsto [v]$. These two mappings are mutually inverse, up to isomorphism of the set of roles.

We summarize this in the following remark.

Remark 9.0.2. For each partition there is a unique associated equivalence relation and a unique associated role assignment and the same holds for all other
combinations.

For the remainder of this chapter, definitions for vertex partitions translate
to associated equivalence relations and role assignments.

9.0.2   Role Graph

The image set of a role assignment can be supplied naturally with a graph
structure. We define that roles are adjacent if there are adjacent vertices playing
these roles:

Definition 9.0.3. *Let $G = (V, E)$ be a graph and $r\colon V \to W$ a role assignment.
The* role graph $R = (W, F)$ *is the graph with vertex set W (the set of roles) and
edge set $F \subseteq W \times W$ defined by*

$$F := \{(r(u), r(v))\,;\ \exists u, v \in V\ \text{such that}\ (u, v) \in E\}\ .$$

R is also called quotient *of G over r.*

The role graph R models roles and their relations. It can also be seen as a
smaller model for the original graph G. Thus, a role assignment can be seen as
some form of network compression. Necessarily, some information will get lost
by such a compression. The goal of role analysis is to find role assignments such
that the resulting role graph displays essential structural network properties,
i. e., that not too much information will get lost.

Thus we have two different motivations for finding good role assignments.
First to know which individuals (vertices) are 'similar'. Second to reduce network
complexity: If a network is very large or irregular, we can't capture its structure
on the individual (vertex) level but perhaps on an aggregated (role) level. The
hope is that the role graph highlights essential and more persistent network
structure. While individuals come and go, and behave rather irregularly, roles
are expected to remain stable (at least for a longer period of time) and to display
a more regular pattern of interaction.

9.1   Structural Equivalence

As mentioned in the introduction, the goal of role analysis is to find meaningful
vertex partitions, where 'meaningful' is up to some notion of compatibility with
the edges of the graph. In this section the most simple, but also most restrictive
requirement of compatibility is defined and investigated. Lorrain and White
[394] proposed that individuals are role equivalent if they are related to the
same individuals.

Definition 9.1.1. *Let $G = (V, E)$ be a graph, and $r\colon V \to W$ a role assignment.
Then, r is called* strong structural *if equivalent vertices have the same (out- and
in-)neighborhoods, i. e., if for all $u, v \in V$*

$$r(u) = r(v) \implies N^+(u) = N^+(v)\ \text{and}\ N^-(u) = N^-(v)\ .$$

Remember Remark 9.0.2: Definitions for role assignments translate to associated partitions and equivalence relations.

Remark 9.1.2. By Definition 9.0.3 it holds for any role assignment r that, if (u,v) is an edge in the graph, then $(r(u),r(v))$ is an edge in the role graph. If r is strong structural, then the converse is also true. This is even an equivalent condition for a role assignment to be strong structural [579]. That is, a role assignment r is strong structural if and only if for all $u,v \in V$, it holds that $(r(u),r(v))$ is an edge in the role graph if and only if (u,v) is an edge in the graph.

We present some examples for strong structural equivalences. The identity mapping id: $V \to V$; $v \mapsto v$ is strong structural for each graph $G = (V,E)$ independent of E. Some slightly less trivial examples are shown in Figure 9.1. For the star, the role assignment that maps the central vertex onto one role and all other vertices onto another, is strong structural. The bipartition of a complete bipartite graph is strong structural. The complete graph without loops has no strong structural role assignment besides id, since the neighborhood of each vertex v is the only one which does not contain v.

Fig. 9.1. Star (left), complete bipartite graph (middle) and complete graph (right)

We note some elementary properties. A class of strong structurally equivalent vertices is either an independent set (induces a subgraph without edges) for the graph or a clique with all loops. In particular, if two adjacent vertices u,v are strong structurally equivalent, then both (u,v) and (v,u) are edges of the graph, and both u and v have a loop.

The undirected distance of two structurally equivalent (non-isolated) vertices is at most 2. For if u and v are structurally equivalent and u has a neighbor w then w is also a neighbor of v. Thus, structural equivalence can only identify vertices that are near each other.

Although in most irregular graph there won't be any non-trivial structural equivalence, the set of structural equivalences might be huge. For the complete graph with loops, every equivalence is structural. In Section 9.1.2, we investigate a partial order on this set.

Variations of structural equivalence. The requirement that strong structurally equivalent adjacent vertices must have loops has been relaxed by some authors.

Definition 9.1.3 ([191]). *An equivalence $\sim$ on the vertex set of a graph is called* structural *if for all vertices $u \sim v$ the transposition of u and v is an automorphism of the graph.*

White and Reitz [579] gave a slightly different definition, which coincides with Definition 9.1.3 on loopless graphs.

9.1.1 Lattice of Equivalence Relations

The set of equivalence relations on a set V is huge. Here we show that this set naturally admits a partial order, which turns out to be a lattice. (For more on lattice theory, see e. g., [261].) This section is preliminary for Sections 9.1.2 and 9.2.2.

Equivalence relations on a set V are subsets of $V \times V$, thus they can be partially ordered by set-inclusion ($\sim_1 \leq \sim_2$ iff $\sim_1 \subseteq \sim_2$). The equivalence relation $\sim_1$ is then called *finer* than $\sim_2$ and $\sim_2$ is called *coarser* than $\sim_1$. This partial order for equivalences translates to associated partitions and role assignments (see remark 9.0.2).

In partially ordered sets, two elements are not necessarily comparable. In some cases we can at least guarantee the existence of lower and upper bounds.

Definition 9.1.4. *Let X be a set that is partially ordered by $\leq$ and $Y \subseteq X$.*

$y^* \in X$ is called an upper bound *(a* lower bound*) for Y if for all $y \in Y$, $y \leq y^*$ $(y^* \leq y)$.*

$y^* \in X$ is called the supremum *(infimum) of Y, if it is an upper bound (lower bound) and for each $y' \in X$ that is an upper bound (lower bound) for Y, it follows $y^* \leq y'$ $(y' \leq y^*)$. The second condition ensures that suprema and infima (if they exist) are unique.*

The supremum of Y is denoted by $\sup(Y)$ the infimum by $\inf(Y)$. We also write $\sup(x, y)$ or $\inf(x, y)$ instead of $\sup(\{x, y\})$ or $\inf(\{x, y\})$, respectively.

A lattice *is a partially ordered set L, such that for all $a, b \in L$, $\sup(a, b)$ and $\inf(a, b)$ exist. $\sup(a, b)$ is also called the* join *of a and b and denoted by $a \vee b$. $\inf(a, b)$ is also called the* meet *of a and b and denoted by $a \wedge b$.*

If $\sim_1$ and $\sim_2$ are two equivalence relations on V, then their intersection (as sets) is the infimum of $\sim_1$ and $\sim_2$. The supremum is slightly more complicated. It must contain all pairs of vertices that are equivalent in either $\sim_1$ or $\sim_2$, but also vertices that are related by a chain of such pairs: The *transitive closure* of a relation $R \subseteq V \times V$ is defined to be the relation $S \subseteq V \times V$, where for all $u, v \in V$

$$uSv \Leftrightarrow \exists k \in \mathbb{N}, \exists w_1, \ldots, w_k \in V \text{ such that}$$
$$u = w_1, v = w_k, \text{ and } \forall i = 1, \ldots, k-1 \text{ it is } w_i R w_{i+1} \ .$$

The transitive closure of a symmetric relation is symmetric, the transitive closure of a reflexive relation is reflexive and the transitive closure of any relation is transitive.

It follows that, if $\sim_1$ and $\sim_2$ are two equivalence relations on V, then the transitive closure of their union is the supremum of $\sim_1$ and $\sim_2$.

We summarize this in the following theorem.

Theorem 9.1.5. *The set of equivalence relations is a lattice.*

The interpretation in our context is the following: Given two equivalence relations identifying vertices that play the same role, there exists a uniquely defined smallest equivalence identifying all vertices which play the same role in either one of the two original equivalences. Moreover, there exists a uniquely defined greatest equivalence distinguishing between actors which play a different role in either one of the two original equivalences.

9.1.2 Lattice of Structural Equivalences

It can easily be verified that if $\sim_1$ and $\sim_2$ are two strong structural equivalences for a graph, then so are their intersection and the transitive closure of their union.

Proposition 9.1.6. *The set of strong structural equivalences of a graph is a sublattice of the lattice of all equivalence relations.*

In particular there exist always a maximum structural equivalence (MSE) for a graph.

The property of being strong structural is preserved under refinement:

Proposition 9.1.7. *If $\sim_1 \leq \sim_2$ and $\sim_2$ is a strong structural equivalence, then so is $\sim_1$.*

Although the above proposition is very simple to prove, it is very useful, since it implies that the set of all structural equivalences of a graph is completely described by the MSE. In the next section we present a linear time algorithm for computing the MSE of a graph.

9.1.3 Computation of Structural Equivalences

Computing the maximal strong structural equivalence for a graph $G = (V, E)$ is rather straight-forward. Each vertex $v \in V$ partitions V into 4 classes (some of which may be empty): Vertices which are in $N^+(v)$, in $N^-(v)$, in both, or in none.

The basic idea of the following algorithm 21 is to compute the intersection of all these partitions by looking at each edge at most twice. This algorithm is an adaption of the algorithm of Paige and Tarjan [459, Paragraph 3] (see Section 9.2.3) for the computation of the regular interior, to the much simpler problem of computing the MSE.

The correctness of algorithm 21 follows from the fact that it divides exactly the pairs of vertices with non-identical neighborhoods.

An efficient implementation requires some datastructures, which will be presented in detail since this is a good exercise for understanding the much more complicated algorithm in Section 9.2.3.

Algorithm 21: Computation of the maximal strong structural equivalence (MSE) of a graph

---

Input: a graph $G = (V, E)$

begin

    maintain a partition $\mathcal{P} = \{C_1, \ldots, C_k\}$ of V, which initially is the complete partition $\mathcal{P} = \{V\}$

    // at the end, $\mathcal{P}$ will be the MSE of G

    foreach $v \in V$ **do**

        foreach *class C to which a vertex $u \in N^+(v)$ belongs to* **do**

            create a new class C' of $\mathcal{P}$

            move all vertices in $N^+(v) \cap C$ from C to C'

            if *C has become empty* **then**

                └ remove C from $\mathcal{P}$

        foreach *class C to which a vertex $u \in N^-(v)$ belongs to* **do**

            create a new class C' of $\mathcal{P}$

            move all vertices in $N^-(v) \cap C$ from C to C'

            if *C has become empty* **then**

                └ remove C from $\mathcal{P}$

end

- A graph $G = (V, E)$ must permit access to the (out-/in-)incidence list of a vertex v in time proportional to the size of this list.
- Scanning all elements of a list must be possible in linear time.
- An edge must permit access to its source and its target in constant time.
- A partition must allow insertion and deletion of classes in constant time.
- A class must allow insertion and deletion of vertices in constant time.
- A vertex must permit access to its class in constant time.

The requirements on partitions and classes are achieved if a partition is represented by a doubly linked list of its classes and a class by a doubly linked list of its vertices.

One refinement step (the outer loop) for a given vertex v is performed as follows.

1. Scan the outgoing edges of v. For each such edge (v, u), determine the class C of u and create an associated block C' if one does not already exist. Move u from C to C'.
2. During the scanning, create a list of those classes C that are split. After the scanning process the list of split classes. For each such class C mark C' as no longer being associated with C and eliminate C if C is now empty.
3. Scan the incoming edges of v and perform the same steps as above.

A loop for a given v runs in time proportional to the degree of v, if v is non-isolated and in constant time else. An overall running time of $\mathcal{O}(|V| + |E|)$ follows, which is also an asymptotic bound for the space requirement.

Conclusion. Structural equivalence is theoretically and computationally very simple. It is much too strict to be applied to irregular networks and only vertices that have distance at most 2, can be identified by a structural equivalence. Nevertheless, structural equivalence is the starting point for many relaxations (see Chapter 10).

9.2   Regular Equivalence

Regular equivalence goes back to the idea of *structural relatedness* of Sailer [501], who proposed that actors play the same role if they are connected to role-equivalent actors – in contrast to structural equivalence, where they have to be connected to identical actors. Regular equivalence has first been defined precisely by White and Reitz in [579]. Borgatti and Everett (e. g., [191]) gave an equivalent definition in terms of colorings (here called role assignments). A coloring is regular if vertices that are colored the same, have the same colors in their neighborhoods. If $r: V \to W$ is a role assignment and $U \subseteq V$ then $r(U) := \{r(u)\,;\, u \in U\}$ is called the *role set* of U.

Definition 9.2.1. *A role assignment* $r: V \to W$ *is called* regular *if for all* $u, v \in V$

$$r(u) = r(v) \quad \Longrightarrow \quad r(N^+(u)) = r(N^+(v)) \text{ and } r(N^-(u)) = r(N^-(v)) \ .$$

The righthand side equations are equations of sets. There are many more equivalent definitions, (see e. g., [579, 90]).

Regular role assignments are often considered as *the* class of role assignments. The term *regular* is often omitted in literature.

Regular equivalence and bisimulation. Marx and Masuch [408] pointed out the close relationship between regular equivalence, bisimulation, and dynamic logic. A fruitful approach to find good algorithms for regular equivalence is to have a look at the bisimulation literature.

9.2.1   Elementary Properties

In this section we note some properties of regular equivalence relations.

The *identity* mapping $\mathrm{id}: V \to V;\ v \mapsto v$ is regular for all graphs. More generally, every structural role assignment is regular.

The next proposition characterizes when the complete partition, which is induced by the constant role assignment $J: V \to 1$ is regular. A *sink* is a vertex with zero outdegree, a *source* is one with zero indegree.

Proposition 9.2.2 ([82]). *The complete partition of a graph* $G = (V, E)$ *is regular if and only if* G *contains neither sinks nor sources or* $E = \emptyset$.

Proof. If: If $E = \emptyset$ then the righthand side in definition 9.2.1 is simply $\emptyset = \emptyset$, thus each role assignment is regular. If G has neither sinks nor sources, then, for all $v \in V$, $J(N^+(v)) = J(N^-(v)) = \{1\}$ and the equations in Definition 9.2.1 are satisfied for all $u, v \in V$.

Only if: Suppose $E \neq \emptyset$ and let $v \in V$ be a sink. Since $E \neq \emptyset$ there exists $u \in V$ with non-zero outdegree. But then

$$J(N^+(v)) = \emptyset \neq \{1\} = J(N^+(u)) \ ,$$

but $J(u) = 1 = J(v)$, thus J is not regular. The case of G containing a source is treated analogously. $\square$

The identity and the complete partition are called *trivial* role assignments. The next lemma is formulated in [190] for undirected connected graphs, but it has a generalization to strongly connected (directed) graphs.

Lemma 9.2.3. *Let G be a strongly connected graph. Then in any non-trivial role assignment r of G, neither $\{r(v)\} = r(N^+(v))$ nor $\{r(v)\} = r(N^-(v))$ holds for any vertex v.*

Proof. If for some vertex v it is $\{r(v)\} = r(N^+(v))$, then the same would need to be true for each vertex in $N^+(v)$. Hence each vertex in successive out-neighborhoods would be assigned the same role and since G is strongly connected it follows that $r(V) = \{r(v)\}$ contradicting the fact that the role assignment is non-trivial. The case of $\{r(v)\} = r(N^-(v))$ for some vertex v is handled equally. $\square$

A graph with at least 3 vertices whose only regular role assignments are trivial is called *role primitive*. The existence of directed role primitive graphs is trivial: For every directed path only the identity partition is regular. Directed graphs which have exactly the identity and the complete partition as regular partitions are for example directed cycles of prime length, since every non-trivial regular equivalence induces a non-trivial divisor of the cycle length.

The existence of undirected role primitive graphs is non-trivial.

Theorem 9.2.4 ([190]). *The graph in Figure 9.2 is role primitive.*

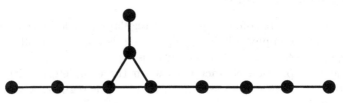

Fig. 9.2. A role-primitive undirected graph

The proof goes by checking that all possible role assignments are either non regular or trivial, where one can make use of the fact that the pending paths of the graph in Figure 9.2 largely diminish the possibilities one has to follow. The proof is omitted here.

A graph in which any role assignment is regular is called *arbitrarily role-assignable*. The next lemma is formulated in [190] for undirected connected graphs.

Lemma 9.2.5. *A strongly connected graph* $G = (V, E)$ *is arbitrarily role-assignable if and only if it is a complete graph, possibly with some but not necessarily all loops.*

Proof. Let $G = (V, E)$ be a graph satisfying the condition of the lemma and let r be any role assignment. We have to show that for all vertices $u, v \in V$

$$r(u) = r(v) \Longrightarrow r(N^+(u)) = r(N^+(v)) \text{ and } r(N^-(u)) = r(N^-(v)) .$$

If $u = v$ this is trivial. Otherwise u and v are connected by a bidirected edge, i.e., the role sets of their in- and out- neighborhoods contain $r(u)$. These role sets also contain all other roles since u and v are connected to all other vertices. So the role sets of the in- and out- neighborhoods of both vertices contain all roles, whence they are equal.

Conversely, let $G = (V, E)$ be a graph with two vertices u and v, such that $u \neq v$ and $(u, v) \notin E$. We assign $V \setminus \{v\}$ one role and v a different one. This is a non-trivial role assignment (note that $n > 2$, since G is connected) with $r(u) = r(N^+(u))$. So by Lemma 9.2.3 this role assignment can't be regular. □

9.2.2 Lattice Structure and Regular Interior

We have seen that the set of regular equivalences of a graph might be huge. In this section we prove that it is a lattice. See the definition of a lattice in Section 9.1.1.

Theorem 9.2.6 ([82]). *The set of all regular equivalences of a graph G forms a lattice, where the supremum is a restriction of the supremum in the lattice of all equivalences.*[1]

Proof. By Lemma 9.2.7, which will be shown after the proof of this theorem, it suffices to show the existence of suprema of arbitrary subsets. The identity partition is the minimal element in the set of regular equivalences, thus it is the supremum for the empty set. Hence we need only to consider the supremum for non-empty collections of regular role assignments. Since the set of all equivalences of a graph is finite, it even suffices to show the existence of the supremum of two regular equivalences.

So let $\sim_1$ and $\sim_2$ be two regular equivalences on G. Define $\equiv$ to be the transitive closure of the union of $\sim_1$ and $\sim_2$.

As mentioned in Section 9.1.1, $\equiv$ is the supremum of $\sim_1$ and $\sim_2$ in the lattice of all equivalences, so it is an equivalence relation and it is a supremum of $\sim_1$

---

[1] For the infimum see proposition (9.2.9).

and $\sim_2$ with respect to the partial order (which is the same in the lattice of all equivalences and in the lattice of regular equivalences). Therefore it remains to show that $\equiv$ is regular.

For this suppose that $u \equiv v$ and let $x \in N^+(u)$ for $u, v, x \in V$. Since $u \equiv v$ there exists a sequence $u, w_2, \ldots, w_{k-1}, v \in V$ where $u \sim_{j_1} w_2, j_1 \in \{1, 2\}$. Since $\sim_{j_1}$ is regular and $x \in N^+(u)$, there exists an $x_2 \in V$ such that $x_2 \in N^+(w_2)$ and $x_2 \sim_{j_1} x$. Iterating this will finally produce an x_k such that $x_k \in N^+(v)$ and $x \equiv x_k$, which shows the condition for the out-neighborhood. The case $x \in N^-(u)$ is handled analogously. $\square$

For the proof of Theorem 9.2.6 we need the following lemma (see e. g., [261]).

Lemma 9.2.7. *Let $(X, \leq)$ be a partially ordered set. If $\sup H$ exists for any subset $H \subseteq X$, then $(X, \leq)$ is a lattice.*

Proof. All we have to show is that for $x, y \in X$ there exists $\inf(x, y)$. Let $H := \{z \in X ; z \leq x$ and $z \leq y\}$. Then one can easily verify that $\sup H$ is the infimum of $\{x, y\}$. $\square$

Corollary 9.2.8. *If G is a graph then there exists a maximum regular equivalence and there exists a minimum regular equivalence for G.*

Proof. The maximum is simply the supremum over all regular equivalences. Dually, the minimum is the infimum over all regular equivalences. Or easier: The minimum is the identity partition which is always regular and minimal. $\square$

Although the supremum in the lattice of regular equivalences is a restriction of the supremum in the lattice of all equivalences, the infimum is not.

Proposition 9.2.9 ([82]). *The lattice of regular equivalences is not a sublattice of the lattice of all equivalences.*

Proof. We show that the infimum is not a restriction of the infimum in the lattice of all equivalences (which is simply intersection). Consider the graph in Figure 9.3 and the two regular partitions $\mathcal{P}_1 := \{\{A, C, E\}, \{B, D\}\}$ and $\mathcal{P}_2 := \{\{A, C\}, \{B, D, E\}\}$. The intersection of $\mathcal{P}_1$ and $\mathcal{P}_2$ is $\mathcal{P} = \{\{A, C\}, \{B, D\}, \{E\}\}$, which is not regular. $\square$

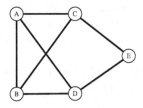

Fig. 9.3. Meet is not intersection

The fact that the supremum in the lattice of regular equivalences is a restriction of the supremum in the lattice of all equivalences implies the existence of a maximum regular equivalence which lies below a given (arbitrary) equivalence.

Definition 9.2.10. *Let G be a graph and $\sim$ an equivalence relation on its vertex set. An equivalence relation $\sim_1$ is called the* regular interior *of $\sim$ if it satisfies the following three conditions.*

1. *$\sim_1$ is regular,*
2. *$\sim_1 \leq \sim$, and*
3. *for all $\sim_2$ satisfying the above two conditions it holds $\sim_2 \leq \sim_1$.*

Corollary 9.2.11. *Let G be a graph and $\sim$ an equivalence relation on its vertex set. Then the regular interior of $\sim$ exists.*

On the other hand there is no minimum regular equivalence above a given equivalence in general (which would have been called a regular closure *or* regular hull*).*

Proof. For the first part, let $G = (V, E)$ be a graph and $\sim$ be an (arbitrary) equivalence on the node set. Then the supremum over the set of all regular equivalence relations that are finer than $\sim$ is the regular interior of $\sim$.

For the second part recall the example in the proof of Prop. 9.2.9 shown in Figure 9.3). It is easy to verify that the regular partitions $\mathcal{P}_1 := \{\, \{A, C, E\}, \{B, D\}\,\}$ and $\mathcal{P}_2 := \{\, \{A, C\}, \{B, D, E\}\,\}$ are both above the (non-regular) partition $\mathcal{P} := \{\, \{A, C\}, \{B, D\}, \{E\}\,\}$ and are both minimal with this property.   □

The regular interior is described in more detail in [90]; its computation is treated in Section 9.2.3. The infimum (in the lattice of regular equivalence relations) of two regular equivalence relations $\sim_1$ and $\sim_2$ is given by the regular interior of the intersection of $\sim_1$ and $\sim_2$.

9.2.3 Computation of Regular Interior

The regular interior (see Definition 9.2.10) of an equivalence relation $\sim$ is the coarsest regular refinement of $\sim$. It can be computed, starting with $\sim$, by a number of refinement steps in each of which currently equivalent vertices with non-equivalent neighborhoods are split, until all equivalent vertices have equivalent neighborhoods. For an example of such a computation see Figure 9.4. The running time of this computation depends heavily on how these refinement steps are organized.

In this section we present two algorithms for the computation of the regular interior. CATREGE [83] is the most well-known algorithm in the social network literature. It runs in time $\mathcal{O}(n^3)$. Tarjan and Paige [459] presented a sophisticated algorithm for the *relational coarsest partition problem*, which is essentially equivalent to computing the regular interior. Their algorithm runs in $\mathcal{O}(m \log n)$ time and is well-known in the bisimulation literature. See [408] for the relationship between bisimulation and regular equivalence.

Fig. 9.4. Computation of the regular interior: initial partition (left), first step (middle) second and final step (right)

CATREGE. In [83], Borgatti and Everett proposed CATREGE as an algorithm for computing the maximal regular equivalence of a graph, or more generally for computing the regular interior of an equivalence relation. CATREGE runs in $\mathcal{O}(n^3)$. On a high-level view CATREGE proceeds as follows:

- CATREGE maintains in each refinement step a current partition $\mathcal{P}$, which is initially set to the complete partition (or alternatively to an arbitrary input partition).
- In each refinement step it tests, for each pair of equivalent vertices (w.r.t. $\mathcal{P}$), whether their neighborhoods are equivalent (w.r.t. $\mathcal{P}$). If so, then these vertices remain equivalent, otherwise they will be non-equivalent after this refinement step.
- The algorithm terminates if no changes happen.

The number of refinement steps is bounded by n, since in each refinement step (except the last) the number of equivalence classes grows by at least one. The running time of one refinement step is in $\mathcal{O}(n^2)$.

The Relational Coarsest Partition Problem. This section is taken from [459], although we translate the notation into the context of graphs.

Problem definition. The RELATIONAL COARSEST PARTITION PROBLEM (RCPP) has as input a (directed) graph $G = (V, E)$ and a partition $\mathcal{P}$ of the vertex set V.

For a subset $S \subseteq V$ we write $E(S) := \{v \in V ; \exists u \in S \text{ such that } uEy\}$ and $E^{-1}(S) := \{u \in V ; \exists v \in S \text{ such that } uEy\}$. For two subsets $B \subseteq V$ and $S \subseteq V$, B is called *stable* with respect to S if either $B \subseteq E^{-1}(S)$, or $B \cap E^{-1}(S) = \emptyset$. If $\mathcal{P}$ is a partition of V, $\mathcal{P}$ is called *stable* with respect to S if all of its blocks are stable with respect to S. $\mathcal{P}$ is called *stable* if it is stable with respect to each of its own blocks.

The RCPP is the problem of finding the coarsest stable refinement for the initial partition $\mathcal{P}$.

In the language of role assignments this condition means that for each two roles, say r_1 and r_2, either no vertex, or all vertices assigned r_1 has/have an out-going edge to a vertex assigned r_2. This is the 'out-part' in Definition 9.2.1.

The algorithm of Paige and Tarjan [459] runs in time $\mathcal{O}(m \log n)$ and space $\mathcal{O}(m + n)$. Especially for sparse graphs this is a significant improvement over CATREGE.

Paige and Tarjan already pointed out that it is possible to generalize their algorithm to handle a bounded number of relations. This generalization can be realized in such a way that it yields asymptotically the same running time (see e. g., [207]). Having done this one can apply the algorithm to compute the coarsest stable refinement with respect to E and E^{T} to obtain the regular interior (see Definition 9.2.10).

The SPLIT *function.* The algorithm uses a primitive refinement operation. For each partition $\mathcal{Q}$ of V and subset $S \subseteq V$, let SPLIT$(S, \mathcal{Q})$ be the refinement of $\mathcal{Q}$ obtained by replacing each block B of $\mathcal{Q}$ such that $B \cap E^{-1}(S) \neq \emptyset$ and $B \setminus E^{-1}(S) \neq \emptyset$ by the two blocks $B' := B \cap E^{-1}(S)$ and $B'' := B \setminus E^{-1}(S)$. We call S a *splitter* of $\mathcal{Q}$ if SPLIT$(S, \mathcal{Q}) \neq \mathcal{Q}$. Note that $\mathcal{Q}$ is unstable with respect to S if and only if S is a splitter of $\mathcal{Q}$.

We note the following properties of SPLIT and consequences of stability. Let S and Q be two subsets of V, and let $\mathcal{P}$ and $\mathcal{R}$ be two partitions of V. The following elementary properties are stated without proof.

Property 9.2.12.   1. Stability is *inherited* under refinement; that is, if $\mathcal{R}$ is a refinement of $\mathcal{P}$ and $\mathcal{P}$ is stable with respect to a set S, then so is $\mathcal{R}$.

 2. Stability is *inherited* under union; that is, a partition that is stable with respect to two sets is also stable with respect to their union.

 3. Function SPLIT is *monotone* in its second argument; that is, if $\mathcal{P}$ is a refinement of $\mathcal{R}$ then SPLIT$(S, \mathcal{P})$ is a refinement of SPLIT$(S, \mathcal{R})$.

 4. Function SPLIT is *commutative* in the sense that the coarsest refinement of $\mathcal{P}$ stable with respect to both S and Q is

$$\text{SPLIT}(S, \text{SPLIT}(Q, \mathcal{P})) = \text{SPLIT}(Q, \text{SPLIT}(S, \mathcal{P})) \ .$$

Basic algorithm. We begin by describing a naive algorithm for the problem. The algorithm maintains a partition $\mathcal{Q}$ that is initially $\mathcal{P}$ and is refined until it is the coarsest stable refinement. The algorithm consists of repeating the following step until $\mathcal{Q}$ is stable:

REFINE: Find a set S that is a union of some of the blocks of $\mathcal{Q}$ and is a splitter of $\mathcal{Q}$; replace $\mathcal{Q}$ by SPLIT$(S, \mathcal{Q})$.

Some observations. Since stability is inherited under refinement, a given set S can be used as a splitter in the algorithm only once. Since stability is inherited under the union of splitters, after sets are used as splitters their unions cannot be used as splitters. In particular, a stable partition is stable with respect to the union of any subset of its blocks.

Lemma 9.2.13. *The algorithm maintains the invariant that any stable refinement of $\mathcal{P}$ is also a refinement of the current partition $\mathcal{Q}$.*

Proof. By induction on the number of refinement steps. The lemma is true initially by definition. Suppose it is true before a refinement step that refines partition $\mathcal{Q}$ using a splitter S. Let $\mathcal{R}$ be any stable refinement of $\mathcal{P}$. Since S is a union of blocks of $\mathcal{Q}$ and $\mathcal{R}$ is a refinement of $\mathcal{Q}$ by the induction hypothesis, S is a union of blocks of $\mathcal{R}$. Hence $\mathcal{R}$ is stable with respect to S. Since SPLIT is monotone, $\mathcal{R} = \text{SPLIT}(S, \mathcal{R})$ is a refinement of $\text{SPLIT}(S, \mathcal{Q})$.   □

The following theorem gives another proof for the existence of the regular interior (see Corollary 9.2.11).

Theorem 9.2.14. *The refinement algorithm is correct and terminates after at most $n - 1$ steps, having computed the unique coarsest stable refinement.*

Proof. The assertion on the number of steps follows from the fact that the number of blocks is between 1 and n. Once no more refinement steps are possible, $\mathcal{Q}$ is stable, and by Lemma 9.2.13 any stable refinement is a refinement of $\mathcal{Q}$. It follows that $\mathcal{Q}$ is the unique coarsest stable refinement.   □

The above algorithm is more general than is necessary to solve the problem: There is no need to use unions of blocks as splitters. Restricting splitters to blocks of $\mathcal{Q}$ will also suffice. However, the freedom to split using unions of blocks is one of the crucial ideas needed in developing a fast version of the algorithm.

Preprocessing. In an efficient implementation of the algorithm it it useful to reduce the problem instance to one in which $|E(\{v\})| \geq 1$ for all $v \in V$ (that is only to vertices having out-going edges). To do this we preprocess the partition $\mathcal{P}$ by splitting each block B into $B' := B \cap E^{-1}(V)$ and $B'' := B \setminus E^{-1}(V)$. The blocks B'' will never be split by the refinement algorithm; thus we can run the refinement algorithm on the partition $\mathcal{P}'$ consisting of the set of blocks B'. $\mathcal{P}'$ is a partition of the set $V' := E^{-1}(V)$, of size at most m. The coarsest stable refinement of $\mathcal{P}'$ together with the blocks B'' is the coarsest stable refinement of $\mathcal{P}$. The preprocessing and postprocessing take $\mathcal{O}(m+n)$ time if we have available the preimage set $E^{-1}(v)$ of each element $v \in V$. Henceforth, we shall assume $|E(\{v\})| \geq 1$ for all $v \in V$. This implies $m \geq n$.

Running time of the basic algorithm. We can implement the refinement algorithm to run in time $\mathcal{O}(mn)$ by storing for each element $v \in V$ its preimage set $E^{-1}(v)$. Finding a block of $\mathcal{Q}$ that is a splitter of $\mathcal{Q}$ and performing the appropriate splitting takes $\mathcal{O}(m)$ time. (Obtaining this bound is an easy exercise in list processing.) An $\mathcal{O}(mn)$ time bound for the entire algorithm follows.

Improved algorithm. To obtain a faster version of the algorithm, we need a good way to find splitters. In addition to the current partition $\mathcal{Q}$, we maintain another partition $\mathcal{X}$ such that $\mathcal{Q}$ is a refinement of $\mathcal{X}$ and $\mathcal{Q}$ is stable with respect to every block of $\mathcal{X}$ (in Section 9.3.4, $\mathcal{Q}$ will be called a *relative regular equivalence* w. r. t. $\mathcal{X}$). Initially $\mathcal{Q} = \mathcal{P}$ and $\mathcal{X}$ is the complete partition (containing V as its single block). The improved algorithm consists of repeating the following step until $\mathcal{Q} = \mathcal{X}$:

REFINE: Find a block $S \in \mathcal{X}$ that is not a block of $\mathcal{Q}$. Find a block $B \in \mathcal{Q}$ such that $B \subseteq S$ and $|B| \leq |S|/2$. Replace S within $\mathcal{X}$ by the two sets B and $S \setminus B$; replace $\mathcal{Q}$ by $\text{SPLIT}(S \setminus B, \text{SPLIT}(B, \mathcal{Q}))$.

The correctness of this improved algorithm follows from the correctness of the original algorithm and from the two ways given previously in which a partition can inherit stability with respect to a set.

Special case: If E is a function. Before discussing this algorithm in general, let us consider the special case in which E is a function, i.e., $|E(\{v\})| = 1$ for all $v \in V$. In this case, assume that $\mathcal{Q}$ is a partition stable with respect to a set S that is a union of some of the blocks of $\mathcal{Q}$, and $B \subseteq S$ is a block of $\mathcal{Q}$. Then $\text{SPLIT}(B, \mathcal{Q})$ is stable with respect to $S \setminus B$ as well. This holds, since if B_1 is a block of $\text{SPLIT}(B, \mathcal{Q})$, $B_1 \subseteq E^{-1}(B)$ implies $B_1 \cap E^{-1}(S \setminus B) = \emptyset$, and $B_1 \subseteq E^{-1}(S) \setminus E^{-1}(B)$ implies $B_1 \subseteq E^{-1}(S \setminus B)$. It follows that in each refinement step it suffices to replace $\mathcal{Q}$ by $\text{SPLIT}(B, \mathcal{Q})$, since $\text{SPLIT}(B, \mathcal{Q}) = \text{SPLIT}(S \setminus B, \text{SPLIT}(B, \mathcal{Q}))$. This is the idea underlying Hopcroft's 'process the smaller half' algorithm for the functional coarsest partition problem. The refining set B is at most half the size of the stable set S containing it.

Back to the general case. In the more general relational coarsest partition problem, stability with respect to both S and B does *not* imply stability with respect to $S \setminus B$, and Hopcroft's algorithm cannot be used. This is a serious problem since we cannot afford (in terms of running time) to scan the set $S \setminus B$ in order to perform one refinement step. Nevertheless, we are still able to exploit this idea by refining with respect to both B and $S \setminus B$ using a method that explicitly scans only B.

A preliminary lemma. Consider a general step in the improved refinement algorithm.

Lemma 9.2.15. *Suppose that partition $\mathcal{Q}$ is stable with respect to a set S that is a union of some of the blocks of $\mathcal{Q}$. Suppose also that partition $\mathcal{Q}$ is refined first with respect to a block $B \subseteq S$ and then with respect to $S \setminus B$. Then the following conditions hold:*

1. *Refining $\mathcal{Q}$ with respect to B splits a block $D \in \mathcal{Q}$ into two blocks $D_1 = D \cap E^{-1}(B)$ and $D_2 = D - D_1$ iff $D \cap E^{-1}(B) \neq \emptyset$ and $D \setminus E^{-1}(B) \neq \emptyset$.*
2. *Refining $\text{SPLIT}(B, \mathcal{Q})$ with respect to $S \setminus B$ splits D_1 into two blocks $D_{11} = D_1 \cap E^{-1}(S \setminus B)$ and $D_{12} = D_1 - D_{11}$ iff $D_1 \cap E^{-1}(S \setminus B) \neq \emptyset$ and $D_1 \setminus E^{-1}(S \setminus B) \neq \emptyset$.*
3. *Refining $\text{SPLIT}(B, \mathcal{Q})$ with respect to $S \setminus B$ does not split D_2.*
4. *$D_{12} = D_1 \cap (E^{-1}(B) \setminus E^{-1}(S \setminus B))$.*

Proof. Conditions 1 and 2 follow from the definition of SPLIT.

Condition 3: Form Condition 1 it follows that if D is split, it is $D \cap E^{-1}(B) \neq \emptyset$. Since D is stable with respect to S, and since $B \subseteq S$, then $D_2 \subseteq D \subseteq E^{-1}(S)$. Since by Cond. 1 $D_2 \cap E^{-1}(B) = \emptyset$, it follows that $D_2 \subseteq E^{-1}(S \setminus B)$.

Condition 4: This follows from the fact that $D_1 \subseteq E^{-1}(B)$ and $D_{12} = D_1 \setminus E^{-1}(S \setminus B)$. □

Performing the three-way splitting of a block D into D_{11}, D_{12}, and D_2 as described in Lemma 9.2.15 is the hard part of the algorithm. Identity 4 of Lemma 9.2.15 is the crucial observation that we shall use in our implementation. Remember that scanning the set $S \setminus B$ takes (possibly) too long to obtain the claimed running time. We shall need an additional datastructure to determine $D_1 \setminus E^{-1}(S \setminus B) = (D \cap E^{-1}(B)) \setminus E^{-1}(S \setminus B)$ by scanning only B.

Running time of the improved algorithm. A given element of V is in at most $\log_2 n + 1$ different blocks B used as refining sets, since each successive such set is at most half the size of the previous one. We shall describe an implementation of the algorithm in which a refinement step with respect to block B takes $\mathcal{O}(|B| + \sum_{u \in B} |E^{-1}(\{u\})|)$ time. From this an $\mathcal{O}(m \log n)$ overall time bound for the algorithm follows by summing over all blocks B used for refinement and over all elements in such blocks.

Datastructures. (See Section 9.1.3 for an example of a much simpler algorithm which already uses some of the ideas of this algorithm.)

Graph $G = (V, E)$ is represented by the sets V and E. Partitions $\mathcal{Q}$ and $\mathcal{X}$ are represented by doubly linked lists of their blocks.

A block S of $\mathcal{X}$ is called *simple* if it contains only a single block of $\mathcal{Q}$ (equal to S but indicated by its own record) and *compound* if it contains two or more blocks of $\mathcal{Q}$.

The various records are linked together in the following ways. Each edge uEv points its source u. Each vertex v points to a list of incoming edges uEv. This allows scanning the set $E^{-1}(\{v\})$ in time proportional to its size. Each block of $\mathcal{Q}$ has an associated integer giving its size and points to a doubly linked list of the vertices in it (allowing deletion in $\mathcal{O}(1)$ time). Each vertex points to the block of $\mathcal{Q}$ containing it. Each block of $\mathcal{X}$ points to a doubly linked list of the blocks of $\mathcal{Q}$ contained in it. Each block of $\mathcal{Q}$ points to the block of $\mathcal{X}$ containing it. We also maintain a set C of compound blocks of $\mathcal{X}$. Initially C contains the single block V, which is the union of the blocks of $\mathcal{P}$. If $\mathcal{P}$ contains only one block (after the preprocessing), $\mathcal{P}$ itself is the coarsest stable refinement and we terminate the algorithm here.

To make three-way splitting (see Lemma 9.2.15) fast we need one more collection of records. For each block S of $\mathcal{X}$ and each element $v \in E^{-1}(S)$ we maintain an integer $\text{COUNT}(v, S) := |S \cap E(\{v\})|$. Each edge uEv with $v \in S$ contains a pointer to $\text{COUNT}(u, S)$. Initially there is one count per vertex (i. e., $\text{COUNT}(v, V) = |E(\{v\})|$) and each edge uEv points to $\text{COUNT}(u, V)$.

This COUNT function will help to determine the set $E^{-1}(B) \setminus E^{-1}(S \setminus B)$ in time proportional to $|\{uEv \,;\, v \in B\}|$ (see step 5 below).

Both the space needed for all the data structures and the initialization time is $\mathcal{O}(m)$.

The refinement algorithm consists of repeating refinement steps until C is empty.

Performing one refinement step. For clarity we divide one refinement step into 7 substeps.

1. (select a refining block). Remove some block S from C. (Block S is a compound block of $\mathcal{X}$.) Examine the first two blocks in the list of blocks of $\mathcal{Q}$ contained in S. Let B be the smaller one. (Break a tie arbitrarily.)

2. (update $\mathcal{X}$). Remove B from S and create a new (simple) block S' of $\mathcal{X}$ containing B as its only block of $\mathcal{Q}$. If S is still compound, put S back into C.

3. (compute $E^{-1}(B)$). Copy the vertices of B into a temporary set B'. (This facilitates splitting B with respect to itself during the refinement.) Compute $E^{-1}(B)$ by scanning the edges uEv such that $v \in B$ and adding each vertex u in such an edge to $E^{-1}(B)$ if it has not already been added. Duplicates are suppressed by marking vertices as they are encountered and linking them together for later unmarking. During the same scan compute $\text{COUNT}(u, B) = |\{v \in B \,;\, uEv\}|$, store this count in a new integer and make u point to it. These counts will be used in step 5.

4. (refine $\mathcal{Q}$ with respect to B). For each block D of $\mathcal{Q}$ containing some element (vertex) of $E^{-1}(B)$, split D into $D_1 = D \cap E^{-1}(B)$ and $D_2 = D \setminus D_1$. Do this by scanning the elements of $E^{-1}(B)$. To process an element $u \in E^{-1}(B)$, determine the block D of $\mathcal{Q}$ containing it and create an associated block D' if one does not already exist. Move u from D to D'.

During the scanning, construct a list of those blocks D that are split. After the scanning, process the list of split blocks. For each such block D with associated block D', mark D' as no longer being associated with D (so that it will be correctly processed in subsequent iterations of Step 4). Eliminate the record for D if D is now empty and, if D is nonempty and the block of $\mathcal{X}$ containing D and D' has been made compound by the split, add this block to C.

5. (compute $E^{-1}(B) \setminus E^{-1}(S \setminus B)$). Scan the edges uEv with $v \in B'$. To process an edge uEv, determine $\text{COUNT}(u, B)$ (to which u points) and $\text{COUNT}(u, S)$ (to which uEv points). If $\text{COUNT}(u, B) = \text{COUNT}(u, S)$, add u to $E^{-1}(B) \setminus E^{-1}(S \setminus B)$ if it has not been added already.

6. (refine $\mathcal{Q}$ with respect to $S \setminus B$). Proceed exactly as in Step 4 but scan $E^{-1}(B) \setminus E^{-1}(S \setminus B)$ (computed in Step 5) instead of $E^{-1}(B)$.

7. (update counts). Scan the edges uEv such that $v \in B'$. To process and edge uEv, decrement $\text{COUNT}(u, S)$ (to which uEv points). If this count becomes zero, delete the COUNT record, and make uEv point to $\text{COUNT}(u, B)$ (to which u points). After scanning all the appropriate edges, discard B'.

Note that in step 5 only edges terminating in B' are scanned. Step 5 is correct (computes $E^{-1}(B) \setminus E^{-1}(S \setminus B)$) since for each vertex u in $E^{-1}(B)$, it holds that u is in $E^{-1}(B) \setminus E^{-1}(S \setminus B)$ iff u is not in $E^{-1}(S \setminus B)$ iff all edges starting at u and terminating in S terminate in B iff $\text{COUNT}(u, B) = \text{COUNT}(u, S)$.

The correctness of this implementation follows in a straightforward way from our discussion above of three-way splitting. The time spent in a refinement step is $\mathcal{O}(1)$ per edge terminating in B plus $\mathcal{O}(1)$ per vertex of B, for a total of $\mathcal{O}(|B| + \sum_{v \in B} |E^{-1}(\{v\})|)$ time. An $\mathcal{O}(m \log n)$ time bound for the entire algorithm follows as discussed above. It is possible to improve the efficiency of the algorithm by a constant factor by combining various steps, which have been kept separate for clarity.

Adaptation to Related Problems. The above algorithm turns out to be the key to efficiently solve several partition refinement problems that arise in this chapter. We will briefly sketch this generality.

Computing the maximal strong structural equivalence (as described in Section 9.1.3) or the relative regular equivalence (see Section 9.3.4) is much simpler than computing the regular interior. Nevertheless we can use the idea of iteratively splitting blocks according to intersection with certain neighborhoods. (See algorithm 21 and the comments in Section 9.3.4.) These problems can be solved by algorithms that run in $\mathcal{O}(m + n)$.

Computing the coarsest *equitable* (see Section 9.3.1) has been solved earlier than the problem of computing the regular interior (see [110] for an $\mathcal{O}(m \log n)$ algorithm and the comments in [459]).

Refining a partition w. r. t. *multiple relations* (see Definition 9.4.1) is also possible in $\mathcal{O}(m \log n)$ (if the number of relations is bounded by a constant). This extension of the algorithm can be used to compute the regular interior w. r. t. incoming and out-going edges. Shortly, a partition can be refined w. r. t. multiple relations by performing steps 3–7 (see above) for fixed B and S successively for all relations, one at a time. (See e. g., [207].)

9.2.4   The Role Assignment Problem

In this section we investigate the computational complexity of the decision problem whether a given graph admits a regular role assignment with prespecified role graph, or with prespecified number of equivalence classes. In this section we consider only undirected graphs.

The most complete characterization is from Fiala and Paulusma [209]. Let $k \in \mathbb{N}$ and R be an undirected graph, possibly with loops.

Problem 9.2.16 (k-Role Assignment (k-RA)). Given a graph G.
Question: Is there a regular equivalence for G with exactly k equivalence classes?

Problem 9.2.17 (R-Role Assignment (R-RA)). Given a graph G.
Question: Is there a regular role assignment $r : V(G) \rightarrow V(R)$ with role graph R?

Note that we require role assignments to be surjective mappings.

Theorem 9.2.18 ([209]). *k-RA is polynomially solvable for $k = 1$ and it is $\mathcal{NP}$-complete for all $k \geq 2$.*

Theorem 9.2.19 ([209]). *R-RA is polynomially solvable if each component of R consists of a single vertex (with or without a loop), or consists of two vertices without loops and it is $\mathcal{NP}$-complete otherwise.*

We give the proof of one special case of the R-Role Assignment Problem.

Theorem 9.2.20 ([493]). *Let R_0 be the graph in Figure 9.5. Then R_0-RA is $\mathcal{NP}$-complete.*

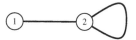

Fig. 9.5. Role graph R_0

Proof. It is easy to see that R-RA is in $\mathcal{NP}$ since one can easily check in polynomial time whether a given function $r : V \to \{1, 2\}$ is a 2-role assignment with role graph R_5.

We will show that the 3-satisfiability problem (3SAT) is polynomially transformable to R_0-RA. So let $U = \{u_1, \ldots, u_n\}$ be a set of variables and $C = \{c_1, \ldots, c_m\}$ be a set of clauses (each consisting of exactly three literals). We will construct a graph $G = (V, E)$ such that G is 2-role assignable with role graph R_0 if and only if C is satisfiable.

The construction will be made up of two components, truth-setting components and satisfaction testing components (see Figure 9.6).

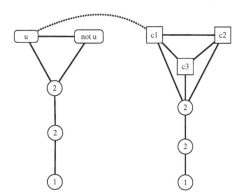

Fig. 9.6. Truth-setting component for variable u (left); satisfaction testing component for clause $\{c_1, c_2, c_3\}$ (right) and communication edge if literal c_1 equals u (dashed). The roles of the vertices in the pending paths are uniquely determined (as indicated by the labels 1 resp. 2) if the role assignment should be regular with role graph R_0

For each variable $u_i \in U$, there is a truth-setting component $T_i = (V_i, E_i)$ with

$$V_i := \{u_i, \overline{u}_i, a_{i1}, a_{i2}, a_{i3}\} \ ,$$
$$E_i := \{u_i\overline{u}_i, u_ia_{i3}, \overline{u}_ia_{i3}, a_{i1}a_{i2}, a_{i2}a_{i3}\} \ .$$

Note that, although we write $u_i\overline{u}_i$ for the edge $\{u_i, \overline{u}_i\}$, the graph is undirected.

The intuition behind the construction of T_i is the following: If a graph containing T_i as a subgraph (such that the a_{ij} are adjacent only to the vertices in V_i as specified above) admits a regular role assignment r with role graph R_0, then necessarily $r(a_{i1}) = 1$, since a_{i1} has degree one and a vertex which is assigned 2 must have degree ≥ 2. Then $r(a_{i2}) = 2$, since a 1-vertex is adjacent to a 2-vertex and $r(a_{i2}) = 2$, since a 2-vertex is adjacent to a 2-vertex. Finally exactly one of u_i or $\overline{u}_i$ is assigned 2, meaning that variable u_i is set to *true* or *false*, respectively. Thus component T_i ensures that a variable gets either *true* or *false*.

For each clause $c_j \in C$, let vertices c_{j1}, c_{j2}, and c_{j3} be three vertices corresponding to the three literals in the clause c_j. Then there is a satisfaction testing component $S_j = (V'_j, E'_j)$ with

$$V'_j := \{c_{j1}, c_{j2}, c_{j3}, b_{j1}, b_{j2}, b_{j3}\} \ ,$$
$$E'_j := \{c_{j1}c_{j2}, c_{j1}c_{j3}, c_{j2}c_{j3}, c_{j1}b_{j3}, c_{j2}b_{j3}, c_{j3}b_{j3}, b_{j1}b_{j2}, b_{j2}b_{j3}\} \ .$$

The intuition behind the construction of S_j is the following: If a graph containing S_j as a subgraph (such that the b_{jl} are adjacent only to the vertices in V_j as specified above) admits a regular role assignment r with role graph R_0, then necessarily $r(b_{j1}) = 1$, $r(b_{j2}) = r(b_{j3}) = 2$, which ensures that one of the vertices c_{j1}, c_{j2}, c_{j3} is assigned 1, thus ensuring that every adjacent vertex of this 1-vertex must be assigned 2. This will be crucial later.

The construction so far is only dependent on the number of variables and clauses. The only part of the construction that depends on which literals occur in which clauses is the collection of communication edges. For each clause $c_j = \{x_{j1}, x_{j2}, x_{j3}\} \in C$ the communication edges emanating from S_j are given by

$$E''_j := \{c_{j1}x_{j1}, c_{j2}x_{j2}, c_{j3}x_{j3}\} \ .$$

(The x_{jl} are either variables in U or their negations.) Notice that for each c_{jk}, there is exactly one vertex that is adjacent to c_{jk} in E''_j, which is the corresponding literal vertex for c_{jk} in the clause c_j.

To complete the construction of our instance of R_0-RA, let $G = (V, E)$ with V being the union of all V_is and all V'_js and E the union of all E_is, all E'_js and all E''_js.

As mentioned above, given a regular role assignment for G with role graph R_0, for each $j = 1, \ldots, m$ there is a vertex c_{jk} such that $r(c_{jk}) = 1$ implying that the corresponding adjacent literal is assigned 2. Setting this literal to *true* will satisfy clause c_j.

Thus we have shown that the formula is satisfiable if G is regularly R_0 assignable.

Conversely, suppose that C has a satisfying truth assignment. We obtain an assignment $r: V \to \{1, 2\}$ as follows. For each $i = 1, \ldots, n$ set $r(u_i)$ to 2 (and

$r(\overline{u_i})$ to 1) if and only if variable u_i is *true* and set the role of the vertices a_{ik} and b_{jk} as implied by the fact that r should be regular (see above). Moreover, for each $j = 1, \ldots, m$ let c_{jk}, $k \in \{1, 2, 3\}$, be some vertex whose corresponding literal in the clause c_j is *true* – such a k exists since the truth assignment is satisfying for C. Set $r(c_{jk}) := 1$ and $r(c_{jl}) := 2$ for $l \in \{1, 2, 3\}$, $l \neq k$.

The proof is complicated a bit by the fact that more than one literal in a clause might be true, but setting $r(c_{jk}) = 1$ is allowed for only one $k \in \{1, 2, 3\}$. Since a 2-vertex may be adjacent to another 2-vertex, this does not destroy the regularity of r.   □

9.2.5   Existence of k-Role Assignments

We have seen in the previous section that the decision whether a graph admits a regular equivalence with exactly k equivalence classes is $\mathcal{NP}$-complete for general graphs. Nevertheless, there are easy-to-verify sufficient, if not necessary, conditions that guarantee the existence of regular k-role assignments. Briefly, the condition is that the graph differs not too much from a regular graph.

Theorem 9.2.21 ([474]). *For all $k \in \mathbb{N}$ there is a constant $c_k \in \mathbb{R}$ such that for all graphs G with minimal degree $\delta = \delta(G)$ and maximal degree $\Delta = \Delta(G)$ satisfying*

$$\delta \geq c_k \log(\Delta) \;,$$

there is a regular equivalence for G with exactly k equivalence classes.

To exclude trivial counterexamples we assume in the following that all graphs in question have at least k vertices.

For the proof we need a uniform version of the LOVASZ LOCAL LEMMA.

Theorem 9.2.22 ([25, Chapter 5 Corollary1.2]). *Let $A_i, i \in I$, be events in a discrete probability space. If there exists M such that for every $i \in I$*

$$|\{A_j \,;\; A_j \text{ is not independent of } A_i\}| \leq M \;,$$

and if there exists $p > 0$ such that $\mathrm{Pr}(A_i) \leq p$ for every $i \in I$, then

$$ep(M + 1) \leq 1 \Longrightarrow \mathrm{Pr}\left(\bigcap_{i \in I} \overline{A_i}\right) > 0 \;,$$

where e is the EULER number $e = \sum_{i=0}^{\infty} 1/i!$.   □

Proof (of Theorem 9.2.21). Define $r : V \rightarrow \{1, \ldots, k\}$ as follows: For every $v \in V$ choose $r(v)$ uniformly at random from $\{1, \ldots, k\}$.

For $v \in V$, let A_v be the event that $r(N(v)) \neq \{1, \ldots, k\}$. It is

$$\mathrm{Pr}(A_v) \leq k\left(\frac{k-1}{k}\right)^{d(v)} \leq k\left(\frac{k-1}{k}\right)^{\delta(G)} \;.$$

Because all $r(w)$ are chosen independently and for a fixed value i, the probability that i is not used for any of the vertices adjacent to v is $\left(\frac{k-1}{k}\right)^{d(v)}$, and there are k choices for i.

Also note that A_v and A_w are not independent if and only if $N(v) \cap N(w) \neq \emptyset$. Hence, A_v with $M := \Delta(G)^2$ and $p := k \left(\frac{k-1}{k}\right)^{\delta(G)}$ satisfies the conditions of the LOVASZ LOCAL LEMMA. Therefore,

$$ek \left(\frac{k-1}{k}\right)^{\delta(G)} (\Delta(G)^2 + 1) \leq 1 \Rightarrow \Pr\left(\bigcap_{v \in V} \overline{A_v}\right) > 0 . \qquad (9.1)$$

If the righthand side of (9.1) holds, there exists at least one r such that $r(N(v)) = \{1, \ldots, k\}$ for every $v \in V$, that is, there exists at least one regular k-role assignment. In order to finish the proof we note that the lefthand side of (9.1) is equivalent to

$$\delta(G) \geq \frac{\log(ek(\Delta(G)^2 + 1))}{\log\left(\frac{k}{k-1}\right)} .$$

Clearly, there exists a constant c_k such that $c_k \log(\Delta(G))$ is greater than the righthand side of the above inequality. □

Conclusion. Regular equivalences are well investigated in computer science. Results indicate that many regular equivalences exist even in irregular graphs, but it is unclear how to define and/or compute the best, or at least a good one. Fast algorithms exist for the computation of the maximal regular equivalence or for the regular interior of an a priori partition. The maximal regular equivalence could be meaningful for directed graphs (for undirected it is simply the division into isolates and non-isolates). Also, the regular interior could be a good role assignment if one has an idea for the partition to be refined. Specifying the number of equivalence classes or the role graph yields $\mathcal{NP}$-hard problems, in the general case. Optimization approaches for these problems are presented in Section 10.1.7 in the next chapter.

9.3　Other Equivalences

In this section we briefly mention other (than structural or regular) types of role equivalences.

9.3.1　Exact Role Assignments

In this section we define a class of equivalence relations that is a subset of regular equivalences. These equivalences will be called *exact*. The associated partitions are also known as *equitable partitions* in graph theory, they have first been defined as *divisors* of graphs.

While for regular equivalences only the occurrence or non-occurrence of a role in the neighborhood of a vertex matters, for exact equivalences, the number of occurrence matters.

The graph model of this section are undirected multigraphs.

Definition 9.3.1. *A role assignment r is called* exact *if for all $u, v \in V$*

$$r(u) = r(v) \quad \Longrightarrow \quad r(N(u)) = r(N(v)) \ ,$$

where the last equation is an equation of multi-sets, *i. e., vertices, that have the same role, must have the same number of each of the other roles in their neighborhoods.*

The coloring in Figure 9.7 defines an exact role assignment for the shown graph.

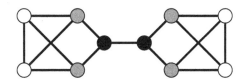

Fig. 9.7. An exact role assignment

While an equivalence is regular for a multigraph if and only if it is regular for the induced simple graph (each edge at most once), for exact equivalences the multiplicity of an edge matters.

It is straightforward to see that exact role assignments are regular, the converse is not true.

An equivalent definition is the following.

Definition 9.3.2 ([247]). *A partition $\mathcal{P} = \{C_1, \ldots, C_k\}$ of the vertex set V of an undirected (multi-)graph $G = (V, E)$ is called* equitable *if there are integers b_{ij}, $i, j = 1, \ldots, k$, such that each vertex in class C_i has exactly b_{ij} neighbors in class C_j. The matrix $B = (b_{ij})_{i,j=1,\ldots,k}$ defines a (directed) multi-graph, which is called the* quotient *of G modulo $\mathcal{P}$, denoted by $G/\mathcal{P}$.*

A partition is equitable if and only if the associated role assignment is exact. The above definition also extends the definition of the quotient or role graph (see Section 9.0.2) to multigraphs. Note that this is possible only for exact role assignments.

Note that even if the graph is undirected the quotient is possibly directed, meaning that the multiplicity of an edge may differ from the multiplicity of the reversed edge. This happens always if two 'adjacent' equivalence classes are of different size.

Exact role assignments are compatible with algebraic properties of a graph.

Theorem 9.3.3 ([247]). *Let G be a graph, $\mathcal{P}$ an equitable partition. Then, the characteristic polynomial of the quotient $G/\mathcal{P}$ divides the characteristic polynomial of G.* □

This theorem implies that the spectrum of the quotient $G/\mathcal{P}$ is a subset of the spectrum of G.

The set of all exact role assignments of a graph forms a lattice [191]. The maximal exact role assignment of a graph can be computed by an adaption of the algorithm in Section 9.2.3. (See [110] and the comments in [459].)

Many problems around exact role assignments are $\mathcal{NP}$-complete as well. For example the problem of deciding if a graph G admits an exact role assignment with quotient R is $\mathcal{NP}$-complete if both G and R are part of the input, or for some fixed R. This holds, since the $\mathcal{NP}$-complete problem of deciding whether a 3-regular graph has a perfect code [370], can be formulated as the problem of deciding whether G has an exact role assignment with quotient

$$R = \begin{bmatrix} 0 & 3 \\ 1 & 2 \end{bmatrix} .$$

The quotient over an equitable partition has much more in common with the original graph than, e.g., the role graph over a regular equivalence. Exact role assignments also ensure that equivalent vertices have the same degree, which is not true for regular role assignments.

Conclusion. Exact role assignments, also called equitable partitions are well investigated in algebraic graph theory. While some problems around equitable partitions are $\mathcal{NP}$-complete, there are efficient algorithms to compute the maximal equitable partition of a graph, or to compute the coarsest equitable refinement of an a priori partition. These algorithms could be used to compute role assignments, but, due to irregularities, the results contain in most cases too many classes and miss the underlying (possibly perturbed) structure. Brandes and Lerner [97] introduced a relaxation of equitable partitions that is tolerant against irregularities.

9.3.2   Automorphic and Orbit Equivalence

Automorphic equivalence expresses interchangeability of vertices.

Definition 9.3.4 ([191]). *Let $G = (V, E)$ be a graph, $u, v \in V$. Then u and v are said to be* automorphically equivalent *if there is an automorphism φ of G with $\varphi(u) = v$.*

Automorphically equivalent vertices cannot be distinguished only in terms of the graph structure. Therefore it could be argued that at least automorphically equivalent vertices should be considered to play the same role.

It is easy to see that structurally equivalent vertices are automorphically equivalent.

A partition of the vertex set which has the property that each pair of equivalent vertices is automorphically equivalent is not necessarily a regular equivalence. However we have the following result.

Proposition 9.3.5 ([190]). *Let $G = (V, E)$ be a graph with automorphism group $A(G)$, and $H < A(G)$ be a subgroup of $A(G)$. Then assigning roles according to the orbits of H defines an exact role assignment for G. Such a partition is called an* orbit partition.

Proof. Let r be a role assignment as in the formulation of the proposition. If $r(u) = r(v)$ then there exists $\varphi \in H$ such that $\varphi(u) = v$. If $x \in N^+(u)$, then $\varphi(x) \in N^+(\varphi(u)) = N^+(v)$. Furthermore $r(x) = r(\varphi(x))$ by definition. It follows that $r(N^+(u)) \subseteq r(N^+(v))$ (as multisets). The other inclusion and the corresponding assertion for the in-neighborhoods is shown similar.   □

In particular, orbit equivalences are regular.

For example, the coloring in Figure 9.7 defines the orbit partition of the automorphism group of the shown graph.

The set of orbit equivalences forms a proper subset of the set of all exact equivalences, which can be proved by any regular graph which is not vertex-transitive. For example, the complete partition for the graph in Figure 9.7 is exact but not an orbit partition.

The above proposition can also be used to prove that every undirected role primitive graph (see Section 9.2.1) is a graph with trivial automorphism group [190]. This is not true for directed graphs as can be seen by directed cycles of prime length.

Orbit equivalence has the nice feature that its condition is invariant w.r.t. a shift to the complement graph. This does not hold neither for regular nor for exact equivalence.

The computation of orbit equivalences is related to the problem of computing the automorphism group which has open complexity status.

Conclusion. Automorphically equivalent vertices cannot be distinguished in terms of graph structure, but only by additional labels or attributes. It could therefore be argued that at least automorphically equivalent vertices play the same role. Computation of automorphic equivalence seems to be hard, but, in irregular networks, there won't be any significant automorphisms anyway.

9.3.3 Perfect Equivalence

Perfect equivalence is a restriction of regular equivalence. It expresses the idea that there must be a reason for two vertices for being *not* equivalent.

Definition 9.3.6 ([191]). *A role assignment r defines a* perfect equivalence *if for all $u, v \in V$*

$$r(u) = r(v) \quad \Longleftrightarrow \quad r(N^+(u)) = r(N^+(v)) \text{ and } r(N^-(u)) = r(N^-(v)).$$

A regular equivalence is perfect if and only if the induced role graph has no strong structural equivalent vertices (see Section 9.1).

The set of perfect equivalence relations of a graph is a lattice [191], which is neither a sublattice of all equivalence relations (Section 9.1.1) nor of the lattice of regular equivalence relations (Section 9.2.2). A *perfect interior* of an equivalence relation $\sim$ would be a coarsest perfect refinement of $\sim$ (compare Definition 9.2.10). In contrast to the regular interior, the perfect interior does not exist in general.

Theorem 9.3.7. *In general, the transitive closure (see Section 9.1.1) of the union of two perfect equivalence relations is not perfect. In particular, for some equivalences there is no perfect interior.*

Fig. 9.8. Graph for the proof of Theorem 9.3.7. Supremum of two perfect equivalences is not perfect

Proof. Consider the graph in Figure 9.8 and the two perfect partitions $\mathcal{P}_1 = \{\{1,5\}, \{2,6\}\{3,4\}\}$ and $\mathcal{P}_2 = \{\{1,2\}, \{5,6\}\{3\}, \{4\}\}$. The transitive closure of $\mathcal{P}_1$ and $\mathcal{P}_2$ is $\mathcal{P} = \{\{1,2,5,6\}, \{3,4\}\}$, which is not perfect.

For the second statement, note that $\mathcal{P}_1$ and $\mathcal{P}_2$ are both perfect refinements of $\mathcal{P}$ and are both maximal w.r.t. this property.   □

The second statement has a more trivial proof: For a graph with two strong structurally equivalent vertices, the identity partition has no perfect refinement.

Some decision problems concerning perfect equivalence are $\mathcal{NP}$-complete as well. This can be seen by Theorems 9.2.18 and 9.2.19, restricted to role graphs without strong structurally equivalent vertices.

Although perfect equivalences rule out some trivial regular equivalences, there is no evidence why roles shouldn't be strong structurally equivalent.

Conclusion. Perfect equivalence is a restriction of regular equivalence, but it doesn't seem to yield better role assignments. Some mathematical properties of regular equivalences get lost and there are examples where the condition on perfect equivalence rules out good regular role assignments.

9.3.4   Relative Regular Equivalence

Relative regular equivalence expresses the idea that equivalent vertices have equivalent neighborhoods in a coarser, predefined measure.

Definition 9.3.8 ([90]). *Let* $G = (V, E)$ *be a graph and* $r: V \to W$ *and* $r_0: V \to W_0$ *be two role assignments. Then,* r *is called* regular relative to r_0 *if* $r \leq r_0$ *(see Section 9.1.1 for the partial order on the set of role assignments) and for all* $u, v \in V$

$$r(u) = r(v) \Rightarrow r_0(N^+(u)) = r_0(N^+(v)) \text{ and } r_0(N^-(u)) = r_0(N^-(v)) \ .$$

A typical application [90] of relative regular equivalence is given by a network of symmetric friendship ties which a priori is divided into two disjoint friendship cliques A and B. Assume that within each clique every member has at least one tie to some other member of the same clique. The partition into these two cliques would be regular if either there is no tie between the two cliques or each actor would have, in addition to the intra-group ties, at least one tie to a member of the other group. But lets assume that some, but not all, actors have friendship ties to members of the other group. The partition into A and B is no longer regular. Now we can split each group into those actors having ties to some member of the other group and those who don't. Say we obtain the partition into A_1, A_2, B_1, and B_2. Neither is this partition (in general) regular: There might be some actors in, say, A_1 having intra-group ties only with members of A_1, some only with members of A_2, some with both; they don't have equivalent neighborhoods. But they have equivalent neighborhoods with respect to the coarse partition into A and B. Thus, the partition into A_1, A_2, B_1 and B_2 is regular relative to the partition into A and B.

Relative regularity below a fixed equivalence is preserved under refinement. (Compare Prop. 9.1.7 for a similar proposition for structural equivalence.)

Proposition 9.3.9. *Let* $\sim$, $\sim_1$, *and* $\sim_2$ *be equivalence relations on* V *such that* $\sim_1 \leq \sim_2$ *and* $\sim_2$ *is regular relative to* $\sim$. *Then so is* $\sim_1$.

Similar to Prop. 9.1.7, this proposition implies that the set of equivalences that are regular relative to a fixed equivalence $\sim$ is a sublattice of all equivalences and is completely described by the maximum of this set, denoted here by MRRE($\sim$).

Computing the MRRE($\sim$) is possible in linear time by an adaptation of the algorithm 21 for computing the maximal structural equivalence: Instead of splitting equivalence classes from the point of view of single vertices, classes are split from the point of view of the classes of $\sim$ (compare the algorithm in Section 9.2.3). Note that the classes of $\sim$ are fixed and the MRRE($\sim$) has been found after all classes of $\sim$ have been processed once.

Each refinement step in the CATREGE algorithm (see Section 9.2.3) computes an equivalence that is regular relative to the previous one, but the running time of one step is in $\mathcal{O}(n^2)$, which is worse than the above described algorithm on sparse graphs.

Conclusion. Relative regular equivalence is computationally simple but it needs an a priori partition of the vertices and, since its compatibility requirement is only local, is not expected to represent global network structure. It has most been applied in connection with multiple and composite relations (see, e.g., Winship-Pattison Role Equivalence in Section 9.5.1).

9.4   Graphs with Multiple Relations

Actors in a social network are often connected by more than one relation. For example, on the set of employees of a company there might be two relations GIVESORDERSTO and ISFRIENDOF. It is often insufficient to treat these relations separately one at a time since their interdependence matters.

In this section we generalize the graph model to graphs with multiple relations, that is, collections of graphs with common vertex set.

Definition 9.4.1. *A graph with multiple relations* $\mathcal{G} = (V, \mathcal{E})$ *consists of a finite vertex set V, and a finite set of relations (finite set of edge sets) $\mathcal{E} = \{E_i\}_{i=1,\ldots,p}$, where $p \in \mathbb{N}$ and $E_i \subseteq V \times V$.*

For the remainder of this section we often write 'graph' meaning 'graph with multiple relations'. A graph is identified with the one resulting from deleting duplicate relations, where we say that two relations are equal if they consist of the same pairs of vertices. That is relations don't have 'labels' but are distinguished by the pairs of vertices they contain.

The role graph of a graph with multiple relations is again a graph with (possibly) multiple relations. (Compare Definition 9.0.3 of the role graph of a graph with one relation.)

Definition 9.4.2. *Let $\mathcal{G} = (V, \mathcal{E})$ be a graph with multiple relations, and $r \colon V \to W$ be a role assignment. The role graph of $\mathcal{G}$ over r is the graph $\mathcal{R} = (W, \mathcal{F})$, where $\mathcal{F} = \{F_i \,; \ i = 1, \ldots, p\}$, where $F_i = \{(r(u), r(v)) \,; \ (u,v) \in E_i\}$.*

Note that F_i may be equal to F_j even if $E_i \neq E_j$ and that duplicate edge relations are eliminated ($\mathcal{F}$ is a set).

From the above definition we can see that role assignments are actually mappings of vertices and relations. That is $r \colon V \to W$ defines uniquely a mapping of relations $r_{\mathrm{rel}} \colon \mathcal{E} \to \mathcal{F}$. Note that r_{rel} does not map edges of $\mathcal{G}$ onto edges of $\mathcal{R}$ but relations, i.e. edge sets, onto relations.

Having more then one relation, the possibilities for defining different types of role assignments explode. See [579, 471] for a large number of possibilities. We will sketch some of them.

The easiest way to translate definitions for different types of vertex partitions (see Sections 9.1, 9.2, and 9.3) to graphs with multiple relations is by the following generic definition.

Definition 9.4.3. *A role assignment $r \colon V \to W$ is said to be of a specific type t for a graph $\mathcal{G} = (V, \mathcal{E})$ with multiple relations, if for each $E \in \mathcal{E}$, r is of type t for the graph (V, E).*

We illustrate this for the definition of regular equivalence relations.

Definition 9.4.4 ([579]). *Let $\mathcal{G} = (V, \mathcal{E})$ be a graph. A role assignment $r \colon V \to W$ is called* regular *for $\mathcal{G}$ if for each $E \in \mathcal{E}$, r is regular for graph (V, E).*

Besides this natural translation of role assignments from graphs to graphs with multiple relations there is a weaker form (e.g. *weak regular network homomorphism* [579]), which makes use of the mapping of relations r_{rel}.

Theorems for certain types of vertex partitions (see Sections 9.1, 9.2, and 9.3) mostly translate to the case of multiple relations if we apply Definition 9.4.3.

Next we introduce a stronger form of compatibility with multiply relations. Regular role assignments as defined in Definition 9.4.4 make sure that equivalent vertices have, in each of the graphs relations identical ties to equivalent counterparts. Sometimes it is considered as desirable that they have the same combinations of relations to equivalent counterparts. That is, if we consider the example at the beginning of this section, it matters whether an individual gives orders to someone and is the friend of another individual or whether he gives orders to a friend.

Definition 9.4.7 formalizes this. First we need some preliminary definitions:

Definition 9.4.5 ([579]). *Given a graph $\mathcal{G} = (V, \mathcal{E})$ and $u, v \in V$, we define the* bundle (of relations) *from u to v as*

$$B_{uv} = \{E \in \mathcal{E} \,;\, (u, v) \in E\} \ .$$

These bundles define a new graph with multiple relations.

Definition 9.4.6 ([191, 579]). *Let $\mathcal{G} = (V, \mathcal{E})$ be a graph and $\mathcal{B}$ be the set of all non-empty bundles. For each bundle $B \in \mathcal{B}$ defines a graph with vertex set V and edge set M_B where $(u, v) \in M_B$ if and only if $B_{uv} = B$. M_B is called a* multiplex *relation induced by the graph $\mathcal{G} = (V, \mathcal{E})$. Let $\mathcal{M} = \{M_B\}_{B \in \mathcal{B}}$, then $MPX(\mathcal{G}) := (V, \mathcal{M})$ is called the* multiplex graph *of $\mathcal{G}$.*

For each pair of vertices (u, v) there is a unique bundle associated with it. This bundle may be either empty or a member of $\mathcal{B}$ (the set of all non-empty bundles). This implies that either (u, v) is a member of no M_B or has only one such multiplex relation. Thus, the multiplex graph of a graph can be viewed as a graph with a single relation, but with edge-labels. We call such a graph a *multiplex graph* [579]. That is, a multiplex graph is a graph $\mathcal{G} = (V, \mathcal{M})$ such that for each pair of relations $M_1, M_2 \in \mathcal{M}$ either $M_1 \cap M_2 = \emptyset$ or $M_1 = M_2$ holds.

For example, the multiplex graph $MPX(\mathcal{G})$ of a graph $\mathcal{G}$, is a multiplex graph.

Now we can define the type of equivalence relation which ensures that equivalent vertices have the same bundles of relations to equivalent counterparts.

Definition 9.4.7 ([191]). *Let $\mathcal{G} = (V, \mathcal{E})$ be a graph with multiple relations. A role assignment $r \colon V \to W$ that is regular for $MPX(\mathcal{G})$ is called* multiplex regular *for $\mathcal{G}$.*

As in the above definition one might define *multiplex strong structural* role assignments, but one can easily verify that a strong structural role assignment on a graph (with multiple relations) is necessarily strong structural on the corresponding multiplex graph.

Remark 9.4.8. An equivalent definition of multiplex regular role assignments is given in [83]: Let $\mathcal{G} = (V, \mathcal{E})$ be a graph, where $\mathcal{E} = \{E_1, \ldots, E_p\}$. Let

$$\mathcal{M} := \left\{ \bigcap_{i \in I} E_i \, ; \; I \subseteq \{1, \ldots, p\}, \, I \neq \emptyset \right\} \; .$$

Then the regular role assignments of $(V, \mathcal{M})$ are exactly the multiplex regular role assignments of $\mathcal{G}$.

Regular role assignments of a graph are in general not multiplex regular. Regularity however is preserved in the opposite direction.

Proposition 9.4.9 ([579]). *If $\mathcal{G} = (V, \mathcal{E})$ is a graph, $C := MPX(\mathcal{G})$, and $r : V \to W$ a role assignment then the following holds.*

1. *If r is regular for C then it is regular for $\mathcal{G}$.*
2. *If r is strong structural for C then it is strong structural for $\mathcal{G}$.*

Proof. For the proof of 1 and 2 let $E \in \mathcal{E}$ be a relation of $\mathcal{G}$ and let $u, v, u' \in V$ with $(u, v) \in E$ and $r(u) = r(u')$. Let B_{uv} be the bundle of relations of u and v (in particular $E \in B_{uv}$) and let $M := \{(w, w') \, ; \; B_{ww'} = B_{uv}\}$ be the corresponding multiplex relation (in particular $(u, v) \in M$).

1. If we assume that r is regular for C, there exist $v' \in V$ such that $r(v') = r(v)$ and $(u', v') \in M$, in particular it is $(u', v') \in E$ which shows the out-part of regularity for $\mathcal{G}$.
2. If we assume that r is strong structural for C, then $(u', v) \in M$, in particular it is $(u', v) \in E$ which shows the out-part of the condition for r being strong structural for $\mathcal{G}$.

The in-parts are treated analogously. $\square$

9.5   The Semigroup of a Graph

Social relations also have an indirect influence: If A and B are friends and B and C are enemies then this (probably) has some influence on the relation between A and C.

In this section we want to formalize such higher-order relations and highlight the relationship with role assignments.

The following definitions and theorems can be found, essentially, in [579], but have been generalized here to graphs with multiple relations (see Section 9.4).

Labeled paths of relations (like ENEMYOFAFRIEND) are formalized by composition of relations; beware of the order.

Definition 9.5.1. *If Q and R are two binary relations on V then the (Boolean) product of Q with R is denoted by QR and defined as*

$$QR := \{(u, v) \, ; \; \exists w \in V \text{ such that } (u, w) \in Q \text{ and } (w, v) \in R\} \; .$$

Boolean multiplication of relations corresponds to Boolean multiplication of the associated adjacency matrices, where for two $\{0, 1\}$ matrices A and B the Boolean product AB is defined as

$$(AB)_{ij} = \bigvee_{k=1}^{n} A_{ik} \wedge B_{kj} \ .$$

It is also possible to define *real* multiplication of weighted relations or multi-edge sets by real matrix multiplication (this has been advocated e. g., in [89]).

Definition 9.5.2. *Let* $\mathcal{G} = (V, \mathcal{E})$ *be a graph (with multiple relations). Then, the* semigroup *induced by* $\mathcal{G}$ *is defined to be*

$$S(\mathcal{G}) := \{E_1 \ldots E_k \,;\ k \in \mathbb{N},\ E_1, \ldots, E_k \in \mathcal{E}\} \ .$$

We also write $S(\mathcal{E})$ *for* $S(\mathcal{G})$.

Note that two elements in $S(\mathcal{G})$ are equal if and only if they contain the same set of ordered pairs in $V \times V$.

Furthermore, note that $S(\mathcal{G})$ is indeed a semigroup since the multiplication of relations is associative, i. e., $(AB)C = A(BC)$ holds for all relations A, B, and C.

In general, $S(\mathcal{G})$ has no neutral element, relations have no inverse and the multiplication is not commutative.

Although the length of strings in the definition of $S(\mathcal{G})$ is unbounded, $S(\mathcal{G})$ is finite since the number of its elements is bounded by $2^{(|V|^2)}$, the number of all binary relations over V.

The interesting thing about composite relations is the identities satisfied by them. For example we could imagine that on a network of individuals with two relations FRIEND and ENEMY, the identities FRIENDFRIEND=FRIEND and FRIENDENEMY=ENEMYFRIEND=ENEMY hold. At least the fact whether these identities hold or not gives us valuable information about the network. In all cases identities exist necessarily since $S(\mathcal{G})$ is finite but the set of all strings $\{E_1 \ldots E_k \,;\ k \in \mathbb{N},\ E_i \in \mathcal{E}\}$ is not.

Role assignments identify individuals. Thus they introduce more identities on the semigroup of the graph. The remainder of this section investigates the relationship between role assignments and the identification of relations.

A role assignment on a graph induce a mapping on the induced semigroup.

Definition 9.5.3 ([579]). *Let* $\mathcal{G} = (V, \mathcal{E})$ *be a graph with multiple relations and* $r : V \to W$ *a role assignment. For* $Q \in S(\mathcal{G})$, $r_{\mathrm{rel}}(Q)$ *(compare Section 9.4) is the relation on* W *defined by* $r_{\mathrm{rel}}(Q) := \{(r(u), r(v)) \,;\ (u, v) \in Q\}$ *called the* relation induced by Q and r. *Thus* r *induces a mapping* r_{rel} *on the semigroup* $S(\mathcal{G})$.

Note that in general $r_{\mathrm{rel}}(S(\mathcal{G}))$ is not the semigroup of the role graph of $\mathcal{G}$ over r, however, this is true if r is regular. Role assignments do not necessarily preserve composition, i. e., r_{rel} is not a semigroup homomorphism. One of the

main results (see Theorem 9.5.6) of this section is that regular role assignments
have this property.

Lemma 9.5.4 ([579]). *Let $\mathcal{G} = (V, \mathcal{E})$ be a graph and $r: V \to W$ a role assignment which is regular with respect to Q and $R \in S(\mathcal{G})$. Then, $r_{\text{rel}}(QR) = r_{\text{rel}}(Q)r_{\text{rel}}(R)$.*

Proof. Let $w, w' \in W$ with $(w, w') \in r_{\text{rel}}(QR)$. By the definition of $r_{\text{rel}}(QR)$
there exist $v, v' \in V$ such that $f(v) = w$, $f(v') = w'$, and $(v, v') \in QR$. Therefore
there is a vertex $u \in V$ with $(v, u) \in Q$ and $(u, v') \in R$ implying $(w, r(u)) \in r_{\text{rel}}(Q)$ and $(r(c), w') \in r_{\text{rel}}(R)$, whence $(w, w') \in r_{\text{rel}}(Q)r_{\text{rel}}(R)$. We conclude
$r_{\text{rel}}(QR) \subseteq r_{\text{rel}}(Q)r_{\text{rel}}(R)$. Note that this holds without the assumption of r
being regular.

Conversely, let $w, w' \in W$ with $(w, w') \in r_{\text{rel}}(Q)r_{\text{rel}}(R)$. Then there is a
$z \in W$ such that $(w, z) \in r_{\text{rel}}(Q)$ and $(z, w') \in r_{\text{rel}}(R)$. By the definition of
r_{rel} there are $v, v', u_1, u_2 \in V$ with $r(v) = w$, $r(v') = w'$, $r(u_1) = r(u_2) = z$,
$(v, u_1) \in Q$, and $(u_2, v') \in R$. Since r is regular and $r(u_1) = r(u_2)$ there is a
vertex $v'' \in V$ with $r(v'') = f(v')$ and $(u_1, v'') \in R$. It follows that $(v, v'') \in QR$
whence $(w, w') = (r(v), r(v'')) \in r_{\text{rel}}(QR)$, implying $r_{\text{rel}}(Q)r_{\text{rel}}(R) \subseteq r_{\text{rel}}(QR)$.
$\square$

The next theorem shows that regular or strong structural on the set of generator relations $\mathcal{E}$ implies regular resp. strong structural on the semigroup $S(\mathcal{E})$.
This is the second step in proving Theorem 9.5.6.

Theorem 9.5.5 ([579]). *Let $\mathcal{G} = (V, \mathcal{E})$ be a graph. If $r: V \to W$ is regular
(strong structural) with respect to $\mathcal{E}$ then r is regular (strong structural) for any
relation in $S(\mathcal{G})$.*

Proof. By induction on the string length of a relation in $S(\mathcal{G})$ written as a
product of generating relations (see definition 9.5.2), it suffices to show that
if r is regular (strong structural) with respect to two relations $Q, R \in S(\mathcal{G})$,
then it is regular (strong structural) for the product QR. So let $Q, R \in S(\mathcal{G})$ be
two relations and $u, v \in V$ such that $(r(u), r(v)) \in r_{\text{rel}}(QR)$. By Lemma 9.5.4,
this implies $(r(u), r(v)) \in r_{\text{rel}}(Q)r_{\text{rel}}(R)$, whence there is a $w \in W$ such that
$(r(u), w) \in r_{\text{rel}}(Q)$ and $(w, r(v)) \in r_{\text{rel}}(R)$. Since r is surjective, there exists
$u_0 \in V$ with $r(u_0) = w$, and it is $(r(u), r(u_0)) \in r_{\text{rel}}(Q)$ and $(r(u_0), r(v)) \in r_{\text{rel}}(R)$.

Now, suppose that r is *regular* with respect to Q and R. We have to show
the existence of $c, d \in V$ such that $(c, v) \in QR$, $(u, d) \in QR$, $r(c) = r(u)$ and
$r(d) = r(v)$. Since r is regular with respect to Q and $(r(u), r(u_0)) \in r_{\text{rel}}(Q)$
there exists $u_1 \in V$ such that $r(u_1) = r(u_0)$ and $(u, u_1) \in Q$. Similarly, since r
is regular with respect to R and $(r(u_0), r(v)) \in r_{\text{rel}}(R)$, there exists $d \in V$ such
that $r(d) = r(v)$, and $(u_1, d) \in R$. Since $(u, u_1) \in Q$ and $(u_1, d) \in R$ it follows
$(u, d) \in QR$, which is the first half of what we have to show. The proof of the
second half can be done along the same lines.

Now, suppose that f is *strong structural* with respect to Q and R. Then $(r(u), r(u_0)) \in r_{\mathrm{rel}}(Q)$ and $(r(u_0), r(v)) \in r_{\mathrm{rel}}(R)$ immediately implies $(u, u_0) \in Q$ and $(u_0, v) \in R$, whence $(u, v) \in QR$.   □

The next theorem might be seen as the main result of this section. It states that regular role assignments induce homomorphisms on the induced semigroups.

Theorem 9.5.6 ([579]). *Let $\mathcal{G} = (V, \mathcal{E})$ be a graph with multiple relations. If $r : V \to W$ is a regular role assignment with role graph $\mathcal{R}$, then $r_{\mathrm{rel}} : S(\mathcal{G}) \to S(\mathcal{R})$ is a surjective semigroup homomorphism.*

Proof. We know from Lemma 9.5.4 that the identity $r_{\mathrm{rel}}(QR) = r_{\mathrm{rel}}(Q) r_{\mathrm{rel}}(R)$ holds whenever r is regular with respect to Q and R. Theorem 9.5.5 states that r is regular with respect to all relations in $S(\mathcal{G})$. Thus the image of $S(\mathcal{G})$ under r_{rel} is equal to $S(\mathcal{R})$ (the images of the generator relations $\mathcal{E}$ are the generator relations of the semigroup of the role graph $S(\mathcal{R})$) and r_{rel} is a semigroup homomorphism.   □

The condition that r be regular, is not necessary for r_{rel} being a semigroup homomorphism. Kim and Roush [355] gave a more general sufficient condition. Also compare [471].

The next theorem shows that the role graph of a strong structural role assignment has the same semigroup as the original graph.

Theorem 9.5.7 ([579]). *Let $\mathcal{G} = (V, \mathcal{E})$ be a graph with multiple relations. If $r : V \to W$ is a strong structural role assignment with role graph $\mathcal{R}$, then $r_{\mathrm{rel}} : S(\mathcal{G}) \to S(\mathcal{R})$ is a semigroup isomorphism.*

Proof. By Theorem 9.5.6 r_{rel} is a surjective semigroup homomorphism. It remains to show that r_{rel} is injective. So let $Q, R \in S(\mathcal{G})$ with $r_{\mathrm{rel}}(Q) = r_{\mathrm{rel}}(R)$. Then, for all $u, v \in V$ if holds $(u, v) \in Q$ iff $(r(u), r(v)) \in r_{\mathrm{rel}}(Q)$ (since r is strong) iff $(r(u), r(v)) \in r_{\mathrm{rel}}(R)$ iff $(u, v) \in R$ (since r is strong).   □

Do Semigroup-Homomorphisms Reduce Networks? The above theorems give the idea to an alternative approach to find role assignments: In Theorem 9.5.6 it has been shown that role assignments introduce new identities on the semigroup of (generator and compound) relations of a network. Conversely, one could impose identities on relations that are almost satisfied, or that are considered to be reasonable. Now the interesting question is: *Does identification of relations imply identification of vertices of the graph which generated the semigroup?* (See [73].)

That is, given a graph $\mathcal{G}$ with semigroup $S(\mathcal{G})$ and a surjective semigroup homomorphism $S(\mathcal{G}) \to S'$ onto some semigroup S', *is there a graph $\mathcal{G}'$ and a graph homomorphism $\mathcal{G} \to \mathcal{G}'$ such that S' is the semigroup generated by $\mathcal{G}'$?*

This would be the counterpart of Theorem 9.5.6, which states that role assignments on graphs induce, under the condition of regularity, reductions of the induced semigroups, (i.e., surjective semigroup homomorphisms).

The answer is in general *no*, simply for the reason that not every semigroup is a semigroup of relations. But *under what conditions on S' and on the semigroup* *homomorphism would we get a meaningful role graph and a meaningful role assignment?*

Although the question is open for the general case some examples can be found in [89] and [471].

9.5.1 Winship-Pattison Role Equivalence

The condition for regular equivalent vertices is: *equivalent vertices have the same ties to equivalent counterparts.* In this section the phrase *to equivalent counterparts* is replaced by the weaker requirement *to some vertices.* As mentioned in Remark 9.5.9 the four equivalences defined in this section, are special cases of relative regular equivalence (see Section 9.3.4).

Definition 9.5.8. *Let $\mathcal{G} = (V, \mathcal{E})$ be a graph and $\sim$ an equivalence on V. Then $\sim$ is said to be a* weak role equivalence *for $\mathcal{G}$ if for all $u, v, w \in V$ and $E \in \mathcal{E}$, $u \sim v$ implies both*

– uRw implies there exists x such that vRx,
– wRu implies there exists x such that xRv.

Note that in contrast to the definition of regular equivalence one does not consider the role of x. So weak role-equivalent vertices don't share the same relations *to equivalent counterparts*, but they only share the same relations. If the graph has one single relation, the maximal weak role equivalence is simply the partition into isolates, sinks, sources, and vertices with positive in- and out-degree.

The indifference in regard to the role of adjacent vertices makes weak role equivalence a much weaker requirement than e. g., regular or strong structural equivalences.

Weak role equivalence could have been defined using relative regular equivalence (see Section 9.3.4).

Remark 9.5.9. Weak role equivalences are exactly the equivalences which are regular relative to the complete partition. This remark immediately generalizes to the next three definitions.

Weak role equivalence can be tightened in two directions: to include multiplexity, which leads to Definition 9.5.11, or to include composition of relations, which leads to Definition 9.5.10.

Definition 9.5.10. *Let $\mathcal{G} = (V, \mathcal{E})$ be a graph, $S := S(\mathcal{G})$ its semigroup, and $\sim$ an equivalence on V. Then $\sim$ is called a* compositional equivalence *of $\mathcal{G}$ if it is a weak role equivalence of (V, S) (see Definition 9.5.8).*

Note that in contrast to regular equivalences, where an equivalence is regular with respect to $\mathcal{E}$ if and only if it is regular with respect to $S(\mathcal{E})$, it makes a difference whether we require $\sim$ to be a weak role equivalence of $\mathcal{G}$ or of (V, S). Compositional equivalences are weak role equivalences.

Definition 9.5.11 ([579]). *Let* $\mathcal{G} = (V, \mathcal{E})$ *be a graph,* $C = (V, \mathcal{M}) := MPX(\mathcal{G})$ *its multiplex graph (see Definition 9.4.6) and* $\sim$ *an equivalence on* V*. Then,* $\sim$ *is called a* bundle equivalence *of* $\mathcal{G}$ *if it is a weak role equivalence (see Definition 9.5.8) of* C*.*

Bundle equivalences are weak role equivalences.

Winship-Pattison role equivalence is most often defined in terms of the *role-set* of an actor (see [471, p. 79ff]): Two actors are equivalent if they have the same role-sets (also compare [82, p. 81]). We restate the definitions given there in our terminology.

Definition 9.5.12. *Let* $\mathcal{G} = (V, \mathcal{E})$ *be a graph. An equivalence relation* $\sim$ *on* V *is called a* local role equivalence *or* Winship-Pattison role equivalence *if* $\sim$ *is a bundle equivalence (see Definition 9.5.11) of the graph* $(V, S(\mathcal{G}))$*.*

Local role equivalences are both bundle and compositional equivalences. Local role equivalences are, in general, not regular, which immediately implies the same for the three other (weaker) equivalences defined in this section: Let vertices u and v be connected by a bidirected edge and v have an out-going edge to a third vertex w. Then u and v are locally role equivalent but not regularly equivalent.

Conclusion. The semigroup of a graph is a possibility to describe the interaction of multiple and compound relations. An idea to use identification of relations in order to get role assignments has been sketched. This approach seems to be rather hard, both theoretically and computationally.

9.6 Chapter Notes

Vertex partitions that yield role assignments have first been introduced by Lorrain and White [394], who defined structural equivalence.

Sailer [501] pointed out that structural equivalence is to restrictive to meet the intuitive notion of social role. He proposed that actors play the same role if they are connected to *role-equivalent* actors (in contrast to *identical* actors, as structural equivalence demands). His idea of structural relatedness has been formalized as regular equivalence by White and Reitz in the seminal paper [579]. In this work, they gave a unified treatment of structural, regular, and other equivalences for graphs with single or multiple relations. Furthermore, they developed conditions for graph homomorphisms to induce (structural or regular) vertex partitions and to be compatible with the composition of relations.

Borgatti and Everett [82, 83, 190, 191] established many properties of the set of regular equivalences, including lattice structure, and developed the algorithm CATREGE to compute the maximal regular equivalence of a graph. Furthermore they introduced other types of vertex partitions to define roles in graphs. Boyd and Everett [90] further clarified the lattice structure and defined relative regular equivalence.

Marx and Masuch [408] commented that regular equivalence is already known, under the name of bisimulation in computer science. Their report has been the reason that we found the algorithm of Paige and Tarjan [459], which can compute the maximal regular equivalence and is much faster than CATREGE.

Roberts and Sheng [493] first showed that there are $\mathcal{NP}$-complete problems stemming from regular role assignments. A more complete treatment is from Fiala and Paulusma [209].

Role assignments for graphs with multiple and composite relations are already treated in [394, 579]. The possibilities to define role assignments in graphs with multiple relations are abundant. We could sketch only few of them in this chapter. Additional reading is, e.g., Kim and Roush [355] and Pattison [471] who found many conditions for vertex partitions to be compatible with the composition of relations. In the latter book, the algebraic structure of semigroups of relations is presented in detail. Boyd [89] advocated the use of real matrix multiplication to define semigroups stemming from graphs. These semigroups often admit sophisticated decompositions, which in turn, induce decompositions or reductions of the graphs that generated these semigroups.

In order to be able to deal with the irregularities of empirical networks, a formalization of role assignment must – in addition to choosing the right compatibility criterion – provide some kind of relaxation. (See Wasserman and Faust [569] for a more detailed explanation.) Relaxation has not been treated in this chapter, which has been focused on the 'ideal' case of vertex partitions that satisfy strictly the different compatibility constraints. Possibilities to relax structural equivalence, optimizational approaches for regular equivalence, and stochastic methods for role assignments are presented in Chapter 10 about blockmodels. Brandes and Lerner [97] introduced a relaxation of equitable partitions to provide a framework for role assignments that are tolerant towards irregularities.

10 Blockmodels

Marc Nunkesser and Daniel Sawitzki

In the previous chapter we investigated different types of vertex equivalences which lead us to the notion of a *position* in a social network. We saw algorithms that compute the sets of equivalent actors according to different notions of equivalence. However, which of these notions are best suited for the analysis of concrete real world data seems to depend strongly on the application area.

Practical research in sociology and psychology has taken another way: Instead of applying one of the equivalences of the previous chapter, researchers often use heuristical role assignment algorithms that compute approximations of strong structural equivalence. More recently, statistical estimation methods for stochastic models of network generation have been proposed.

Typically, researchers collect some relational data on a group of persons (the *actor set*) and want to know if the latter can be partitioned into positions with the same or at least similar relational patterns. The corresponding area of network analysis is called *blockmodeling*. Relational data is typically considered as a directed loopless graph G consisting of a node set $V = \{v_1, \ldots, v_n\}$ and R edge sets $E_1, \ldots, E_R \subseteq V^2 \setminus \{(v,v) \mid v \in V\}$. The following definition of a blockmodel sums up most of the views that can be found in the literature.

Definition 10.0.1. *A* blockmodel *$BM = (\mathcal{P}, B_1, \ldots, B_R)$ of G consists of two parts:*

1. *A partition $\mathcal{P} = (P_1, \ldots, P_L)$ of V into L disjoint subsets called the* positions *of G. For $v \in V$, the position number k with $v \in P_k$ is denoted by $P(v)$.*
2. *Matrices $B_r = (b_{k,\ell,r})_{1 \le k, \ell \le L} \in \{0,1\}^{L \times L}$, $1 \le r \le R$, called* image matrices *that represent hypotheses on the relations between the positions with respect to each relation.*

Thus, a blockmodel is a simplified version of G whose basic elements are the positions. If we demand that nodes of the same position have exactly the same adjacencies, the equivalence classes of the structural equivalence relation introduced in Definition 9.1.3 (denoted by $\simeq \in V^2$ in this chapter) give us a unique solution $\mathcal{P}_\simeq$ to our partitioning problem.

Because the field of blockmodeling is concerned with processing real world data possibly collected in experiments, it is assumed that there is some 'true' blockmodel underlying the observed graph which may not be reflected cleanly by G. This may be caused by measurement errors or natural random effects. $\mathcal{P}_\simeq$

U. Brandes and T. Erlebach (Eds.): Network Analysis, LNCS 3418, pp. 253–292, 2005.
© Springer-Verlag Berlin Heidelberg 2005

does not catch these deviations, and is therefore expected to contain too many positions hiding the true blockmodel.

Hence, blockmodeling is much about building relaxations of structural equivalence which are able to tolerate random distortions in the data up to an appropriate degree. Corresponding blockmodels are expected to have a minimal number of positions while tolerating only small deviations from the assumption of structural equivalence. Historically, the first methods used in blockmodeling have been heuristic algorithms which were believed to give good trade-offs between these two criterions.

In blockmodeling, graphs are often viewed from an adjacency matrix point of view. Let $A_r = (a_{i,j,r})_{i,j}$ denote the adjacency matrix of E_r, i. e., $a_{i,j,r} = 1 \Leftrightarrow (v_i, v_j) \in E_r$. Then, a blockmodel is represented by a permuted version of A which contains nodes of the same position in consecutive rows and columns.

Definition 10.0.2. *The* $\mathcal{P}$*-permuted adjacency matrix* $A_r^* := \left(a_{i,j,r}^*\right)_{i,j} := \left(a_{\pi^{-1}(i),\pi^{-1}(j),r}\right)_{i,j}$ *is obtained by reordering rows and columns of* A_r *with respect to the permutation* $\pi \in \Sigma_n$ *defined by*

$$\pi(i) < \pi(j) :\Leftrightarrow \left[P(v_i) < P(v_j)\right] \vee \left[(P(v_i) = P(v_j)) \wedge (i < j)\right]$$

for all $1 \leq i < j \leq n$.
For $1 \leq k, \ell \leq L$, *the* $|P_k| \times |P_\ell|$*-submatrix*

$$A_r^{k,\ell} := \left(a_{\pi^{-1}(i),\pi^{-1}(j),r}\right)_{(v_i,v_j) \in P_k \times P_\ell}$$

is called a block *and contains the connections between positions* k *and* ℓ *with respect to relation* r.

That is, the rows and columns of A_r^* are lexicographically ordered with respect to position and index. Nodes of the same position have consecutive rows and columns. A block $A_r^{k,\ell}$ represents the part of A_r^* that corresponds to the relation between P_k and P_ℓ. The entry $b_{k,\ell,r}$ of B_r should contain this information in distilled form. For $\mathcal{P}_{\simeq}$, each block is solely filled with ones or zeros (except the diagonal elements), and it makes sense to set $b_{P(v_i),P(v_j),r} := a_{i,j,r}$.

In blockmodels obtained by heuristic algorithms, the nodes of a position P_k do not necessarily have equal neighborhoods. Nevertheless, the adjacencies of nodes of the same position should be very similar in a good model, and $b_{k,\ell,r}$ should express trends existing in $A_r^{k,\ell}$. Therefore, a variety of methods has been employed to derive the image matrices from the blocks according to $\mathcal{P}$. If we consider B_r as an adjacency matrix of a graph having the positions as nodes, we obtain the so called *reduced graph* of E_r (compare Definition 9.0.3).

Figure 10.1(a) gives an example of a network G, its adjacency matrix A, and its three positions due to the structural equivalence relation $\simeq$. Figure 10.1(b) shows both the corresponding permuted adjacency matrix A^* and the reduced graph with positions as nodes.

E	1	2	3	4	5	6	7	8	9
1	-	0	0	1	0	0	0	0	1
2	0	-	0	0	1	0	1	0	0
3	0	1	-	0	1	1	1	1	0
4	1	0	0	-	0	0	0	0	1
5	0	1	0	0	-	0	1	0	0
6	0	1	1	0	1	-	1	1	0
7	0	1	0	0	1	0	-	0	0
8	0	1	1	0	1	1	1	-	0
9	1	0	0	1	0	0	0	0	-

(a) G's adjacency matrix A with corresponding graph and positions P_1, P_2, and P_3 due to the structural equivalence relation.

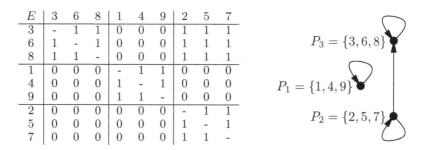

E	3	6	8	1	4	9	2	5	7
3	-	1	1	0	0	0	1	1	1
6	1	-	1	0	0	0	1	1	1
8	1	1	-	0	0	0	1	1	1
1	0	0	0	-	1	1	0	0	0
4	0	0	0	1	-	1	0	0	0
9	0	0	0	1	1	-	0	0	0
2	0	0	0	0	0	0	-	1	1
5	0	0	0	0	0	0	1	-	1
7	0	0	0	0	0	0	1	1	-

$P_3 = \{3, 6, 8\}$

$P_1 = \{1, 4, 9\}$

$P_2 = \{2, 5, 7\}$

(b) $\mathcal{P}$-permuted adjacency matrix A^* and the corresponding reduced graph. Blocks in A^* are separated by lines. Due to the structural equivalence, they contain solely ones or zeros in non-diagonal positions.

Fig. 10.1. Example network $G = (V, E)$ and its blockmodel due to the structural equivalence relation

Contents. This chapter gives a survey on selected blockmodeling approaches which are either well-established and have been widely used, or which seem to be promising and to give novel perspectives on this quite old field in network analysis. We will restrict ourselves to graphs $G = (V, E)$ containing only one edge set corresponding to one actor relation. Most approaches can be easily adapted to the case of several (sometimes weighted) relations.

Section 10.1 presents blockmodeling approaches that are mainly based on heuristic assumptions on positional interplay without a concrete model of network generation. In contrast to these so called deterministic models, which include some of the oldest algorithms used in blockmodeling, Section 10.2 presents approaches based on stochastic models. They assume that the positional structure influences a randomized process of network generation and try to estimate the parameters of the corresponding probability distribution. In this way, we can both generate and evaluate hypotheses on the network positions. Conclusions on both kinds of methods and an overview of the relevant literature are given in Section 10.3

10.1   Deterministic Models

In this section, well-established blockmodeling approaches are presented which are mainly based on heuristic assumptions on positional interplay without a concrete stochastic model of the process that generates the network. Instead, certain relaxations of the structural equivalence relation are used to decide whether two nodes share the same position. Because the decision criterions are based upon static network properties, we call these approaches deterministic models.

In order to weaken the structural equivalence, we need to measure to what extend two nodes are equivalent. Therefore, Section 10.1.1 is devoted to two of the most popular measures. These need not to be metrics, but the techniques for multidimensional scaling discussed in Section 10.1.2 can be used to embed actors in a low-dimensional Euclidian space. Having pair-wise distance values for the actors, clustering based methods like Burt's algorithm (see Section 10.1.3) are popular ways to finally partition the actor set V into positions $\mathcal{P}$. Section 10.1.4 presents the CONCOR algorithm that is an alternative traditional method to obtain $\mathcal{P}$.

The methods up to this point have been mainly introduced in the 70's and represent classical approaches. They are only used to compute a partition $\mathcal{P}$ of the actor set; the image matrix B is typically obtained by applying some standard criterions to $\mathcal{P}$ discussed in Section 10.1.5. In Section 10.1.6 we discuss different goodness-of-fit indices that are obtained by comparing the $\mathcal{P}$-permuted adjacency matrix A^* with the image matrix B of a concrete blockmodel. Finally, Section 10.1.7 introduces a generalized blockmodeling framework which integrates the steps of partitioning the actor set, computing B, and evaluating the resulting blockmodel. It represents the most recent blockmodeling approach in this section on deterministic models.

10.1.1   Measuring Structural Equivalence

We have already noted in the introduction that relations between actors in observed real-world networks may reflect an eventual underlying positional structure only in a distorted and inexact way. Therefore, blockmodeling algorithms have to tolerate a certain deviation from perfect structural equivalence, whose idea of equal neighborhoods seems to be reasonable in principle. Hence, it does not suffice to know if two nodes v_i and v_j are equivalent w. r. t. $\simeq$—we also want to know some value $\delta_{i,j}$ describing how close a node pair (v_i, v_j) is to equivalence. In the following, $\delta_{i,j}$ will always denote a symmetric distance measure between the adjacency relations of node v_i and v_j with the properties $\delta_{i,i} = 0$ and $\delta_{i,j} = \delta_{j,i}$. Superscripts identify special measures.

In order to apply geometrical distance measures, we consider the concatenation of the ith row and ith column of A as a point in the $2n$-dimensional space $\mathbb{R}^{2n}$. Burt [107] was the first who proposed to use the *Euclidian distance* in blockmodeling:

Definition 10.1.1. *The* Euclidian distance $\delta^e_{i,j}$ *between actors v_i and v_j is defined by*

$$\delta^e_{i,j} := \sqrt{\sum_{k \neq i,j} (a_{i,k} - a_{j,k})^2 + \sum_{k \neq i,j} (a_{k,i} - a_{k,j})^2} \qquad (10.1)$$

for $1 \leq k \leq n$.

Note that $\delta^e_{i,i} = 0$, $\delta^e_{i,j} = \delta^e_{j,i}$, and $0 \leq \delta^e_{i,j} \leq \sqrt{2(n-2)}$.

A second widely used measure of structural equivalence is the *correlation coefficient*, also known as *product-moment coefficient*. In contrast to the Euclidian distance, it does not directly compare entries in A, but their deviations from mean values of rows and columns.

Definition 10.1.2. *Let* $\bar{a}_{i,\cdot} := \sum_{1 \leq k \leq n} a_{i,k}/(n-1)$ *resp.* $\bar{a}_{\cdot,i} := \sum_{1 \leq k \leq n} a_{k,i}/(n-1)$ *be the mean of the values of the ith row resp. ith column of A. The* correlation coefficient *(or product-moment coefficient) $c_{i,j}$ is defined by*

$$\frac{\sum\limits_{k \neq i,j} (a_{i,k} - \bar{a}_{i,\cdot})(a_{j,k} - \bar{a}_{j,\cdot}) + \sum\limits_{k \neq i,j} (a_{k,i} - \bar{a}_{\cdot,i})(a_{k,j} - \bar{a}_{\cdot,j})}{\sqrt{\sum\limits_{k \neq i,j} [(a_{i,k} - \bar{a}_{i,\cdot})^2 + (a_{k,i} - \bar{a}_{\cdot,i})^2]} \sqrt{\sum\limits_{k \neq i,j} [(a_{j,k} - \bar{a}_{j,\cdot})^2 + (a_{k,j} - \bar{a}_{\cdot,j})^2]}}$$

$$(10.2)$$

for $1 \leq k \leq n$. The matrix $C = (c_{i,j})_{i,j}$ is called the correlation matrix *of A.*

That is, its numerator is the sum of products of v_i's and v_j's deviations from their respective row and column mean values. In the denominator, these deviations are squared and summed separately for v_i and v_j before their respective square roots are taken and the results are multiplied.

Note that $c_{i,j} \in [-1, 1]$. In statistics, the correlation coefficient is used to measure to what degree two variables are linearly related; an absolute correlation value of 1 indicates perfect linear relation, while a value of 0 indicates no linear relation. Especially, $c_{i,i} = 1$ and $c_{i,j} = c_{j,i}$. On the other hand, $|c_{i,j}| = 1$ does not imply $v_i \simeq v_j$, and $c_{i,j} = 0$ does not mean that the ith and jth row/column of A are not related at all—they are just not linearly related.

In order to derive a measure that fulfills the property $\delta_{i,i} = 0$, we normalize $c_{i,j}$ to $\delta^c_{i,j} := 1 - |c_{i,j}|$.

Comparison of Euclidian Distance and Correlation Coefficient. Let us compare the two measures $\delta^e_{i,j}$ and $\delta^c_{i,j}$. We have already seen that the Euclidian distance $\delta^e_{i,j}$ is directly influenced by the difference between the entries for v_i and v_j in A, while the normalized correlation coefficient $\delta^c_{i,j}$ also incorporates the mean values $a_{i,\cdot}$, $a_{\cdot,i}$, $a_{j,\cdot}$, and $a_{\cdot,j}$. Thus, $\delta^e_{i,j}$ measures the absolute similarity between the neighborhoods of v_i and v_j, while $\delta^c_{i,j}$ measures the similarity of the mean deviations.

In order to make the formal relationship between $\delta^e_{i,j}$ and $c_{i,j}$ better understandable, we temporarily assume that both (10.1) and (10.2) contain only the row-related sums.

Property 10.1.3. Let $\sigma_{i,\cdot} := \sqrt{\sum_{k \neq i} (a_{i,k} - \bar{a}_{i,\cdot})^2 / (n-1)}$ resp. $\sigma_{\cdot,i} := \sqrt{\sum_{k \neq i} (a_{k,i} - \bar{a}_{\cdot,i})^2 / (n-1)}$, $1 \leq k \leq n$, be the standard deviation of the ith row resp. ith column of A. Then, it holds

$$\left(\delta^e_{i,j}\right)^2 = (n-2)\left[\left(\bar{a}_{i,\cdot} - \bar{a}_{\cdot,i}\right)^2 + \sigma^2_{i,\cdot} + \sigma^2_{j,\cdot} - 2c_{i,j}\sigma_{i,\cdot}\sigma_{j,\cdot}\right] \ .$$

That is, the Euclidian distance grows with increasing mean difference $|\bar{a}_{i,\cdot} - \bar{a}_{j,\cdot}|$ and variance difference $|\sigma^2_{i,\cdot} - \sigma^2_{j,\cdot}|$, while these are filtered out by $c_{i,j}$. If the used blockmodeling method leaves freedom in choosing a measure, structural knowledge about G should influence the decision: If the general tendency of an actor to be related to others is assumed to be independent of his position, the use of $\delta^c_{i,j}$ is expected to give a better insight in the positional structure than $\delta^e_{i,j}$. For example, if the relational data was obtained from response rating scales, some actors may tend to give consistently higher ratings than others.

In the following sections, we will see how such symmetric measures $\delta_{i,j}$ are used in computing the actor set partition $\mathcal{P}$.

10.1.2   Multidimensional Scaling

Blockmodels and MDS. In the previous section we saw that the deterministic blockmodeling problem is connected to (dis-)similarity measures between the rows and columns of the adjacency matrix A that correspond to the actors. After we have decided upon a particular dissimilarity measure, we get for the set of actors a set of pairwise dissimilarities, from which we might want to deduce the positions and the image matrix of the blockmodel. This in turn can be considered as a reduced graph, which we already saw in the introduction. This process can be seen as one of information reduction from the initial dissimilarities to an abstract representation. Clearly, this is not the only way to represent the blockmodel. In this section, we will discuss in detail a slightly different approach, where the abstract representation maps the actors to points in the plane. The distances between the points should roughly correspond to the dissimilarities between the actors. Points that are close to each other with respect to the other points could then again be interpreted as positions. The underlying general problem is called *multidimensional scaling* (MDS): Given a set of dissimilarities of actors, find a 'good' representation as points in some space (two-dimensional Euclidean space for our purposes). It has been used as an intermediate step for blockmodeling, where clustering algorithms are run on the points produced by the MDS algorithm (see Section 10.1.3), and it is also considered a result in itself that needs no further postprocessing. The result can then be seen as a concise visual representation of the positional structure of a social network. Let us define the problem formally:

Problem 10.1.4 (Multidimensional Scaling Problem (MDS)). Given n objects by their $n \times n$ dissimilarity matrix δ, a dimension d and a loss function $\ell\colon \mathbb{R}^{n \times n} \times \left\{S \subset \mathbb{R}^d \,\middle|\, |S| = n\right\} \to \mathbb{R}^+$, construct a transformation $f\colon \{1, \ldots, n\} \to \mathbb{R}^d$ such that the loss $\ell(\delta, P)$ is minimal, for $P = f(\{1, \ldots, n\})$.

The loss function $\ell(\delta, P)$ measures how much the dissimilarity matrix of the objects $\{1, \ldots, n\}$ is distorted by representing them as a point set P in d dimensional Euclidean space. Obviously, different loss functions lead to different MDS problems. In this whole section, we set $d = 2$ for ease of presentation even if the first approach to be presented can be easily extended to higher dimensions. When discussing solutions to multidimensional scaling problems we will often directly talk about the point set $P = \{(p_x^1, p_y^1), \ldots, (p_x^n, p_y^n)\}$ that implicitly defines a possible transformation f. Then we also write $\delta[p, q]$ for the dissimilarity $\delta[f^{-1}(p), f^{-1}(q)]$ of the preimages of p and q for some points $p = p^i = (p_x^i, p_y^i)$ and $q = p^j = (p_x^j, p_y^j)$. We call any candidate set of points P for a solution a *configuration* and write $P = f(\delta)$ abusing notation slightly. In the next two sections we have selected out of the multitude of different MDS-approaches two algorithms that are particularly interesting: Kruskal's MDS algorithm and a recent algorithm with quality guarantee by Bădoiu. Kruskal's algorithm is probably the one that has been used most frequently in the blockmodeling context because it has become relatively established. On the other hand we are not aware of any study in blockmodeling in which Bădoiu's algorithm has been used. We present it here, because it is an algorithmically interesting method and has appealing properties like the quality guaranty.

Kruskal's MDS Algorithm. Historically, Kruskal's algorithm was among the first that gave a sound mathematical foundation for multidimensional scaling. Kruskal called his approach *nonmetric multidimensional scaling* to set it apart from earlier approaches that fall into the class of *metric scaling*. The latter approach tries to transform the dissimilarity matrix into distances by some class of parametric functions and then finds the parameters that minimize the loss function. This scenario is very similar to the classical estimation task and can be solved by least squares methods. In contrast to this parametric approach nonmetric multidimensional scaling makes no parametric assumptions about the class of legal transformations; the only condition that the transformation f should fulfill best-possible is the *monotonicity constraint* (MON)

$$\delta[p, q] < \delta[r, s] \Rightarrow \|p - q\|_2 \leq \|r - s\|_2 \tag{10.3}$$

for all $p, q, r, s \in P$. This constraint expresses that if a pair of objects is more similar than another pair then the corresponding pair of points must have a smaller (or equal) distance than the distance of the other pair of points. In (10.3), the only necessary information about the dissimilarities is their relative order.

The Stress. The key to Kruskal's algorithm is the right choice of a loss function that he calls *stress*. It is best introduced via *scatter diagrams*. Given a dissimilarity matrix and a candidate configuration of points P, we can plot the distances $d_{ij} = \|p^i - p^j\|_2$ versus the dissimilarities δ in a scatter diagram like in Figure 10.2(a).

Obviously, the configuration in Figure 10.2(a) does not fulfill the monotonicity constraint, because when we trace the points in the order of increasing dissimilarity, we sometimes move from larger to smaller distances, i.e. we move left. Let us call the resulting curve the *trace* of the configuration. A trace of a configuration that fulfills MON must be a *monotone curve* like in Figure 10.2(b). The idea is now to take the minimum deviation of the trace of a configuration from a monotone curve as the loss function. Clearly, if the trace itself is monotone this deviation is zero. More precisely, we define the *raw stress* of a configuration as

$$\min \left\{ \sum_{i<j} (d_{ij} - \hat{d}_{ij})^2 \middle| (\hat{d}_{ij})_{ij} \text{ fulfill MON} \right\} .$$

This means that the error is measured only at y-coordinates of points in the scatter diagram. At these y-coordinates, we search for points $\hat{d}_{ij}$ that together fulfill MON and minimize the squared error of distances to the corresponding points of the configuration. The raw stress has some disadvantages, for example it is not invariant under uniform stretching or shrinking of the dissimilarities. Therefore, stress is defined as follows.

Definition 10.1.5. *Given a dissimilarity matrix δ and a configuration of points P, the* stress *of P is defined by*

$$S(P) = \min \left\{ \frac{\sum_{i<j} (d_{ij} - \hat{d}_{ij})^2}{\sum_{i<j} d_{ij}^2} \middle| (\hat{d}_{ij})_{ij} \text{ fulfill MON} \right\} . \tag{10.4}$$

Note that the values of δ do not enter in (10.4); however, their order occurs implicitly via MON. In Figure 10.2(c) there is an example of a configuration together with a monotone curve that minimizes the stress, in Figure 10.2(d) the corresponding values of $\hat{d}_{ij}$ are shown as squares.

The Algorithm. To complete the description of the algorithm we need to know two further details: How is the stress computed and how is a configuration with minimum stress found? Assume we have answered the first question such that we have a procedure to compute the stress of any given configuration. For a

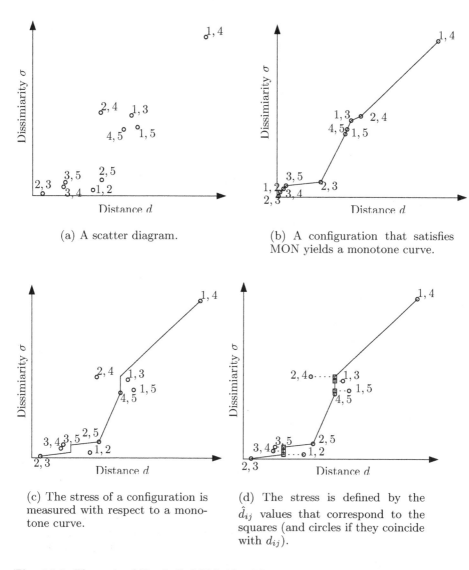

(a) A scatter diagram.

(b) A configuration that satisfies MON yields a monotone curve.

(c) The stress of a configuration is measured with respect to a monotone curve.

(d) The stress is defined by the $\hat{d}_{ij}$ values that correspond to the squares (and circles if they coincide with d_{ij}).

Fig. 10.2. Elements of Kruskal's MDS-Algorithm

given configuration it returns the correct values $\hat{d}_{ij}$. These values correspond to a *local stress function*

$$S_\ell(P) = S_\ell((p_x^1, p_y^1), \ldots, (p_x^n, p_y^n)) = \sum_{i<j}(d_{ij} - \hat{d}_{ij})^2 \Big/ \sum_{i<j} d_{ij}^2 \;, \qquad (10.5)$$

where we have still $d_{ij} = \|p^i - p^j\|_2$. The problem of finding a configuration with minimum local stress turns out to be the numerical problem of minimizing a function of $2n$ variables with respect to a given objective function, the stress. There-fore, any standard method for function minimization can be used. Kruskal pro-poses the *method of steepest descent* that starts with an arbitrary point in search space, computes the gradient of the local stress $(\partial S_\ell/\partial p_x^1, \partial S_\ell/\partial p_y^1, \dots, \partial S_\ell/\partial p_y^n)$, and moves towards the negative direction of the gradient. Then, it recomputes the local stress in the new configuration and iterates until a local minimum is found. This need not be the global minimum. In this sense the algorithm is a heuristic without performance guarantee (just as any other general algorithm for minimization of non-convex functions). To understand that the algorithm is really as straight-forward as it sounds, observe that it is indeed possible to calculate the partial derivatives of a local stress function. In general, also other methods for function minimization could be used.

As for the computation of the $\hat{d}_{ij}$ we will briefly sketch the algorithm. It relies on the following observation.

Observation 10.1.6. *The $\hat{d}_{ij}$ that minimize the stress for a given configuration have the following form: The ordered list of dissimilarities can be partitioned into consecutive blocks $\{b_1, \dots, b_k\}$ such that within each block, $\hat{d}_{ij}$ is constant and equals the average of the d_{ij} values in the block.*

Note that the $\hat{d}_{ij}$ values in Figure 10.2(d) have this form. From this observation it is clear that the problem can be solved by finding the correct partition. This is achieved by starting from the finest possible partition (each point in one block) and then iteratively joining an arbitrary pair of neighboring blocks for which the monotonicity constraint is violated.

MDS with Quality Guarantee. In this section we present a relatively new approach to multidimensional scaling by Bădoiu [105] that relies more on the combinatorial structure of the problem. As before the algorithm constructs an embedding of a given dissimilarity matrix into the plane. In this case the dissimi-larity matrix is also called distance matrix, because Bădoiu's algorithm searches for a point set that not only qualitatively mirrors the order relation on the distances/dissimilarities, but also its objective is that the distances in the em-bedding should approximate the distances given by the matrix δ as precisely as possible. It is an approximation algorithm in the sense that the loss of the con-structed embedding is bounded by $c\varepsilon$ if an optimal embedding has loss ε. Note that this is not the same as having a constant loss with respect to the original dissimilarity measure which is impossible in general. This algorithm is a quite recent result and was the first to give such guarantees. Its success stems from a clever choice of the loss function combined with beautiful insights into the com-binatorial nature of the problem. Unfortunately, it is slightly too complicated to be presented here in its entirety. However, we will see the important parts and explain the ideas for the missing parts. All missing proofs can be found in [105].

The Loss Function. The loss function that is employed here is one that uses the L_∞-norm to measure the distance in $\mathbb{R}^2$. Remember that the infinity norm of a vector its component with maximum absolute value. The loss is the maximum deviation of embedded distances from original distances.

$$\ell(\delta, f(\delta)) = \max_{1 \leq i < j \leq n} \left\{ \left| \delta[i, j] - \| f(i) - f(j) \|_\infty \right| \right\} \tag{10.6}$$

$$= \max_{p, q \in P} \left\{ \left| \delta[p, q] - \| p - q \|_\infty \right| \right\} \tag{10.7}$$

The first equation is in terms of the objects, the second in terms of the configuration $P = f(\delta)$. We call this loss function *distortion*. It measures the maximum additive error of the embedding. Let us have a closer look at the properties of this loss function. Assume we know the distortion $\varepsilon^\star = \min_f \{\ell(\delta, f(\delta))\}$ of the optimal solution and search the corresponding point set $P^\star$. Then, for each pair of points $p, q \in P^\star$ it must hold that

$$-\varepsilon^\star \leq \delta[p, q] - \max \left\{ |p_x - q_x|, |p_y - q_y| \right\} \leq \varepsilon^\star .$$

Note that the infinity norm destroys the symmetry suggested by the absolute value in (10.6) in the following sense. For the lower bound it must hold that

$$-\varepsilon^\star \leq \delta[p, q] - |p_x - q_x| \quad \text{and} \quad -\varepsilon^\star \leq \delta[p, q] - |p_y - q_y| , \tag{10.8}$$

whereas for the upper bound it must hold that

$$\delta[p, q] - |p_x - q_x| \leq \varepsilon^\star \quad \text{or} \quad \delta[p, q] - |p_y - q_y| \leq \varepsilon^\star . \tag{10.9}$$

We sum these two equations up in the following simple observation.

Observation 10.1.7. *Let $P^\star$ be the point set with minimum distortion $\varepsilon^\star$. For any two points $p, q \in P^\star$ the lower bound $-\varepsilon^\star \leq \delta[p, q] - |p_z - q_z|$ must hold for both x- (z = x) and y-coordinate (z = y). The upper bound $\delta[p, q] - |p_z - q_z| \leq \varepsilon^\star$ must hold for either x- or y-coordinates.*

The observation also suggests that x- and y-coordinates can be treated independently to a certain extend.

The Algorithm. The general idea of the algorithm is to do the following:

1. Guess $\varepsilon^\star = \min_f \{\ell(\delta, f(\delta))\}$
2. Find x-coordinates of an embedding with distortion $\varepsilon' \leq c_1 \cdot \varepsilon^\star$.
3. For these x-coordinates find y-coordinates such that the resulting point set P has distortion no more than $\varepsilon'' \leq c_2 \cdot \varepsilon'$.

We will see that the resulting point set P has distortion $\varepsilon'' \leq 30\varepsilon^\star$. Guessing the right $\varepsilon^\star$ is done by a binary search in the end. The most interesting part of the algorithm is how the y-coordinates are found. For this reason we will discuss this part in detail. Then we will sketch how the x-coordinates are found.

The y-coordinates. Let us assume that we are given x-coordinates $X = \{p_x^1, \ldots, p_x^n\}$ of a point set P with the property that for all $p, q \in P$ it holds

$$-\varepsilon' \leq \delta[p, q] - \max\{|p_x - q_x|, |p_y - q_y|\} \leq \varepsilon' . \tag{10.10}$$

Let us call this assumption the *quality assumption*. It will not be possible to exactly recover the y-coordinates in P. But we will construct y-coordinates such that the resulting point set P' has the property

$$-5\varepsilon' \leq \delta[p, q] - \max\{|p_x - q_x|, |p_y - q_y|\} \leq 5\varepsilon'$$

for all $p, q \in P'$. We call such a solution a 5-approximation solution. In doing so, we see finding the y-coordinates as a problem in its own right, i.e. we only want to know how the distortion grows with respect to ε'.

From Observation 10.1.7 it is clear that all x-coordinate pairs (p_x, q_x) have to fulfill the lower bound. In the special case where all such pairs fulfill also the upper bound, it follows by the same observation that it suffices to find y-coordinates such that all y-coordinate pairs (p_y, q_y) fulfill the lower bound. In terms of the absolute value this means $|p_y - q_y| \leq \delta[p, q] + \varepsilon'$. It is easy to express this condition as linear constraints because $|x| \leq c$ is equivalent to $x \leq c$ and $-x \leq c$. The linear constraints become

$$-\varepsilon' - \delta[p, q] \leq p_y - q_y \leq \delta[p, q] + \varepsilon' \tag{10.11}$$

for all $p, q \in P'$. Note that in this special case we actually recover the 'correct' y-coordinates that fulfill (10.10). It follows that we only need to care about pairs (p_x, q_x) that do not fulfill the upper bound. We introduce the notion of edges between such points that model how bad a pair (p_x, q_x) exceeds the upper bound.

Definition 10.1.8. *If for a pair (p_x, q_x) it holds*

$$\delta[p, q] - |p_x - q_x| > 3\varepsilon' , \tag{10.12}$$

there is a strong edge *between p and q. If*

$$3\varepsilon' \geq \delta[p, q] - |p_x - q_x| > \varepsilon' , \tag{10.13}$$

there is a weak edge *between p and q. We denote the set of all strong edges by E_s, the set of all weak edges by E_w.*

In the special case where all edges are weak edges we can again find y-coordinates via linear programming with constraints of Type (10.11). The result is then at least a 3-approximation.

The set of strong edges E_s together with the points P form (a drawing of)[1] a graph G. For the correctness of the algorithm, it is important that the connected components of G can be separated by vertical lines, i.e., they do not overlap. The graph G does not have this property. Therefore, we define the edge set E' and claim that the resulting graph G' has the desired property.

---

[1] We will simply refer to the drawing of the graph G as *the graph* G because it will always be clear that we are discussing embeddings in the plane, and it will also be clear which drawing we refer to, namely the one given by P.

Definition 10.1.9. *Let $\mathcal{C} = \{C_1, \ldots, C_k\}$ be the connected components of G and l_i (r_i) be the leftmost (rightmost) point in C_i. Let $\tilde{E}_w \subset E_w$ be the set of weak edges that have exactly one endpoint in some component C_i and the other one between l_i and r_i: $\tilde{E}_w = \{\{p,q\} \in E_w \mid \exists i: p \in C_i, q \notin C_i, l_i \le q_x \le r_i\}$. We define E' as $E_s \cup \tilde{E}_w$.*

The resulting graph G' has the desired property:

Claim 10.1.10. The connected components of G' can be separated by vertical lines that do not intersect any vertex. Moreover, every weak edge in G' is adjacent to at least one strong edge (see Figure 10.3).

The (easy) proof uses the definitions of strong and weak edges and the triangle inequality.

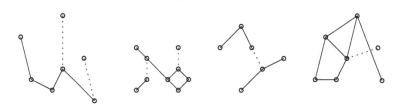

Fig. 10.3. Structure of G'. Solid lines represent strong edges, dotted lines weak edges. The four connected components do not overlap. Each weak edge is adjacent to at least one strong edge

Now that we know the structure of the graph G' that is constructed from the strong and weak edges it is interesting to see how exactly these edges can help to find an embedding. From Observation 10.1.7 we know that for a strong edge $\{p,q\}$ the y-coordinates p_y and q_y have to fulfill both the upper and the lower bound. If we try to express the upper bound similarly to (10.11), we run into the problem that $|x| \ge c$ is equivalent to $x \ge c$ or $-x \ge c$, which we cannot express as linear constraints of a linear program,[2] which have to be fulfilled simultaneously. But if we know whether $q_y \ge p_y$ or $p_y > q_y$, this problem vanishes and we can again use linear programming.

Definition 10.1.11. *For an edge $e = \{p,q\} \in E'$, $p_x \le q_x$ we say that e is oriented up if $q_y \ge p_y$, we say that it is oriented down if $p_y > q_y$.*

Lemma 10.1.12. *If we know the orientation of all strong edges, we can compute a 3-approximation via linear programming.*

Proof. We construct the following linear program.[3]

---

[2] More generally, the constraint $|x| \ge c$ is non-convex, because the function $-|x|$ is not convex. On the other hand only convex optimization problems can be solved efficiently, which is a hint that we cannot express it in any way in a linear program. See [91] for more information on convex optimization.

[3] In the original paper [105] there are inconsistencies both in the definition of orientation and in the linear program.

min δ
s.t.

$$-\delta \leq \delta[p,q] - (q_y - p_y) \leq \delta \qquad \begin{cases} \text{if } \{p,q\} \in E \\ \text{is oriented up} \end{cases} \qquad (10.14)$$

$$-\delta \leq \delta[p,q] - (p_y - q_y) \leq \delta \qquad \begin{cases} \text{if } \{p,q\} \in E \\ \text{is oriented down} \end{cases}$$

$$-\delta[p,q] - \delta \leq \quad q_y - p_y \quad \leq \delta[p,q] + \delta \text{ if } (p,q) \notin E$$

By the quality assumption there is a solution P that fulfills (10.10). This solution leads to a solution to the linear program with $\delta = \varepsilon'$. On the other hand, a solution with optimal value $\delta \leq \varepsilon'$ to the linear program is only guaranteed to have distortion lower than $3\varepsilon'$: For all edges $\{p,q\} \in E'$ the first two inequalities guarantee that the distortion is at most δ. For all pairs $\{p,q\} \notin E_w$ and $\{p,q\} \notin E_s$ the third inequality bounds the distortion by δ; but for all $\{p,q\} \in E_w \setminus E'$ the only guarantee for an upper bound is via the weakness of the edges. Thus, the guaranteed upper bound is $3\varepsilon'$ (see Definition 10.1.8). □

After Lemma 10.1.12 it is clear that it is useful to find out the orientation of the edges in E'. The following lemma states how to perform this task for one connected component of E'.

Lemma 10.1.13. *By fixing the orientation of one arbitrary edge in a connected component of G' we also fix the orientation of all other edges in this connected component.*

Proof. We show that the orientation of an edge $e = \{v,w\}$ fixes the orientation of all adjacent strong edges or of all edges if e itself is strong. Without loss of generality let v,w be oriented up. As $\{v,w\} \in E'$, both the upper and the lower bound must hold for the y-coordinate. Thus it holds (by the quality assumption) that

$$w_y - v_y + \varepsilon' \geq \delta[v,w] \geq w_y - v_y - \varepsilon' . \qquad (10.15)$$

Furthermore, $w_y - v_y > w_x - v_x$ because $u,w \in E'$ and the upper bound must be established. For an adjacent edge $\{w,t\}$ that is oriented up, we get

$$\delta[v,t] \overset{\text{Obs 10.1.6}}{\geq} (t_y - v_y) - \varepsilon' = (t_y - w_y) + (w_y - v_y) - \varepsilon'$$

$$\overset{(10.15)}{\geq} \delta[v,w] + \delta[w,t] - 3\varepsilon' . \qquad (10.16)$$

As $\{w,t\}$ is a strong edge $\delta[w,t] - 3\varepsilon' > 0$, it holds

$$\delta[v,t] > \delta[v,w] . \qquad (10.17)$$

As $\{v,w\}$ is (at least) a weak edge and $\{w,t\}$ is a strong edge, we get by combining Equations (10.12), (10.13), and (10.16)

$$\delta[v,t] > (w_x - v_x) + |t_x - w_x| + \varepsilon' \geq |t_x - v_x| + \varepsilon' . \qquad (10.18)$$

In the other case where $\{w, t\}$ is oriented down we get

$$\delta[v, t] \leq \|t - v\|_\infty + \varepsilon' \leq \max\{|t_x - v_x| + \varepsilon', |(w_y - v_y) - (w_y - t_y)| + \varepsilon'\}$$
$$\leq \max\{|t_x - v_x| + \varepsilon', \delta[v, w] + \varepsilon' - \delta[w, t] + \varepsilon' + \varepsilon'\}$$
$$\leq \max\{|t_x - v_x| + \varepsilon', \delta[v, w]\} \ . \quad (10.19)$$

Where the first inequality follows by the quality assumption, the second by the orientation of the edges, the third by the fact that both are edges and Observation 10.1.7, and the fourth because $\{w, t\}$ is a strong edge.

Equations (10.17) and (10.18) together contradict (10.19); therefore it is possible to find out the orientation of edge $\{w, t\}$. A similar argument shows that in the case where $\{w, t\}$ is a weak edge and $\{v, w\}$ is strong we can find out the orientation of $\{w, t\}$. As in a connected component each weak edge is connected to a strong edge, we can iteratively find the orientation of all edges in it by fixing one.   □

The previous two lemmata together already yield a 3-approximation solution if G' consists of a single connected component.

If G' consists of more than one connected component the algorithm arbitrarily fixes the orientation of one edge in each connected component. In the case where all these relative orientations are accidentally chosen correctly, we still have a 3-approximation. Surprisingly, even if the relative orientations are chosen incorrectly we still have a 5-approximate solution. The intuition behind this result is that between connected components there are no strong edges (but potentially weak edges) and therefore by choosing the wrong relative orientation between the components not too much distortion is created. The following lemma makes this statement precise.

Lemma 10.1.14. *There is a 5-approximate solution for every relative orientation between the edges of the components.*

Sketch of Proof. The idea of the proof is to show how we can transform the optimal solution (i.e. the solution in which the orientations are chosen optimally) into a solution with arbitrary relative orientation. To this end, we scan through the components $\{C_1, \ldots, C_k\}$ from left to right and flip in the ith step components $\{C_i, \ldots, C_k\}$ by an appropriately chosen horizontal line if the orientations in C_i in the arbitrary and the optimal solution disagree. For this choice to be possible it is necessary that the components can be separated by vertical lines. Then it needs to be established that this (clever) choice of flips does not create too much additional distortion.   □

The x-coordinates. We will see a sketch of the method to find the x-coordinates. We start again by stating the *quality assumption* (q. a.) that the optimal embedding has error $\varepsilon^\star$. Now let the diameter be given by the points p and q and assume it is defined by $q_x - p_x$. As the origin of the coordinate system is arbitrary, we can fix p at $(0, 0)$.

Let A be the set of points $v \in P \setminus \{p, q\}$ with $\delta[p, q] + k\varepsilon^\star \geq \delta[p, v] + \delta[v, q]$ for some constant k.

$$A = \{v \in P \setminus \{p, q\} \mid \delta[p, q] + k\varepsilon^\star \geq \delta[p, v] + \delta[v, q]\}$$

Points in A fulfill the following two inequalities

$$v_x \overset{\text{q. a.}}{\leq} \delta[p, v] + \varepsilon^\star \,, \tag{10.20}$$

$$v_x \overset{2 \times \text{q. a.}}{\geq} \delta[p, q] - \delta[v, q] - 2\varepsilon^\star \overset{v \in A}{\geq} \delta[p, v] - (k+2)\varepsilon^\star \,. \tag{10.21}$$

If we fix v_x at the arithmetic mean of the two right hand sides $v_x = (2\delta[p, v] - (k+1)\varepsilon^\star)/2 = \delta[p, v] - ((k+1)\varepsilon^\star)/2$, the additive error with respect to the optimal value for v_x is bounded by $(k+3)\varepsilon^\star/2$. If all points $v \in P \setminus \{p, q\}$ are in A, the problem is solved. In the case $P \setminus A \neq \emptyset$, the algorithm makes a (lengthy) case distinction that we will not present in detail. The general idea is to partition the set $P \setminus A$ into finer sets B, C, and D. Then, similar to the case of the problem with the y-coordinates, equations are derived that hold under the assumptions that a point p' is in B, C, or D. As the equations are again contradictory, it is possible to find out to which of the sets p' belongs. From this membership it is then possible to find a good approximation of the x-coordinate.

This completes the presentation of Bădoiu's algorithm. To sum up, it achieves its goal of guaranteeing a constant loss with respect to the optimal embedding by connecting the MDS-problem to a discrete structure—the graph G' together with an orientation on it. This makes possible the use of a combinatorial algorithm. Note that on bad instances the distortion of the constructed embedding can still be very high if even the optimal embedding has high distortion, see the bibliography for references on this problem.

10.1.3   Clustering Based Methods

In the preceding sections, we have discussed how to derive measures of structural equivalence from the adjacency matrix A of G and how to refine them by multidimensional scaling. We will now investigate clustering based methods which use such a measure $\delta_{i,j}$ for computing an actor set partition $\mathcal{P}$ that hopefully corresponds to the true positional structure of G.

Having a symmetric distance measure $\delta_{i,j}$, we could of course apply general clustering techniques in order to identify subsets of actors which are expected to represent one position. Chapter 8 gives an overview over this broad area of network analysis. Nevertheless, in the area of blockmodeling a rather simple clustering heuristic has been implemented and applied by most researchers.

In general, we speak of *hierarchical clustering* if the clustering algorithm starts with clusters $P_1, \ldots, P_n$ with $P_i := \{v_i\}$ before it iteratively joins pairs of clusters with minimal distance $d(P_k, P_\ell)$. Different measures $d\colon \mathcal{P}(V) \times \mathcal{P}(V) \to \mathbb{R}$ for the inter-cluster distance have been proposed. This clustering framework generates a hierarchy of subsets and finally results in a single cluster containing all actors of V. Then, the researcher has to select a minimum distance β that has to be between two clusters resp. positions.

Formally, we start with a partition $\mathcal{P}_1 = \{\{v_1\}, \ldots, \{v_n\}\}$. In general, we have a current partition $\mathcal{P}_x$ and compute $\mathcal{P}_{x+1}$ by joining two different clusters $P_{k^*}, P_{\ell^*} \in \mathcal{P}_x$, i.e., $\mathcal{P}_{x+1} := (\mathcal{P}_x \setminus \{P_{k^*}, P_{\ell^*}\}) \cup \{P_{k^*} \cup P_{\ell^*}\}$ for $(P_{k^*}, P_{\ell^*}) :=$ arg $\min_{P_k, P_\ell \in \mathcal{P}_x} d(P_k, P_\ell)$. The result is a sequence $\mathcal{P}_1, \ldots, \mathcal{P}_n$ of partitions. The researcher has to choose a threshold value β which is used to discard cluster unions incorporating cluster pairs of larger distance than β. After having pruned the hierarchy in this way, the resulting actor subsets are taken as the positions of the blockmodeling analysis.

Cluster Distance Measures. There are four popular ways how to define the cluster distance d (see [18]). All of them have been justified by successful analyses of positional structures and may be selected depending on the relational data of G. However, the single linkage hierarchical clustering is not considered to be very good because of chaining effects. Nevertheless, it is able to discover well-separated shape clusters.

Single linkage $d^{\mathrm{s}}(P_k, P_\ell)$. In case of single linkage, we set $d^{\mathrm{s}}(P_k, P_\ell) := \min \{\delta_{i,j} \mid v_i \in P_k,\ v_j \in P_\ell\}$. That is, the smallest distance between two members v_i of P_k and v_j of P_ℓ is taken as distance between the clusters P_k and P_ℓ.

Complete linkage $d^{\mathrm{c}}(P_k, P_\ell)$. In case of complete linkage, we demand that every pair $(v_i, v_j) \in P_k \times P_\ell$ has at most distance $d^{\mathrm{c}}(P_k, P_\ell)$, i.e., $d^{\mathrm{c}}(P_k, P_\ell) := \max \{\delta_{i,j} \mid v_i \in P_k,\ v_j \in P_\ell\}$.

Average linkage $d^{\mathrm{a}}(P_k, P_\ell)$. In contrast to the maximum or minimum actor-wise distances of the two previous measures, the average linkage takes the average actor distances into account. The average linkage distance d^{a} is defined by

$$d^{\mathrm{a}}(P_k, P_\ell) := \frac{1}{|P_k| \cdot |P_\ell|} \cdot \sum_{v_i \in P_k, v_j \in P_\ell} \delta_{i,j} \ .$$

Average group linkage $d^{\mathrm{g}}(P_k, P_\ell)$. Finally, the average group linkage considers the average distance between all actor-pairs of the join of P_k and P_ℓ. This result $P_k \cup P_\ell$ contains $\binom{|P_k|+|P_\ell|}{2}$ actor pairs, and it is

$$d^{\mathrm{g}}(P_k, P_\ell) := \sum_{v_i \in P_k, v_j \in P_\ell} \delta_{i,j} \bigg/ \binom{|P_k| + |P_\ell|}{2} \ .$$

Burt's Algorithm. We finally want to mention a special well-established hierarchical clustering approach to blockmodeling that was presented by Burt [107] in 1976. Basically, he uses the Euclidian distance $\delta_{i,j}^{\mathrm{e}}$ together with the single linkage cluster distance d^s. Furthermore, Burt assumes that the vector $\delta_{i,\cdot}^e :=$ $\left(\delta_{i,j}^e\right)_j$ of the observed actor distances between v_i and the other actors is composed mainly of two components: First, a position-dependent vector $p_k \in \mathbb{R}^n$ which contains the hypothetical distances of an ideal member of position $k :=$ $P(v_i)$ to all other actors. Second, $\delta_{i,\cdot}^e$ is influenced by an additive error component $w_i \in \mathbb{R}^n$ as small as possible which is (besides of the covariance) used to explain the deviations of $\delta_{i,\cdot}^e$ from p_k. In detail, Burt's model states

$$\delta_{i,\cdot}^e = \mathrm{cov}\left(\delta_{i,\cdot}^e, p_k\right) \cdot p_k + w_i \ ,$$

where $k := P(v_i)$ and $\mathrm{cov}\left(\delta_{i,\cdot}^e, p_k\right)$ is the covariance between $\delta_{i,\cdot}^e$ and p_k. That is, vectors $\delta_{i,\cdot}^e$ and $\delta_{j,\cdot}^e$ for $P(v_i) = P(v_j)$ may only differ by their mean, while the remaining deviation w_i resp. w_j should be small for a good blockmodel.

Burt gives methods to compute the unknown components p_k, $1 \leq k \leq L$ and w_i, $1 \leq i \leq n$, from the distances $\delta_{i,j}$ minimizing the error components w_i. These results can then be used for further interpretation of the blockmodel or to evaluate its plausibility by means of the magnitudes of the error components.

10.1.4 CONCOR

Besides clustering based methods, the CONCOR algorithm represents the most popular method in traditional blockmodeling. It was presented by Breiger, Boorman, and Arabie [99] in 1975 and has been extensively used in the 70's and 80's.

CONCOR is a short form of *convergence of iterated correlations*. This stems from the observation in sociological applications that the iterated calculation of correlation matrices of the adjacency matrix A typically converges to matrices of special structure. In detail, the algorithm computes the symmetric *correlation matrix* $C_1 := \left(c_{i,j}^{(1)}\right)_{i,j} := (c_{i,j})_{i,j}$ of A corresponding to Definition 10.1.2. Then, it iteratively computes the correlation matrix $C_{s+1} := \left(c_{i,j}^{(s+1)}\right)_{i,j}$ of $C_s := \left(c_{i,j}^{(s)}\right)_{i,j}$. This process is expected to converge to a matrix $\mathcal{R} := (r_{i,j})_{i,j}$ consisting solely of -1 and $+1$ entries. Furthermore, it has been observed that there typically exists a permutation $\pi \in \Sigma_n$ on the set of actor indices $\{1, \ldots, n\}$ and an index i^* such that the rows and columns of $\mathcal{R}$ can be permuted to a matrix $\mathcal{R}^* := \left(r_{i,j}^*\right)_{i,j} := \left(r_{\pi(i),\pi(j)}\right)_{i,j}$ with $r_{i,j}^* = 1$ for $(i,j) \in$ $(\{1, \ldots, i^*\} \times \{1, \ldots, i^*\}) \cup (\{i^*+1, \ldots, n\} \times \{i^*+1, \ldots, n\})$ and $r_{i,j}^* = -1$ for $(i,j) \in (\{i^*+1, \ldots, n\} \times \{1, \ldots, i^*\}) \cup (\{1, \ldots, i^*\} \times \{i^*+1, \ldots, n\})$ (see Figure 10.1.4).

$$\mathcal{R}^* = \left(\begin{array}{c|c} +1 & -1 \\ \hline -1 & +1 \end{array}\right)$$

Fig. 10.4. Layout of matrix $\mathcal{R}^*$

Assume that the actor set partition $\mathcal{P} = \{P_1, \ldots, P_L\}$ reflects the true positional structure of G. Let $\mathcal{P}_1$ and $\mathcal{P}_2$ be disjoint subsets of $\mathcal{P}$ with $\mathcal{P}_1 \cup \mathcal{P}_2 = \mathcal{P}$ such that actors of different positions $P_k \in \mathcal{P}_x$ and $P_\ell \in \mathcal{P}_x$, $x \in \{1, 2\}$, are more similar to each other than actors of positions $P_{k'} \in \mathcal{P}_1$ and $P_{\ell'} \in \mathcal{P}_2$. That is, we assume that $\mathcal{P}_1$ and $\mathcal{P}_2$ are the result of some clustering method dividing $\mathcal{P}$ into two subsets.

The CONCOR algorithm is based on the assumption that the correlation coefficient $c_{i,j}^{(s)}$ between actors v_i, v_j of the same part $\mathcal{P}_x$ converges to 1, while this index converges to -1 if v_i and v_j are placed in different halves of $\mathcal{P}$. Therefore, the algorithm is iterated until for some s^* matrix C_{s^*} is close enough to $\mathcal{R}$; then, the actors V are divided into $V_1 := \{v_{\pi(1)}, \ldots, v_{\pi(i^*)}\}$ and $V_2 := \{v_{\pi(i^*+1)}, \ldots, v_{\pi(n)}\}$. Now, V_x, $x \in \{1, 2\}$, should correspond to $\bigcup_{k \in \mathcal{P}_x} P_k$.

In order to finally obtain the positional subsets $P_1, \ldots, P_L$ of V, CONCOR is recursively applied to the induced subgraphs $G_x := (V_x, E \cap (V_x \times V_x))$ for $x \in \{1, 2\}$, until the user decides to stop. One criterion for this could be the speed of convergence to $\mathcal{R}$; in most papers reporting on applications of CONCOR, this decision is taken heuristically depending on G. That is, we get a subdivision tree of V (often called *dendrogram*) whose leaves correspond to the final output $\mathcal{P} = \{P_1, \ldots, P_L\}$.

Criticism. Although this method is well-established and was applied in many blockmodel analyses of social data, there has also been a lot of criticism of CONCOR. Doreian [160], Faust [199], and Sim and Schwartz [520] applied it to several hypothetical networks whose positional structure was known and experienced CONCOR to be unable to recover the correct blockmodel.

Breiger, Boorman, and Arabie proposed the CONCOR algorithm without a mathematical justification for its procedure or an idea what it exactly computes. In [508], Schwartz approaches this problem by investigating CONCOR's mathematical properties. Experiments show that for most input graphs G, the result matrix C_{s^*} has rank 1. It can be easily proved that this property implies the special structure of $\mathcal{R}$ (that is, $r_{i,j} \in \{-1, 1\}$ and the existence of π and i^*). Schwartz also gives concrete counterexamples for which this does not hold. Furthermore, the only eigenvector of such a rank 1 matrix seems almost always to correspond to the *first principal component* obtained by the statistical method of principal component analysis (PCA) [335]. That is why there seems to be no substantial reason to use CONCOR instead of a PCA, whose properties are well-understood.

10.1.5 Computing the Image Matrix

It was already mentioned that the partition $\mathcal{P}_\simeq$ of the structural equivalence classes causes the $\mathcal{P}_\simeq$-permuted adjacency matrix $A^* = \left(a_{i,j}^*\right)_{i,j}$ to consist solely of 0- and 1-blocks $A^{k,\ell}$, $1 \leq k, \ell \leq L$. It has also been argued that $\simeq$ is not suited to retrieve the hidden positional structure of real-world graphs G. Therefore, heuristic methods based on some relaxation of $\simeq$ have been introduced in the preceding sections.

Let $\mathcal{P}$ be an actor set partition produced by such a heuristic blockmodeling method. The corresponding $\mathcal{P}$-permuted matrix A^* is expected to consist of blocks $A^{k,\ell}$ containing both zeros and ones. In order to decide if position P_k is adjacent to P_ℓ in the reduced graph represented by the image matrix $B = (b_{i,j})_{i,j} \in \{0,1\}^{L \times L}$, several criterions have been proposed in the literature. We describe the three most popular ones for the case $k \neq \ell$.

Zeroblock Criterion. The zeroblock criterion corresponds to the assumption that two positions $P_k, P_\ell \in \mathcal{P}$ are only non-adjacent if the k–ℓ block $A^{k,\ell}$ of the $\mathcal{P}$-permuted matrix A^* solely contains zeros, i.e., $b_{k,\ell} = 0 :\Leftrightarrow \forall (v_i, v_j) \in P_k \times P_\ell : (v_i, v_j) \notin E$. If the zeroblock criterion is used, the image matrix B corresponds to the adjacency matrix of the role graph introduced in Definition 9.0.3.

Oneblock Criterion. In contrast to the zeroblock criterion, the oneblock criterion corresponds to the assumption that two positions $P_k, P_\ell \in \mathcal{P}$ are only adjacent if $A^{k,\ell}$ solely contains ones, i.e., $b_{k,\ell} = 1 :\Leftrightarrow \forall (v_i, v_j) \in P_k \times P_\ell : (v_i, v_j) \in E$.

α-Density Criterion. In most cases, we do not assume that a single entry in a block $A^{k,\ell}$ decides about the relation between positions $P_k, P_\ell \in \mathcal{P}$. We would rather accept small deviations from perfect 0- or 1-blocks and, therefore, want to know to which block type $A^{k,\ell}$ is more similar.

First, we define a supporting identifier for the number of non-diagonal elements of a block.

Definition 10.1.15. *The* block cardinality $S_{k,\ell}$ *of block* $A^{k,\ell}$ *is defined by* $S_{k,\ell} := |P_k| \cdot |P_\ell|$ *if* $k \neq \ell$ *and* $S_{k,\ell} := |P_k| \cdot (|P_\ell| - 1)$ *if* $k = \ell$.

Definition 10.1.16. *The* block density $\Delta_{k,\ell}$ *of block* $A^{k,\ell}$ *is defined by*

$$\Delta_{k,\ell} := \frac{1}{S_{k,\ell}} \cdot \sum_{v_i \in P_k,\, v_j \in P_\ell} a_{i,j} \ .$$

This definition excludes the diagonal elements of A. Using the α-density criterion, we set $b_{k,\ell}$ to zero iff $\Delta_{k,\ell}$ is smaller than a certain threshold value α, i.e., $b_{k,\ell} = 0 :\Leftrightarrow \Delta_{k,\ell} < \alpha$. Often, the over-all density of the adjacency matrix A is used as threshold α, i.e., $\alpha := \sum_{1 \leq i,j \leq n} a_{i,j} / (n(n-1))$. That is, two positions P_k and P_ℓ are decided to be connected if the relative edge number in their induced subgraph is at least as high as the relative edge number of the whole graph G.

10.1.6 Goodness-of-Fit Indices

Due the heuristical nature of the algorithms discussed in this section on deterministic models, it makes sense to apply several different methods on the same data and to compare the results. In order to decide which result to accept as the best approximation of the true positional structure of the input graph G, quality or *goodness-of-fit* indices are needed to evaluate the plausibility of a blockmodel.

So let us assume that $B = (b_{i,j})_{i,j}$ is an image matrix produced by some blockmodeling method for graph G with adjacency matrix A and corresponding actor set partition $\mathcal{P}$.

Density Error. A blockmodel can be evaluated by comparing $A = (a_{i,j})_{i,j}$ with a hypothetical ideal adjacency matrix induced by B. Such an ideal matrix would have only 1-entries in blocks $A^{k,\ell}$ with $b_{k,\ell} = 1$ resp. 0-entries if $b_{k,\ell} = 0$ (excluding diagonal elements $a_{i,i}$). In detail, we compute the sum of error differences between the block densities $\Delta_{k,\ell}$ and the elements of B.

Definition 10.1.17. *The* density error e_d *of image matrix B is defined by*

$$e_\mathrm{d} := \sum_{1 \leq k, \ell \leq L} |b_{k,\ell} - \Delta_{k,\ell}| \ .$$

It is $e_\mathrm{d} \in [0, L^2]$. The smaller e_d, the more structural equivalent are actors of same position. Therefore, the blockmodel with the smallest density error is expected to be a better representation of the positional structure of G.

Carrington-Heil-Berkowitz Index. A second widely used goodness-of-fit index is the Carrington-Heil-Berkowitz index [112], which is tailored for evaluating blockmodels that have been created using the α-density criterion (see Section 10.1.5). Remember that we define $b_{i,j} = 0$ iff the block density $\Delta_{k,\ell}$ is smaller than the threshold value α. The choice of $b_{k,\ell}$ seems to be more reliable if the difference $|\Delta_{k,\ell} - \alpha|$ is large. The best possible difference for $b_{k,\ell} = 0$ is α, while it is $1 - \alpha$ for $b_{k,\ell} = 1$.

The Carrington-Heil-Berkowitz index is the normalized weighted sum of squared ratios of the observed difference $|\Delta_{k,\ell} - \alpha|$ to the ideal one α resp. $1 - \alpha$. Again, let $S_{k,\ell}$ be the block cardinality of block $A^{k,\ell}$ defined in Definition 10.1.15.

Definition 10.1.18. *Let $t_{k,\ell} := 1$ for $b_{k,\ell} = 0$ and $t_{k,\ell} := 1/(1-\alpha)$ for $b_{k,\ell} = 1$. The* Carrington-Heil-Berkowitz index e_b *of image matrix B is defined by*

$$e_\mathrm{b} := \sum_{1 \leq k, \ell \leq L} \left(\frac{\Delta_{k,\ell} - \alpha}{t_{k,\ell} \cdot \alpha} \right)^2 \cdot \frac{S_{k,\ell}}{n(n-1)} \ .$$

That is, the summand for block $A^{k,\ell}$ is weighted by the ratio $S_{k,\ell}/(n(n-1))$ it contributes to the whole matrix A. It is $e_\mathrm{b} \in [0, 1]$, and a value of 1 indicates perfect structural equivalence. A value close to 0 stems from all $\Delta_{k,\ell}$s being close to α. Then, many values $b_{k,\ell}$ are expected to be wrong because the α-density criterion classified them just due to little random deviations of $\Delta_{k,\ell}$ around α. Hence, the corresponding blockmodel is assumed to be bad.

10.1.7   Generalized Blockmodeling

Batagelj, Ferligoj, and Doreian [42, 45, 206] present an approach called *generalized blockmodeling*. They consider blockmodeling as an optimization problem on the set of partitions $\Pi := \{\mathcal{P} = (P_1, \ldots, P_L) \mid V = P_1 \uplus \cdots \uplus P_L\}$, where L is part of the input.

In the classical blockmodeling framework, the entry $b_{k,\ell} \in \{0, 1\}$ of the image matrix B represents a hypothesis on the existence of a connection between positions P_k and P_ℓ. That is, the blocks of a good blockmodel are assumed to be filled mainly either with ones or with zeros. In contrast, generalized blockmodeling is not just based on a relaxation of structural equivalence, but allows positions to be related by a variety of different connection types $\mathcal{T}$. Hence, the entries $b_{k,\ell}$ of B now take values in $\mathcal{T}$.

The optimization problem that has to be solved is defined by an error measure $D \colon \Pi \to \mathbb{R}$ with $D(\mathcal{P}) := \sum_{1 \le k, \ell \le L} d(P_k, P_\ell)$ summing up blockwise errors $d \colon \mathcal{P}^2 \to \mathbb{R}$. The final result is an optimal partition $\mathcal{P}^* := \arg\min_{\mathcal{P} \in \Pi} \{D(\mathcal{P})\}$. The authors use a local search heuristic to find $\mathcal{P}^*$. Starting from an initial (possibly random) partition, a current partition $\mathcal{P}$ is iteratively improved by replacing it with the best partition $\mathcal{P}' \in \mathcal{N}(\mathcal{P})$ from the *neighborhood* $\mathcal{N}(\mathcal{P})$ of $\mathcal{P}$. This neighborhood is defined by all partitions resulting from one of two operations applied on $\mathcal{P}$:

1. A *transition* moves some node v from its position P_k to another position P_ℓ.
2. A *transposition* exchanges the positional membership of two distinct nodes $v \in P_k$ and $v_j \in P_\ell$, $k \ne \ell$.

This optimization method does not depend on D and leaves freedom for the definition of the blockwise error measure d. It is assumed that each connection type $T \in \mathcal{T}$ has a set $I(T)$ of ideal blocks that fit T perfectly. For any block $A^{k,\ell}$ according to the current partition $\mathcal{P}$, the type-specific error measure $\delta(A^{k,\ell}, T) \in \mathbb{R}$ gives the minimal distance of $A^{k,\ell}$ to any block of $I(T)$. Then, we assume that P_k and P_ℓ are related by some connection type T with minimal distance to $A^{k,\ell}$, that is, $b_{k,\ell} := \arg\min_{T \in \mathcal{T}} \{\delta(A^{k,\ell}, T)\}$. Some priority order on $\mathcal{T}$ can be used to determine $b_{k,\ell}$ if the nearest connection type is not unique. Alternatively, $\mathcal{P}$ can be optimized for a pre-defined image matrix B or a whole class of image matrices (see, e.g., [44, 162]). From B, the blockwise errors are obtained by $d(P_k, P_\ell) := \delta(A^{k,\ell}, b_{k,\ell})$.

Some Proposed Connection Types. The generalized blockmodeling framework can be used with arbitrary user-defined connection types. In [45], Batagelj, Ferligoj, and Doreian propose a set of nine types motivated from different existing blockmodeling approaches, which will be briefly discussed in the following. In order to simplify the descriptions, we assume that blocks contain no diagonal elements of A.

Complete and null. This corresponds to the zeroblock- and oneblock-criterion in classical blockmodeling. An ideal complete (null) block $A^{k,\ell}$ contains solely ones (zeros) and represents the case that all nodes P_k are connected to all nodes P_ℓ (no node of P_k is connected to any node of P_ℓ). If all blocks are either ideal complete or ideal null, the partition corresponds to the equivalence classes of a structural equivalence relation.

Row-dominant and col-dominant. An ideal row-dominant (col-dominant) block contains at least one row (column) entirely filled with ones. That is, there is at least one actor in the row position connected to all of the other group (there is at least one actor in the column position to which every actor from the row position is connected).

Row-regular, col-regular, and regular. An ideal row-regular (col-regular) block contains at least one one in each row (column). That is, every actor in the row position is connected to at least one of the column position (every actor in the column position is connected from at least one of the row position). A block is called regular if it is both row-regular and col-regular. If all blocks are ideal regular, the partition corresponds to a regular role assignment (see Definition 9.0.3).

Row-functional and col-functional. An ideal row-functional (col-functional) block contains exactly one one in each column (row). That is, every actor in the column position is connected to exactly one of the row position (every actor in the row position is connected from exactly one of the column position).

Figure 10.1.7 illustrates ideal subgraphs of each connection type, while Table 10.1.7 lists the definitions of deviation functions $\delta(P_k, P_\ell, T)$ for each of the nine types. These sum up elements of rows resp. column which do not fit the ideal block of a particular connection type.

Generalized blockmodeling has been successfully applied to several networks (see, e. g., [163]). The method seemed to be limited to networks of at most hundreds actors.

10.2   Stochastic Models

Recently, many researchers have advocated the use of stochastic models instead of deterministic models because they make explicit the assumptions on the model and enable us to make precise statements on the validity of hypotheses about social networks. In Section 10.2.1 we present the p_1 model that was the first stochastic model to become established in social network analysis. Then we investigate in the context of the p_1 model how hypothesis testing can be done for stochastic models in Section 10.2.2. In Section 10.2.3 we explore the use of the p_1 model for blockmodeling. We also describe stochastic models that are more adapted to the specific setting of blockmodeling in Section 10.2.4. Finally, we present an advanced stochastic model in Section 10.2.5.

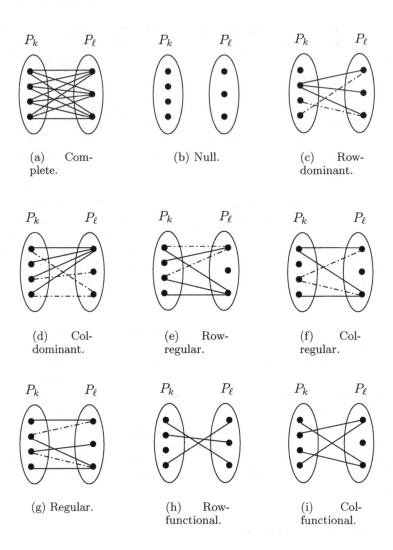

Fig. 10.5. Examples of ideal subgraphs for the different block types of generalized blockmodeling as proposed by Batagelj, Ferligoj, and Doreian. Dashed lines are not necessary for the blocks to be ideal

10.2.1   The p_1 Model

If we want to understand blockmodeling from a statistical point of view, we need to make an assumption on a model that generates the data. In the setting of parameterized statistics, this is a parameterized probability distribution. As the data in blockmodeling is a directed graph, we need to understand suitable distributions on graphs. Historically, the p_1 model was one of the first such

Table 10.1. Deviation functions for the connection types of generalized blockmodeling as proposed by Batagelj, Ferligoj, and Doreian. (For blocks containing diagonal elements, the formulas have to be slightly modified)

T	$\delta(P_k, P_\ell, T)$
complete	$|P_k| \cdot |P_\ell| - c$
null	c
row-dominant	$(|P_\ell| - M_r) \cdot |P_k|$
col-dominant	$(|P_k| - M_c) \cdot |P_\ell|$
row-regular	$(|P_k| - N_r) \cdot |P_\ell|$
col-regular	$(|P_\ell| - N_c) \cdot |P_k|$
regular	$(|P_k| - N_r) \cdot |P_\ell| + (|P_\ell| - N_c) \cdot |N_r|$
row-functional	$c - N_r + (|P_k| - N_r) \cdot |P_\ell|$
col-functional	$c - N_c + (|P_\ell| - N_c) \cdot |P_k|$

c Number of ones in $A^{k,\ell}$.
N_r Number of non-null rows in $A^{k,\ell}$.
N_c Number of non-null column in $A^{k,\ell}$.
M_r Maximal row-sum in $A^{k,\ell}$.
M_c Maximal column-sum in $A^{k,\ell}$.

distributions that has been used in social network analysis. Its main advantages are its intuitive appeal and its simplicity.

Generally, we want to express for each graph x on n nodes with an $n \times n$ adjacency matrix A the probability that it is drawn from the set of all possible graphs on n nodes $\mathcal{G}_n$. If we define a random variable X that assumes values in $\mathcal{G}_n$, we could express any distribution by defining $\Pr[X = x]$ explicitly for all $x \in \mathcal{G}_n$. Of course, this direct approach becomes infeasible already for moderately big n. We are interested in an simple, 'intuitive' distribution. Therefore, it is natural to try to connect $\Pr[X = x]$ to the presence or absence of individual edges x_{ij} in x, which we express by a $\{0, 1\}$-random variable:

$$X_{ij} = \begin{cases} 1 & \text{if edge } x_{ij} \text{ present in } x, \\ 0 & \text{otherwise.} \end{cases}$$

Note that in contrast to the part on deterministic models concrete graphs are called x in this part and the edges are referred to as x_{ij}. The reason for this change in notation is that graphs are now seen as an outcome of a draw from a distribution that is represented by a random variable X. Probably one of the easiest ways to specify a distribution is to set $\Pr[X_{ij} = 1] = 1/2$ for all $i, j \in \{1, \ldots, n\}, i \neq j$ and to assume that all X_{ij} are independent. This is equivalent to giving all graphs in $\mathcal{G}_n$ the same probability, i.e. $\Pr[X = x] = 2^{-n(n-1)}$. Obviously, this model is too simple to be useful. It is not possible to infer anything from it as the distribution is not parameterized. A very simple parameterization is to set $\Pr[X_{ij} = 1] = a_{ij}$. If we assume independence of all

X_{ij} we get

$$\Pr[X = x] = \prod_{1 \leq i,j \leq n} a_{ij}^{x_{ij}} (1 - a_{ij})^{1-x_{ij}} .$$

One reason why this closed form is so simple is that we have assumed independence. On the other hand this model has serious drawbacks: First, by the independence assumption it is impossible to infer how likely it is that a relation from a to b is reciprocated. Unfortunately, this question is at the heart of many studies in social network analysis. Second, the model has too many parameters, which cannot be estimated from a single observation (i.e. the observed social network), this problem is often referred to as this model not being *parsimonious*.

The p_1 model that we derive now from a first 'tentative' distribution overcomes these drawbacks. In order to model reciprocation effects let us assume statistical dependence of the variables X_{ij} and X_{ji} for all $1 \leq i < j \leq n$ which are together called the *dyad* $D_{ij} := X_{ij} \times X_{ji}$. Let the rest of the variables still be independent, i.e. the probability of an edge from a to b is only dependent on the existence of an edge from b to a. The resulting distribution, which we call p_t (for tentative), is easy to specify in terms of dyads. We set

$$\Pr[D_{ij} = (1,1)] = m_{ij} ,$$
$$\Pr[D_{ij} = (1,0)] = a_{ij} ,$$
$$\Pr[D_{ij} = (0,0)] = n_{ij} ,$$

with $m_{ij} + a_{ij} + a_{ji} + n_{ij} = 1$. Here, m_{ij} stands for the probability of a mutual relation, a_{ij} for an asymmetric relation and n_{ij} for no relation between actors i and j. For the probability of a given graph x we get

$$p_t(x) = \Pr[X = x] = \prod_{i<j} m_{ij}^{x_{ij}(1-x_{ij})} \prod_{i \neq j} a_{ij}^{x_{ij}(1-x_{ji})} \prod_{i<j} n_{ij}^{(1-x_{ij})(1-x_{ji})} .$$

This formula completely specifies p_t. We have still the problem of too many variables, which we will address soon. From a statistical point of view, it is desirable to find out into which class of distributions p_t falls, so that the standard theory can be applied. For p_t we show that it belongs to the *exponential family* of distributions:

Definition 10.2.1. *A distribution of a random variable X belongs to the s-dimensional exponential family iff its probability density or frequency function can be written as*

$$f(x, \eta) = \exp \left[\sum_{i=1}^{s} \eta_i T_i(x) - A(\eta) \right] h(x) ,$$

where the η_i are parameters, A is a real-valued function of the parameters, the T_i are real-valued statistics, and the factor $h(x)$ is any function depending only on x.

To see that p_t has indeed this form we first transform it by taking logarithms and exponentiating:

$$p_t(x) = \exp\left[\sum_{i<j} \rho_{ij} x_{ij} x_{ji} + \sum_{i \neq j} \theta_{ij} x_{ij}\right] \prod_{i<j} n_{ij} \ , \tag{10.22}$$

where $\rho_{ij} = \ln\left[(m_{ij} n_{ij})/(a_{ij} a_{ji})\right]$ and $\theta_{ij} = \ln\left[a_{ij}/n_{ij}\right]$. The distribution p_t is in the exponential family: the η are all θ and ρ parameters, the statistics are the x_{ij} and the $x_{ij} x_{ji}$, the function $A(\eta)$ is $\sum_{i<j} \log n_{ij}$ and finally $h(x)$ is just the constant 1. The dimension is $2n^2$. Equation (10.22) is a reparameterization of p_t that is now expressed in terms of ρ_{ij} and θ_{ij}. The parameters ρ and θ are so-called *log-odds ratios*. Intuitively, $\exp(\rho_{ij})$ divides the symmetric cases by the asymmetric cases and therefore ρ_{ij} measures the tendency for reciprocation. The odds ratio $\exp(\theta_{ij})$ divides a case where there is an edge from i to j by a case where there is no edge. Therefore, θ_{ij} is an indicator of the probability of an edge from i to j.

To overcome the problem of too many parameters (that can be read off from the high dimension of p_t) we constrain the parameters in the following way:

$$\rho_{ij} = \rho \quad \forall i < j$$

and

$$\theta_{ij} = \theta + \alpha_i + \beta_j \quad \forall i \neq j \tag{10.23}$$

with $\sum_i \alpha_i = \sum_i \beta_i = 0$. The constraints imply that a global reciprocation parameter ρ is assumed and that the density from i to j is split up into three additive components: θ, a global density parameter, α_i, actor i's *expansiveness* (or *productivity*), and β_j, actor v_j's *attractiveness*. The resulting distribution is the p_1 distribution:

$$p_1(x) = \exp\left[\rho m' + \theta \sum_{i,j} x_{ij} + \sum_i \alpha_i \sum_j x_{ij} + \sum_j \beta_j \sum_i x_{ij}\right] \cdot \prod_{i<j} n_{ij}$$
$$= \exp\left[\rho m' + \theta e + \sum_i \alpha_i \Delta_{\text{out}}(i) + \sum_j \beta_j \Delta_{\text{in}}(j)\right] \cdot \prod_{i<j} n_{ij} \ . \tag{10.24}$$

Also p_1 belongs to the exponential family: The statistics are the number of mutual edges m', the total number e of edges, and $\Delta_{\text{in}}(i)$ and $\Delta_{\text{out}}(i)$, i.e. for all i the in- and out-degrees of node i. The dimension is $2n + 2$, significantly lower than for p_t. Equation 10.24 shows that all statistics except for m' can be expressed as so-called *margins*, i.e. as a sum over the variables where some indices are fixed and others go over the complete range.

After having deduced the p_1 model, the most natural question is how we can estimate the parameters $\underline{\theta} = (\rho, \theta, \alpha_1, \dots, \alpha_n, \beta_1, \dots, \beta_n)$ from an observed graph. The standard estimation procedure for p_1 is maximum likelihood (ML-) estimation which yields the parameters that maximize the probability $p_1(x \mid \theta)$ for the observed x. The general approach to find the ML-estimator is to differentiate the probability density function for the parameters and to search for maxima. In this context the density function is called *likelihood function* $\ell_x(\theta)$ because it is seen as a function in the parameters and not in the data values. The theory of exponential families (that is beyond the scope of this book, see [385] for details) directly gives the result that the maximum likelihood estimation can be found as the solution of the *likelihood equations* in which the (sufficient) statistics are equated to their expected values.[4] In our case the sufficient statistics are all statistics that define p_1. Therefore we get

$$m' \overset{!}{=} E[m'] = \sum_{i<j} m_{ij} \ , \tag{10.25}$$

$$\Delta_{\text{in}}(i) \overset{!}{=} E[\Delta_{\text{in}}(i)] = \sum_{j}(m_{ij} + a_{ij}) \ \forall i \in \{1, \dots, n\} \ , \tag{10.26}$$

$$\Delta_{\text{out}}(j) \overset{!}{=} E[\Delta_{\text{out}}(j)] = \sum_{i}(m_{ij} + a_{ij}) \ \forall j \in \{1, \dots, n\} \ . \tag{10.27}$$

Note that for ease of presentation the variables θ and ρ_{ij} have been transformed back. Theoretically, any standard method that solves such a system of linear equations (like the Newton-method) can be applied. However, the structure of these equations can lead to nontrivial convergence problems. Therefore, specific algorithms have been developed; one of them is the *generalized iterative scaling* algorithm. In fact after a transformation of the variables also standard iterative scaling can be used. As this transformation is also needed in the next section it is presented here. Let

$$Y_{ijk\ell} = \begin{cases} 1 & \text{if } X_{ij} = k, X_{ji} = \ell \text{ for } k, \ell \in \{0, 1\}, \\ 0 & \text{otherwise}. \end{cases}$$

With this representation *all* statistics in the p_1 model can be expressed as margins of the variables. In particular, $m' = 1/2 \sum_{i,j} y_{ij11}$. For a single dyad we get

$$\Pr[Y_{ijk\ell} = 1] = \exp\left[k\alpha_i + k\beta_j + \ell\alpha_j + \ell\beta_i + (k + \ell)\theta + k\ell\rho + \lambda_{ij}\right] \ ,$$

where the λ_{ij} are chosen such that $\sum_{k,\ell} Y_{ijk\ell} = 1$ and $\sum_i \alpha_i = \sum_i \beta_i = 0$. It can be verified that this is equivalent to the p_1 model by expressing the original parameters m_{ij}, a_{ij} and n_{ij} in terms of the new parameters. Besides, a little calculation reveals that indeed

$$\prod_{i<j,k,\ell} \Pr[Y_{ijk\ell} = 1] = p_1(x) \ .$$

---

[4] Roughly, this result can be obtained by maximizing $\ell(\cdot)$ in setting $\partial\ell/\partial\theta$ to zero and observing that p_1 is a convex function being in the exponential family.

The new representation allows to apply the theory of *generalized linear models* and *categorical data analysis*[5] to p_1.

The p_1 model incorporates the possibility to do goodness-of-fit tests and general hypothesis testing.

10.2.2   Goodness-of-Fit Indices

One of the major advantages of statistical models over the 'ad-hoc' deterministic models is the (at least theoretical) possibility to make precise statements on both how appropriate a model is for the observed data and how justified hypothesis on the social network are.

We review basic facts from statistics that are necessary to understand this. Whether we want to evaluate the goodness-of-fit of the model or whether we are interested in verifying claims about the social network, we are always in a similar setting in which we have two alternative hypothesis, the *null hypothesis* H_0 and the *alternative hypothesis* H_A. Already the names suggests that we usually treat these two hypothesis asymmetrically, which will become clear later in this section. To give an example H_0 might state that the observed social network is from a p_1 distribution with a given parameter set $\{\theta = \theta', \rho = \rho', \alpha_1 = \alpha_1', \ldots, \alpha_n = \alpha_n', \beta_1 = \beta_1', \ldots, \beta_n = \beta_n'\}$, whereas H_A could state that this is true except for the reciprocation parameter, which is different: $\{\theta = \theta', \rho \neq \rho', \alpha_1 = \alpha_1', \ldots, \alpha_n = \alpha_n', \beta_1 = \beta_1', \ldots, \beta_n = \beta_n'\}$ In this example H_0 is called a *simple hypothesis* because it completely specifies the distribution, whereas H_A does not specify ρ and is therefore called a *composite hypothesis*. In general composite hypotheses specify that the parameters can come from a subset of all possible parameters. A *test statistic* T is a random variable that maps the observed data x to a value $T(x)$, often with $T(x) \in [0,1]$. The set of values of T for which H_0 is accepted (rejected) is denoted by *acceptance region* (resp. *rejection region*). Often the rejection region is of the form $\{x \mid T(x) < c\}$ or $\{x \mid T(x) > c\}$, then the value c that separates the rejection region from the acceptance region is called the *critical value*. In the ideal case all x for which H_0 holds are mapped to values in the acceptance region and all other x are mapped to values in the rejection region. In almost all nontrivial cases errors occur. These errors can be of two types:

1. H_0 is true, but the test rejects it. This is called a *type I error*, its probability α is called the *significance level* of the test.
2. H_0 is false, but is accepted. This (usually less detrimental) error is called a *type II error*. Let its probability be β, then we call $1 - \beta$ (the probability that H_0 is false and rejected by the test) the *power* of the test.

The asymmetry of H_0 and H_A is reflected in the usual test procedure: A significance level α is fixed (typically at small values like $0.01, 0.05$ or 0.1) and an appropriate test T is chosen. Obviously, tests with higher power for the fixed

---

[5] To be more precise the transformation shows that p_1 is a *loglinear model of homogeneous association* or of *no three-factor interaction*, see [3, 213].

significance level are preferable. The choice of the significance level reflects how detrimental the researcher assesses a type I error. The critical value c is set according to this significance level and finally $T(x)$ is computed on the observed data x. If $T(x) > c$ the null hypothesis is rejected, otherwise it is accepted. In order to set the critical value c according to the significance level we need to find a c for which $\Pr[T(x) > c] \leq \alpha$ under the assumption that the null hypothesis is true. Therefore, it is in general necessary to know the distribution of the test statistic under the null hypothesis. This distribution is the so-called *null distribution*.

Finding a good test, i.e. finding a test T with high or even maximum power among all possible tests, is a complicated problem beyond the scope of this book. However we present a paradigm from which many tests are constructed: Given two hypothesis H_0 and H_A expressed by the subsets of parameters ω_0 and ω_A to which they restrict the likelihood function $\ell(\theta)$, then the statistic

$$\Lambda^* = \frac{\sup_{\theta \in \omega_0} \ell_x(\theta)}{\sup_{\theta \in \omega_A} \ell_x(\theta)}$$

is called the *likelihood ratio test statistic*. High values of Λ^* suggest that H_0 should be accepted, low values suggest it should be rejected. The *likelihood ratio test* rejects for values below some critical value c and accepts above it. One reason why this test is often used is that it can be shown to be optimal for simple hypotheses (in this case the supremum is over a single value $\underline{\theta_0}$ resp. $\underline{\theta_A}$).

Lemma 10.2.2 (Neyman-Pearson). *Let H_0 and H_A be simple hypotheses given by the two parameter vectors $\underline{\theta_0}$ resp. $\underline{\theta_A}$. If the likelihood ratio test that rejects H_0 for $\frac{\ell_x(\theta_0)}{\ell_x(\theta_A)} < c$ and rejects it otherwise has significance level α then any other test statistic with significance level $\alpha' \leq \alpha$ has power less than or equal to that of the likelihood ratio test.*

Note that in the case of composite hypothesis nominator and denominator are the ML-estimates from the respective restricted parameter sets ω_0 and ω_A. For distributions involving exponentiation like the exponential family it is often easier to work with the ratio of the logarithms. In this case we get the statistic G^2 that is called *log likelihood ratio statistic*:

$$G^2 = -2 \log \lambda \ .$$

The factor of -2 has the reason that with this definition, G^2 has an approximate chi-square distribution in many cases.

Testing with p_1. We now investigate how to apply the general setting above to p_1 models. For goodness-of-fit evaluation we would state H_0 as "the data is generated by a p_1 model". Intuitively, H_A should express that "the data is generated by some other (more complicated) model". Making this statement precise is difficult, we need to define a family of distributions p_s that is a meaningful superset of p_1 with two properties: First, for the likelihood ratio tests we need to be able to do ML-estimation in p_s. Second, we need to determine the null distribution of the likelihood ratio test statistic, in order to set a meaningful critical value. For p_1 both problems are nontrivial. One possibility to extend p_1 to a more general distribution is to allow for *differential reciprocity*, i.e. instead of setting $\rho_{ij} = \rho$ every actor gets a reciprocity parameter ρ_i and we set $\rho_{ij} = \rho + \rho_i + \rho_j$. Let us ignore the estimation problems for this model and assume that we can calculate the ML-estimates for given data. Then the likelihood ratio is the maximum likelihood of the p_1 model over the maximum likelihood of this extended model (which cannot be smaller because it contains the p_1 model). The value of this ratio indicates how justified the assumption of a global reciprocity parameter in the p_1 model is.

10.2.3 Blockmodels and p_1

The p_1-model has been extensively used for blockmodels. Recall that the p_1-model estimation yields—apart from the global density and reciprocation estimates θ and ρ—an expansiveness and an attractiveness estimate α_i respectively β_i for each actor.

One prominent approach is from Anderson, Faust, and Wasserman [30]. They propose to interpret the stochastic equivalence of two actors as them having the same α_i and β_i values . From this they derive the following blockmodeling procedure.

1. Fit a p_1-model to the observed digraph G, giving a set of parameters $\{\theta, \rho, \alpha_1, \dots, \alpha_n, \beta_1, \dots, \beta_n\}$.
2. Attribute the point $q_i = (\alpha_i, \beta_i)$ to each actor $i \in \{1, \dots, n\}$.
3. Cluster the points into k clusters and return the clusters as a partition $\mathcal{P}$ for the blockmodel.

Alternatively Anderson, Faust, and Wasserman suggest to take the points as a result of the blockmodel. The parameter k is an input parameter to the blockmodeling procedure. For the clustering any of the clustering methods from Section 10.1.3 or Chapter 8 can be used. Once the partition $\mathcal{P}$ has been found we can test its quality by the testing methods of the previous section: Let the null-hypothesis H_0 be that $\mathcal{P}$ is indeed the partition and therefore all actors in each $P_k \in \mathcal{P}$ are stochastically equivalent and have $\alpha_i = \alpha_j \, \forall i, j \in P_k$. For a maximum likelihood ratio test we need to evaluate G^2 which can be shown to be

$$2 \sum_{i<j,k,\ell} y_{ijk\ell} \log \frac{y_{ijk\ell}}{\hat{y}_{ijk\ell}^{\mathcal{P}}} \, ,$$

where $y_{ijk\ell}$ are the observed data and $\hat{y}_{ijk\ell}^{\mathcal{P}}$ are the ML-estimates for $\Pr[Y_{ijk\ell} = 1]$ under the side constraints given by $\mathcal{P}$. The null distribution of G^2 is a chi-squared distribution, the degrees of freedom of which are a function of the number of partitions.

The above model has some serious drawbacks that will become clear in a typical example, in which the social network consists of a class of pupils in a primary school. The pupils are asked for their best friends. The answers are encoded in the directed graph, i.e. an edge from pupil i to pupil j means that pupil i sees pupil j as one of his best friends. Typically, this setting results in a graph with two 'clusters', the boys and the girls. Both among the boys and among the girls a high 'within'-density of directed edges can be observed. Between the two clusters there are usually considerably less edges, corresponding to a low 'between'-density. From the discussion about the different types of equivalences it should be clear that the two groups should be taken into consideration for the blockmodel and that the partition should be into boys and girls. Unfortunately, the p_1-model attributes a single expansiveness and attractiveness parameter to each boy and girl and is thus unable to model the difference between 'within'- and 'between'-densities. This is a serious drawback because the different densities reflect the blockmodel. To overcome these shortcomings Wang and Wong proposed a refinement of the p_1-model [567]. In particular, if the partition $\mathcal{P}$ is already known in advance (like in the school class example) we can define indicator variables $d_{ijk\ell}$ that represent $\mathcal{P}$ as follows.

$$d_{ijk\ell} = \begin{cases} 1 & \text{if actor } i \text{ is in } P_k \text{ and actor } j \text{ is in } P_\ell, \\ 0 & \text{otherwise.} \end{cases}$$

Recall that in the derivation of p_1 we set θ_{ij} as in Eq. (10.23). We incorporate the knowledge about $\mathcal{P}$ in the new model by setting

$$\theta_{ij} = \theta + \alpha_i + \beta_j + \sum_{k,\ell} d_{ijk\ell}\lambda_{k\ell} \ \forall i \neq j \ .$$

Here, λ_{ij} are the newly introduced *block parameters* that model the deviations from the average in the expansiveness and attractiveness between two specific partitions P_k and P_ℓ. In the school class example we would get a negative λ between the boys and girls and positive λs within the two partitions. Maximum likelihood estimation in this model can again be done via generalized iterative scaling, a transformation into a loglinear model of homogeneous association like for p_1 is not known however. For reasons of parsimony it is often preferably to restrict subsets of the λ_{ij}s to be equal.

10.2.4   Posterior Blockmodel Estimation

In the model of Wang and Wong that we saw in the previous section we had to content ourselves with blockmodels that needed the partition of actors as input and served only as a means of testing hypotheses on this partition. Such an approach is called *a priori* blockmodeling, because the partition constitutes a priori knowledge. It is justified whenever it is clear from the nature of the sociological question or by attributes of the actors (gender, age etc.) what the partition of interest is. In this section we consider a stochastic approach by Nowicki and Snijders to *a posteriori* blockmodeling, where the partition is unknown

[453]. This model does not have the drawbacks of other a posteriori approach we saw by Anderson, Faust, and Wasserman.

As in the p_1-model Nowicki and Snijders consider dyads, i.e., ordered pairs of actors (vertices) between which relations are given. Here a relation can be the presence or absence of directed edges between two actors, but more generally, the model allows the relation from vertex v_i to vertex v_j to take on any value from a finite set A, which is similar to allowing multiple edge sets as in Definition 10.0.1. Therefore, dyads (v_i, v_j) can take values x_{ij} in a set $\mathcal{A} \subset A \times A$, the values of all dyads together form the *generalized adjacency matrix*. For ease of presentation we will continue with directed graphs, thus we assume $\mathcal{A} = \{0, 1\}^2$; for example $x_{ij} = (0, 1)$ stands for the asymmetric dyad (v_i, v_j) with an edge from v_j to v_i. The crucial concept that models the partition is that vertices v_i have *colors* c_i from a set $\chi = \{1, \ldots, L\}$ that are not observed (latent). The authors call this model a *colored relational structure*. It is given by a generalized adjacency matrix x and a coloring $c = (c_1, \ldots, c_n)$.

The stochastic model that generates the data is now defined as follows. We model the coloring by independent identically distributed (i. i. d.) random variables C_i for each vertex $v_i \in V$. Thus we set

$$\Pr[C_i = k] = \theta_k$$

for each color $k \in \chi$. The joint distribution of a given coloring $c = \{c_1, \ldots, c_n\}$ is

$$\Pr[C_1 = c_1, \ldots, C_n = c_n] = \prod_{1 \leq i \leq n} \theta_{c_i} \ .$$

As we have seen in the discussion on different types of equivalences, we assume in blockmodeling that all actors in one block behave similarly. Therefore, it is assumed here that the type of relation between two actors i and j depends only on their colors:

$$\Pr[X_{ij} = a \mid C = c] = \eta(i, j, a) \ ,$$

where the η parameterize the distribution (we have to require $\forall i, j \in C \colon \sum_{a \in \mathcal{A}} \eta(i, j, a) = 1$). We obtain the following conditional distribution of relations x and colors c given the parameters:

$$\Pr[x, c \mid \theta, \eta] = \prod_{i=1}^{n} \theta_{c_i}$$

$$\cdot \prod_{a,b} \left(\prod_{1 \le i < j \le L} \eta(i, j, (a, b))^{|\{x_{k\ell} \mid x_{k\ell} = (a,b), c_k = i, c_\ell = j\}|} \right)$$

$$\cdot \prod_{a=b} \left(\prod_{1 \le i \le L} \eta(i, i, (a, b))^{|\{x_{k\ell} \mid x_{k\ell} = (a,b), c_k = c_\ell = i\}|/2} \right)$$

$$\cdot \prod_{a \ne b} \left((\prod_{1 \le i \le L} \eta(i, i, (a, b))^{|\{x_{k\ell} \mid x_{k\ell} = (a,b), c_k = c_\ell = i, k < \ell\}|} \right)$$

$$\forall a, b \in \{0, 1\} \ .$$
$$(10.28)$$

Note that this formula basically multiplies for each vertex the probability of its color θ_i and between all color classes the probabilities for the observed relations between the two classes. The first double product does the multiplication for different color classes, the last two double products do this for all monochromatic dyads. From a statistical point of view such a model falls into the class of *mixture models*.

We will briefly describe one way of statistical inference for such models. Assume some black box allows us to get a sample of values (θ, η, x) from the distribution given by the density function $f(\theta, \eta, c \mid x)$. Thus we have at our disposal a set of triplets $\{(\theta^0, \eta^0, x^0), (\theta^1, \eta^1, x^1), \ldots, (\theta^{K-1}, \eta^{K-1}, x^{K-1})\}$. This sample provides us with information about the underlying model parameters and hidden data. For example, the value

$$\frac{1}{K} \sum_{k=0}^{K-1} [x_i^k = x_j^k]$$

indicates how likely it is that actors v_i and v_j are in the same color class. Indeed, if the sample consists of *independent* draws from the distribution given by $f(\theta, \eta, c \mid x)$ it follows directly from the law of large numbers that the above value is a meaningful approximation of the random indicator variable $[C_i = C_j]$.

$$\Pr[C_i = C_j] = E[[C_i = C_j]] \approx \frac{1}{K} \sum_{k=0}^{K-1} [c_i^k = c_j^k] \qquad (10.29)$$

For convenience we restate the (weak)[6] law of large numbers, which also makes precise the sense in which the $\approx$ symbol is to be understood.

---

[6] The type of convergence shown here is called *convergence in probability*. Other versions of this theorem exist in which stronger types of convergence are shown.

Theorem 10.2.3 (Weak Law of Large Numbers). *Let* $X_1, X_2, \ldots, X_n$ *be a sequence of independent random variables with* $E[X_i] = \mu$ *and* $\mathrm{Var}[X_i] = \sigma^2$. *Then for any* $\varepsilon > 0$,

$$\Pr\left[\left|\mu - \frac{1}{n}\sum_{i=1}^{n} X_i\right| > \varepsilon\right] \to 0$$

for $n \to \infty$.

The proof follows straight forward by one application of the Chebyshev inequality and can be found in any textbook on probability theory, for example [492].

This theorem makes no statement on the speed of convergence. In the case of independent random variables, the central limit theorem makes such a statement. Unfortunately we will see that our black box does not give us independent samples. With the same approach as above estimates for θ and η can be obtained. The value

$$\frac{1}{K}\sum_{k=0}^{K-1} \theta_{c_i}^k \tag{10.30}$$

gives an estimate of the probability of the color class containing actor i. Finally,

$$\frac{1}{K}\sum_{k=0}^{K-1} \eta(c_i, c_j, a) \tag{10.31}$$

is an estimate of the probability that between actor v_i's color class and actor v_j's color class a relation of type a holds.

These slightly awkward constructions are necessary, because there is an identifiability problem in the estimation process: It is not meaningful to talk about color class i because arbitrary permutations of the color class labels can lead to identical results. Similarly, it is not meaningful to estimate the probability $\Pr[C_i = j]$ (instead of 10.29). All functions in (10.29), (10.30), and (10.31) are invariant under permutations and therefore circumvent the identifiability problems.

Up to now we have gently ignored the question how we get the sample from $f(\theta, \eta, c \mid x)$. To this end a method called *Gibbs sampling* can be used. While the method itself is easy to describe its precise mathematical foundations are beyond the scope of this book. The general approach for Gibbs sampling from a distribution $f(x_1, \ldots, x_d)$ with prior distributions $\pi(x_i)$ for all variables is to start with a random point $(x_1^0, x_2^0, \ldots, x_d^0)$ as the first point of the sample. The next point is identical to the first, except in the first coordinate. The first coordinate is drawn from the full conditional distribution $f(x_1 \mid x_2 = x_2^0, \ldots, x_d = x_d^0)$. Usually it is much easier to get a sample from such a full conditional distribution then from the general one. In the ith step, the new point is the same as the last, except for the $(i \bmod d)$th coordinate, which is drawn from the distribution $f(x_{i \bmod d} \mid x_1, \ldots, x_{(i \bmod d)-1}, x_{(i \bmod d)+1}, x_d)$. Often only every dth point is taken, so that the next point potentially differs in all coordinates from the present one. In our case the Gibbs sampler works as follows: Given values (x^t, θ^t, η^t) the next values are determined as

1. $(\theta^{t+1}, \eta^{t+1})$ is drawn from $f(\theta, \eta \mid x^t, y)$.
2. For each value $i \in \{1, \ldots, n\}$ the color x_i^{t+1} is drawn from

$$f(x_i \mid \theta^t, \eta^t, x_1^t, \ldots, x_{i-1}^t, x_{i+1}^t, \ldots, x_d^t) \ .$$

It can be verified that the full conditional distributions used here have a comparatively easy form. The Gibbs sampler has the property that the sample $\{(x^0, \theta^0, \eta^0), \ldots, (x^{K-1}, \theta^{K-1}, \eta^{K-1})\}$ approximates the distribution $f(\theta, \eta, c \mid x)$ for large K. It is obvious from this description that the sample points are highly dependent, because the values $(x^{t+1}, \theta^{t+1}, \eta^{t+1})$ are constructed from (x^t, θ^t, η^t). The sequence of samples forms a Markov chain. Fortunately, the general theory of Markov chains comprises the so-called *ergodic theorem* that is in a sense the counterpart of the law of large numbers for dependent samples that are produced by a Markov chain. For a precise statement of the theorem too much terminology for Markov chains is required, therefore we leave the presentation at that intuitive level and refer the interested reader to the bibliography.

To sum up, Nowicki and Snijders propose to see blockmodeling as actors getting colors from a distribution defined by the θ parameters. The probabilities of relations between actors are influenced by their colors. As the colors are latent variables, the task is to predict them from the observations (namely the relations between the actors) and to estimate the parameters η that govern how the actors in the color classes relate to each other. The prediction and estimation is done by Gibbs sampling from the conditional distribution $f(\theta, \eta, c \mid x)$ and then evaluating invariant functions on the sample from which information about the coloring and the parameters can be inferred.

10.2.5 p^* Models

In Sects. 10.2.1 and 10.2.3, we have seen how a stochastic model of social network generation can be used to evaluate an a priori blockmodel and to compute an a posteriori blockmodel. The simple structure of the node-wise parameters α and β allows to define stochastic equivalence and, therefore, to express a blockmodel in terms of a restricted parameter space of the p_1 model. Moreover, the parameters of such simple models can be estimated exactly and efficiently.

On the other hand, we made quite strong assumptions on the network generation process. Even the basic assumption that the dyads are drawn independently of each other has been heavily criticized since the p_1 model was proposed in social network analysis. In 1996, Wasserman and Pattison [570] introduced a more powerful family of random graph distributions called p^* *models*, whose applications to blockmodeling will be discussed in the following. The aim of p^* models is to have greater flexibility in expressing the dependencies among the relations X_{ij} between the actors. To state this more formally we make the following definition.

Definition 10.2.4 (Conditional Independence). *Let X, Y and Z be (sets of) random variables. We say that X is* conditional independent *of Y given Z if*

$$\Pr[X \mid Y, Z] = \Pr[X \mid Z] \text{ for } \Pr[Z] > 0 .$$

This is written as $X \amalg Y \mid Z$.

In the p_1 model we have $\{X_{ij}\} \amalg X \setminus \{X_{ij}\} \mid \{X_{ji}\}$, i.e. the edge from i to j only depends on the edge from j to i. In order to model more complicated dependencies than this, we introduce a graph that represents these dependencies.

Definition 10.2.5. *Let $W := \{X_{ij} \mid 1 \leq i, j \leq n\}$ with $X_{ij} \in \{0, 1\}$ be the set of random variables of the edges of a random graph X on n nodes. Thus as before X is the random variable that assumes values in $\mathcal{G}_n$.*

- *A distribution on X is called* random field *if all graphs $x \in \mathcal{G}_n$ get positive probability.*
- *The undirected graph $\mathcal{I}_X = (W, F)$, $F \subseteq W^2$, is called the* dependency graph *of X if for all $X_{ij} \in W$ it holds that*

$$\{X_{ij}\} \amalg W \setminus \{X_{ij}\} \mid \mathcal{N}(X_{ij})$$

where $\mathcal{N}(X_{ij})$ are all random variables adjacent to X_{ij} in $\mathcal{I}_X$.
- *A random field that can be expressed via a dependency graph is called a* Markov field.

The p_1 model can be seen as a Markov field (or *Markov graph*) with a dependency graph that consists of all edges $\{X_{ij}, X_{ji}\}$. The idea of p^* is to try to find explicit distributions for arbitrary dependency graphs. The Hammersley-Clifford Theorem [59, 589] states that this is always possible in the sense that for each Markov field there is a distribution that can be expressed by an (almost) closed form. We state the theorem in a simplified version:

Theorem 10.2.6 (Hammersley-Clifford). *Let $\mathcal{I}_X = (W, F)$ be the dependency graph of a Markov graph X. Let $\mathcal{C}$ be the set of cliques of $\mathcal{I}_X$. Then, there exist potentials $\{\lambda_c \in \mathbb{R} \mid c \in \mathcal{C}\}$ such that*

$$\Pr[X = x] = \frac{\exp\left(\sum_{c \in \mathcal{C}} \lambda_c \cdot \prod_{X_{ij} \in c} x_{ij}\right)}{\kappa} ,$$

where $\kappa := \sum_{z \in \mathcal{G}_n} \exp\left(\sum_{c \in \mathcal{C}} \lambda_c \cdot \prod_{X_{ij} \in c} x_{ij}\right)$ is a normalization constant.

Observe that the products over the cliques are one if and only if all edges in the clique in the dependency graph are present, otherwise they equal zero. Given a dependency graph $\mathcal{I}_X$ that expresses our assumptions about independence between relations in the observed graph x, we get a minimal parameter set for the graph distribution consisting of the potentials λ_c of all cliques of $\mathcal{I}_X$. Distributions of this kind are called p^* *models*.

Estimating the Potentials. Estimation in p^* models has been a topic of vivid research discussions in the last years. Several estimation methods have been proposed, the most prominent ones being the *pseudolikelihood method* and the *Markov Chain Monte Carlo (MCMC) method* which we saw briefly already in Section 10.2.4. Both methods are mathematically involved and have serious drawbacks as discussed in [529] and references therein, therefore we will not present them here. See the bibliography for detailed references on the methods.

Using p^* Models for Blockmodeling. Up to now, we have only seen a stochastic model for graph generation. Due to the clique-wise potentials, there is no obvious counterpart to the stochastic equivalence in p_1 blockmodels. We present an approach proposed in [472].

Consider an a priori blockmodel with actor set partition $\mathcal{P} := \{P_1, \ldots, P_L\}$ for an observed graph x. Let $\mathcal{C}$ be the set of cliques of the dependency graph $\mathcal{I}_X = \{W, F\}$. We call the subgraph of x on the nodes of a clique $c \in \mathcal{C}$ the *configuration $C(c, x)$.*

Definition 10.2.7. *Two configurations $C(a, x)$ and $C(b, x)$, $a, b \in \mathcal{C}$, are called* isomorphic *if there is a bijective map $\phi: a \to b$ satisfying*

$$\phi(X_{ij}) = X_{i'j'} \Leftrightarrow (x_{ij} = x_{i'j'}) \wedge (x_{ji} = x_{j'i'})$$
$$\wedge (P(v_i) = P(v_{i'})) \wedge (P(v_j) = P(v_{j'})) \qquad \forall X_{ij} \in W .$$

We can incorporate the blockmodel into the parameter estimation by forcing $\lambda_a = \lambda_b$ for isomorphic configurations $C(a, x)$ and $C(b, x)$. Then, the plausibility of a blockmodel can be scored by computing the likelihood ratio statistic G^2 using ML-estimates for both the unrestricted and restricted parameter spaces (see Sects. 10.2.1 and 10.2.3).

10.3   Chapter Notes

We have seen that the tight concepts of roles and equivalences discussed in Chapter 9 are not suited to analyze real-world data occurring in psychology and sociology. Therefore, a variety of methods has been developed since the 70's that realize some kind of relaxation of the strict structural equivalence.

Traditional methods in blockmodeling are mainly based on measures of similarity of actor relationships, which are then used to compute the partition of actors into positions. These measures can be turned into metrics using techniques for multidimensional scaling in order to refine the relational data or, alternatively, to enable a visual interpretation. Often, clustering based methods are used to compute the actor set partition. We have seen also the popular but heavily criticized CONCOR algorithm, which works with iterated correlations of adjacency matrices. Afterwards, different criterions may be used to decide on relations between the positions and, therefore, to obtain a simplified representation of the original data. With generalized blockmodeling, an integrated optimizational approach has been presented which solves both the partitioning problem and the image matrix computation by minimizing a common error function.

Second, stochastic models have been introduced which assume certain kinds of stochastic generation processes for the observed relational data. They represent the more recent developments in blockmodeling. Both simple models offering exact and efficient estimation methods and more complex, realistic models have been presented. For the latter, different approaches to parameter estimation have been discussed which do not offer both exactness and efficiency, but which have been successfully applied to social network data. We have seen that the adaptation and application to blockmodeling follows the introduction of a new stochastic model originally proposed as general explanation for observed data.

Finally, we conclude that the area of blockmodeling seems to be strongly application driven. Researchers from psychology and sociology are in need of methods to analyze the positional structure of observed networks and serve themselves from different scientific areas like computer science and statistics to obtain methods giving them the desired analytic results. Hence, the approaches and techniques are quite heterogenous. At the moment, most researchers in this area seem to use rather the traditional methods discussed in Section 10.1 than the more recent methods of Section 10.2.

Further information on the properties of the correlation coefficient and its relationship to the Euclidean distance can be found in [492, 528].

Kruskal's Multidimensional Scaling algorithm was published in two seminal articles as early as 1964 [372, 373]. Cox and Cox discuss Multidimensional Scaling in a recent book [134] from a statistical point of view, it also contains among other topics a presentation of Kruskal's algorithm and its relation to other MDS-methods. The approximation algorithm for metric embedding is by Bădoiu [105]. The part of the algorithm that finds the x-coordinates is in the appendix of the paper and can be found at http://theory.lcs.mit.edu/~mihai/. More on the related metric embedding problems can be found in [320] and the references therein.

A variety of applications of the CONCOR algorithm to social network data can be found in [31, 98, 100, 230, 364, 424, 434].

An implementation of generalized blockmodeling is included in the Pajek software available via http://vlado.fmf.uni-lj.si/pub/networks/pajek/default.htm.

Exponential models are discussed in [385]. The p_1 model was introduced by Holland and Leinhardt in [302]. The goodness-of-fit test against differential reciprocity is advocated by Fienberg and Wasserman in [213]. In this paper the authors also show how to understand p_1 as a special case of a so-called general linear model. Loglinear models of homogeneous association can be found in the textbook by Agresti [3]. The Neyman-Pearson Lemma is proved in [384]. A gentle introduction into testing and statistics in general can be found in the book by Rice [492].

First applications of p_1 to blockmodeling can be found in [30, 212, 301]. The refinement of the p_1-model presented here is by Wang and Wong [567]. The a posteriori blockmodeling approach is by Nowicki and Snijders [453, 530]. In these papers, the identifiability problems are discussed in more detail. Furthermore, methods to test the adequacy of the obtained class structure are handled therein.

A proof of the strong law of large numbers can be found in [492]. The ergodic theorem is discussed in [245]. More information on the related Markov Chain Monte Carlo methods, Gibbs sampling, and mixture models can be found in [243, 245].

The p^* models can be seen as an application of Markov random graphs to social sciences. Markov random graphs were introduced by Frank and Strauss [222]. They were made popular in social network analysis by a sequence of papers by Pattison, Robins, and Wasserman [472, 494, 570]. Recently, Snijders has analyzed them in detail and pointed out estimation problems together with categorical problems which call into question the appropriateness of p^* to many social network problems [529]. More information on the two estimation methods for p^* can be found in [245, 529, 589]. Finally, [29, 200] contain more social network analyses using p^* models.

11 Network Statistics

Michael Brinkmeier and Thomas Schank

Owing to the sheer size of large and complex networks, it is necessary to reduce the information to describe essential properties of vertices and edges, regions, or the whole graph. Usually this is done via *network statistics*, i.e., a single number, or a series of numbers, catching the relevant and needed information. In this chapter we will give a list of statistics which are not covered in other chapters of this book, like distance-based and clustering statistics. Based on this collection we are going to classify statistics used in the literature by their basic types, and describe ways of converting the different types into each other.

Up to this point *network statistic* has been a purely abstract concept. But one has quite good ideas what a statistic should do:

A network statistic ...

... should describe essential properties of the network.
This is the main task of network statistics. A certain property should be described in a compact and handy form. We would like to forget the exact structure of the underlying graph and concentrate on a restricted set of statistics.

... should differentiate between certain classes of networks.
A quite common question in network analysis regards the type of the 'measured' network and how to generate models for it. This requires the decision whether a generated or measured graph is similar to another one. In many situations this may be done by identifying several statistics, which are invariant in the class of networks of interest. Using these statistics an arbitrary graph can be tested for membership in a specific class, by determining its statistics and comparing them with some references.

... may be useful in algorithms and applications.
Some network statistics may be used for algorithms or calculations on the graph. Or they might indicate which graph elements have certain properties regarding the application.

To which degree a certain statistic fulfills one or more of these tasks obviously depends on the application and the network. Therefore we will not go into detail about the interpretation, and restrict ourselves to the description of types of statistics, common constructions, and several examples.

U. Brandes and T. Erlebach (Eds.): Network Analysis, LNCS 3418, pp. 293–317, 2005.

11.1   Degree Statistics

The most common and computationally easy statistic is the vertex degree. Depending on the underlying network and its application, it may be a simple measure for the strength of connection of a specific vertex to the graph, or – as in the
case of indegrees – a measure for the relevance. But usually, instead of using this
statistic directly, the main interest lies in the absolute number or the fraction
of vertices of a given in-, out-, or total degree. It has been discovered that the
distribution of degrees in many naturally occurring graphs significantly differs
from that of classical random graphs.

In a classical undirected random graph $G_{n,p}$ the fraction of vertices of degree
k is expected to be

$$\binom{n-1}{k} p^k (1-p)^{n-1-k} \qquad \text{(Binomial distribution)}$$

if the number of vertices n is small, or approximately

$$\frac{(np)^k}{k!} e^{-np} \qquad \text{(Poisson distribution)}$$

if n is large. But in many natural graphs the degrees seem to follow a *power law*,
i.e.

$$c\,k^{-\gamma} \qquad \text{with } \gamma > 0 \text{ and } c > 0.$$

The power law is a good example for a parameter elimination
by means of a functional description, as described in Section 11.7.1. Since
the degree distribution can be described as $ck^{-\gamma}$, it suffices to determine the
constant exponent γ. This can easily be done with linear regression of the log-
log-plot of the distribution. The scaling constant c is then dictated by the fact
that the sum over all k is either the number of edges (in the absolute case) or
1 (in the relative case). Of course, the exponent is only meaningful if the degree
distribution has the appropriate form.

Examples for graphs and networks whose degree distribution seems to follow
a power law include

– the actor collaboration graph ($\gamma \approx 2.3$) [40],
– the World Wide Web ($\gamma_{\text{in}} \approx 2.1, \gamma_{\text{out}} \approx 2.45/2.72$)[1][16, 40, 102],
– the power grid of the United States of America ($\gamma \approx 4$) [40],
– the Internet (router and autonomous systems) ($\gamma \approx 2.2$) [197].

More details about the power law, and models for generating graphs satisfying
it, can be found in Chapter 13.

Experiments and measurements indicate that the power law and the resulting
exponents are good statistics for the classification of graphs. They are especially
useful for the decision whether a specific model generates graphs which are close
to the naturally occurring ones.

---

[1] In the examples shown in Figures 11.4 and 11.5 near the end of this chapter, we
have $\gamma_{\text{in}} \approx 2$ and $\gamma_{\text{out}} \approx 3.25$.

11.2 Distance Statistics

Another basic, but computationally more complex statistic is the distance between two vertices, defined as $d(u,v) = \min\{|P| \mid P \text{ is a path from } u \text{ to } v\}$. Arranging the distances leads to a $V \times V$-matrix D, whose columns and rows are indexed by the vertices of the graph, with

$$D = (d(u,v))_{u,v \in V}$$

containing the distance $d(u,v)$ in row u and column v.

For arbitrary edge weights $w \colon E \to \mathbb{R}$ the problem of finding a shortest path is $\mathcal{NP}$-hard (see [240]). But for more special (but also more common) cases the distance can be calculated in polynomial time by solving the all-pairs shortest-path problem (APSP). Before we go into detail about the algorithmic aspects in Section 11.2.5 (see also Section 4.1.1), we will describe related statistics that are commonly used in the literature. As we will see, often not the distances themselves but more 'condensed' statistics are used.

11.2.1 Average or Characteristic Distance

The *average* or *characteristic distance* $\bar{d}$ is the arithmetic mean of all distances in the graph, i.e.,

$$\bar{d} := \frac{1}{|V|^2 - |V|} \sum_{u \neq v \in V} d(u,v).$$

For disconnected graphs we obviously have $\bar{d} = \infty$. Hence it might be useful to restrict ourselves to all connected pairs, leading to the *average connected distance*

$$\bar{d} := \frac{1}{k} \sum_{\substack{u \neq v \in V \\ 0 < d(u,v) < \infty}} d(u,v)$$

where k is the number of connected pairs $u \neq v$. In abuse of notation we denote both statistics with $\bar{d}$, since usually only the second is meaningful.

If the distance-matrix D is known, the average distance can be calculated in $\mathcal{O}(n^2)$ time.

11.2.2 Radius, Diameter and Eccentricity

By fixing one argument of the distance, the number of parameters of this statistic is reduced, leading to the *eccentricity* $\varepsilon(u)$ of a vertex u. This is the maximal distance of another node from u, i.e.,

$$\varepsilon(u) := \max\{d(u,v) \mid v \in V\}.$$

The *radius* rad(G) is the minimal eccentricity of all vertices, i.e.

$$\mathrm{rad}(G) := \min\{\varepsilon(u) \mid u \in V\}.$$

Maximizing over both arguments of the distance, one obtains the *diameter* $\text{diam}(G)$ of a graph G as the maximal distance between two arbitrary (connected) vertices, i.e.,

$$\text{diam}(G) := \max\{d(u,v) \mid u,v \in V\}.$$

As with the average distances, the diameter and the eccentricity can be calculated from the distance matrix in $\mathcal{O}(n^2)$ time. The eccentricity of a single vertex may also be calculated via the single-source shortest-paths problem (SSSP), see Sections 4.1.1 and 11.2.5.

11.2.3   Neighborhoods

The *h-neighborhood* $\text{Neigh}_h(v)$ *of a vertex* v is the set of all vertices u with distance less than or equal to h from v, i.e.,

$$\text{Neigh}_h(v) := \{u \in V \mid d(v,u) \le h\}.$$

The sizes of the h-neighborhoods form a parameterized statistic

$$N(v,h) := |\text{Neigh}_h(v)|.$$

The *(absolute) hop plot* $P(h)$ eliminates the dependence on the vertex by assigning the number of pairs (u,v) with $d(u,v) \le h$ to each parameter h, i.e.,

$$P(h) := \left|\{(u,v) \in V^2 \mid d(u,v) \le h\}\right| = \sum_{v \in V} N(v,h).$$

The *(relative) hop plot* $p(h)$ is the fraction of pairs with a distance less than or equal to h, i.e.,

$$p(h) := \frac{P(h)}{n^2} = \frac{1}{n^2} \sum_{v \in V} N(v,h).$$

The *average h-neighborhood size* $\overline{\text{Neigh}}(h)$ is defined as

$$\overline{\text{Neigh}}(h) := \frac{1}{n} \sum_{v \in V} N(v,h) = \frac{P(h)}{n} = np(h).$$

Again the absolute and relative hop plots, and the neighborhood sizes $N(v,h)$, can be calculated from the distance matrix in $\mathcal{O}(n^2)$ time and space.

The absolute hop plot of the Internet is examined by Faloutsos et al. in [197]. They observe that it follows a power law with an exponent around 4.7.

The h-neighborhoods that are necessary for the hop plot can be approximated using the ANF-algorithm of Palmer et al. presented in [462]. We will go into detail about that in Section 11.2.6.

11.2.4   Effective Eccentricity and Diameter

The *effective eccentricity* ε_{eff} and the *effective diameter* diam_{eff} are obtained via an interesting construction from the eccentricity and the diameter. The former measures the minimal distance at which a specified fraction r of all nodes lies from a specific source v, i.e.,

$$\varepsilon_{\text{eff}}(v, r) := \min \{h \mid N(v, h) \geq rn\} .$$

The latter does the same without specifying a concrete source, i.e.,

$$\text{diam}_{\text{eff}}(r) := \min \{h \mid P(h) \geq rn^2\} = \min \{h \mid p(h) \geq r\} .$$

If N and P are known, both statistics can be calculated in $\mathcal{O}(\log \text{diam}(G))$ using binary search on $N(v, h)$ and $P(h)$.

The effective diameter and the effective eccentricity occur in [462] for a fixed value of $r = 0.9$.

11.2.5   Algorithmic Aspects

As mentioned at the beginning of this section, the problem of finding a shortest path between two vertices in a network with arbitrary edge weights $w \colon E \to \mathbb{R}$ is $\mathcal{NP}$-hard (see [240]). But if we restrict ourselves to networks without cycles of negative weight, the problem can be solved in polynomial time by well-known algorithms described in the following.

The Path Algebra. Assume that $G = (V, E)$ is a weighted graph (directed/undirected) without cycles of negative weight. Then the most general approach, which may be adapted to several other problems, for the calculation of the distance matrix is given by matrix multiplication over the *path algebra*.

Let $d_i(u, v)$ be the weight of a shortest path (i.e. a path of minimal weight) from u to v using at most i edges. This implies

$$d_0(u, v) = \begin{cases} 0 & \text{if } u = v \\ \infty & \text{otherwise.} \end{cases}$$

Since a path with at most $i + 1$ edges either has at most i edges or consists of a path of length $\leq i$ to a predecessor v' of v and the edge (v', v), we have

$$d_{i+1}(u, v) = \min_{v' \in V} (d_i(u, v), d_i(u, v') + w(v', v)) .$$

Here it is important to keep in mind that G does not contain cycles of negative weight.

The distance matrix may be calculated via the summation and multiplication of an adapted adjacency matrix A over the *commutative semi-ring* $(\bar{\mathbb{R}}, \min, +)$, with $\bar{\mathbb{R}} = \mathbb{R} \cup \{\infty\}$. The entries of A are given by

$$a_{vu} = \begin{cases} 0 & \text{if } u = v \\ w(u,v) & \text{if there exists an edge from } u \text{ to } v \text{ .} \\ \infty & \text{otherwise} \end{cases}$$

Explicitly, this means that summation is replaced by the minimum and multiplication by the addition. Then the distance matrix D is $A^{\text{diam}(G)}$, or, written as an iteration, D is the limit of the iteration $D_{i+1} = D_i \cdot A$ and $D_0 = A$.

Fortunately it suffices to iterate $\text{diam}(G)$-times, where the diameter is taken on the graph without weights. This is due to the fact that each simple path has at most length $\text{diam}(G)$.

If $T(n)$ is the time needed for a multiplication of two $n \times n$-matrices, the iteration leads to a running time of $\mathcal{O}(T(n)\text{diam}(G))$. Using the iteration $D_{2i} = D_i \cdot D_i$ instead, one obtains a running time of $\mathcal{O}(T(n)\log(\text{diam}(G)))$.

Hence, the time needed for the calculation of the distance matrix via matrix multiplication is dominated by the time $T(n)$ needed for the multiplication of two $n \times n$-matrices. For the naive matrix multiplication we have $T(n) = \mathcal{O}(n^3)$. In [595] Zwick described an algorithm with $T(n) = \mathcal{O}(n^{2.575})$.

Single-Source Shortest-Paths. The distance matrix may be calculated by solving the single-source shortest-paths problem (SSSP) for all sources. Depending on the type of weight function, several algorithms are known.

- If the network is unweighted, i.e., $w(e) = 1$ for all edges $e \in E$, then SSSP can be solved via *breadth-first-search* in $\mathcal{O}(m)$ (see [133]).

- If G has only edges of nonnegative weights then the algorithm of *Dijkstra* solves SSSP in $\mathcal{O}(n \log n + m)$ time when using Fibonacci Heaps (see [133] or Algorithm 4 in Section 4.1.1.).

- If G contains no cycles of negative weight, then the *Bellman-Ford* algorithm solves SSSP in $\mathcal{O}(nm)$ (see [133]).

The runtime of the algorithm of Dijkstra may be improved if more sophisticated data structures and strategies are used. In [547] Thorup describes an algorithm which calculates the SSSP for a given source on an undirected graph with positive integers as weights in $\mathcal{O}(m)$ time and space, leading to $\mathcal{O}(nm)$ time for the distance matrix. He uses an alternative strategy for the choice of the next node in the Dijkstra algorithm, following [152], and realizes the priority queue using buckets.

For further results about shortest paths, see [224, 225, 546, 548, 487, 12].

All-Pairs Shortest-Paths. An alternative to the use of matrix multiplications over the path algebra is the *Floyd-Warshall* algorithm for the solution of the all-pairs shortest-paths problem (APSP). Again the algorithm can only be applied to networks without cycles of negative weight. It requires $\mathcal{O}(n^3)$ time and can be found in textbooks (e.g., [133]) or in this volume (Algorithm 5 in Section 4.1.2).

Another approach to solving the APSP consists of running an SSSP algorithm for each of the n vertices of the given graph. This leads to the following runtimes for APSP:

1. unweighted: $\mathcal{O}(nm)$
2. non-negative weights: $\mathcal{O}(nm + n^2 \log n)$
3. no cycles of negative weight: $\mathcal{O}(n^2 m)$

An improved solution to the APSP on a weighted graph without cycles of negative weight is achieved by Johnson's algorithm (details can be found in [133]). It first calculates the distances from an artifical source to all vertices in the graph using the Bellman-Ford algorithm. If the graph does not contain cycles of negative weight, it recalculates the edge weights using the results of the Bellman-Ford algorithm and then determines the distances of all pairs by calling the Dijkstra algorithm n times. If the graph contains cycles of negative length it simply terminates. In total this leads to a runtime of $\mathcal{O}(n^2 \log n + nm)$.

11.2.6 ANF – The Approximate Neighborhood-Function

In [461] Palmer et al. described an algorithm for the estimation of the hop plot of an unweighted graph $G = (V, E)$ using approximative counting as introduced by Flajolet and Martin in [215]. The basis of the algorithm is the observation that the set $\text{Neigh}_h(u) = \{v \mid d(u, v) \leq h\}$ of nodes v with distance at most h from u can be described as

$$\text{Neigh}_h(u) = \text{Neigh}_{h-1}(u) \cup \bigcup_{(u,v) \in E} \text{Neigh}_{h-1}(v).$$

Therefore it suffices to approximately count the number of elements in an iteratively increasing set. The main idea is to represent the sets $\text{Neigh}_h(u)$ by bitmasks, such that each bit represents a subset of vertices. Then the union corresponds to a logical or of the bitmasks. The resulting algorithm is shown as Algorithm 22.

In the last loop the estimate for $N(u, h)$ is calculated from the lowest position $R[u, h]$ of a 0-bit in the bitmask $B[u, h]$ starting at 0, i.e. $B[u, h]$ is of the form $\{0, 1\}^* 01^{R[u,h]}$. Following [215] the expected value of $R[u, h]$ is $\log(\varphi N(u, h))$ with $\varphi \approx 0.77351$.

If this procedure is repeated z times with different position assignments $k[u]$, one can use the average $\bar{R}[u, h]$ over all lowest 0-bit positions to obtain an even better estimate with

$$N(u, h) = \frac{2^{\bar{R}[u,h]}}{0.77351\,(1 + 0.31/z)},$$

where the additional factor $(1 + 0.31/z)$ is caused by a bias of the expected value of $N[u, h]$ in this situation. Details can be found in [215].

Experiments conducted by Palmer, Gibbons and Faloutsos in [461] indicate a high accuracy of the estimations (less than 10% error for $z = 64$ trials) while

---

Algorithm 22: The ANF-Algorithm

Input: A graph $G = (V, E)$ and a natural number r
Data: Bitmasks $B[u, h]$ for each $u \in V$, $h = 1, \ldots, n$ of length $\log n + r$
Output: Estimations $N(u, h)$ for the number of nodes within distance h of u
for each vertex u

begin
  $l \leftarrow \log n + r$
  foreach $v \in V$ **do**
    Set $k[v] = i$ with probability $\frac{1}{2^{i+1}}$ for $0 \leq i < l - 1$ and probability $\frac{1}{2^{l-1}}$
    for $i = l - 1$
  end
  foreach $v \in V$ **do**
    Set position $k[v]$ of $B[v, 0]$ to 1
  end
  for $h \in \{1, \ldots, n\}$ **do**
    foreach $u \in V$ **do**
      $B[u, h] \leftarrow B[u, h - 1]$
    end
    foreach $(u, v) \in E$ **do**
      $B[u, h] \leftarrow B[u, h] \vee B[v, h - 1]$
    end
  end
  foreach $u \in V$ and $h \in \{0, \ldots, n\}$ **do**
    Let $R[u, h]$ be the lowest position of a 0-bit in $B[u, h]$
    $N(u, h) \leftarrow 2^{R[u,h]}/0.77351$
  end
end

---

providing higher speed and lower space requirements than other approximation methods based on random intervals and sampling. More exact results involving the standard deviation of the estimate can be found in [215].

11.3 The Number of Shortest Paths

A statistic closely related to the distance is the *number $c(u, v)$ of distinct shortest paths* between two vertices

$$c(u, v) := |\{P \mid P \text{ is a shortest path from } u \text{ to } v\}|.$$

Since the general APSP is $\mathcal{NP}$-hard, the same obviously holds for the calculation of the number of distinct shortest paths. But again it becomes polynomially solvable if the network does not contain cycles of negative or zero length. The $c(u, v)$ can be calculated by a modified Floyd-Warshall-algorithm (see Algorithm 23). In a similar way the Dijkstra algorithm may be modified if all edge weights are positive.

Closely related to $c(u, v)$ is the number of *disjoint* shortest paths. But this problem is known to be $\mathcal{NP}$-hard [346].

Alternatively, one can calculate the number of distinct shortest paths between two arbitrary vertices, using a specific edge, i.e.,

---

Algorithm 23: Counting the number of distinct shortest paths

Input: A graph $G = (V, E)$ with edge weights w and without cycles of negative or zero length.

Output: The distances $d(u, v)$ and the number of distinct shortest paths between each pair u, v of vertices

begin

    foreach $v \in V$ **do**

        foreach $u \in V$ **do**

$$d(u, v) \leftarrow \begin{cases} 0 & \text{if } u = v \\ w(u, v) & \text{if } (u, v) \in E \\ \infty & \text{otherwise} \end{cases}$$

$$c(u, v) \leftarrow \begin{cases} 1 & \text{if } (u, v) \in E \\ 0 & \text{otherwise} \end{cases}$$

    foreach $v' \in V$ **do**

        foreach $u \in V$ **do**

            foreach $v \in V$ **do**

                if $d(u, v') + d(v', v) = d(u, v)$ **then**

                    $c(u, v) \leftarrow c(u, v) + c(u, v')c(v', v)$

                else

                    if $d(u, v') + d(v', v) < d(u, v)$ **then**

                        $d(u, v) \leftarrow d(u, v') + d(v', v)$

                        $c(u, v) \leftarrow c(u, v')c(v', v)$

end

$$c(e) := |\{P \mid P \text{ is a shortest path including } e\}|.$$

Since each shortest path passing through the edge $e = (u, v)$ from u' to v' consists of a shortest path from u' to u, the edge e, and a shortest path from v to v', the statistic $c(e)$ can be calculated from the $c(u, v)$:

$$c(e) = \sum_{d(u', v') = d(u', u) + 1 + d(v, v')} c(u', u)c(v, v').$$

11.4 Distortion and Routing Costs

Consider a spanning tree T of $G = (V, E)$, i.e., an acyclic, connected subgraph of G containing all vertices $v \in V$. Then the *distortion* $D(T)$ of T is the average length of paths in T between two adjacent vertices in G. More precisely

$$D(T) := \frac{1}{|E|} \sum_{\{u,v\} \in E} d_T(u, v)$$

where $d_T(u, v)$ is the distance from u to v in T.

The global *distortion* $D(G)$ *of* G is the minimum distortion over all spanning trees of G, i.e.,

$$D(G) := \min \{D(T) \mid T \text{ is a spanning tree of } G\}.$$

The distortion is closely related to the *communication costs* of a spanning tree T of a network, as introduced by Hu in [317]. In this setting a requirement $r_{u,v}$ is given for each pair of vertices. The communication cost of T is given as the sum of all distances between vertices in T, multiplied by the requirement values. By setting $r_{u,v} = \frac{1}{m}$ if $\{u,v\} \in E$ and 0 otherwise, one obtains the distortion. In [333] Johnson et al. proved that the calculation of the minimal communication costs of a network with $r_{u,v} = 1$ for all pairs of vertices is $\mathcal{NP}$-hard.

11.5   Clustering Coefficient and Transitivity

The clustering coefficient introduced by Watts and Strogatz [573] in the year 1998 has become a frequently used tool in network analysis. For a node v the clustering coefficient $c(v)$ is supposed to represent the likeliness that two neighbors of v are connected. The clustering coefficient $C(G)$ of a graph is the average of $c(v)$ taken over all nodes. The latter seems to be a very popular index in network analysis.

In 2000 Barrat and Weigt [41] used an 'alternative formulation' for this average, claiming that their redefinition does not alter the 'physical significance' for certain generated small world networks. In the year 2002 the so-called transitivity was introduced by Newman, Strogatz and Watts [447], again as an alternative formulation of the clustering coefficient of a graph. It really turned out to be equivalent to the formulation of Barrat and Weigt, but not at all equivalent to the original clustering coefficient.

11.5.1   Definitions

We will define both indices in terms of triangles and triples of a node and a graph, respectively. Let G be a simple and undirected graph. A *triangle* $\triangle = \{V_\triangle, E_\triangle\}$ is a complete subgraph of G with exactly three nodes, see Figure 11.1. We use

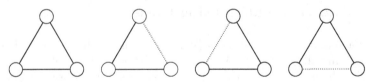

Fig. 11.1. A triangle and its three triples

$\lambda(G)$ for the number of triangles of a graph G. Accordingly we define $\lambda(v) = |\{\triangle \mid v \in V_\triangle\}|$ as the number of triangles of a node. Note that $\lambda(G) = 1/3 \sum_{v \in V} \lambda(v)$.

A *triple* is a subgraph of G (not necessarily an induced subgraph) with three nodes and two edges. A triple is a *triple at node v* if v is incident with both edges of the triple. The number of triples at a node v can be formulated in dependence of its degree $d(v)$ as

$$\tau(v) = \binom{d(v)}{2} = \frac{d(v)^2 - d(v)}{2}.$$

The number of triples for the whole graph is $\tau(G) = \sum_{v \in V} \tau(v)$. In the above terms the *clustering coefficient* of a node v with $\tau(v) \neq 0$ is defined as

$$c(v) = \frac{\lambda(v)}{\tau(v)}.$$

With $V' = \{v \in V \mid d(v) \geq 2\}$ we define the clustering coefficient of the whole graph as

$$C(G) = \frac{1}{|V'|} \sum_{v \in V'} c(v).$$

Note that there is some variation in the literature in how nodes of degree less than two are handled, e.g., $c(v)$ is defined to be zero or one and included in the averaging process.

The *transitivity* for a graph is defined as

$$T(G) = \frac{3\lambda(G)}{\tau(G)}.$$

As there are exactly three triples in each triangle, see Figure 11.1, $3\lambda(G) \leq \tau(G)$ holds and consequently $T(G)$ is a rational number between zero and one.

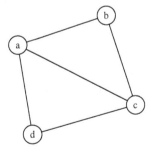 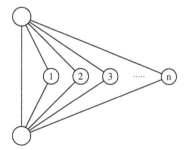

Fig. 11.2. On the left: Graph with clustering coefficients: $c(a) = c(c) = 2/3$, $c(b) = c(d) = 1$, $C(G) = \frac{1}{4}(2 + 4/3) \approx 0.83$ and transitivity $T(G) = 3 \cdot 2/8 = 0.75$. On the right: family of graphs where $T(G) \rightarrow 0$, $C(G) \rightarrow 1$ for $n \rightarrow \infty$.

11.5.2   Relation Between C and T

The transitivity $T(G)$ was first introduced as an alternative formulation for the clustering coefficient $C(G)$ in [447]. The left hand side of Figure 11.2 shows a small graph where the two values differ. The right hand side shows a family of graphs where $T(G)$ approaches zero while $C(G)$ approaches one, for increasing n. The equation

$$T(G) = \frac{\sum_{v \in V'} \tau(v)c(v)}{\sum_{v \in V'} \tau(v)}$$

given by Bollobás and Riordan [68] shows a formal relation between the two indices: The transitivity is equal to the triple weighted clustering coefficient. Hence $T(G)$ equals $C(G)$, e.g., if all nodes have the same degree or all clustering coefficients are equal.

11.5.3   Computation

To compute the clustering coefficients we need to compute the number of triples $\tau(v)$ and the number of triangles $\lambda(v)$, for each node v. It is straight-forward to compute the number of triples $\tau(v)$ in linear time. This leaves us with the task of computing the number of triangles. For the transitivity it suffices to compute $\lambda(G)$ for the whole graph. It is not known whether there is an algorithm that is asymptotically faster in computing the triangles globally vs. locally.

The standard method is to iterate over all nodes and check whether the edge between any two neighbors is present. This algorithm has running time in $\mathcal{O}(nd_{\max}^2)$ where $d_{\max} = \max\{d(v) \mid v \in V\} = \Delta(G)$.

We assumed above that it is possible to test for edge existence in constant time. This could be done with a $n \times n$ matrix by using the 'indirection trick', see e.g. Exercise 2.12 in [4]. A more useful approach, requiring only linear space, is to use hashing. If each node is assigned a random bit vector of appropriate length, and the hash function uses a combination of two of these, we will get a randomized algorithm with expected testing time in $\mathcal{O}(1)$.

The second approach is to use matrix multiplication. Note that if A is the adjacency matrix of graph G, then the diagonal elements of A^3 contain two times the number of triangles of the corresponding node. This gives an algorithm with running time in $\mathcal{O}(n^\gamma)$, where γ is the matrix multiplication coefficient. It is currently known that $\gamma \leq 2.376$ [132].

Itai and Rodeh [321] proposed an algorithm with running time in $\mathcal{O}(m^{3/2})$. This is an improvement for sparse graphs, compared to the mentioned methods.

We will now discuss the algorithm of Alon, Yuster, and Zwick [26]. It improves the running time by using fast matrix multiplication, and still expresses only dependence on m in the running time. If used with standard matrix multiplication ($\gamma = 3$) it will achieve the same bound as the algorithm of Itai and Rodeh.

The pseudocode of the algorithm is listed in Algorithm 24. Informally the algorithm splits the node set into low degree vertices $V_{\mathrm{low}} = \{v \in V : \mathrm{d}(v) \leq \beta\}$

Algorithm 24: AYZ triangle algorithm

Input: Graph G with adjacency array representation and hashed edge set
matrix multiplication parameter γ
Output: number of triangles $\lambda(v)$ for each node

1 $\beta \longleftarrow m^{(\gamma-1)/(\gamma+1)}$
2 **for** $v \in V$ **do**
    $\lambda(v) \leftarrow 0$
    if $d(v) \leq \beta$ **then**
      $\lfloor$ $V_{\text{low}} \leftarrow V_{\text{low}} \cup \{v\}$
    else
      $\lfloor$ $V_{\text{high}} \leftarrow V_{\text{high}} \cup \{v\}$

3 **for** $v \in V_{low}$ **do**
    for *all pairs of neighbors* $\{u, w\}$ *of* v **do**
4   **if** *edge between* u *and* w *exists* **then**
5   **if** $u, w \in V_{low}$ **then**
          for $z \in \{v, u, w\}$ **do**
            $\lfloor$ $\lambda(z) \leftarrow \lambda(z) + 1/3$
6   **else if** $u, w \in V_{high}$ **then**
          for $z \in \{v, u, w\}$ **do**
            $\lfloor$ $\lambda(z) \leftarrow \lambda(z) + 1$
7   **else**
          for $z \in \{v, u, w\}$ **do**
            $\lfloor$ $\lambda(z) \leftarrow \lambda(z) + 1/2$

8 A $\leftarrow$ adjacency matrix of node induced subgraph of V_{high}
9 M $\leftarrow A^3$
10 **for** $v \in V_{high}$ **do**
11   $\lfloor$ $\lambda(v) \leftarrow \lambda(v) + M(i, i)/2$ where i is index of v

---

and high degree vertices $V_{\text{high}} = V \setminus V_{\text{low}}$, where $\beta = m^{\gamma - 1/\gamma + 1}$. It then performs
the standard method on the low degree nodes, and uses fast matrix multiplication
on the subgraph induced by the high degree nodes.

Lemma 11.5.1. *Algorithm 24 computes the triangles $\lambda(v)$ for each node and
can be implemented to have running time in $\mathcal{O}(m^{2\gamma/(\gamma+1)})$, or $\mathcal{O}(m^{1.41})$.*

Proof. The correctness of the algorithm can be easily seen by checking the case
distinction for different types of triangles consisting of exactly three (line 5), two
(line 7), one (line 6), or zero (line 11) low degree nodes.

To prove the time complexity, we first note that the lines 1, 2, and 10 can
clearly be implemented to run in linear time. We prove that the time required
for the loop beginning at line 3 is in $\mathcal{O}(m^{2\gamma/(\gamma+1)})$. Line 4 requires a test for
edge existence in constant time. We have discussed this in the context of the
standard method above. The following tests in line 5, 6 and 7 are clearly in
constant time. The bound follows then from $\sum_{v \in V_{\text{low}}} \binom{d(v)}{2} \leq m\beta$. The running
time of line 8 is less than that of line 9. To bound line 9 we have to show that

$\mathcal{O}\left(n_{\text{high}}^{\gamma}\right) \subset \mathcal{O}\left(m^{2\gamma/(\gamma+1)}\right)$. Utilizing the hand shaking lemma $\sum_{v \in V} d(v) = 2m$, we get $n_{\text{high}}\beta \leq 2m$, which yields $n_{\text{high}} \leq 2m^{2/(\gamma+1)}$.   □

11.5.4 Approximation

For processing very large networks, linear or sublinear running time is desired. We outline now how we can achieve approximations with sublinear running time using random sampling. A more detailed description can be found in [502].

Let X_i be independent real random variables bounded by $0 \leq X_i \leq M$ for all i. With k denoting the number of samples, and ϵ some error bound, Hoeffding's bound [299] states:

$$\Pr\left(\left|\frac{1}{k}\sum_{i=1}^{k}X_i - \mathbb{E}\left[\frac{1}{k}\sum_{i=1}^{k}X_i\right]\right| \geq \epsilon\right) \leq e^{\frac{-2k\epsilon^2}{M^2}} \tag{11.1}$$

We assume that the graph is in an appropriate data structure in working memory. Specifically, we require that testing whether an edge between two nodes exists is in constant time. When giving the running time we regard the error bound ϵ and probability of correctness to be constants. To approximate the clustering coefficient for a node v, we estimate $\lambda(v)$ by checking for the required number (determined by Eq. 11.1) of neighbor pairs whether they are connected. This leads to an $\mathcal{O}(n)$-time algorithm for approximating $c(v)$ for all nodes. The clustering coefficient for the graph $C(G)$ can be approximated in a similar fashion. A short computation shows that it is sufficient to choose a random node v and then two random neighbors for each sample. This gives an algorithm running in constant time. To approximate the transitivity one can proceed in a similar fashion, however, the nodes have to be chosen with weights corresponding to $\tau(v)$. This can be done in time $\mathcal{O}(n)$.

Lemma 11.5.2. *Consider the error bound ϵ and the probability of correctness to be constants. Then there exist algorithms that approximate the clustering coefficients for each node $c(v)$ and the transitivity $T(G)$ in time $\mathcal{O}(n)$. The clustering coefficient $C(G)$ can be approximated in time in $\mathcal{O}(1)$.*

11.6 Network Motifs

In molecular biology a functional domain of a molecule is called a *motif*. Thus motifs are building blocks for molecules on a higher level than atoms. Milo et al. [423] introduced the term *motifs* for networks, following the idea of building blocks on a higher level than nodes and edges. They enumerated the occurrences of small subgraphs in a network G. A connected subgraph that occurs in G significantly more often than in a random network of same size and degree distribution is called a motif of G.

The authors considered several networks from biology (food webs, neuronal networks, gene regulation), engineering (electrical circuits), and also a domain of the World Wide Web. They looked for weakly connected subgraphs up to four nodes, and discovered for each of the networks distinctive motifs.

11.6.1   Algorithmic Aspects

A very basic algorithm to enumerate the occurrences of a subgraph with k nodes is to check all $\binom{n}{k}$ possible combinations of k nodes. The comparison itself can be done by permuting the k nodes in $k!$ steps, where in each step all potential $2\binom{k}{2}$ edges are considered. To do this the number of possible automorphisms of the subgraph has to be known, see Section 12.1.

Unfortunately the original work of Milo et al. [423] does not describe the algorithm, and the 'supplementary materials' remain very vague in that respect, too. The 'supplementary materials', some of the considered networks, and an implementation of the algorithm can be downloaded from an author's website.[2] However, it has been reported that the published results produced by that program are incorrect.[3]

Algorithm 24 of Alon, Yuster, and Zwick [26] for enumerating all triangles in an undirected graph can be modified to count directed and connected subgraphs of three nodes without incurring additional costs in the asymptotic running time.

In [363] the basic idea of the triangle counting algorithm of Alon, Yuster, and Zwick was refined by Kloks, Kratsch and Müller to achieve an algorithm for counting all K_4's in a graph in time $\mathcal{O}(m^{\frac{\gamma+1}{2}})$. Here $\gamma \leq 2.376$ [132] is the matrix multiplication coefficient. Further, for the undirected case, they were able to derive an algorithm that counts all subgraphs with at most four nodes in time $\mathcal{O}(n^{\gamma} + m^{\frac{\gamma+1}{2}})$. It is not known whether this can be generalized to the directed case whilst obeying the same bound in running time.

11.7   Types of Network Statistics

Browsing through the literature, and by examination of the previous examples, one can identify four basic types of statistics which can be described by two pairs of exclusive attributes:

single-valued vs. distribution   and   global vs. local

Even though the attributes are intuitively clear, we give a formal description of the four resulting types. Let $\mathcal{G}$ be a class of graphs (e.g., (un-)weighted, (un-)directed, (un-)connected etc.), P a set of parameters, and Y a set of *values*. Usually we have, e.g., $P, Y = \mathbb{N}, \mathbb{Z}, \mathbb{R}$ or products of them. Furthermore let X_G be a set of not specified *graph elements* in G, which may consist of vertices, edges, subgraphs, paths etc.

Global (Single-valued) Statistics. A *global (single-valued) statistic* γ assigns a single value $\gamma_G \in Y$ to each graph $G \in \mathcal{G}$.

> Examples: Number of vertices/edges, diameter (Section 11.2.2), clustering coefficient of a graph (Section 11.5), edge- and vertex-connectivity (Chapter 7).

---

[2] http://www.weizmann.ac.il/mcb/UriAlon/
[3] Falk Schreiber, personal communication regarding MFinder 1.1

Global Distributions. A *global distribution* Γ assigns a map $\Gamma_G \colon P \to Y$ to each graph $G \in \mathcal{G}$. We usually denote the value by $\Gamma_G(t)$, where t is the parameter. For example for $P = \mathbb{N}$ a global distribution Γ_G is a sequence $(\Gamma_G(0), \Gamma_G(1), \dots)$ of values for each graph in $\mathcal{G}$.

Examples: absolute/relative distribution of in/out-degrees (Section 11.1), hop plot (Section 11.2.3).

Local (Single-valued) Statistics. A *local (single-valued) statistic* λ_G assigns a single value $\lambda_G(x) \in Y$ to a certain graph element x of a given graph $G \in \mathcal{G}$, where x may be a vertex, an edge, a set of vertices or edges, a subgraph, or whatever may be seen as a local graph element.
More formally $\lambda_G \colon X_G \to Y$ is a map from a set X_G of graph elements in G to the set of values Y.

Examples: in/out-degree (Section 11.1), weight/capacity/length of edges, distance (Section 11.2), clustering coefficient of a vertex (Section 11.5) , various centralities (Chapters 3 and 5).

Local Distributions. A *local distribution* Λ assigns a map $\Lambda_G \colon X_G \times P \to Y$ from the cartesian product of the appropriate set of graph elements and a parameter set P to Y to each graph $G \in \mathcal{G}$. As in the global case we write $\Lambda_G(x, t)$ for the value at x with parameter t.

Examples: Sizes of neighborhood (Section 11.2.3), effective eccentricity and diameter (Section 11.2.4).

Since we already used the term *distribution* for multiple valued statistics, we will adapt our nomenclature slightly. The term *statistic* refers to a single-valued, and *distribution* to multiple-valued, network statistics. Nonetheless both types are included in the notion *network statistics*.

11.7.1   Transformation of Types of Statistics

The four types are not isolated from each other. In the literature several techniques can be found to transform one into another. In fact, it usually is possible to classify network statistics by their 'underlying' statistic, from which they are deduced in one or more of the constructions we are going to describe. A rough scheme of the possible transformations is presented in Figure 11.3. A more detailed description is given in the following sections.

Of course, all constructions may be *composed*, e.g., after going from a global statistic to a local distribution, one may construct a local statistic from the result. In this section we restrict ourselves to the operations which seem to be more or less independent of each other. In the literature and in the examples below, many network statistics are obtained by the application of two or more of the steps described here.

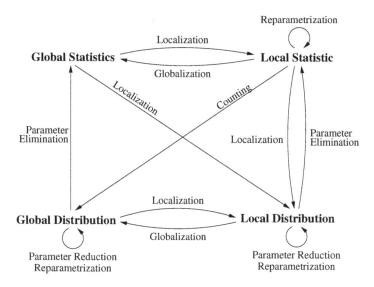

Fig. 11.3. Transformations of types of statistics

Some of the operations do not occur in the literature (at least as far as we know). They are mainly presented for completeness, and because they seem to be very natural possibilities. Furthermore we do not discuss the intuition behind the transformations, since this depends on the concrete application and interpretation of the values. We only give a formal description of the techniques.

Globalization. As the name says, globalization eliminates the dependence of a local statistic or distribution on graph elements. This is often done by calculating, choosing, or constructing a single value γ_G from the values $\lambda_G(x)$ for all graph elements $x \in X_G$. The most common examples are:

– the maximum
$$\gamma_G := \max\{\lambda_G(x) \mid x \in X_G\},$$

– the minimum
$$\gamma_G := \min\{\lambda_G(x) \mid x \in X_G\},$$

– the summation
$$\gamma_G := \sum_{x \in X_G} \lambda_G(x),$$

– and averaging
$$\gamma_G := \frac{1}{|X_G|} \sum_{x \in X_G} \lambda_G(x).$$

For distributions, the same constructions may be applied parameterwise, e.g.,
$$\Gamma_G(t) := \max_{x \in X_G}\{\lambda_G(x,t)\}.$$

Examples are the average (connected) distance (Section 11.2.1), the diameter and the radius (Section 11.2.2), the relative and absolute hop plot (Section 11.2.3), and the global distortion (Section 11.4).

Counting. To transform a local statistic λ into a global distribution Γ, one may count the number of graph elements x such that $\lambda_G(x)$ lies in a specific range of values. For discrete statistics (e.g., $\lambda_G(x) \in \mathbb{N}, \mathbb{Z}$), the absolute number of occurrences of the value t, i.e.,

$$\Gamma_G(t) := |\{x \in X_G \mid \lambda_G(x) = t\}|,$$

or the relative number of occurrences, i.e.

$$\Gamma_G(t) := \frac{|\{x \in X_G \mid \lambda_G(x) = t\}|}{|X_G|}$$

are very common. Similarly, for continuous statistics (e.g., $\lambda_G(x) \in \mathbb{R}$), the absolute or relative number of elements with $\lambda_G(x) \leq t$, i.e.,

$$\Gamma_G(t) := |\{x \in X_G \mid \lambda_G(x) \leq t\}| \qquad \text{or}$$

$$\Gamma_G(t) := \frac{|\{x \in X_G \mid \lambda_G(x) \leq t\}|}{|X_G|}$$

is widely used.

Examples are the absolute, relative, and complementary cumulative degree distributions (Section 11.1), the absolute and relative hop plot (Section 11.2.3).

Parameter Elimination and Reduction. Perhaps the widest class of type changes is the elimination of parameters. In some respect nearly all types of changes can be interpreted in this way. But we restrict ourselves to parameters, i.e., values which are not directly related to the examined network.

Similar to the transformation from local to global statistics and distributions, one may calculate a single value λ_G from a sequence/map of values given by a distribution Λ_G. But instead of using the graph elements as variables, one uses the parameter of the distribution. Examples are

– the maximum

$$\lambda_G(x) := \max \{\Lambda_G(x, t) \mid t \in P\},$$

– the minimum

$$\lambda_G(x) := \min \{\Lambda_G(x, t) \mid t \in P\},$$

– summation

$$\lambda_G(x) := \sum_{t \in P} \Lambda_G(x, t),$$

– averaging (only sensible if P is finite)

$$\lambda_G(x) := \frac{1}{|P|} \sum_{t \in P} \Lambda_G(x,t),$$

– and the projection onto a certain parameter $t_0 \in P$

$$\lambda_G(x) := \Lambda_G(x, t_0).$$

For single-valued statistics one simply has to neglect the argument x. The eccentricity (Section 11.2.2) is an example for this transformation.

If a local or global distribution has more than one parameter, i.e., if its set of parameters is the cartesian product $P = P' \times P''$, one may use only one parameter for the calculations, leading to a *parameter reduced* distribution, e.g.,

$$\Lambda'_G(x, t') := \max_{t'' \in P''} \{ \Lambda_G(x, t', t'') \}.$$

If the local or global distribution allows the description as a function of the parameters, some may be eliminated using a different technique. Assume that the local distribution $\Lambda_G(x, t)$ can be described by a function f_r with parameter $r(x)$, i.e., $\Lambda_G(x, t) = f_{r(x)}(t)$. Then the parameters $r(x)$ form a local statistic which indirectly describes the local distributions. If the graph element x is omitted, the same technique can be applied for the construction of a global statistic from a global distribution.

An example for this transformation is the power law exponent (Section 11.1). The degree distributions are reduced to a single value.

Reparametrization. Instead of eliminating a parameter, one can change it. For a local distribution $\Lambda_G(x, t)$ with parameter $t \in P$, we usually can write the reparameterized distribution $\tilde{\Lambda}_G$ with parameter $r \in P'$ as

$$\tilde{\Lambda}_G(x, r) := f_r(x, (\Lambda_G(x, t))_{t \in P})$$

where f_r is a function with parameter r, using all values of $\Lambda_G(x, t)$ for a given x.

For example one may chose $P = P' = \mathbb{N}$,

$$f_r(x; y_0, y_1, \dots) = \max\{y_r, y_{r+1}, \dots\}$$

leading to

$$\tilde{\Lambda}_G(x, r) := \max \{ \Lambda_G(x, t) \mid t \geq r \}.$$

Another example is $f_r(x; y_0, y_1, \dots) = \sum_{t \geq r} \Lambda_G(x, t)$ leading to

$$\tilde{\Lambda}_G(x, r) := \sum_{t \geq r} \Lambda_G(x, t).$$

Concrete examples are the effective eccentricity and diameter (Section 11.2.4).

Another type of reparametrization affects the locality of local statistics and distributions. In some situations it may useful to change the set of graph elements on which a certain statistic is defined. For example one may change a vertex-based statistic $\lambda_G(v)$ to an edge-based statistic $\lambda'_G(e)$ via the general relation

$$\lambda'_G(e) := f(\lambda_G(u), \lambda_G(v))$$

where $e = \{u, v\}$, and f is an arbitrary function with two arguments (like sum, average, product etc.).

An example for this transformation is the eccentricity (Section 11.2.2), where the notion of locality is transformed from 'a pair of vertices' to 'a single vertex'.

Localization. To construct a local distribution Λ from a global statistic γ, one may choose a subgraph $H(x, t) \subseteq G$ for each graph element $x \in X_G$ and each parameter $t \in P$. Then set

$$\Lambda_G(x, t) := \gamma_{H(x,t)}.$$

Of course this requires that the statistic γ is defined on the subgraph $H(x, t)$. An example for the choice of $H(x, t)$ is:

- The ball of radius t around x, i.e., the subgraph induced by the t-neighborhood $\text{Neigh}_t(x)$ (all vertices v with $d(x, v) \leq t$) of x ([486, 540]).
- The largest/smallest subgraph of G containing x and satisfying a certain criterion, possibly depending on t, e.g. the largest subgraph among all subgraphs containing x and having edge connectivity $\geq t$.

Similarly, a local statistic λ can be derived from a global statistic γ by choosing a subgraph $H(x)$ depending only on x and setting

$$\lambda_G(x) := \gamma_{H(x)}.$$

An example for the choice of $H(x)$ is:

- The union of the cliques containing x that have largest size among all cliques containing x.

The localization of a global distribution Γ can be done in a similar manner to that described above. If the subgraph $H(x)$ is chosen solely based on the graph element x, this leads to the local distribution

$$\Lambda_G(x, t) := \Gamma_{H(x)}(t).$$

The last type of localization constructs a local distribution from a local statistic. Again a subgraph $H(x, t)$ is chosen for each graph element $x \in X_G$. But this time it additionally depends on a new parameter $t \in P$. This leads to

$$\Lambda_G(x, t) := \lambda_{H(x,t)}(x).$$

Transformations of these types can be found in [486, 540].

11.7.2   Visualization

A single-valued global statistic γ is simply a value. Hence its visualization is trivial. It becomes more interesting if several graphs of a class $\mathcal{G}$ are examined. If the graphs depend on a parameter t, then one may examine the distribution $\gamma_{G(t)}$ where $G(t)$ is a graph with parameter t. This distribution may be treated like the distributions on a single graph, as described in the following.

Global distributions $\Gamma\colon P \rightarrow Y$ are maps from the set of parameters to the set of values. Hence they may be visualized as functions in the usual way. Furthermore this interpretation allows the application of techniques like interpolation or regression to obtain a functional description based on a set of parameters. Since we assume that the reader is familiar with these notions and techniques, we do not go into detail about them here.

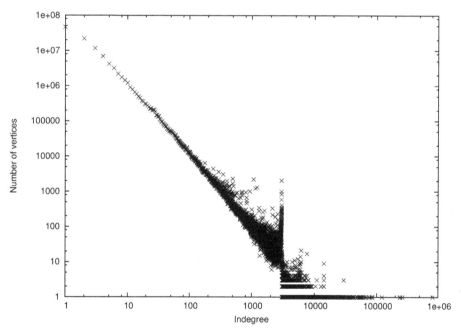

Fig. 11.4. The absolute indegree distribution of the 2001 Crawl of WebBase [566] with logarithmic scale

Figures 11.4 and 11.5 show two examples of visualizations of this type. They show the absolute in- and outdegree distributions of the 2001 WebBase Crawl [566]. Both diagrams have logarithmic scale and show a linear relation, indicating a power law (see Section 11.1).

For local statistics and distributions the visualization is more tricky. For a single-valued statistic $\lambda_G(x)$, one could choose an order $x_1, x_2, \ldots$ on the graph

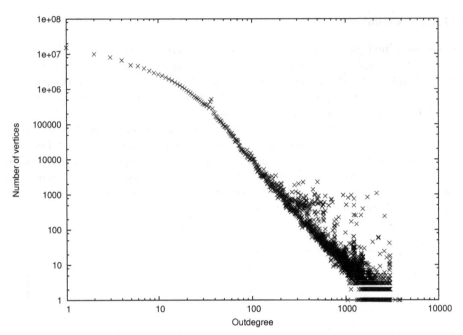

Fig. 11.5. The absolute outdegree distribution of the 2001 Crawl of WebBase [566] with logarithmic scale

elements and plot the resulting pairs $(i, \lambda_G(x_i))$. But the resulting diagram heavily depends on the chosen order. A better approach would be counting or averaging, and then using the global distributions or statistics, leading to a global distribution.

Different local statistics of the same graph, based on the same set X_G of graph elements, may be visualized using a *scatter plot*. Let λ_G and λ'_G be two local statistics over the same set X_G of graph elements. Then the *scatter plot* of λ_G and λ'_G consists of the pairs $(\lambda_G(x), \lambda'_G(x))$ for all graph elements $x \in X_G$. The scatter plot visualizes the relation between λ and λ' in the underlying graph G. If the resulting diagram is an unstructured cloud, the two statistics seem to be unrelated. If, on the other hand, the resulting diagram allows a functional description (obtained via regression, interpolation or similar techniques), the two statistics seem to be closely related – at least in the given graph.

Figure 11.6 shows the scatter plot of the absolute in- and outdegrees of the 2001 WebBase Crawl in logarithmic scale. Each cross represents one or more vertices having the corresponding combination of in- and outdegree. Since the resulting diagram is a very dense cloud, the two types of degree do not seem to be directly related.

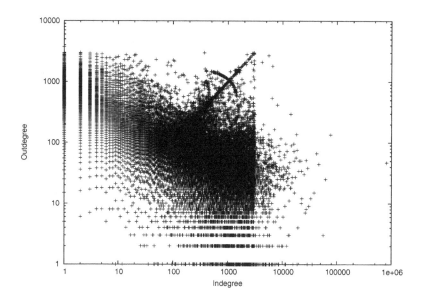

Fig. 11.6. The scatter plot of the in- and outdegrees of the 2001 Crawl of WebBase [566] with logarithmic scale

11.7.3   Sampling of Local Statistics

As we have seen, many network statistics are computationally expensive on large graphs, even though they require only polynomial space and time. Hence it seems sensible to use approximations.

Assume that σ_G is an arbitrary local or global statistic derived from a local statistic λ_G based on graph elements $x \in X_G$. Then one very common and accepted technique for the approximation of σ_G is *sampling*. Instead of using the values $\lambda_G(x)$ for all graph elements $x \in X_G$ for the calculation of σ, only the values for some *samples* $x \in X_{\text{samples}} \subset X_G$ are used.

In general, the sampling of a (global) statistic σ_G from a local statistic $\lambda_G \colon X_G \to Y$ has the following form:

1. Choose a set X_{samples} of samples in X_G. (This may be done completely at random or using a specific strategy.)
2. Calculate $\lambda_G(x)$ for each sample $x \in X_{\text{samples}}$.
3. Calculate σ_G from the set $\{\lambda_G(x) \mid x \in X_{\text{sample}}\}$.

Of course, this description is very rough. Depending on the concrete situation, adaptations might be necessary or useful. For example the strategy for the choice of the samples often determines the quality of the approximation. On the other hand, the calculation of the approximation of σ_G from the samples may require changes in the definition or calculation of σ_G to improve the quality or to obtain reasonable values.

In any case the quality of this approximation has to be certified either by experiments or by analysis of the specific graph models. In general it is clear that an increase of the number of samples directly results in a better approximation.

For example the relative hop plot of a graph can be estimated quite easily using sampling (see Algorithm 25). For $l = n$ one obtains the exact values $\bar{P}$. The runtime of this algorithm is $\mathcal{O}(ln^2 \log n)$ and the space requirements are $\mathcal{O}(n)$ (in addition to the graph). Unfortunately, we do not know of any results about the quality of the estimates.

Algorithm 25: A simple example for sampling: The relative hop plot

---

Input: A graph $G = (V, E)$.

begin

    Set $P(h) = 0$ for $h = 0, 1, 2, \ldots, \mathrm{diam}(G)$

    for $i \in \{1, \ldots, l\}$ **do**

        Choose an **unused** vertex u_i

        Calculate the distances $d(u_i, v)$ for all $v \in V$ by solving the SSSP for source u_i

        for $v \in V$ **do**

            for $h \in \{d(u_i, v), \ldots, \mathrm{diam}(G)\}$ **do**

                $P(h) \leftarrow P(h) + 1$;

    $p(h) \leftarrow \frac{P(h)}{ln}$

end

11.8 Chapter Notes

The first articles mentioning the power law distribution of the degrees of large 'naturally' occurring networks are [16, 40, 197]. Following that, more detailed studies were done. During these and the initial studies a power law was found for several distributions. These include

- the hop plot of the Internet (see Section 11.2.3) [197],
- the distribution of the eigenvalues of the Internet (see Chapter 14) [197],
- the size of certain components of the WWW [102].

Other examples of power law distributions can be found in [405, 147].

Recently, the seeming universality of the power law was questioned (e.g. [117]). Furthermore Bu and Towsley observed in [104] that the *complementary cumulative distribution of degrees* of the Internet seems to fit better with the power law than the degree distributions did. More precisely, this means that

$$D(k) := |\{v \mid d(v) \leq k\}| = ck^\lambda$$

for some constants c and λ.

The average (connected) distance of networks is used in [573, 16, 102, 104].

The eccentricity of the Internet on the level of autonomous systems is examined by Magoni and Pansiot in [405]. They observe a mean eccentricity of 7, while the radius is 5.

The eccentricity also appears in [540], where it is called *node diameter*.

In [486, 540] Tangmunarunkit et al. use the average of absolute and relative hop plots over all vertices in the graph. They call their statistic *expansion*.

In [197] Faloutsos et al. give an alternative definition of the effective diameter, which relies on a power law distribution of the hop plot. They observe that the absolute hop plot $P(h)$ satisfies $P(1) = n + 2m$ (distance 1 is given by the edges) and therefore $P(h) = (n + 2m)h^{\mathcal{H}}$ for a constant exponent $\mathcal{H}$. Then their effective diameter δ_{eff} is the value such that $P(\delta_{\text{eff}}) = (n + 2m)\delta_{\text{eff}}^{\mathcal{H}} = n^2$, leading to $\delta_{\text{eff}} = \left(\frac{n^2}{n+2m}\right)^{1/\mathcal{H}}$. Therefore their effective diameter estimates the maximal distance under the assumption that the hop plot satisfies the power law exactly.

The number of distinct shortest paths is used in [405]. There the fraction of all pairs of vertices with a certain number of distinct shortest paths is measured.

Tangmunarunkit et al. apply the distortion in [486, 540] to analyze the network of autonomous systems of the Internet. They sample the distortion over a collection of spanning trees, generated by unspecified heuristics using the number of all-pairs-shortest-paths traversing a specific link. This global statistic is localized with balls of radius h around a vertex v, leading to a local distribution. This in turn is globalized by averaging over all center nodes.

The term *clustering coefficient* might be somewhat misleading. Clustering in networks, see Chapter 8, is based on many concepts. There are graphs with high clustering coefficient that are contrary to most of these concepts. A relatively high *clustering coefficient* together with a small diameter is known as the *small world* property of a network [573], see also Section 13.1.2 for details.

Also the term *transitivity* as used in Section 11.5 is somewhat misleading. In a directed graph the edge (u, w) is called transitive if there is a path consisting of the two edges (u, v) and (v, w). In this sense the term *transitivity ratio* was defined by Harary and Kommel in 1979 [280]. The equivalent *index of transitivity* was defined in the context of linguistics even earlier, in the year 1957 [282]. The *transitivity* of Newman, Watts and Strogatz [447] is an undirected version of the above terms. Interestingly the term *triple* also appeared already in its directed version in [280].

12 Network Comparison

Michael Baur and Marc Benkert

A fundamental question in comparative network analysis is whether two given networks have the same structure. To formalize what to relate to structural equivalence, the following definition was made:

Definition 12.0.1. *Two undirected simple graphs $G_1 = (V_1, E_1)$ and $G_2 = (V_2, E_2)$ are isomorphic (denoted by $G_1 \simeq G_2$) if there is an edge-preserving bijective vertex mapping $\phi : V_1 \longrightarrow V_2$, i.e. a bijection ϕ with*

$$\forall u, v \in V_1 : \{u, v\} \in E_1 \Longleftrightarrow \{\phi(u), \phi(v)\} \in E_2.$$

The graph isomorphism problem (GI) is to determine whether two given graphs are isomorphic. Figure 12.1 shows an example of two – differently embedded – isomorphic graphs. However, in practice it will be extremely rare that two graphs are isomorphic. We can deal with this fact, as in most cases it is comparatively easy to recognize two graphs as non-isomorphic. We simply have to check necessary conditions: trivially, the number of vertices and edges has to match. For each degree value the number of vertices having this degree has to match, the two graphs must form the same number of connected components, the diameter has to match, and so on. We can also use more complicated properties like those from other chapters in this book: if it should be possible that two graphs are isomorphic, their spectra should be equal, all centrality indices have to match, etc. One could give an ever increasing list of necessary conditions, but thus far no one has succeeded in giving a sufficient condition that is polynomially computable. More details are given in Section 12.1, especially in the overview and in Section 12.1.3.

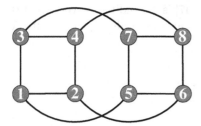 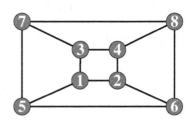

Fig. 12.1. Two isomorphic graphs. The labeling indicates a possible isomorphism; it is not part of the graph

U. Brandes and T. Erlebach (Eds.): Network Analysis, LNCS 3418, pp. 318–340, 2005.
© Springer-Verlag Berlin Heidelberg 2005

Indeed, even in the case of two graphs being non-isomorphic, we want to make a statement as to how similar the graphs are. For example, in chemistry it is often desired to determine the similarity between two molecular structures. Several approaches were made to give such similarity measures; we will present the important ones in Section 12.2. One can also ask if one graph is a part of another; this leads to the *Subgraph Isomorphism Problem*: Determine for two given graphs H and G whether there is a subgraph $H' \subset G$ with $H \simeq H'$. This problem is $\mathcal{NP}$–complete [240] and can probably only be solved in time exponential in the number of vertices of the subgraph.

12.1 Graph Isomorphism

Although the graph isomorphism problem has been studied since the Seventies [489], its complexity status is still unknown. Clearly GI$\in \mathcal{NP}$, but it is not known that GI is polynomially solvable or that it is $\mathcal{NP}$-complete. Also the relationship of GI to co-$\mathcal{NP}$ is not known. Unless $\mathcal{P} = \mathcal{NP}$ there are problems whose complexity status is intermediate, that means their complexity class lies between $\mathcal{P}$ and $\mathcal{NPC}$. The widespread conjecture is that GI is such an intermediate problem. Indications for that conjecture are on one hand that, in spite of enduring research, no polynomial algorithm has been found. And on the other hand, it is known that the counting version of the problem (determine the number of all isomorphisms) is equally difficult as the decision version itself [409]. This together with a theorem proven by Boppana, Håstad and Zachos [80] and Schöning [504] indicates that GI is unlikely to be $\mathcal{NP}$-complete. The theorem makes a statement on the collapse of the polynomial time hierarchy as a consequence of GI$\in \mathcal{NPC}$, which is considered to be very unlikely. One approach in complexity theory to follow this conjecture is to define a special complexity class *isomorphism-complete* which contains GI and all problems as hard as GI. However, Lubiw described in 1981 $\mathcal{NP}$-complete problems similar to GI [398].

Nevertheless, GI is in $\mathcal{P}$ for many graph classes, and graph classes for which GI is really difficult seem to be rare. GI is in $\mathcal{P}$ for trees [4], planar graphs [313], [314], graphs with bounded degree [402], circular-arc graphs [316] and interval graphs (as a subclass of circular-arc graphs). Recently, Cvetkovic, Rowlinson and Simic [136] showed that GI is in $\mathcal{P}$ for graphs with eigenvalues of bounded multiplicity. On the other hand, isomorphism-completeness is maintained on bipartite graphs, line graphs [4], chordal graphs [77] and regular graphs [76]. Most of the positive results are mainly of theoretical interest as the introduced algorithms are of little practical use.

A powerful approach to solve GI is to consider the automorphism group $\mathrm{Aut}(G_1)$ of a given graph G_1, or at least the computable information about $\mathrm{Aut}(G_1)$. Clearly, if $\mathrm{Aut}(G_1)$ is known, $G_1 \simeq G_2$ can be decided by testing $\phi(G_1) = G_2$ for all $\phi \in \mathrm{Aut}(G_1)$. Even if we cannot compute $\mathrm{Aut}(G_1)$ explicitly, we can restrict the number of possible isomorphisms between two graphs by grouping their vertices in equivalence classes. For this, vertex invariants are used. A vertex invariant is a function *inv* defined on the vertex set of a graph with the following property: if there is an isomorphism between G_1 and G_2 that maps v to w then $inv(v) = inv(w)$. The simplest vertex invariant, and in many cases the most powerful, is the degree of a vertex. We can immediately recognize two graphs to be non-isomorphic if their degree sequences are different. If the sequences are equal, but the cardinality of equivalence classes is small, the number of possible isomorphisms is restricted, e.g. if in each graph there are only three vertices of degree d and all other degrees appear only once, the number of possible isomorphisms is 6. In general, we can solve GI polynomially on graph classes for which the automorphism group is polynomially computable, or at least the vertices can be grouped in equivalence classes such that the number of possible isomorphisms between two graphs is polynomial. This raises the question for which graph classes this approach does not work, as these might be difficult to solve. The degree sequence, for example, yields no restriction for regular graphs, but also more elaborate properties may fail. In Section 12.1.3 we give two comparatively small examples that show the hardness of solving GI polynomially using invariants. It gets difficult if the graph does not allow a meaningful vertex grouping because the graph structure is very regular. For more details see Section 4 of [221].

To solve the problem (on general graphs) in practice, there are mainly two methods. Naturally, the direct one: take the two graphs that are to be compared and try to compute an isomorphism. This has the advantage that if there are many isomorphisms, only one has to be found. The second method is to define – independently from the comparison of two specific graphs – a canonical label C, which is a function on the set of all graphs, such that G_1 and G_2 are isomorphic if and only if $C(G_1) = C(G_2)$. This has the advantage that already computed information can be recycled for new comparisons. *McKay's nauty algorithm* grabs this second idea and has become the most practical algorithm for GI. We will elaborate on it later in Section 12.1.2, but refer to [415] for full details. However, first we will have a look at a simple backtracking algorithm that follows the first method.

12.1.1   A Simple Backtracking Algorithm

For the first method we give an algorithm that uses vertex invariants in order to find an isomorphism. The more powerful the invariant, the less the number of functions tested to be isomorphisms from the $n!$ possible ones. Let $\mathcal{R}$ be a set with a linear order '$<$'. Let $inv : V \to \mathcal{R}$ denote some vertex invariant, e.g. $inv(v) = d(v)$ and $\mathcal{R} = \mathbb{N}$. Let $\Pi(V, inv) = (V_1, \dots, V_k)$ be the ordered vertex partition of V with respect to inv, i.e. $\forall v, w \in V_i : inv(v) = inv(w)$ and for all $v \in V_i, w \in V_j$ with $i < j : inv(v) < inv(w)$.

Let $G_1 = (V = \{v_1, \dots, v_n\}, E_1)$ and $G_2 = (W = \{w_1, \dots, w_n\}, E_2)$ denote the two graphs that are checked for isomorphism. The output of the algorithm will be a permutation ϕ of $\{1, \dots, n\}$, such that $v_i \to w_{\phi(i)}$, $1 \le i \le n$, is an isomorphism between G_1 and G_2, or 'NON-isomorphic', if no isomorphism exists. The algorithm will extend isomorphisms between subgraphs of G_1 and G_2 step-by-step and either stop if an isomorphism can be extended to the whole graphs or

---

Algorithm 26: ISOMORPH$(G_1, G_2, (V_1, \ldots, V_m), (W_1, \ldots, W_m), \phi')$

Input: Graphs $G_1 = \big(V = \{v_1, \ldots, v_n\}, E_1 \big)$, $G_2 = \big(W = \{w_1, \ldots, w_n\}, E_2 \big)$,
vertex partitions $(V_1, \ldots, V_m), (W_1, \ldots, W_m)$ with $\bigcup V_i \subset V, \bigcup W_i \subset W$
and $|V_i| = |W_i|$, and an isomorphism ϕ' between the subgraphs induced
by $V \setminus \bigcup V_i$ and $W \setminus \bigcup W_i$.
Output: ϕ, if ϕ' is extensible to an isomorphism ϕ between G_1 and G_2,
'NON-isomorphic' otherwise.

if $(V_1, \ldots, V_m) = \emptyset$ **then return** ϕ'
compute $V_i \in \big\{ V_j \mid |V_j| \leq |V_\ell|, \, 1 \leq \ell \leq m \big\}$
let $V_i = \{v_{i1}, v_{i2}, \ldots\}, W_i = \{w_{i1}, w_{i2}, \ldots\}$
for $j = 1, \ldots, |V_i|$ **do**
> **if** ϕ' extended by $i1 \to ij$ is an isomorphism between the subgraphs induced
> by $V \setminus \bigcup V_k \cup \{v_{i1}\}$ and $W \setminus \bigcup W_k \cup \{w_{ij}\}$ **then**
>> $branch = $ ISOMORPH$(G_1, G_2, (V_1, \ldots, V_i \setminus v_{i1}, \ldots V_m),$
>> $\qquad\qquad (W_1, \ldots, W_i \setminus w_{ij}, \ldots W_m), \phi' \cup \{i1 \to ij\})$
>> **if** $branch \neq$ '*NON-isomorphic*' **then return** $branch$

return 'NON-isomorphic'

---

if all possibilities have been checked unsuccessfully. Isomorphisms on subgraphs will be denoted by ϕ'. Note that any ϕ' is a bijection between two subsets of $\{1, \ldots, n\}$. Initially $\Pi(V, inv) = (V_1, \ldots, V_k)$ and $\Pi(W, inv) = (W_1, \ldots, W_{k'})$ are computed. If $k \neq k'$ or $|V_i| \neq |W_i|$ for any $1 \leq i \leq k$, the two graphs cannot be isomorphic because each possible mapping does not preserve inv. Let us assume that we have checked $k = k'$ and $|V_i| = |W_i|$ successfully in the preprocessing; then ISOMORPH$\big(G_1, G_2, (\Pi(V, inv), \Pi(W, inv), \emptyset\big)$ is called, see Algorithm 26.

First, the vertex subset V_i with minimum cardinality among all subsets of the partition is determined; obviously W_i has the same cardinality. Any isomorphism ϕ between G_1 and G_2 has to map the vertices of V_i to the vertices of W_i. Thus it is sufficient to fix a mapping between a vertex of V_i and W_i and to go on. The smallest cell is chosen in the hope of detecting 'NON-isomorphism' as fast as possible. Now, in the *for*–loop we determine the mapping. If there is an isomorphism ϕ, then $\phi(v_{i1}) \in W_i$, and checking all mappings $v_{i1} \to w_{ij}$ is sufficient in order to obtain an isomorphism. If it is now still possible to extend $\phi' \cup \{v_{i1} \to w_{ij}\}$ to an isomorphism ϕ, we check the mappings of the remaining unmapped vertices. This is done by a recursive call of ISOMORPH.

12.1.2 McKay's Nauty Algorithm

An example of the approach to compute a canonical label that has been implemented is *McKay's nauty algorithm*. In which *nauty* stands for NO AUTOMorphisms YeT?

We first explain McKay's idea to define a canonical label. For an undirected graph $G = (V, E)$ with $V = \{v_1, v_2, \ldots, v_n\}$ let $\mathrm{Adj}(G, \delta)$ be the adjacency matrix of G with respect to the vertex order $v_{\delta(1)}, v_{\delta(2)}, \ldots, v_{\delta(n)}$, where δ is a permutation of $\{1, \ldots, n\}$. Then C_{adj} defined by

$$C_{\text{adj}}(G) = \min_{\delta \in S_n} \text{Adj}(G, \delta)$$

is a canonical label, where $\text{Adj}(G, \delta)$ is interpreted as a n^2-bit binary number derived by concatenation of all rows. Two labels $C_{\text{adj}}(G_1)$ and $C_{\text{adj}}(G_2)$ are equal if and only if G_1 and G_2 are isomorphic. This is because the minimum adjacency matrix is uniquely defined, and two graphs are isomorphic only if there are vertex orders that yield equal adjacency matrices. The naive approach to compute $C_{\text{adj}}(G)$ would look at all $n!$ vertex orders and for each order compare two adjacency matrices of size $n \times n$. However, even for comparatively small values of n this would not be feasible in acceptable time. To speed up this approach McKay uses various techniques in his nauty algorithm in order to compute a label $C(G)$. In general, $C(G)$ will be different from $C_{\text{adj}}(G)$ as the nauty algorithm does not look at all $n!$ orders but at a special sample and computes the minimum matrix among them. The *Refinement Procedure* will determine these samples. The number of samples depends on the structure of the graph, but usually the sample size is significantly smaller than $n!$. To compute all vertex orders that will be checked, the nauty algorithm uses a search tree $\mathcal{T}$ in which each leaf corresponds to a vertex order. The algorithm traverses $\mathcal{T}$ and examines all adjacency matrices that are induced by the vertex orders of visited leaves. Now, the next trick comes into play: not all leaves are visited. Group theory, more precisely the information about the automorphism group $\text{Aut}(G)$ already known, allows to exclude subtrees of $\mathcal{T}$ from the traversal. A subtree is pruned if it is known that it contains only vertex orders that lead to adjacency matrices not smaller than the best one found so far. Using algebra, mainly group theory, it is shown that the label C derived by this approach is indeed canonical, see Theorem 12.1.2. There is another technique to prune $\mathcal{T}$, but, as it is very abstract, we will mention it only briefly in the chapter notes (Section 12.3). Next, we introduce some basics from group theory that we need in the sequel.

Basics in Group Theory

We denote the permutation group of n elements by S_n. An element $\delta \in S_n$ is simply a bijection between the sets $\{1, \ldots, n\}$ and $\{1, \ldots, n\}$. Obviously, there are $n!$ such bijections. The product of two elements f and g in a group of functions is defined by composition, i.e. $f \cdot g = f \circ g$. For a finite group $\mathcal{G}$ and a subset of elements $\mathcal{F} \subseteq \mathcal{G}$ the group product of $\mathcal{F}$ in $\mathcal{G}$ is the subgroup $\langle \mathcal{F} \rangle$ defined by

$$\langle \mathcal{F} \rangle = \{f \in \mathcal{G} \mid \exists m \exists f_1, \ldots, f_m \in \mathcal{F} : f = f_1 \cdot \ldots \cdot f_m\}.$$

The elements of $\mathcal{F}$ are called *generators* of $\langle \mathcal{F} \rangle$.

A group $\mathcal{G}$ *operates* on a set $\mathcal{M}$ with respect to a function $\sigma : \mathcal{G} \times \mathcal{M} \to \mathcal{M}$, if for the neutral element $e \in \mathcal{G}$ and all $f, g \in \mathcal{G}$ and $x \in \mathcal{M}$ it holds that $\sigma(e, x) = x$ and $\sigma(f \cdot g, x) = \sigma(f, \sigma(g, x))$. Then, $\mathcal{G}$ and σ induce an equivalence relation on $\mathcal{M}$ in the following way:

$$x \sim y \iff \exists f \in \mathcal{G} : \sigma(f, x) = y.$$

We call the equivalence class of x, i.e. the set $\{\sigma(f, x) \mid f \in \mathcal{G}\}$, the *orbit* of x. The set of all equivalence classes of $\mathcal{M}$ with respect to $\mathcal{G}$ and σ is called the *orbit partition*. In our case a subgroup $\Phi \subseteq \mathrm{Aut}(G)$ of the automorphism group of a graph $G = (V, E)$ will operate on V. For an automorphism $\phi \in \Phi$ and a vertex $v \in V$ the function σ is simply defined by $\sigma(\phi, v) = \phi(v)$.

The Search Tree $\mathcal{T}$

In the following $G = (V, E)$ is the undirected graph whose label $C(G)$ we want to compute. Let the cardinality of V be n. We first fix an initial indexing of the vertices $V = \{v_1, \ldots, v_n\}$. We now give a formal definition of what we mean by *vertex partition*.

Definition 12.1.1 (Vertex partition). *A* vertex partition *of G is an ordered list $\Pi = (V_1, \ldots, V_r)$ of vertex subsets $V_i \subseteq V$, the so-called* cells, *with*

1. $V_i \cap V_j = \emptyset$, $1 \leq i \neq j \leq r$
2. $\bigcup_{i \in \{1, \ldots, r\}} V_i = V$
3. $|V_i| \geq 1$, $1 \leq i \leq r$.

The number r of vertex subsets of Π is denoted by $|\Pi|$. A vertex partition Π is called unit partition *if $r = 1$ and* discrete partition *if $r = n$.*

From now on, by vertex partition we will always mean a vertex partition of G. Any node of $\mathcal{T}$ corresponds to a vertex partition with which we will identify that node. To specify these vertex partitions, we have to introduce a refinement procedure f in advance. For a vertex partition Π, $f(\Pi)$ will be a refinement of Π, i.e. for each cell V' in $f(\Pi)$ there will be a cell V in Π with $V' \subseteq V$. The refinement is arranged such that vertices that have 'equal' adjacencies are grouped together. For a vertex $v \in V$ and a vertex set $W \subset V$ let $d(v, W)$ be the number of vertices in W that are adjacent to v. For simplicity assume that we want to compute the refinement $f(\Pi)$ of the unit partition $\Pi = (V)$. In the first refinement step, the number $d(v, V)$ is computed for each vertex v, which simply means the degree of v. Then, the vertices are partitioned according to their degrees, i.e. the result of this first refinement step is a partition $\Pi_1 = (W_1, \ldots, W_j)$ in which any two vertices of each cell are of same degree and for a vertex $v \in W_k$ and a vertex $w \in W_\ell$ it holds that $d(v, V) < d(w, V)$ if and only if $k < \ell$. Next, each cell of Π_1 is refined with respect to Π_1. We proceed in basically the same manner as before. For each vertex v of a cell W_i its number $\eta(v) = \big(d(v, W_1), \ldots, d(v, W_j)\big)$ is computed and the vertices of W_i are partitioned according to these numbers. (Two vectors are compared according to their lexicographical order.) Doing this for all cells results in a refined partition Π_2. Partition Π_3 is then the refined partition of Π_2 and so on. This is done as long as Π_{i+1} is a real refinement of Π_i, see Algorithm 27.

Note that the partition $f(\Pi) = (V_1, \ldots, V_{r'})$ fulfills the following property: for any two (not necessarily distinct) cells V_i, V_j of $f(\Pi)$ and for any two vertices

---

Algorithm 27: REFINEMENT PROCEDURE $f(\Pi)$

Input: A vertex partition $\Pi = (V_1, \ldots, V_r)$.
Output: The refined vertex partition $f(\Pi)$.

$\Pi_{\text{new}} = \Pi$
repeat
$\quad$ $\Pi_{\text{old}} = \Pi_{\text{new}}$
$\quad$ let $\Pi_{\text{old}} = (V_1, \ldots, V_{r'})$
$\quad$ **for** $i = 1$ **to** r' **do**
$\quad\quad$ **for** *each* $v \in V_i$ **do**
$\quad\quad\quad$ **compute** $\eta(v) = (d(v, V_1), \ldots, d(v, V_{r'}))$
$\quad\quad$ **partition** V_i into $W_1, \ldots, W_j$ such that for $v \in W_k$, $w \in W_\ell$:
$$\eta(v) < \eta(w) \iff k < \ell$$
$\quad\quad$ **replace** V_i in Π_{new} by $W_1, \ldots, W_j$
until $\Pi_{\text{new}} = \Pi_{\text{old}}$
return Π_{new}

$v, w \in V_i$ it holds that $d(v, V_j) = d(w, V_j)$. A partition that satisfies this property has been called *equitable* in Section 9.3.1, where the same method has been discussed. We say that two vertices are *structurally equivalent* (w.r.t. $f(\Pi)$) if they lie in the same cell of $f(\Pi)$.

Now, we can precisely describe the nodes of $\mathcal{T}$. All nodes will correspond to equitable partitions. The root $\Pi = (V_1, \ldots, V_r)$ corresponds to the refinement of the unit partition $f((V))$. If Π is already a discrete partition, Π has no descendants and $\mathcal{T}$ consists of just one node, otherwise the descendants of Π are derived as follows: let $V_i = \{v_1', \ldots, v_m'\}$ be the first *non-trivial* cell of Π, i.e. the first cell that contains more than one vertex. Then, Π has m descendants, namely $f(\Pi \setminus v_1'), \ldots, f(\Pi \setminus v_m')$, where $f(\Pi \setminus v_j')$ is short hand for $f((V_1, \ldots, V_{i-1}, \{v_j'\}, V_i \setminus \{v_j'\}, V_{i+1}, \ldots V_r))$. This means that we take each vertex $v' \in V_i$ out of V_i once, define $\{v'\}$ artifically as new cell and refine this partition in order to get the descendant $f(\Pi \setminus v')$. This makes sense as Π was equitable before, i.e. any two vertices of one cell of Π were structurally equivalent, and we now check each possibility to refine Π by removing each vertex v' out of V_i and make it an artificial cell.

For any other node $\Pi' \in \mathcal{T}$ that does not correspond to a discrete partition, the descendants are derived in exactly the same manner as for Π. Hence, all leaves of $\mathcal{T}$ correspond to discrete partitions. The order of such a discrete partition $(\{v_{\delta(1)}\}, \ldots, \{v_{\delta(n)}\}), \delta \in S_n$ determines the adjacency matrix of the corresponding leaf. Recall that the purpose of f is to make $\mathcal{T}$ as small as possible by means of structurally equivalent vertices with respect to the current partition. However, the real size of $\mathcal{T}$ depends on the structure of G. For the example graph in Figures 12.2, $\mathcal{T}$ has only three nodes, while the search tree $\mathcal{T}$ of the example graph in Figure 12.3, which contains somewhat more regular structures than the graph of Figure 12.3, is much bigger.

McKay now defines the label $C(G)$ as minimum adjacency matrix found among all leaves of $\mathcal{T}$. This is indeed a canonical label:

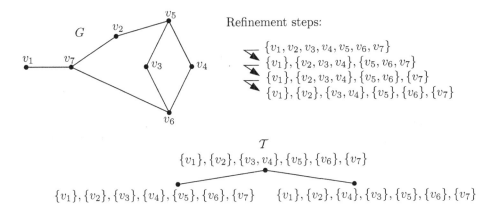

Fig. 12.2. A graph G and the corresponding search tree $\mathcal{T}$. At the beginning only v_3 and v_4 are structurally equivalent

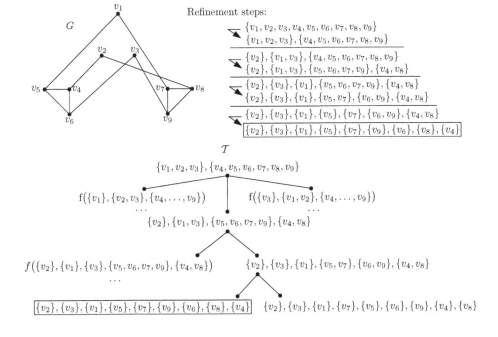

Fig. 12.3. A graph G, an extract of the corresponding search tree $\mathcal{T}$ and the refinement steps for the emphasized path in $\mathcal{T}$. At the beginning the vertices in $\{v_1, v_2, v_3\}$ and in $\{v_4, \ldots, v_9\}$ are structurally equivalent

Theorem 12.1.2. *Let G_1 and G_2 be two undirected graphs. Let $C(G_1)$ and $C(G_2)$ be the labels that were derived from the corresponding search trees. It holds that*

$$C(G_1) = C(G_2) \Longleftrightarrow G_1 \simeq G_2.$$

On one hand it is clear that non-isomorphic graphs G_1 and G_2 cannot have the same label as each adjacency matrix of G_1 is different from each adjacency matrix of G_2 (otherwise the graphs would be isomorphic). In the other direction the clear prescript to generate the search trees gives a hint that two isomorphic graphs really get the same label. Of course, this has to be proven exactly. However, the proof is very technical. We refer the interested reader to [415, Theorem 2.19].

Using Automorphisms to Prune $\mathcal{T}$

The nauty algorithm does not compute $\mathcal{T}$ explicitly. Instead the algorithm parses $\mathcal{T}$ in a special early-to-late order and tries to exclude as many subtrees from the search. Actually, the partition that corresponds to a node is not computed until the node is visited by the search. When the algorithm reaches a leaf ℓ, the adjacency matrix A_ℓ induced by ℓ is computed. During the traversal the algorithm maintains the minimum adjacency matrix $A_{\min}$ it has found so far. When the algorithm reaches the first leaf ℓ_1, $A_{\min}$ is initialized by A_{ℓ_1}. When another other leaf ℓ is reached, it is tested whether $A_\ell < A_{\min}$ and, if so, $A_{\min}$ is set to A_ℓ. Thus, at the end $A_{\min}$ contains the label $C(G)$. Additionally, the algorithm maintains the subgroup $\Phi_t(G)$ of the automorphism group of G that has been computed so far. We will denote this group by $\Phi_t(G)$. It holds that $\Phi_t(G) = \langle \phi_1, \ldots, \phi_{i(t)} \rangle$, where $\phi_1, \ldots, \phi_{i(t)}$ are all automorphisms that we know at time t. An automorphism ϕ is found when two leaves induce equal adjacency matrices: let $w_1, \ldots, w_n$ and $w'_1, \ldots, w'_n$ be the vertex orders of the two leaves. Then, $\phi : w_i \to w'_i$ for $i = 1, \ldots, n$ is an automorphism, see Figure 12.4.

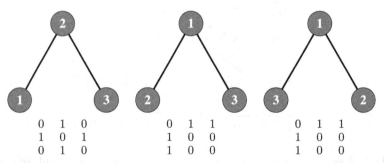

Fig. 12.4. Recognizing automorphisms: the matrix below each graph is the adjacency matrix induced by the particular labeling. If and only if two matrices are equal, the mapping that matches identically labeled vertices is an automorphism

To see how $\mathcal{T}$ can be pruned, we need some more definitions. First a linear order on the nodes in $\mathcal{T}$ is introduced to establish the early-to-late order. Let Π be an inner vertex of $\mathcal{T}$. We denote the subtree rooted at a descendant $f(\Pi \setminus v_i)$ of Π by $\mathcal{T}(\Pi \setminus v_i)$.

Definition 12.1.3 (Linear order on the nodes of $\mathcal{T}$). *Let Π_1, Π_2 be two different nodes of $\mathcal{T}$ and let Π be the least common ancestor of Π_1 and Π_2 in $\mathcal{T}$. We define Π_1 '<' Π_2 if $\Pi_1 = \Pi$ or if for the vertices v_i and v_j in the first non-trivial cell of Π with $\Pi_1 \in \mathcal{T}(\Pi \setminus v_i)$ and $\Pi_2 \in \mathcal{T}(\Pi \setminus v_j)$ it holds that $i < j$. Otherwise Π_2 '<' Π_1.*

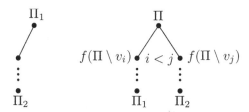

Fig. 12.5. Linear order on the nodes of $\mathcal{T}$: the two cases where Π_1 '<' Π_2

It is easy to see that the relation '<' is a linear order, see also Figure 12.5. The nauty algorithm traverses the nodes of $\mathcal{T}$ with respect to this order. Next, we need an equivalence relation on the nodes of $\mathcal{T}$.

Definition 12.1.4 (Equivalence relation on the nodes of $\mathcal{T}$).
Let $\Pi_1 = (V_1, \ldots, V_m) \in \mathcal{T}$ and $\Pi_2 = (W_1, \ldots, W_m) \in \mathcal{T}$. Then $\Pi_1 \sim \Pi_2$ if and only if there is an automorphism $\phi \in \mathrm{Aut}(G)$ and a permutation $\delta \in S_m$ such that $\phi(V_i) = W_{\delta(i)}$ for $i = 1, \ldots, m$. We say that ϕ witnesses $\Pi_1 \sim \Pi_2$.

Automorphisms that witness the equivalence of two partitions can be thought of as color-preserving automorphisms. For each cell V_i of Π_1 color its vertices in a distinct color and color the vertices of cell $W_{\delta(i)}$ in the same color. Then there exists an automorphism ϕ that preserves the color of each vertex. We can now state the first of two important theorems on the way to prune $\mathcal{T}$.

Theorem 12.1.5. *Let $\Pi_1 \sim \Pi_2 \in \mathcal{T}$ and let $\mathcal{T}_1$ and $\mathcal{T}_2$ be the subtrees of $\mathcal{T}$ rooted at Π_1 and Π_2, respectively. Then for each node $\Pi_1' \in \mathcal{T}_1$ there is a node $\Pi_2' \in \mathcal{T}_2$ with $\Pi_1' \sim \Pi_2'$.*

For the proof we refer to [415], Theorem 2.14. As an immediate consequence of Theorem 12.1.5 we can discard the subtree $\mathcal{T}_2$ rooted at a node $\Pi_2 \in \mathcal{T}$ if we know that there is a node Π_1 with Π_1 '<' Π_2 and $\Pi_1 \sim \Pi_2$. This is due to the fact that each leaf of $\mathcal{T}_2$ is equivalent to a leaf of the subtree rooted at Π_1. Thus, we have already seen all adjacency matrices that would be induced by the leaves of $\mathcal{T}_2$. We have to see how $\Phi_t(G)$ is applied to find equivalent inner nodes. For a vertex $v \in V$, $\{\phi(v) \mid \phi \in \Phi_t(G)\}$ is the orbit of v with respect to

$\Phi_t(G)$. Let Θ_t be the orbit partition of V at time t. The algorithm has access to Θ_t at any time. Initially Θ_t is the discrete partition, i.e. $\Theta_0 = \{v_1\}, \ldots, \{v_n\}$. Every time a new automorphism is discovered, Θ_t is updated. This means Θ_t is getting coarser as the new automorphism can enlarge $\Phi_t(G)$ and thus vertices can become equivalent (w.r.t. $\Phi_t(G)$) that were not equivalent before. We can now detect equivalent descendants of a node $\Pi \in \mathcal{T}$ by means of the following theorem which corresponds to Theorem 2.15. in [415].

Theorem 12.1.6. *Let $\Pi = (V_1, \ldots, V_r) \in \mathcal{T}$ and $V_i = \{v'_1, \ldots, v'_m\}$ be the first non-trivial cell of Π. If there are $v'_i, v'_j \in V_i$ that lie in the same orbit of Θ_t, there is an automorphism $\phi \in \Phi_t(G)$ that witnesses $f(\Pi \setminus v'_i) \sim f(\Pi \setminus v'_j)$.*

This theorem is used to prune $\mathcal{T}$ in two ways. The first is obvious: assume the algorithm reaches a node $\Pi \in \mathcal{T}$ whose first non-trivial cell is V_i. Then, Θ_t induces a partition of V_i into cells such that any two vertices of each cell lie in the same orbit. We denote this partition by $\Theta_t \wedge V_i$. According to Theorem 12.1.6, we have to consider only one descendant $\mathcal{T}(\Pi \setminus v')$ for each cell $\Theta_t \wedge V_i$. Namely v' is the vertex that is minimal in its cell, i.e. has the lowest initial index out of all vertices in its cell. In other words, we have to consider the descendants that are derived by the minimal cell representatives of $\Theta_t \wedge V_i$.

The second way is a bit trickier. Assume that the algorithm reaches a node $\Pi \in \mathcal{T}$ at time t_1. Again let V_i be the first non-trivial cell of Π, and let $v_i, v_j \in V_i$ be vertices that do not lie in the same orbit w.r.t. Θ_{t_1}. This means that the algorithm will examine $\mathcal{T}(\Pi \setminus v_i)$ and $\mathcal{T}(\Pi \setminus v_j)$ by the information that it gets from Θ_{t_1}. W.l.o.g. let v_i have the smaller initial index than v_j, and thus $\mathcal{T}(\Pi \setminus v_i)$ will be examined before $\mathcal{T}(\Pi \setminus v_j)$. The algorithm proceeds, and at time t_2 it finds a new automorphism ϕ' such that now there is an automorphism $\phi \in \Phi_{t_2}(G)$ with $\phi(v_i) = v_j$. Hence, v_i and v_j lie in the same orbit w.r.t. Θ_{t_2}. (Note that ϕ is not necessarily the new automorphism ϕ' itself but a composition of ϕ' and automorphisms that have been found before.) Now, the algorithm has the information that $\mathcal{T}(\Pi \setminus v_j)$ can be pruned. Of course, this cannot be taken into account anymore if t_2 is after the examination of $\mathcal{T}(\Pi \setminus v_j)$ has been completed. Otherwise this examination can be discarded or (if $\mathcal{T}(\Pi \setminus v_j)$ is already being examined) aborted. If the algorithm indeed aborts the examination of a subtree $\mathcal{T}(\Pi \setminus v_j)$ and jumps back to Π, a new automorphism has just been found such that Θ_t allows this step. Now, it might even be possible to jump back to an ancestor of Π because Θ_t now also allows to abort the examination of a subtree in which Π is contained. Actually, when a new automorphism is found, the algorithm immediately checks how far it can jump back in $\mathcal{T}$ by means of the new information.

A challenge is to determine an appropriate number of adjacency matrices to be stored. Storing and comparing adjacency matrices needs a lot of time and space. However, if the algorithm maintains a large number of adjacency matrices, the number of detected automorphisms will also be higher. Thus $\mathcal{T}$ can be pruned more efficiently which in turn will again decrease the running time. McKay claims that the storage of only two adjacency matrices has stood

---

Algorithm 28: NAUTYALGORITHM $\left(G = (V, E), V = \{v_1, \ldots, v_n\}\right)$

Input: A graph $G = (V, E)$ and an initial vertex indexing $V = \{v_1, \ldots, v_n\}$.
Output: The label $C(G)$.
adj.matrix A_{ℓ_1}, vertex_order(A_{ℓ_1}) = nil
adj.matrix $A_{\min}$, vertex_order$(A_{\min})$ = nil
$\Phi(G) = \{\mathrm{id}\}$
$\Theta = \{v_1\}, \ldots, \{v_n\}$
process$\left(f((V))\right)$
return $A_{\min}$.

---

process$\left(\Pi = (V_1, \ldots, V_r)\right)$
if $r = n$ then
  identify $V_1 = \{v_1'\}, \ldots, V_n = \{v_n'\}$ with vertex order $v_1', \ldots, v_n'$
  compute adj. matrix A_Π induced by $v_1', \ldots, v_n'$
  if A_{ℓ_1} = nil then
    A_{ℓ_1} = A_Π, vertex_order(A_{ℓ_1}) = $v_1', \ldots, v_n'$
    $A_{\min} = A_\Pi$, vertex_order$(A_{\min}) = v_1', \ldots, v_n'$

  else
    if $A_{\min} > A_\Pi$ then $A_{\min} = A_\Pi$, vertex_order$(A_{\min}) = v_1', \ldots, v_n'$
    else
      ϕ = nil
      if $A_{\ell_1} = A_\Pi$ then
        compute automorphism ϕ induced
          by vertex_order(A_{ℓ_1}) and $v_1', \ldots, v_n'$
      if $A_{\min} \neq A_{\ell_1}$ and $A_{\min} = A_\Pi$ then
        compute automorphism ϕ induced
          by vertex_order$(A_{\min})$ and $v_1', \ldots, v_n'$
      if $\phi \neq$ nil then
        $\Phi(G) = \langle\{\Phi(G) \cup \phi\}\rangle$
        update Θ
        check jump back

else
  let $V_i = \{v_1', \ldots, v_m'\}$ be the first non-trivial cell of Π
  let $v_1'', \ldots, v_{m'}''$ be the minimum cell representatives of $\Theta \wedge V_i$
  for $j = 1$ **to** m' **do** process$\left(f(\Pi \setminus v_j'')\right)$

---

the test in practice. At any time, the nauty algorithm stores two adjacency matrices, the matrix A_{ℓ_1} of the first visited leaf and $A_{\min}$. We summarize the nauty algorithm in Algorithm 28. For simplification we have omitted a detailed, rather complicated description of the jump-back steps.

12.1.3   The Difficulty of GI or 'How to Trick Nauty'

Recall that the complexity status of GI is not yet known. Assume we strongly believe that there is a polynomial algorithm and we want to try to solve GI polynomially (many people have in fact tried to derive such an algorithm). The obvious way to do it would be to use an idea similar to that of McKay. This section tries to illustrate why it seems to be hard to succeed in solving GI like this; we show that even elaborate approaches fail. In principle we want to proceed as in the nauty algorithm, but to use a different refinement procedure and compute only *one* leaf of the search tree. The label $C(G)$ is then again defined as the adjacency matrix induced by the vertex order of this leaf. As stated before two non-isomorphic graphs G_1 and G_2 are never recognized as isomorphic because each adjacency matrix of G_1 is different from any adjacency matrix of G_2. To make sure that two isomorphic graphs G_1 and G_2 are recognized as isomorphic we want to ensure the following: Let Π_k be the leaf in the search tree of G_1 that has been computed and that defines $C(G_1)$. Let Π'_k be the leaf in the search tree of G_2 that defines $C(G_2)$. Let $\Pi_1, \ldots, \Pi_k$ and $\Pi'_1, \ldots, \Pi'_{k'}$ be the vertex partitions that have been computed in order to get to Π_k and $\Pi'_{k'}$, respectively. Then it should hold that $k = k'$ and for $i = 1, \ldots, k$ each vertex partition Π_i matches Π'_i in terms of number of cells and cardinality of each cell. Finally we need that for $\Pi_i = (V_1, \ldots, V_r)$ and $\Pi'_i = (V'_1, \ldots, V'_r)$ and for each pair $V_j = \{v_1, \ldots, v_m\}, V'_j = \{v'_1, \ldots, v'_m\}$ of cells the following holds: for all $(v, v') \in V_j \times V'_j$ there is an isomorphism ϕ between G_1 and G_2 with $\phi(v) = v'$. This last condition justifies our computing only one leaf of the search tree. Then it is irrelevant which vertices v and v' we take out of the first non-trivial cells of Π_i and Π'_i, define artificially as new equivalence classes and refine according to these new partitions. To see this, note that if later $C(G_1)$ really equals $C(G_2)$, the corresponding vertex orders of the two leaves induce an isomorphism ϕ between G_1 and G_2. Defining $\{v\}$ and $\{v'\}$ as new equivalence classes simply means that we fix $\phi(v) = \phi(v')$ and refine with respect to this information. And if there is really an isomorphism ϕ that maps v onto v', which is guaranteed by the last condition, we will still find it.

To illustrate that it seems difficult to solve GI polynomially in the way described above, we give two counterexamples. First, we look at the refinement procedure f used in the nauty algorithm. The 3-regular graph G in Figure 12.6 proves that f does not help to solve GI polynomially.

For this graph $G = (V, E)$, it holds that $d(v, V)$ equals $d(w, V)$ for any two vertices $v, w \in V$, since G is regular. Thus, the unit partition is not further refined by f. If we now have two copies G_1 and G_2 of G and take v_1 out of V to derive $C(G_1)$ while we take v_2 out of V to derive $C(G_2)$, we will come to the false conclusion that G_1 and G_2 are non-isomorphic. There is no isomorphism

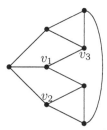

Fig. 12.6. The 3-regular graph G

that maps v_1 onto v_2: from v_1 the distance to any other vertex is 2, while the distance from v_2 to v_3 is 3.

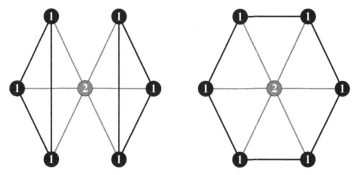

Fig. 12.7. Graph with two components

We now want to see what happens if we apply a different refinement procedure that uses more information than adjacencies to other cells. Recall that the idea of f was to partition the vertex set in equivalence classes as long as it holds that any two vertices of one cell have the same number of neighbors in each other cell. For $v \in V, W \subseteq V$ and $i \in \mathbb{N}$ let now $d_i(v, W)$ be the number of vertices in W of distance i to v. We try to improve f and refine as long as the following holds: for any two vertices v and w of one cell the numbers $d_i(v, W)$ and $d_i(w, W)$ are equal w.r.t. each cell W and each $i \in \mathbb{N}$. The 3-regular graph is no longer a counterexample for this refinement procedure. However, the new method also fails, as the graph in Figure 12.7 shows. The label of each vertex v corresponds to the equivalence class to which v belongs after the refinement of the unit partition. Each 1–vertex has two 1–vertices and one 2–vertex at distance 1 and three 1–vertices at distance 2, while each 2–vertex has six 1–vertices at distance 1. Obviously, there is no isomorphism that maps the 2–vertex of the left component onto the 2–vertex of the right component. For simplification the graph consists of two components, but the graph can be extended, resulting in a connected graph which yields the same result.

12.2   Graph Similarity

The graph isomorphism problem asks if two graphs have identical structure. As this is a very restrictive criterion, one may consider the natural relaxation which tries to specify how similar two graphs are. *Graph similarity*, often called *graph matching*, compares two graphs to give a measure for the similarity, or distance, between them.

There are various applications of this problem, i.e., CAD/CAM, computer vision, and molecule matching. An important advantage of graph similarity over isomorphism is its ability to cope with errors and distortions in the input data, which often occurs when collecting real world data. These errors can change isomorphic graphs to non-isomorphic ones, so a rigorous check for isomorphism is inappropriate. The alternative is an imprecise matching using a graph similarity measure.

Many applications imply a labeling of the vertices or edges, i.e., in molecule matching the labeling is defined by the types of the elements. When labels are present vertices and edges with different labels are either penalized or even not allowed to match. Since we are interested in structural similarity, all graphs are regarded as unlabeled in the following.

There are certain properties a meaningful similarity measure should fulfill. For example, the distance from graph G_1 to graph G_2 should be the same as from G_2 to G_1, and the distance of isomorphic graphs should be 0. An common formalization of such properties is a *graph distance metric*.

Definition 12.2.1. *Let G_1, G_2, and G_3 be graphs. A function $d : G_1 \times G_2 \to \mathbb{R}_0^+$ is called a* graph distance metric *if the following properties hold:*

$$\textbf{\textit{reflexivity:}} \quad d(G_1, G_2) = 0 \Leftrightarrow G_1 \cong G_2 \tag{12.1}$$

$$\textbf{\textit{symmetry:}} \quad d(G_1, G_2) = d(G_2, G_1) \tag{12.2}$$

$$\textbf{\textit{triangle inequality:}} \quad d(G_1, G_2) + d(G_2, G_3) \geq d(G_1, G_3) \tag{12.3}$$

On the other hand, all graph distance metrics are hard to compute since the reflexivity property implies a solution for graph isomorphism. Thus, in practice, one may either relax these properties, or compute an approximation of the measure.

For simplicity, only undirected connected graphs are considered in the following. All statements can be extended to unconnected graphs by considering their connected components, and also to directed (strongly connected) graphs.

We present three types of similarity measures. Two are metrics: one is based on the size of a maximum common subgraph, and the other on the difference in the length of corresponding paths. Another approach defines the distance between two graphs in terms of edit operations needed to transform one into the other. Finally we give a short overview of other methods from literature

12.2.1   Edit Distance

A general and flexible method for matching structural objects is the concept of edit distance. Given a set of allowed edit operations on the objects, the distance between two objects is defined as the minimal number of operations needed to transform one into the other. A well-known example is string edit distance.

In graph edit distance typical operations include the insertion, deletion, and substitution of vertices and edges. There is no general agreement on the set of allowed operations. Instead, a good selection of allowed operations is very application-dependent. Furthermore, non-negative costs can be assigned to operations to better fit special requirements. In this case the distance is defined as the minimum cost taken over all sequences of operations that transform one graph into the other.

Intuitively speaking, for reasonable and meaningful specifications of operations and costs, the problem is hard to solve. For certain combinations of operations and costs the metric properties are satisfied. Recall this implies the problem is at least as hard to solve as GI. On the other hand the distance is efficiently computable only for simple sets of allowed operations. In this case the resulting distance is less significant.

Example 1. The first example illustrates a specification which is easy to handle but does not lead to very meaningful results. The following edit operations are allowed:

– vertex insertion - a new (isolated) vertex is added to the graph
– vertex deletion - a (isolated) vertex is deleted from the graph
– edge insertion - a new edge is added between arbitrary vertices of the graph
– edge deletion - an edge is deleted from the graph

The costs of both vertex operations are one, of both edge operations zero. It is easy to see that the distance defined by this specification is equal to the difference of the number of vertices of the two graphs:

$$d_{\text{exp1}} = ||V(G_1)| - |V(G_2)|| .$$

This means, for example, a path, a star, and a clique of the same number of vertices are equal in terms of this distance.

Example 2. This specification was introduced by Papadopoulos and Manolopoulos [464]. They propose to use three operations, all with cost one:

– vertex insertion – a new (isolated) vertex is added to the graph
– vertex deletion – a (isolated) vertex is deleted from the graph
– edge update – one endvertex of an edge is changed

Insertion or deletion of an edge requires two edge updates in this model. Using these operations on the graphs of Figure 12.8, two operations are required to match G_1 with G_2, namely two edge updates, whereas three operations are required to match G_1 with G_3, namely one vertex insertion and two edge updates. Thus, in this specification, G_1 is more similar to G_2 than to G_3.

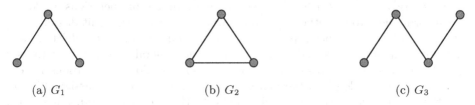

(a) G_1 (b) G_2 (c) G_3

Fig. 12.8. Similarity among graphs: in Example 2, G_1 is more similar to G_2 than to G_3

As already mentioned, the computation of a meaningful graph edit distance is hard. Therefore, the matching condition is relaxed: given two graphs G_1 and G_2, instead of transforming G_1 into G_2, G_1 is transformed into a graph with the same number of vertices and edges and the same degree sequence as G_2. In other words, only the size and the degree sequence of the graphs are considered.

A *degree vector* $x = (x_1, \ldots, x_n)$ of a graph $G = (\{v_1, \ldots, v_n\}, E)$ with n vertices is defined by $x_i := d(v_i)$. A *graph histogram* is a degree vector whose entries are incremented by one and sorted in decreasing order. Given two graphs G_1 and G_2, the distance according to the L_1 metric of the corresponding graph histograms gives the minimum number of operations required to transform G_1 into a graph with the same number of vertices and edges and the same degree sequence as G_2. If the number of vertices of the two graphs differs, zeros are added to the smaller graph histogram.

12.2.2 Difference in Path Lengths

The next similarity measure we present is an example of a graph distance metric [116]. Hence, while the definition is quite simple, its computation is hard. Roughly speaking, the sum of differences of the lengths of corresponding paths for all pairs of vertices is considered. Since this measure is reasonable only for graphs of the same number of vertices, only such graphs are compared in the following.

Definitions. Let G_1, G_2 be isomorphic connected graphs with isomorphism $\phi : V(G_1) \rightarrow V(G_2)$. Two vertices of G_1 are adjacent iff their isomorphic vertices in G_2 are adjacent, in other words:

$$\forall u, v \in V(G_1) : \{u, v\} \in E(G_1) \Leftrightarrow \{\phi(u), \phi(v)\} \in E(G_2) .$$

An equivalent formulation extends this connection property from distance-one vertices to arbitrary pairs of vertices:

$$\forall u, v \in V(G_1) : d_{G_1}(u, v) = d_{G_2}(\phi(u), \phi(v)) . \tag{12.4}$$

Now, let G_1, G_2 be two arbitrary connected graphs of the same number of vertices and $\sigma : V(G_1) \rightarrow V(G_2)$ a bijection. Then, Equation 12.4 does not necessarily hold anymore. Instead, we can use the differences of the path lengths to define the similarity of two graphs with respect to σ.

Definition 12.2.2. *For two connected graphs G_1, G_2 of the same number of vertices and a bijection $\sigma : V(G_1) \to V(G_2)$ we define the σ-distance d_σ by*

$$d_\sigma(G_1, G_2) = \sum_{\{u,v\} \in V(G_1) \times V(G_1)} |d_{G_1}(u, v) - d_{G_2}(\sigma(u), \sigma(v))| \, ,$$

where the sum is taken over all unordered pairs of vertices of G_1.

Since the similarity of two graphs can not depend on a specific mapping between the sets of vertices, the distance is defined as the minimum over all possible bijections between $V(G_1)$ and $V(G_2)$.

Definition 12.2.3. *For two connected graphs G_1, G_2 of the same number of vertices, we define the path distance d_{path} by*

$$d_{path}(G_1, G_2) = \min_{\sigma \in \Lambda} d_\sigma(G_1, G_2) \, ,$$

where Λ is the set of all bijections between $V(G_1)$ and $V(G_2)$.

Example. Let G_1 be the graph shown in Figure 12.9 and let G_2 be a cycle of 4 vertices. At first sight there are $4! = 10$ bijective mappings from $V(G_1)$ to $V(G_2)$. However, because of the highly symmetric structure of the graphs, there are only two inequivalent mappings with respect to path distance. These are depicted in Figure 12.9, where the mappings $\sigma_1, \sigma_2 : V(G_1) \to V(G_2)$ are defined by $\sigma_i j = j$ for $j = 1, \ldots, 4$. Now we determine for each pair of vertices the difference between distance in G_1 and distance of the corresponding images in G_2 and find that

$$d_{\sigma_1}(G_1, G_2) = 2 \quad \text{and} \quad d_{\sigma_2}(G_1, G_2) = 4 \, .$$

Thus $d_{\mathrm{path}}(G_1, G_2) = 2$.

Path Distance Is a Metric. $d_{\mathrm{path}}(G_1, G_2) = d_{\mathrm{path}}(G_2, G_1)$ follows directly from the definition. From Equation 12.4 we get immediately $d_{\mathrm{path}}(G_1, G_2) = 0$ for isomorphic graphs. On the other hand, $d_{\mathrm{path}}(G_1, G_2) = 0$ implies the existence of an isomorphism $\phi : V(G_1) \to V(G_2)$.

The triangle inequality remains to be verified. Let G_1, G_2, and G_3 be connected graphs with $|V(G_1)| = |V(G_2)| = |V(G_3)|$, and $\alpha : V(G_1) \to V(G_2)$ and $\beta : V(G_2) \to V(G_3)$ bijections with $d_\alpha(G_1, G_2) = d_{\mathrm{path}}(G_1, G_2)$ and $d_\beta(G_2, G_3) = d_{\mathrm{path}}(G_2, G_3)$ respectively. Then $\beta \circ \alpha : V(G_1) \to V(G_3)$ is also a bijection and

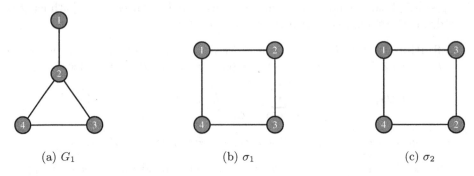

(a) G_1 (b) σ_1 (c) σ_2

Fig. 12.9. Two different mappings of G_1 to a cycle of 4 vertices

$$d_{\text{path}}(G_1, G_3) \leq d_{\beta \circ \alpha}(G_1, G_3)$$

$$= \sum_{\{u,v\} \in V(G_1) \times V(G_1)} |d_{G_1}(u,v) - d_{G_3}((\beta \circ \alpha)(u), (\beta \circ \alpha)(v))|$$

$$\leq \sum_{\{u,v\} \in V(G_1) \times V(G_1)} |d_{G_1}(u,v) - d_{G_2}(\alpha(u), \alpha(v))|$$

$$+ \sum_{\{u,v\} \in V(G_1) \times V(G_1)} |d_{G_2}(\alpha(u), \alpha(v)) - d_{G_3}((\beta \circ \alpha)(u), (\beta \circ \alpha)(v))|$$

$$= d_\alpha(G_1, G_2) + d_\beta(G_2, G_3)$$

$$= d_{\text{path}}(G_1, G_2) + d_{\text{path}}(G_2, G_3) .$$

Therefore, the triangle inequality holds and d_{path} is a graph similarity metric.

Computation of Path Distance. The computation of path distance of two graphs consists of three steps. First, we compute the distance of all pairs of vertices in both graphs. This is exactly the all-pairs shortest path problem (see Section 2.2.2). Then, we can compute the σ-distance for a given bijection σ in time $\mathcal{O}(n^2)$. Finally, we must identify the minimum bijection with respect to path distance.

12.2.3 Maximum Common Subgraphs

In this section we look at a similarity measure based on the size of a maximum common subgraph. The idea to use similar substructures of graphs for graph matching was introduced by Horaud and Skordas [315] and Levinson [391], and refined by Bunke and Shearer [106].

Recall the definition of induced subgraphs in Section 2.1. A graph $G' = (V', E')$ is a *subgraph* of the graph $G = (V, E)$ if $V' \subseteq V$ and $E' \subseteq E$. It is an *induced subgraph* if E' contains all edges $e \in E$ that join vertices in V'.

Definition 12.2.4. *Let G_1, G_2 be undirected graphs. An injective function ϕ : $V(G_1) \rightarrow V(G_2)$ is a* subgraph isomorphism *from G_1 to G_2 if there exists an induced subgraph $G'_2 \subseteq G_2$ such that ϕ is a graph isomorphism between G_1 and G'_2.*

Definition 12.2.5. *Let G_1, G_2 be undirected graphs. A graph S is a* common induced subgraph *of G_1 and G_2 if there exist subgraph isomorphisms from S to G_1 and G_2.*

Definition 12.2.6. *Let G_1, G_2 be undirected graphs. A common induced subgraph S of G_1 and G_2 is* maximum *if there exists no other common subgraph with more vertices than S. We denote such a maximum common induced subgraph (MCIS) by $mcis(G_1, G_2)$.*

A concept closely related to (vertex-)induced subgraphs are edge-induced subgraphs. A graph $G' = (V', E')$ is a *edge-induced subgraph* of the graph $G = (V, E)$ if $E' \subseteq E$ and V' contains only the incident vertices of edges in E'. Note that edge-induced subgraphs contain no isolated vertices. Figure 12.10 shows a comparison of vertex- and edge-induced subgraphs of a simple graph. The prior definitions for induced subgraphs are easily carried over to edge-induced subgraphs.

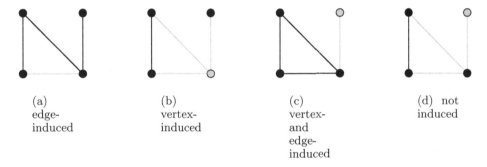

(a)
edge-
induced

(b)
vertex-
induced

(c)
vertex-
and
edge-
induced

(d) not
induced

Fig. 12.10. Comparison of vertex- and edge-induced subgraphs

Definition 12.2.7. *Let G_1, G_2 be undirected graphs. An injective function ϕ : $V(G_1) \rightarrow V(G_2)$ is an* edge subgraph isomorphism *from G_1 to G_2 if there exists an edge-induced subgraph $S \subseteq G_2$ such that ϕ is a graph isomorphism between G_1 and S.*

Definition 12.2.8. *Let G_1, G_2 be undirected graphs. A graph S is a* common edge subgraph *of G_1 and G_2 if there exist edge subgraph isomorphisms from S to G_1 and to G_2.*

Definition 12.2.9. *Let G_1, G_2 be undirected graphs. A common edge subgraph S of G_1 and G_2 is maximum if there exists no other common edge subgraph with more vertices than S. We denote such a maximum common edge subgraph (MCES) by $mces(G_1, G_2)$.*

Note that maximum common subgraphs are neither unique nor connected by definition. Note also that the MCIS or MCES of non-empty graphs consist at least of one vertex or one edge, respectively. Next, induced subgraphs are used to define distance measures for graphs.

Definition 12.2.10. *Let G_1, G_2 be undirected graphs, not both empty. We define the MCIS distance d_{mcis} by*

$$d_{mcis}(G_1, G_2) = 1 - \frac{|V(mcis(G_1, G_2))|}{\max(|V(G_1|, |V(G_2)|)} \tag{12.5}$$

and the MCES distance d_{mces} by

$$d_{mces}(G_1, G_2) = 1 - \frac{|V(mces(G_1, G_2))|}{\max(|V(G_1)|, |V(G_2)|)} . \tag{12.6}$$

MCIS and MCES Distance Are Metrics. Two properties of a graph similarity metric, reflexivity and symmetry, follow directly from the definition. The proof of the triangle inequality consists of a longish case differentiation, so we only give a sketch for MCIS. The complete proof for MCIS is given in [106].

Let G_1, G_2, and G_3 be undirected graphs. For notational convenience, let $n_i = V(G_i)$, $mcis(i, j) = |V(mcis(G_i, G_j))|$, and $\max(i, j) = \max(n_i, n_j)$ for $i, j \in \{1, 2, 3\}$. Using this notation, the triangle inequality is equivalent to

$$1 - \frac{mcis(1, 3)}{\max(1, 3)} \leq 1 - \frac{mcis(1, 2)}{\max(1, 2)} + 1 - \frac{mcis(2, 3)}{\max(2, 3)} .$$

Next, consider a maximum common subgraph of $mcis(G_1, G_2)$ and $mcis(G_2, G_3)$ and denote its number of vertices by $mcis(12, 23)$. Clearly

$$mcis(12, 23) \leq mcis(1, 3) ,$$
$$mcis(12, 23) \leq mcis(1, 2) ,$$
$$mcis(12, 23) \leq mcis(2, 3) ,$$

and

$$mcis(1, 2) + mcis(2, 3) - mcis(12, 23) \leq n_2 .$$

Now distinguish six cases by the possible orderings of n_1, n_2, and n_3 and get the result by combining the above inequalities.

Computation of MCIS and MCES. The detection of a maximum common subgraph is an $\mathcal{NP}$-complete problem [240]. Nevertheless a few exact algorithms have been proposed, based either on an exhaustive search for all subgraphs or on the relation of maximum common subgraph and maximum clique detection.

The first method was proposed by McGregor [414] and is very similar to the search-and-backtrack approach to graph isomorphism. The algorithm identifies common subgraphs by starting from single vertices in each graph and iteratively adding vertices (and incident edges) which do not violate the common subgraph condition. If it is impossible to add any new vertex, the size of the current subgraph is compared to the one previously found and a backtracking is done to test other branches of the search tree. Finally, a largest common subgraph is reported.

The second approach is based on the fact that a MCIS of two graphs corresponds to a maximum clique in their modular product graph. Recall a clique is a completely connected subgraph. A *maximum clique (MC)* is a clique with the largest number of vertices. Note that a MC is not necessarily unique. The *modular product graph* $G_1 \diamond G_2$ of G_1 and G_2 is defined on the vertex set

$$V(G_1 \diamond G_2) = V(G_1) \times V(G_2)$$

and two vertices $(u_i, v_i), (u_j, v_j) \in G_1 \diamond G_2$ being adjacent if either

$$(u_i, u_j) \in E(G_1) \text{ and } (v_i, v_j) \in E(G_2)$$

or

$$(u_i, u_j) \notin E(G_1) \text{ and } (v_i, v_j) \notin E(G_2) .$$

Accordingly, a MCES of two graphs corresponds to a maximum clique in the modular product graph of their line graphs [450].

Exact algorithms for clique detection are based on exhaustive search strategies [240]. This approach is similar to algorithms for MCIS, but takes advantage of a number of upper and lower bounds to prune the search space (e.g., see [466]). Also, many approximation algorithms have been proposed, see [70] for an extensive survey.

12.2.4 Other Methods

RASCAL. This is not a single method but a combination of a fast initial screening process followed by a rigorous MCES detection algorithm [488]. In the initial screening the degree sequence and vertex and edge labels are considered for computing a first approximation of the similarity. Only if it is above a certain threshold is the costly MCES detection executed. The idea is that one does not care about quite different graphs but only about very similar ones. Other benefits of this paper, beside the two phase approach, are a detailed description of the MCES computation including some minor improvements and a good readability.

Motifs. In Section 11.6 the concept of motifs is introduced. Motifs are small connected subgraphs in a graph G that occur in G significantly more often than in a random graph of the same size and degree distribution. Characteristics and quantity of motifs in graphs can be used as indicators for their similarity.

12.3 Chapter Notes

Besides the detection of explicit algorithms (by finding equal adjacency matrices), an automorphism can sometimes be inferred by a special structure of a vertex partition in $\mathcal{T}$. However, this occurs rarely, for details see [415, Lemma 2.25].

McKay uses another trick in order to prune the search tree $\mathcal{T}$: let Λ be a function defined on the set of all vertex partitions. The goal is now to define an indicator function Λ^* on the nodes of $\mathcal{T}$. In the nauty algorithm a node $\Pi_m \in \mathcal{T}$ actually stores all vertex partitions of its ancestors, i.e. the list of refined partitions $f((V)) = \Pi_1, \ldots, \Pi_m$ that were derived in order to get to Π_m. Identify the node from now on with $[\Pi_1, \ldots, \Pi_m]$. The function Λ^* is defined by $\Lambda^*([\Pi_1, \ldots, \Pi_m]) = (\Lambda(\Pi_1), \ldots, \Lambda(\Pi_m))$. McKay's algorithm actually searches the minimum adjacency matrix among the leaves that maximize Λ^*. The algorithm can then prune subtrees as soon at it is clear that all their leaves have a Λ^*–value below the current maximum. This is due to the lexicographical order of Λ^*. The benefit of this method depends eminently on the quality of Λ. For example, if Λ is the identity, Λ has no effect. McKay uses information from the computation of $f(\overline{\Pi}) = \Pi$ to define $\Lambda(\Pi)$, see again [415].

13   Network Models

Nadine Baumann and Sebastian Stiller

The starting point in network analysis is not primarily the mathematically defined object of a graph, but rather almost everything that in ordinary language is called 'network'. These networks that occur in biology, computer science, economy, physics, or in ordinary life belong to what is often called 'the real world'. To find suitable models for the real world is the primary goal here. The analyzed real-world networks mostly fall into three categories.

The biggest fraction of research work is devoted to the Internet, the WWW, and related networks. The HTML-pages and their links (WWW), the newsgroups and messages posted to two or more of them (USENET), the routers and their physical connections, the autonomous systems, and several more are examples from this scope of interest.

In biology, in particular chemical biology, including genetics, researchers encounter numerous structures that can be interpreted as networks. Some of these show their net structure directly, at least under a microscope. But some of the most notorious of the biological networks, namely the metabolic networks, are formed a little more subtly. Here the vertices model certain molecules, and edges represent chemical reactions between these molecules in the metabolism of a certain organism. In the simplest case, two vertices are connected if there is a reaction between those molecules.

Sociological networks often appear without scientific help. We think of cronyism and other (usually malign) networks in politics and economy, we enjoy to be part of a circle of friends, we get lost in the net of administration, and networking has become a publicly acknowledged sport. The trouble – not only but also – for scientists is to get the exact data. How can we collect the data of a simple acquaintance network for a whole country, or even a bigger city? But for some networks the data is available in electronic form. For example, the collaboration of actors in movies, and the co-authorship and the citation in some research communities, partly owe their scientific attraction to the availability of the data.

Many – but not all – of these examples from different areas have some characteristics in common. For example metabolics, the WWW, and co-authorship often form networks that have very few vertices with very high degree, some of considerable degree and a huge number of vertices with very low degree. Unfortunately, the data is sometimes forced to fit into that shape, or even mischievously interpreted to show a so called power law. Often deeper results are

U. Brandes and T. Erlebach (Eds.): Network Analysis, LNCS 3418, pp. 341–372, 2005.
© Springer-Verlag Berlin Heidelberg 2005

not only presented without proof, but also only based on so called experimental observations.

Yet one feature can be regarded as prevalent without any alchemy: Most of the real-world networks are intrinsically historical. They did not come into being as a complete and fixed structure at one single moment in time, but they have developed step by step. They emerged. Therefore, on the one hand, it makes sense to understand the current structure as the result of a process. On the other hand, one is often more interested in the network's future than in one of its single states. Therefore several models have been developed that define a graph, or a family of graphs, via a process in the course of which they emerge.

The mathematical models for evolving networks are developed for three main intentions. First of all, the model should meet or advocate a certain intuition about the nature of the development of the real-world network. Secondly, the model should be mathematically analyzable. A third objective is to find a model that is well suited for computational purpose, i.e., to simulate the future development or generate synthetic instances resembling the real network, for example to test an algorithm.

There are several overviews in particular on models for Internet and WWW networks (see [164, 68] for a more mathematically inclined overview). Some of these papers already exceed this chapter in length. It hardly pays and it is virtually impossible to mention all models, experimental conjectures, and results. We rather intend to endow the reader with the necessary knowledge to spark her own research. We proceed in four sections.

In the first section the founding questions, driving ideas, and predominant models are summarized. Then, in the second section, we compile some methods that are deemed to or have proven to be fruitful in analyzing the structure of a network as a whole. Third, we broaden our scope in order to exemplify the great variety of models for evolving networks. The last section is devoted to the state of the art generators for networks that resemble the Internet.

Up to further notice we consider graphs as directed graphs. Some graph processes may generate multigraphs which should be clear from the context.

13.1 Fundamental Models

13.1.1 The Graph Model ($G_{n,p}$)

First we want to discuss the historical starting point of random graph theory. More precisely, we define the graph model ($G_{n,p}$).

A graph model is a set of graphs endowed with a probability distribution. In this case the graphs under consideration are undirected. The following three graph models stochastically converge to each other as $n \to \infty$:

1. The first way to generate a random graph is to choose a graph uniformly at random among all graphs of given vertex number n and average vertex degree z.

2. Alternatively, choose every edge in a complete graph of n vertices with probability p to be part of $E(G)$, where $\frac{2p\binom{n}{2}}{n} = p(n-1) =: z$ is the expected average degree. This model is denoted by $(G_{n,p})$.

3. In the third method, n vertices v_i are added successively, deciding for each v_i and for each $j < i$ whether to put $\{v_i, v_j\}$ in the edge set or not with probability p.

The last one is an interpretation of the second as a graph process. See Section 13.1.4 for more details about graph processes. The first is of course more restrictive, because the average degree is fixed and not just expected, as in the two other models. Still these models converge. Thereby the first model may be more intuitive, but the second is often more suitable for analysis. These three aspects are also important for the other models we will discuss in this section: Some models capture best our intuition about the real world, others are superior in mathematical tractability. Third, networks in the real world very often are structures which rather emerged from a process than popped up as a whole.

There is a myriad of literature and highly developed theory on the $(G_{n,p})$ and related models. It turns out that a graph chosen according to that distribution, a graph 'generated' by that model, shows a number of interesting characteristics with high probability. On the other hand, this graph model has, precisely because of these characteristics, often been disqualified as a model for real-world networks that usually do not show these characteristics. For example, without deep mathematical consideration one can see that the majority of the vertices will have almost or exactly the average degree. For many networks in the real world this is not the case. Still our interest in this model is more than historical.

We should state at least one fundamental and very illuminating result on $(G_{n,p})$-graphs. Let $G_{n,p}$ denote a fixed graph generated by one of these models.

Theorem 13.1.1. *Let m_ω be the expected number of arcs in a $G_{n,p}$, i.e., $m_\omega = p\binom{n}{2}$. If $m_\omega = \frac{n}{2}(\log n + \omega(n))$, then for $\omega \to -\infty$ $G_{n,p}$ is disconnected with high probability, and for $\omega \to \infty$ $G_{n,p}$ is connected with high probability.*

This chapter will extensively treat degree sequences. Therefore we state the following immediate fact about the probability distribution p of the degree k of a vertex in a $(G_{n,p})$-graph. We use z or z_1 to denote the average degree of a vertex in the graph under consideration.

$$p(k) = \binom{n-1}{k} p^k (1-p)^{n-1-k} \approx \frac{z^k \exp(-z)}{k!}$$

After this classical mathematical model let us turn to a topic strongly inspired by the real-world, the concept of a Small World.

13.1.2   Small World

One of the starting points of network analysis is a sociological experiment conducted to verify the urban legend that anyone indirectly knows each other by just a few other mediators. To scrutinize this assumption Milgram [421] asked several people in the US to deliver a message just by passing it on to people they knew personally. The senders and mediators knew nothing about the recipient but his name, profession, and the town he lived in. These messages reached their destination on average after roughly five or six mediators, justifying the popular claim of six degrees of vicinity. The world, at least in the US, appears to be small.

The notion of 'Small World' has become technical since, usually encompassing two characteristics: First, the average shortest path distances over all vertices in a small world network has to be small. 'Small' is conceptualized as growing at most logarithmically with the number of vertices. In this sense $(G_{n,p})$ graphs (see Section 13.1.1) are small even for small values of p, and the sociological observation would come as no surprise. But in a vicinity-network – like the one the sociological experiment was conducted on – a huge fraction of people one knows personally, also know each other personally. Mathematically speaking a network shows the worldly aspect of a small world if it has a high clustering coefficient. Whereas in an $(G_{n,p})$ graph the clustering coefficient obviously tends to zero. (The clustering coefficient gives the fraction of pairs of neighbors of a vertex that are adjacent, averaged over all vertices of the graph. For a precise definition of the clustering coefficient and a related graph statistic, called transitivity, see 11.5.)

A very popular abstract model of small world networks, i.e., a graph with clustering coefficient bounded from below by a constant and logarithmically growing average path distance, is obtained by a simple rewiring procedure. Start with the kth power of an n-cycle, denoted by C_n^k. The kth power of a cycle is a graph where each vertex is not only adjacent to its direct neighbors but also to its k neighbors to the right and k neighbors to the left. Decide for each edge independently by a given probability p whether to keep it in place or to rewire it, i.e., to replace the edge $\{a, b\}$ by an edge $\{a, c\}$ where c is chosen uniformly at random from the vertex set.

The description contains a harmless ambiguity. Viewing the rewiring process as iteratively passing through all vertices, one may choose an edge to be rewired from both of its vertices. It is not a priori clear how to handle these ties. The natural way to straighten this out is the following: Visit each vertex iteratively in some order, and make the rewiring decisions for each of the *currently* incident edges. Therefore, strictly speaking, the model depends on the order in which the vertex set is traversed. Anyway, the reader should be confident that this does not affect the outcome we are interested in, namely the average shortest path distance and the clustering coefficient C. For small p the clustering coefficient stays virtually that of C_n^k. To be more precise, for small k and p and large n: $C(G_{rewired}) = C(C_n^k)(1 - \frac{p}{2k})$, as the pth fraction of an average of the $2k$ neighbors' contribution is removed from the numerator. On the other hand, the average path distance in such a graph decreases quickly (as p increases) from the original $\frac{n}{4k}$ (on average one has to walk a quarter of the circle by steps of length

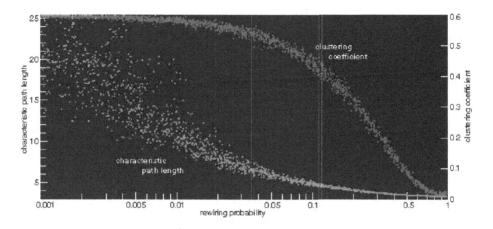

Fig. 13.1. Clustering coefficient and path lengths for the small world model by Watts-Strogats. Found at: http://backspaces.net/PLaw/. Results are from 2,000 random graphs, each with 300 vertices and 900 edges

k, except for maybe the last) to small values, claimed [573] to be in $\mathcal{O}(\log n)$ (compare Figure 13.1).

Unfortunately, these figures were obtained and verified empirically only. The chart suggests that calculation of the second moment of the distributions would be desirable, as the lower cloud of points, i.e., the average shortest path distances, appear far less stable. Maybe the most important problem with such experimental figures is that they can hardly account for the difference between, for example, a logarithmic or a $\sqrt{n}$ behavior.

A weakness of the rewiring model, and thus of the whole definition of small world graphs, is that, by fixing the number of random edges and enlarging k, the clustering coefficient can be kept artificially high, whereas the path distances on average only depend on the number of random edges relative to n. An increase in small deterministic edges does not contribute to the average path distance, except for a constant: On average the number of steps to go from one long-range edge to the other becomes smaller only by a constant factor. Sociologically speaking, having many friends in the neighborhood brings you only a constant closer to the Dalai Lama.

13.1.3 Local Search

Revisiting the sociological experiment, one may not be satisfied that the theoretical explanation only accounts for the existence of short average shortest paths. The fact that the letters reached their destination within a few steps requires short paths not only to exist, but also to be detectable for the ignorant agents in the vicinity network. This led Kleinberg to the idea of a *local algorithm*. Roughly speaking, a local algorithm should act – for example crawl a network – step by step without knowing the whole structure. In each step only a specific, local part of the whole data should be used to reach the current decision. A definition of local algorithm for the specific problem will be given in a moment.

The real-world vicinity network is idealized by a parameterized network model that is easily found to have a high and constant clustering coefficient. It is once again a network comprised of short deterministic and long random edges, modeling the intuition that we know our neighborhood and have some occasional acquaintances. The aim is to determine the parameters under which there exists a local algorithm capable of finding a path with on average logarithmic length for a randomly chosen pair of vertices.

Model for Local Search. The network $G(V, E)$ is parameterized by n, p, q and r. The vertex set V contains the points of a 2-dimensional $n \times n$ lattice. On the one hand, E contains bi-directed arcs between each vertex and its $2p$ closest horizontal and $2p$ closest vertical neighbors. On the other hand, for each vertex, v, there are q directed arcs of the form $(v, x) \in E$, where x is chosen out of $V \setminus \{v\}$ according to the distribution $p(x) = \frac{d^{-r}(v,x)}{\sum_y d^{-r}(v,y)}$, where $d(x, y)$ denotes the minimum number of steps to go from x to y on the grid and $r > 0$ is a constant. We call such a network $G_K(n, p, q, r)$ a Kleinberg-Grid. (Note that for $p = 1$ the clustering coefficient is 0, but for $p > 1$ it is greater than 0 and essentially independent of n.)

Local Algorithm. The following notion of a local algorithm is not very general, but rather tailor-made for the above model. A local algorithm provides a rule giving the subsequent vertex at each vertex of the path to be output in the end, based only on the following types of information:

- Global Knowledge
  - The structure of the underlying grid.
  - The position of the destination vertex in the underlying grid.
- Local Knowledge
  - The positions of the current vertex in the underlying grid and of its neighbors in the whole network (i.e., including its long-range connections).
  - The positions of all vertices visited so far, including their neighbors positions.

Results. The local algorithm Kleinberg analyses is the most natural one – which gives even more explanatory power for the sociological experiment: Every recipient of the message passes it on to that vertex among its neighbors that is closest to the destination in $d(\cdot, \cdot)$. Call this the Kleinberg-Algorithm.

Theorem 13.1.2. *Let* $p, q \in \mathbb{N}$ *be fixed. Then the following holds for every Kleinberg-Grid* $G_K(n, p, q, r)$:

For $r = 0$

every local algorithm finds paths of average length   *in* $\Omega(n^{\frac{2}{3}})$.

For $0 < r < 2$

every local algorithm finds paths of average length   *in* $\Omega(n^{\frac{2-r}{3}})$.

For $r = 2$

the Kleinberg-Algorithm finds paths of average length   *in* $\mathcal{O}(\log^2 n)$.

For $r > 2$

every local algorithm finds paths of average length   *in* $\Omega(n^{\frac{r-2}{r-1}})$.

Sketch of Proof. Though the negative results of Theorem 13.1.2 (that no local algorithm can find a path of the desired length) are the intriguing ones, it is the proof of the positive result that will give us the required insight, and will be sketched here.

In order to estimate the average number of steps which the Kleinberg-Algorithm takes (for $r = 2$ and x the destination vertex) subdivide the vertex space in subsets U_k of vertices v with $2^{k-1} \le d(x, v) < 2^k$. The algorithm always proceeds to a vertex that is closer to x than the current vertex. Thus, if it once reaches a subset U_i it will only advance to U_j where $j \le i$. As the total number of subsets grows logarithmically with n, we are done if we can show that the algorithm needs at most a constant number of steps to leave a subset U_k, independent of k.

As the subset U_k can be very big, we interpret leaving a subset as finding a vertex that has a random edge into $\bigcup_{i<k} U_i$. As the algorithm visits every vertex at most once, we can apply the technique of postponed decisions, i.e., choose the random edge of a vertex v when we reach v. In order to have a constant probability at every level k, the probability for v to have a random contact at distance less than or equal to 2^{k-1} from v, must be constant for all k. This is true for a 2-dimensional lattice if and only if $r = 2$. $\square$

At this point the result of Theorem 13.1.2 seems generalizable to other dimensions, where r should always equal the dimension. This can easily be seen for dimension 1. The details of the proof and the negative results may be more difficult to generalize.

The above proof already gives a hint why the negative results hold for dimension 2. If $r > 2$ the random arcs are on average too short to reach the next section in a constant time, when the algorithm is in a big and far away subset. On the other hand, $r < 2$ distributes too much of the probabilistic mass on long reaching arcs. The algorithm will encounter lots of arcs that bring it far beyond the target, but too rarely one that takes it to a subset closer to the target.

In general, the distribution must pay sufficient respect to the underlying grid structure to allow for a local algorithm to make use of the random arcs, but still need to be 'far-reaching' enough. It seems worthwhile to conduct such an analysis on other, more life-like, models for the vicinity network.

13.1.4   Power Law Models

As already described in Section 11.1 there is a wide interest in finding graphs where the fraction of vertices of a specified degree k follows a power law. That means that the degree distribution p is of the form

$$p(k) = ck^{-\delta} \qquad \delta > 0, c > 0.$$

This mirrors a distribution where most of the vertices have a small degree, some vertices have a medium degree, and only very few vertices have very high degree.

Power laws have not only been observed for degree distributions but also for other graph properties. The following dependencies (according to [197]) can especially be found in the Internet topology:

1. Degree of vertex as a function of the rank, i.e., the position of the vertex in a sorted list of vertex degrees in decreasing order
2. Number of vertex pairs within a neighborhood as a function of the neighborhood size (in hops)
3. Eigenvalues of the adjacency matrix as a function of the rank

A more detailed view to these power laws found in Internet topologies is given in Section 13.4. Since in the literature the most interesting fact seems to be the degree distribution, or equivalently the number of vertices that have a certain degree k, we will focus mostly on this.

In some contexts (protein networks, e-mail networks, etc.) we can observe an additional factor q^k to the power law with $0 < q < 1$ – the so called exponential cutoff (for details see [448]). Trying to fit a degree distribution to this special form, the power law $p(k) = ck^{-\delta}q^k$ obtains a lower exponent δ than would be attained otherwise. A power law with an exponential cutoff allows to normalize the distribution even in the case that the exponent δ lies in $(0, 2]$.

Since the 'strict' power law, i.e., in the form without cutoff, is more fundamental and more explicit in a mathematical way, we will in the following restrict ourselves to some models that construct networks with power laws not considering exponential cutoff. We start by describing the most well-known preferential attachment model and then give some modifications of this and other models.

Preferential Attachment Graphs. In many real life networks we can observe two important facts: *growth* and *preferential attachment*. Growth happens because networks like the WWW, friendships, etc. grow with time. Every day more web sites go online, and someone finds new friends.

An often made observation in nature is that some already highly connected vertices are likely to become even more connected than vertices with small degree. It is more likely that a new website also inserts a link to a well-known website like *google* than to some private homepage. One could argue that someone who already has a lot of friends easily gets more new friends than someone with only a few friends – the so called 'the rich get richer'-phenomenon. This is modeled by a preferential attachment rule.

One of the first models to tackle these two special characteristics is the preferential attachment model presented by Barabási and Albert in [40].

Graph Process. Formally speaking a graph process $(\mathcal{G}^t)$ is a sequence of sets $\mathcal{G}^t$ of graphs (called states of the process $(\mathcal{G}^t)$) each endowed with a probability distribution. Thereby the sets and their distributions are defined recursively by some rule of evolution. More intuitively one thinks of a graph process as the different ways in which a graph can develop over the time states.

In [40] a graph process (G_m^t) is described in this intuitive way as the history of a graph $G = (V, E)$. At every point in time one vertex v with outdegree m is added to the graph G. Each of its outgoing edges connects to some vertex $i \in V$ chosen by a probability distribution proportional to the current degree or indegree of i.

Formally, this description gives a rule how any graph of a certain state of the process is transformed into a graph of the next state. Further, this rule of evolution prescribes for any graph of a state of the graph process the probabilities with which it transforms into a certain graph of the next state. In this way, the sets and distributions of the graph process are recursively defined. Unfortunately, the above description from [40] entails some significant imprecisions which we will discuss now.

The choice of the initial state (which usually contains exactly one graph) is a nonnegligible matter. For example, taking $m = 1$, if the graph is disconnected at the beginning of the sequence then any emerging graph is also disconnected. In contrast, any connected graph stays connected. Moreover, we need at least one vertex to which the m new edges can connect. But it is not defined how to connect to a vertex without any edge, since its probability is zero. Thus, there must be at least one loop at that vertex, or some other rule how to connect to this vertex.

Secondly, one has to spell out that the distribution shall be proportional to the degree. In particular, it has to be clear whether and how the new vertex is already part of the vertex set V. If it is excluded no loops can occur. (Note that for $m = 1$ loops are the only elementary cycles possible.) If it is an element of V it is usually counted as if it had degree m, though its edges are not yet connected to their second vertices. Moreover, if $m > 1$ one has to define a probability distribution on the set of all $\binom{|V|}{m}$ or $\binom{|V|+1}{m}$ possible ways to connect which is not sufficiently defined by requiring proportionality to the degree for each single vertex.

Note that the process (G_1^t) is equivalent to the process (G_m^t) for large t in the following sense: Starting with the process (G_1^{tm}) and always contracting the last m vertices after m states we get the same result as for the process (G_m^t).

With the graph process (G_1^t), the probability that an arbitrary vertex has degree k is $\Pr[k] = k^{-\delta}$, with $\delta = 3$. There are several possibilities to prove this degree distribution. Some of them, and a precise version of the model, are presented in Section 13.2.1.

Other Power-Law Models. There are more models that try to construct a graph that resembles some real process and for which the degree distribution follows a power law. One of them is the *model of initial attractiveness* by Buckley and Osthus (see [68] for details and references). Here the vertices are given a value $a \geq 1$ that describes their initial attractiveness. For example a search engine is already from the start more attractive to be linked to than a specialized webpage for scientists. So the probability that a new vertex is linked to vertex i is proportional to its indegree plus a constant initial attractiveness am.

A different approach to imitate the growing of the world wide web is the *copying model* introduced by Kleinberg and others [375] where we are given a fixed outdegree and no attractiveness. We choose a prototype vertex $v \in V$ uniformly at random out of the vertex set V of the current graph. Let v' be a new vertex inserted into the network. For all edges (v, w) for $w \in V$ edges (v', w) are inserted into the network. In a next step each edge (v', w) is retained unchanged with probability p, or becomes rewired with probability $1 - p$. This simulates the process of copying a web page almost identical to the one the user is geared to and modifying it by rewiring some links. The authors also showed that this model obtains a degree distribution following a power law. One advantage of this model is that it constructs a lot of induced bipartite subgraphs that are often observed in real web graphs (for some further details see Section 3.9.3 about *Hubs & Authorities*). But it does not account for the high clustering coefficient that is also characteristic of the webgraph.

Another model that tries to combine most of the observations made in nature, and therefore does not restrict to only one way of choosing the possibilities for a connection, is the model of Cooper and Frieze [131]. Here we first have the choice between a method NEW and a method OLD, which we choose by a probability distribution α. Method OLD inserts a number of edges starting at a vertex already in the network whereas method NEW inserts first a new vertex and then connects a number of edges to this new vertex. The number of inserted edges is chosen according to probability distribution β. The end vertices to which to connect the edges are chosen either uniformly at random, or depending on the current vertex degree, or by a mixture of both.

13.2   Global Structure Analysis

13.2.1   Finding Power Laws of the Degree Distribution

We would like to have some general approaches to find the exact degree distribution of a given graph model. Since there are no known general methods we will present four different ways of showing the degree distribution of the preferential attachment model. One will be a static representation of one state of the graph process called Linearized Chord Diagrams, introduced by Bollobás [68]. Furthermore we will give three heuristic approaches that yield the same result.

Linearized Chord Diagrams. A Linearized Chord Diagram (LCD) consists of $2n$ distinct points on the x-axis paired off by chords in the upper half-plane.

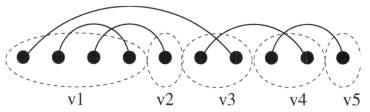

Fig. 13.2. An LCD representing a graph

The goal is now to construct a graph out of this LCD that represents a static state of a graph process.

Reconsider the preferential attachment model by Barabási and Albert (Section 13.1.4). There a graph process (G_m^t) is used. Let us consider the case $m = 1$. Let $\Pr[v]$ be the probability that the vertex v_t inserted at time t is connected to vertex v. We define

$$\Pr[v] = \begin{cases} 1/(2t-1) & \text{if } v = v_t, \\ k_v/(2t-1) & \text{otherwise} \end{cases} \tag{13.1}$$

where k_v denotes the current degree of vertex v before the connection. The normalizing term is $(2t-1)$ because the new edge is understood to be incident to v_t only, until its second endpoint is chosen.

The LCD Model. To construct a Linearized Chord Diagram as defined in the beginning of this section we can use *n-pairings*. An n-pairing L is a partition of the set $S = \{1, 2, \ldots, 2n\}$ into pairs. So there are $\frac{(2n)!}{n!2^n}$ n-pairings. Figure the elements of S in their natural order on the x-axis and represent each pair by connecting its two elements by a chord (compare Figure 13.2). On such a Linearized Chord Diagram the construction of the graph for the pairing L becomes understandable. Construct the graph $\Phi(L)$ by the following rules: starting from the left of the x-axis we identify all endpoints up to and including the first right endpoint of a chord to form vertex v_1. Then we identify all further endpoints until the second right endpoint as vertex v_2 and so on. To form the edges we replace all chords by an edge connecting the vertices associated with the endpoints. Figure 13.2 gives an example of such a Linearized Chord Diagram and the associated graph.

The same can be achieved by choosing $2n$ points at random in the $[0,1]$ interval and associating the points $2i - 1$ and $2i$, $i \in \{1, 2, \ldots, n\}$ as a chord.

LCD's as Static Representation of (G_m^n). For a special point in time $t = n$ we can construct a Linearized Chord Diagram with n chords and build the graph $\Phi(L)$. The obtained graph model is exactly the nth state of the graph process (G_1^t), i.e., G_1^n. To see this observe how the evolution rule of (G_1^t) can be imitated for LCD's. Add one pair to an arbitrary LCD by placing the right point of the pair at the end of the point set and inserting the left point of the pair uniformly at random before any of the $2n + 1$ points. Then the new edge is connected by the same distribution as in the (G_1^t) process.

It can easily be shown that the degree distribution of this 'static' graph follows a power law with exponent $\gamma = -3$ (for details see [69]).

Now we will give three heuristic approaches that work with the preferential attachment model.

Continuum Theory. Let k_i again denote the degree of vertex i. The value k_i increases if a new vertex v enters the network and connects an edge to vertex i. The probability that this happens will be $\frac{k_i}{\sum_{j \in V \setminus \{v\}} k_j}$. Note that this does not yet determine the full probability distribution, but it is sufficient for our argument. In addition we have to specify a start sequence. We want to start with a graph of $m_0 (\geq m)$ vertices and zero edges. As in this case the probability distribution is not defined, we stipulate it to be the uniform distribution for the first step. Obviously after the first step we have a star plus vertices of degree zero which are irrelevant for the further process. Unfortunately, the exact shape of the initial sequence is not given in [15].

We now want to consider k_i as a continuous real variable. Therefore the rate with which k_i changes is proportional to the probability that an edge connects to i. So we can state the following dynamic equation:

$$\frac{\partial k_i}{\partial t} = m \frac{k_i}{\sum_{jj \in V \setminus \{v\}} k_j} \tag{13.2}$$

So we get for the total number of degrees in the network, except for that of the new vertex, $\sum_{j=1}^{N-1} k_j = 2mt - m$.

Thus the above equation changes to $\frac{\partial k_i}{\partial t} = \frac{k_i}{2t-1}$ and, since we consider very large times t, we can approximate it as

$$\frac{\partial k_i}{\partial t} = \frac{k_i}{2t}. \tag{13.3}$$

By construction of the preferential attachment model we know that the initial condition $k_i(t_i) = m$ holds where t_i is the time when vertex i was inserted into the network. Using this initial condition we obtain as a solution of the differential equation (13.3) the following result:

$$k_i(t) = m \left(\frac{t}{t_i} \right)^\beta, \qquad \beta = \frac{1}{2}. \tag{13.4}$$

Our goal is now to determine $p(k)$, the probability that an arbitrary vertex has degree exactly k. Since $p(k) = \frac{\partial \Pr[k_i(t) < k]}{\partial k}$ we firstly have to determine the probability that the degree of vertex i at time t is strictly smaller than k.

By using the solution of the differential equation given above, the following equations arise:

$$\Pr[k_i(t) < k] = \Pr[m\left(\frac{t}{t_i}\right)^\beta < k]$$

$$= \Pr\left[t_i > \frac{m^{\frac{1}{\beta}}t}{k^{\frac{1}{\beta}}}\right]$$

$$= 1 - \Pr\left[t_i \le \frac{m^{\frac{1}{\beta}}t}{k^{\frac{1}{\beta}}}\right]$$

$$= 1 - \int_0^{m^{\frac{1}{\beta}}tk^{-\frac{1}{\beta}}} \Pr[t_i = t]\, dt$$

$$= 1 - \frac{m^{\frac{1}{\beta}}t}{k^{\frac{1}{\beta}}(t + m_0)}$$

The last equation follows from the fact that the probability space over t_i has to sum up to one and the probabilities are assumed to be constant and uniformly distributed, thus $1 = \sum_{i=1}^t \Pr[t_i] \implies \Pr[t_i] = \frac{1}{m_0+t}$.

Differentiating the above equations with respect to k we obtain for $p(k)$:

$$p(k) = \frac{\partial \Pr[k_i(t) < k]}{\partial k} = \frac{2m^{\frac{1}{\beta}}t}{m_0 + t} \cdot \frac{1}{k^{\frac{1}{\beta}+1}}. \tag{13.5}$$

For $t \to \infty$ asymptotically we get $p(k) \sim 2m^{\frac{1}{\beta}}k^{-\gamma}$ with $\gamma = \frac{1}{\beta}+1 = 3$. Note that the exponent is independent of m. So we get a degree distribution that follows a power law where the coefficient is proportional to m^2.

Master Equation Approach. With the master equation approach we want to use recursion to find the shape of the degree distribution. So we are looking for equations that use information from the time steps before, in form of the degree distribution of older time steps. Since we know the initial distribution it is easy to solve this recursion.

This approach to determining the power law of the preferential attachment model was introduced by Dorogovtsev, Mendes, and Samukhin [166].

We study the probability $p(k, t_i, t)$ that a vertex i that entered the system at time t_i has degree k at time t. During the graph process the degree of a vertex i increases by one with probability $\frac{k}{2t}$.

For simplicity of formulas we use the dot notation for the derivative with respect to t.

A master equation for this probability $p(k, t_i, t)$ is of the form:

$$\dot{p}(k, t_i, t) = \sum_{k'}[W_{k' \to k}\, p(k', t_i, t) - W_{k \to k'}\, p(k, t_i, t)] \tag{13.6}$$

Here $W_{k'\to k}$ denotes the probability of changing from state k' to state k. In our model this probability is obviously

$$W_{k'\to k} = \frac{k'}{2t}\delta_{k',k-1} \quad , \text{ where } \quad \delta_{i,j} = \begin{cases} 1 & i=j \\ 0 & \text{otherwise} \end{cases} \tag{13.7}$$

is the Kronecker symbol.

By summing up over all vertices inserted up to time t, we define the probability $P(k,t) := \frac{\sum_{t_i}^t p(k,t_i,t)}{t}$ that some arbitrary vertex has degree k.

As we are interested in a stationary distribution, we are looking for the point where the derivative with respect to time is zero.

$$0 = \dot{P}(k,t) = \frac{t\sum_{t_i}\dot{p}(k,t_i,t) - \sum_{t_i}p(k,t_i,t)}{t^2}$$

$$= \left(\frac{1}{t}\sum_{t_i}\dot{p}(k,t_i,t)\right) - \frac{1}{t}P(k,t)$$

$$= \left(\sum_{k'}\frac{1}{t}[W_{k'\to k}\,p(k',t_i,t) - W_{k\to k'}\,p(k,t_i,t)]\right) - \frac{1}{t}P(k,t)$$

$$= \left(\sum_{k'}[W_{k'\to k}\,P(k',t) - W_{k\to k'}\,P(k,t)]\right) - \frac{1}{t}P(k,t)$$

$$= \left(\sum_{k'}\left[\frac{k'}{2t}\delta_{k',k-1}P(k',t) - \frac{k}{2t}\delta_{k,k'-1}P(k,t)\right]\right) - \frac{2}{2t}P(k,t)$$

$$= \frac{k-1}{2t}P(k-1,t) - \frac{k+2}{2t}P(k,t)$$

There is now a t' so that for every time t greater than t' we get the stationary distribution, $\tilde{P}(k)$. This results in the recursive equation $\tilde{P}(k) = \frac{k-1}{k+2}\tilde{P}(k-1)$ for $k \geq m+1$. For the case $k = m$ the probability directly results from the scaling condition of the probability measure: $\tilde{P}(m) = \frac{2}{m+2}$.

This directly yields the power law of the form $\Pr[k] = \frac{2m(m+1)}{k(k+1)(k+2)}$ which converges to the value of the power law found using the continuum theory, $2m^2\gamma^{-3}$.

This approach can also be used to determine the degree distribution of a more general case of preferential linking. In this model, one new vertex is inserted at every point in time. At the same time we insert m edges with one endpoint at unspecified vertices or from the outside. This can be done since here we only take into consideration the indegree of a vertex. The other endpoints are distributed to existing vertices proportional to $q(s) + A$ where $q(s)$ is the indegree of vertex s, and A an additional attractiveness associated with all vertices.

Rate Equation Approach. In this approach we want to analyze the change over time of the numbers of vertices with degree k – we are looking for the rate at which this number changes.

This approach for the preferential attachment model is due to Krapivsky, Redner, and Leyvraz [369].

We are considering the average number (over all graphs of the state of the process) $N_k(t)$ of vertices that have exactly degree k at time t. Asymptotically we have, by the strong law of large numbers, the following for large t: $N_k(t)/t \sim$ $\Pr[k]$ and $\sum_k k N_k(t)/t \sim 2m$.

If a new vertex enters the network, $N_k(t)$ changes as follows:

$$\Pr[k] = \frac{\partial N_k}{\partial t} = m \frac{(k-1)N_{k-1}(t) - kN_k(t)}{\sum_k kN_k(t)} + \delta_{k,m}. \qquad (13.8)$$

Here the first term of the numerator denotes the total number of edges leaving vertices with degree exactly $k-1$ where new edges connect to those vertices and therefore increase the degree to k. The second term determines the number of edges leaving vertices with degree exactly k where new edges connect to those vertices and therefore increase the degree to a value higher than k. If the newly entered vertex has exactly degree k, i.e., $m = k$, then we have to add a 1 to our rate equation.

Applying the above limits we obtain exactly the same recursive equation as found with the master equation approach, and therefore we have the same power law.

Flexibility of the Approaches. All the approaches mentioned before are very helpful and easy to understand for the case of analyzing the preferential attachment model. Some of them are also applicable for more general versions of the preferential attachment model, as for $m \neq 1$ and others. But it is not clear whether there is a useful application of these approaches to totally different models. For the rate equation approach an adaption to more general evolving networks, as well as for networks with site deletion and link-arrangement, is possible. There is a huge need for approaches that can deal with other models. It would be even more desirable to find a way to treat numerous types of evolving network models with a single approach.

13.2.2 Generating Functions

The power law exemplifies an often faced problem in network-analysis: In many cases all that is known of the network is its degree sequence, or at least the distribution of the degrees. It seems as if one could infer certain other structural features of the network, for example second order neighbors from its degree-sequence. Combinatorics provides a powerful tool to retrieve such insights from sequences: Generating functions. Our goal is to present some basics of generating functions and then develop the method for the special purposes of network analysis.

Ordinary Generating Functions. We are given the distribution of the degree sequence, to be precise a function $p(k)$ mapping each vertex degree k to the probability for a randomly chosen vertex – in a network chosen according to that degree sequence – to be adjacent to k other vertices. (For simplicity we confine ourselves to undirected graphs.) Calculating the expectation of that distribution immediately gives the (expected) average degree z_1, i.e., the average number of neighbors of a random vertex. Can we as easily calculate the probability for a vertex to have k second order neighbors, i.e., vertices whose shortest path distance to it equals exactly 2, from the distribution of the first order neighbors? Trying a direct approach, one might want to average over all degrees of a vertex the average of the degrees of the adjacent vertices. In some sense, one would like to perform calculations that relate the whole distribution to itself. But how to get hold of the entire distribution in a way useful for calculation? A generating function solves exactly this problem: On the one hand, it is an encoding of the complete information contained in the distribution, but on the other hand it is a mathematical object that can be calculated with. We define:

Definition 13.2.1. *For a probability distribution* $p : \mathbb{N} \mapsto [0, 1]$

$$G_p(x) = \sum_k p(k)x^k \tag{13.9}$$

is called the generating function *of p.*

This definition is by no means in its most general form. This particular way of encapsulating p is sometimes called the ordinary generating function.

The formal definition is justified in the light of the following proposition:

Proposition 13.2.2. *Let p be a probability distribution and G_p its generating function:*

1.

$$G_p(1) = 1$$

2.

$$G_p(x) \text{ converges for } x \text{ in } [-1, 1].$$

3.

$$p(k) = \frac{1}{k!} \left. \frac{\partial^k G_p}{\partial x^k} \right|_{(x=0)}$$

4.

$$E(p) := \sum_k kp(k) = G'_p(1)$$

5.

$$\mathrm{Var}(p) := \sum_k k(k-1)p(k) = G''_p(1)$$

The convergence is shown by standard analytic criteria. The other parts of the proposition are immediate from the definition, keeping in mind for the first that $\sum_k p(k) = 1$ for a probability distribution.

Part 3 of the proposition shows that a generating function encodes the information of the underlying distribution. From another perspective a generating function, G_p, is a formal power series that actually converges on a certain interval. Taking powers $(G_p(x))^m$ of it will again result in such a power series. Interpreting this in turn as a generating function amounts to interpreting the coefficient of some x^k in $(G_p(x))^m$. For $m = 2$ this is $\sum_{j+l=k} p(j)p(l)$, in other words, this is the probability that the values of two independent realizations of the random variable with distribution p sum up to k. In general $(G_p(x))^m$ is the generating function for the distribution of the sum of the values of m independent realizations of the random variable distributed according to p.

Generating Functions for Degree Sequences. For $k \in \mathbb{N}$ let D_k be a random variable equal to the number of vertices of degree k. Further $p(k)$ shall be the probability that a randomly chosen vertex has degree equal to k. It holds that $np(k) = E(D_k)$, the expectation of the random variable. To construct a random graph according to the distribution p may mean two slightly different things. We may take D_k to be a constant function for every k, thus, there is a fixed number of vertices with degree k. Alternatively, we only require the *expectation* of D_k to equal that fixed number. The latter model will make the graphs to which the first model is confined only the most probable. Moreover, as the first fixes the degree sequence, only those sequences of fixed values of D_k that are realizable generate a non-empty model. For example the sum of all degrees must not be odd. (The next section will discuss which degree sequences are realizable.) Despite these differences, for a realizable sequence the statistical results we are interested in here are not affected by changing between these two models. We confine our explicit considerations to the second and more general interpretation, where $p(k)$ only prescribes the expectation of D_k.

To justify the technicality of generating functions, some structural features of the network should be easily derived from its degree sequence's distribution. So far the average vertex degree z_1 has been shown to be $G_p'(1)$, which is not a real simplification for computation. Next we ask for the degree distribution of a vertex, chosen by the following experiment: Choose an edge of the graph uniformly at random and then one of its endpoints. The probability f to thereby reach a vertex of degree k is proportional to $kp(k)$. That means the corresponding generating function is $G_f(x) = \sum_k \frac{kp(k)}{\sum_k kp(k)} x^k = x\frac{G_p'(x)}{G_p'(1)}$. Removing the factor x in the right hand term amounts to reducing the exponent of x in the middle term, thus obtaining a generating function, where the coefficient of x^k in $G_f(x)$ becomes the coefficient of x^{k-1} in the new generating function. Hence the new function is the generating function of $f(k-1)$. Interpreting this combinatorially, we look at the distribution of the degrees minus one. In other words, we want to know the probability distribution p^* for the number of edges that are incident to the vertex, not counting the one edge we came from in the above choosing procedure. Its generating function can thus be written nicely, as

$$G_{p^*}(x) = \sum_k \frac{kp(k)}{\sum_k kp(k)} x^{k-1} = \frac{G'_p(x)}{G'_p(1)} \, . \tag{13.10}$$

This distribution p^* is useful to determine the distribution of rth neighbors of a random vertex.

Distribution of rth Neighbors. What is the probability, for a randomly chosen vertex v, that exactly k vertices are at a shortest path distance of exactly r? For $r = 2$, assume that the number of vertices of distance exactly 2 from v is $(\sum_{w \in N(v)} d(w)) - d(v)$, (where $d(v)$ denotes the degree of v), and for general r that the network contains no cycles. This assumption seems to be a good approximation for big, sparse, random graphs, as the number of occasional cycles seems negligible. But its exact implications are left to be studied. For the sake of explanation, assume the network to be directed in the following way: Choose a shortest-path tree from a fixed vertex v and direct each of its edges in the orientation in which it is used by that tree. Give a third orientation, zero, to the remaining edges. In this way the definition is consistent even for non tree-like networks. But assume again a tree structure. Any vertex except for v has exactly one in-edge and p^* is the distribution of the number of out-edges of that in-edge. Now a second assumption is required: For an out-edge $\{x, y\}$ of a vertex x the degree of y shall be distributed independently of that of x's other out-edges' endvertices, and independently of the degree of x. Of course, there are examples of pathological distributions for the degree-sequence where this assumption fails. Again, the assumption seems reasonable in most cases. Again, precise propositions on the matter are left to be desired.

Given these two assumptions, tree structure and independence, the generating function of the distribution of second neighbors is seen to be

$$\sum_k p(k)(G_{p^*}(x))^k = G_p(G_{p^*}(x)), \tag{13.11}$$

recalling that k independent realizations of a random variable amount to taking the kth power of its generating function. Correspondingly, the generating function of the distribution of the rth neighbors $G_{(r)}$ is:

$$G_{(r)} := \underbrace{G_p(G_{p^*}(G_{p^*} \ldots G_{p^*}(x)))}_{r \text{ functions altogether}} \, . \tag{13.12}$$

Taking expectations for second neighbors, i.e., calculating z_2, simplifies nicely:

$$z_2 = [G_p(G_{p^*}(x))]'|_{(x=1)} = G'_p(\underbrace{G_{p^*}(1)}_{=1})G'_{p*}(1) = G''_p(1) \tag{13.13}$$

Recall that the expectation of the first neighbors z_1 is $G'_p(1)$. Note that in general the formula for r-neighbors does not boil down to the rth derivative.

Component Size. For the analysis of the component size, first consider the case without a giant component. A giant component is a component of size in $\Theta(n)$. Thus we assume that all components have finite size even in the limit. Assume again the network to be virtually tree-like. Again the results are subject to further assumptions on the independency of certain stochastic events. And again these assumptions are false for certain distributions and, though likely for large networks, it is unclear where they are applicable. To point out these presuppositions we take a closer look at the stochastic events involved.

Construct a random graph G for a given probability distribution of the vertex degree, p, as always in this section. Next, choose an edge e uniformly at random among the edges of G. Flip a coin to select v, one of e's vertices. The random variable we are interested in is the size of the component of v in $G \setminus e$. Let p° be its distribution, and p^* as above the distribution of the degree of v in $G \setminus e$ found by this random experiment. Then for example $p^\circ(1) = p^*(0)$. In general, for k the degree of v, let $n_1, \ldots, n_k$ be the neighbors of v in $G \setminus e$. Further, we need a laborious definition: $P_k(s-1) := \Pr[\text{The sizes of the components}$ of the k vertices $n_1 \ldots n_k$ in $G \setminus \{e, (v, n_1), \ldots (v, n_k)\}$ sum up to $s-1$.] Then when may write: $p^\circ(s) = \sum_k p^*(k)P_k(s-1)$. How to compute P_k? It does not in general equal the distribution of the component size of a randomly chosen vertex when removing one of its edges. Take into account that in the above experiment a vertex is more likely to be chosen the higher its degree. On the other hand, supposing a tree-like structure, the component size of n_j is the same in $G \setminus \{e, (v, n_1), \ldots (v, n_k)\}$ as in $G \setminus (v, n_j)$. Now, assume that our experiment chooses the edges (v, n_i) independently and uniformly at random among all edges in G, then P_k is distributed as the sum of k random variables distributed according to p°. These assumptions are not true in general. Yet, granted their applicability for a special case under consideration, we can conclude along the following lines for the generating function of p°:

$$G_{p^\circ}(x) = \sum_{s=0}^{n} p^\circ x^s = \sum_{s=0}^{n} x^s \sum_k p^*(k)P_k(s-1)$$

$$= x \sum_k p^*(k) \underbrace{\sum_{s=0}^{n} x^{s-1} P_k(s-1)}_{G_{P_k}(x)}$$

Since we presume P_k as the distribution of the sum of k independent realizations of p°, we have $G_{P_k}(x) = G_{p^\circ}^k(x)$, and $G_{p^\circ}(x) = x \sum_k p^*(k)(G_{p^\circ}(x))^k$. This can be restated as

$$G_{p^\circ}(x) = xG_{p^*}(G_{p^\circ}(x)). \tag{13.14}$$

In a similar way we arrive at a consistency requirement for $p^\bullet$, the distribution of the component size of a randomly chosen vertex:

$$G_{p^\bullet}(x) = xG_p(G_{p^\circ}(x)) \tag{13.15}$$

The assumptions on stochastic independence made here are not true in general. Granted they are verified for a specific degree distribution, the functional

Equations (13.14) and (13.15) still withstand their general solution. Numerical solutions have been carried out for special cases (for details see [448]).

But the expected component size of a random vertex can be computed directly from those equations. The expectation of a distribution is the derivative of its generating function at point 1. Therefore $E(p^\bullet) = G'_{p^\bullet}(1) = 1 + G'_p(1)G'_{p^\circ}(1)$. But, as $G'_{p^\circ}(1) = 1 + G'_{p^*}(1)G'_{p^\circ}(1)$, this becomes:

$$E(p^\bullet) = G'_{p^\bullet}(1) = 1 + \frac{G'_p(1)}{1 - G'_{p^*}(1)} = 1 + \frac{z_1^2}{z_1 - z_2} \qquad (13.16)$$

Giant Component. So far we have excluded graphs with a giant component, i.e., a component that grows linearly with the graph. For a distribution that would generate such a component, the probability for a cycle would of course be no longer negligible. If we, however, still infer a tree-like structure as a good approximation, Formula 13.16 for the expected component size should no longer be independent of n, the number of vertices.

Indeed for $G'_{p^*}(1) \to 1$ equation (13.16) diverges, meaning that the expected component size is not bounded for unbounded n. What can be derived from $G'_{p^*}(1) = 1$?

$$1 = G'_{p^*}(1) \qquad \Longleftrightarrow$$

$$\sum_k k(k-1)p(k)x^{k-2}\bigg|_{(x=1)} = \sum_k kp(k) \qquad \Longleftrightarrow$$

$$\sum_k k(k-2)p(k) = 0$$

This equation marks the phase transition to the occurrence of a giant component, as the sum on the left increases monotonically with the relative number of vertices of degree greater than 2.

How much of the graph is occupied by the giant component? In [448] it is claimed that the above considerations on the component size still apply to the 'non-giant' part of the graph. But $G_{p^\bullet}(1)$ becomes smaller than 1 in these cases. Following the lines of [448], this should in turn give the fraction of the vertex set that is covered by non-giant components. In other words, $n(1 - G_{p^\bullet}(1))$ equals the (expected) number of vertices in the giant component. This is an audacious claim, as we calculate information about the non-giant part insinuating that it shows the same degree distribution as the whole graph. For example high-degree vertices could be more likely to be in the giant-component. Maybe those calculations actually lead to reasonable results, at least in many cases, but we cannot give any mathematical reason to be sure.

Generating Functions for Bipartite Graphs. So far this section has collected several ideas based on generating functions in order to squeeze as much information as possible from the mere knowledge of the degree distribution. Some of them depend on further assumptions, some are less appropriate than others. Some conclusions drawn in the literature are left out due to their questionable validity.

Finally, we become a little more applied. Many real-world networks show a bipartite structure. For example, in [184] we find a graph model used to analyze how infections can spread in a community. The model consists of two sets of vertices, persons and places, and edges from any person to any place the person regularly visits. As in other examples, like the co-author or the co-starring network, we are given bipartite data, but the interest is often mainly in the projection onto one of the vertex sets, mostly the persons'. Suppose we are given two probability distributions of degrees a and b, for the persons and the places, and the fraction ρ between the numbers of persons and places. Make a partition of n vertices according to ρ, realize a and b each in one part of the partition, and choose $\bar{H}$ uniformly at random among all bipartite graphs of n vertices with the same partition and degree sequences on the partition sets. Let H be the projection of $\bar{H}$ on the persons' vertices, and p the corresponding distribution of its degree sequence. Then $G_p = G_b(G_{a^*})$ and $G_{p^*} = G_{b^*}(G_{a^*})$. Now the whole machinery of generating functions can be applied again. In this way generating functions can help to bridge the gap between the bipartite data we are given and the projected behavior we are interested in.

13.2.3   Graphs with Given Degree Sequences

Given a degree sequence, generating functions allow to derive some deeper graph properties. Now we wish to construct a graph of a given degree sequence. At best, the generating algorithm would construct such a graph with uniform probability over all graphs that have a proposed degree sequence $d_1, d_2, \ldots, d_n$.

For reasons of simplicity we assume that $d_1 \geq d_2 \geq \cdots \geq d_n$ are the degrees of vertices $v_1, v_2, \ldots, v_n$.

Definition 13.2.3. *A degree sequence $d_1, d_2, \ldots, d_n$ is called* realizable *if there is a graph G with vertices $v_1, v_2, \ldots, v_n \in V$ with exactly the given degree sequence.*

Erdős and Gallai [180] gave sufficient and necessary conditions for realizability of a simple, undirected graph.

Necessary and Sufficient Conditions. In order to construct a graph with a given degree sequence we should at first verify whether that sequence is realizable at all. Secondly, we are only interested in connected graphs. Thus, we also want to know whether the degree sequence can be realized by a connected graph.

Starting with the first property we can observe the following. A degree sequence $d = (d_1, d_2, \ldots, d_n)$ is realizable if and only if $\sum_{i=1}^{n} d_i$ is even (since the sum of the degrees is twice the number of edges), and for all subsets $\{v_1, v_2, \ldots, v_\ell\}$ of the ℓ highest vertex degrees, the degrees of those vertices can be absorbed within those vertices and with the outside degrees. This means that there are enough edges within the vertex set and to the outside to bind to all the degrees. More formally we can state the following theorem.

Theorem 13.2.4. *A degree sequence* $d = (d_1, d_2, \ldots, d_n)$ *is realizable if and only if* $\sum_{i=1}^{n} d_i$ *is even and*

$$\sum_{i=1}^{\ell} d_i \leq \ell(\ell - 1) + \sum_{i=\ell+1}^{n} \min\{\ell, d_i\} \qquad 1 \leq \ell \leq n. \qquad (13.17)$$

This inequality is intuitively obvious, and therefore one direction of the theorem is trivial to prove. All degrees in the first ℓ degrees of highest order have to be connected first of all to the $(\ell - 1)$ other vertices in this set of vertices. The rest of the open degrees have to be at least as many as there are open degrees in the outside of the chosen set. How many can there be? For each vertex there is the minimum of either ℓ (since no more are needed for the ℓ vertices in the chosen set) or the degree of a vertex i where only vertices $\ell + 1, \ldots, n$ are taken into account.

A more precise theorem about realizability of an undirected, simple graph is given below.

Theorem 13.2.5. *A sequence* $d = (d_1, d_2, \ldots, d_n)$ *is realizable if and only if the sequence* $H(d) = (d_2 - 1, d_3 - 1, \ldots, d_{d_1+1} - 1, d_{d_1+2}, d_{d_1+3}, \ldots, d_n)$ *is realizable.*

Furthermore we are interested not only in a graph with this degree sequence, but in a connected graph. The necessary and sufficient conditions on connectedness are well known, but should be repeated here for completeness.

Theorem 13.2.6. *A graph G is connected if and only if it contains a spanning tree as a subgraph.*

As we neither have a graph nor a spanning tree, we are interested in a property that can give us the information whether a graph with certain degree sequence is constructible. As a spanning trees comprises $(n - 1)$ edges, the sum of degrees must be at least $2(n-1)$. This necessary condition is already sufficient, as it will become clear from the constructing algorithms given below.

If we can fulfill Theorem 13.2.4, and $\sum_{v_i \in V} d_i \geq 2(n - 1)$ holds, there exists a connected graph with the given degree sequence.

Algorithms. There are several easy-to-implement algorithms with linear running time that construct a graph with a given degree sequence. In the following we present two slightly different algorithms; one constructs a graph with a sparse core, the other constructs a graph with a dense core. The reader has to be aware that all of these easy algorithms do not construct a random graph out of all graphs with the desired degree sequence with the same probability. But starting from the graph constructed by one of these algorithms we give a method to generate a random instance that is indeed equiprobable among all graphs with the desired degree sequence. We assume that the sum of all degrees is at least $2(n - 1)$.

For both algorithms we need a subroutine called `connectivity`. This subroutine first of all checks whether the constructed graph is connected. If the graph G is not connected, it finds a connected component that contains a cycle. Such a connected component must exist because of the assumption on the degrees made above. Let uv be an edge in the cycle, and st be an edge in another connected component. We now delete edges uv and st, and insert edges us and vt to the network.

Sparse Core. In this section we want to describe an algorithm that constructs a graph with the given degree sequence that additionally is sparse. We are given a degree sequence $d_1 \geq d_2 \geq \cdots \geq d_n$, and we assign the vertices $v_1, v_2, \ldots, v_n$ to those degrees. As long as there exists a vertex v_i with $d_i > 0$, we choose the vertex v_ℓ with the currently lowest degree d_ℓ. Then we insert d_ℓ edges from v_ℓ to the first d_ℓ vertices with highest degree. After that we update the residual vertex degrees $d_i = d_i - 1$ for $i = 1, \ldots, d_\ell$ and $d_\ell = 0$. Last, but not least, we have to check connectivity and, if necessary, establish it using the above mentioned method `connectivity`.

Dense Core. To construct a graph with a dense core for a certain degree sequence, we only have to change the above algorithm for sparse cores slightly. As long as there exists a vertex v_i with $d_i > 0$ we now choose such a vertex arbitrarily and insert edges from v_i to the d_i vertices with the highest residual degrees. After that we only have to update the residual degrees and establish connectivity if it is not given.

Markov-Process. To generate a random instance from the space of all graphs with the desired degree sequence, we start using an easy to find graph G with the desired realization. In a next step, 2 edges (u, v) and (s, t) with $u \neq v, s \neq t$ such that $(u, s), (v, t) \notin G$ are chosen uniformly at random. The second step is to delete the edges (u, v) and (s, t) and replace them with (u, s) and (v, t).

This process is a standard Markov-chain process often used for randomized algorithms. We can observe that the degree distribution is unchanged by this algorithm. If rewiring two edges would induce a disconnected graph, the algorithm simply does not do this step, and repeats the random choice. The following theorem states that this algorithm constructs a random instance out of the space of all graphs with the desired degree sequence.

Theorem 13.2.7. *Independent of the starting point, in the limit, the above Markov-chain process will reach every possible connected realization with equal probability.*

For practical reasons one has to find a stopping rule so that we can bound the number of steps of the algorithm. Mihail et al. [420] observed that the process levels off in terms of the difference of two sorted lists (at different points in time) of all neighbors (by degree) of nodes with unique degrees. Using this measure they heuristically claim a number of at most 3 times the level-off number of steps to get a good random graph for instances like today's AS-level topology (about 12,000 vertices). They observed the number of steps to level-off to be less than 10,000 for graphs of 3,000 vertices, less than 75,000 for graphs with 7,500 vertices, and less than 180,000 for graphs with 11,000 vertices.

d-Regular Graphs. A special variant of graphs with given degree sequences are d-regular graphs where each vertex has exactly degree d.

There are several algorithms known that can construct an equiprobable d-regular graph. McKay and Wormald [416] gave an algorithm that is also applicable for arbitrary degree sequences. For a given $d \in \mathcal{O}(n^{\frac{1}{3}})$ its expected running time is in $\mathcal{O}(n^2 d^4)$, and furthermore it is very difficult to implement. A modification of this algorithm for only d-regular graphs improves the running time to $\mathcal{O}(nd^3)$, but does not remove the disadvantages. Tinhofer [550] gave a simpler algorithm that does not generate the graphs uniformly at random and, moreover, the resulting probability distribution can be virtually unknown. Jerrum and Sinclair [329] introduced an approximation algorithm where all graphs have only a probability varying by a factor of $(1 + \varepsilon)$, but the d-regular graph can be constructed in polynomial time (in n and ε), and the algorithm works for all possible degrees d.

A very simple model is the pairing model, introduced in the following. The running time is exponential ($\mathcal{O}(nd \exp{(\frac{d^2-1}{4})})$), and the graph can only be constructed in this running time for $d \leq n^{\frac{1}{3}}$.

Pairing Model. A simple model to construct a d-regular graph is the so-called pairing model. There, nd points are partitioned into n groups – clearly every group should include exactly d points. In a first step a random pairing of all points has to be chosen. Out of this pairing we now construct a graph G. Let the n groups be associated with n vertices of the graph. There is an edge (i, j) between vertices i and j in the graph if and only if there is a pair in the pairing containing points in the ith and jth group. This so constructed graph is a d-regular graph if there are no duplicated edges. Furthermore, we have to check a posteriori whether the graph is connected.

13.3   Further Models of Network Evolution

In this section we want to present some further models for evolving networks. Since there is a huge variety of them, we want to consider only some of those network models that include significantly new ideas or concepts.

13.3.1   Game Theory of Evolution

The literature for games *on* a (fixed) network is considerable. But game theoretical mechanisms can also be used to *form* a network, and this falls in our scope of interest. The following example is designed to model economic cooperation.

Vertices correspond to agents, who establish or destroy an edge between each other trying to selfishly maximize their value of a cost revenue function.

The objective function of an agent sums revenues that arise from each other agent directly or indirectly connected to him minus the costs that occur for each edge incident to him: Let c be the fixed costs for an incident edge and $\delta \in (0,1)$. The cost revenue of a vertex v is $u_v(G) = (\sum_{w \in V(G)} \delta^{d(v,w)}) - \deg(v)c$, where G is the current network and $d(v,w)$ is the edge-minimal path distance from v to w in G. (To avoid confusion we denote the degree of a vertex v by $\deg(v)$ here.) Set that distance to infinity for vertices in different components, or confine the index set of the sum to the component of v.

An edge is built when it increases the objective function of at least one of the agents becoming incident and does not diminish the objective function of the other. To delete an edge it suffices that one incident agent benefits from the deletion. In fact the model analyzed is a little more involved. Agents may simultaneously delete any subset of their incident edges, while participating in the creation of a new edge, and consider their cost revenue function after both actions.

To put it formally:

Definition 13.3.1. *A network G is* stable *if for all $v \in V(G)$*

$$\forall e \in E(G) : v \in e \implies u_v(G) \geq u_v(G \setminus e)$$

and

$$\forall w \in V(G), \forall S \subseteq \{e \in E(G) \mid v \in e \vee w \in e\} :$$
$$u_v((G \cup \{v,w\}) \setminus S) > u_v(G) \implies u_w((G \cup \{v,w\}) \setminus S) < u_w(G)$$

This quasi-pareto notion of stability does not guarantee that a stable network is in some sense 'good', namely that at least in total the benefit is maximal. Therefore we define:

Definition 13.3.2. *A network G is* efficient *if*

$$\forall G' : V(G) = V(G') \implies \sum_v u_v(G') \leq \sum_v u_v(G).$$

Theorem 13.3.3. *In the above setting we have:*
For $c < \delta, (\delta - c) > \delta^2$ the complete graph is stable.
For $c < \delta, (\delta - c) \leq \delta^2$ the star is stable.
For $c \geq \delta$ the empty graph is stable.

Theorem 13.3.4. *In the above setting we have:*
For $(\delta - c) > \delta^2$ only the complete graph is efficient.
For $(\delta - c) < \delta^2, c < \delta + (n-2)\delta^2/2$ only a star is efficient.
For $(\delta - c) < \delta^2, c > \delta + (n-2)\delta^2/2$ only the empty graph is efficient.

Through the work of A. Watts [572], this approach received a push towards evolution. Given the parameters c and δ, the question is, which networks will emerge? This remains unclear until the order in which agents may alter their incident part of the edge set (and the initial network) is given. Up to now only the case for an empty network as the initial configuration has been scrutinized. The order, in which the agents can make their decisions, is given in the following random way: At each step t of a discretized time model, one edge e of the complete graph of all agents (whether e is part of the current network or not) is chosen uniformly at random. Then the two incident agents may decide whether to keep or drop or, respectively, establish or leave out the edge e for the updated network. This means for an existing edge e that it is deleted if and only if one of its endvertices benefits from the deletion, and for a non-existing edge e that it is inserted if and only if at least one of the potentially incident vertices will benefit and the other will at least not be worse off. Note that in this model more sophisticated actions, that are comprised of the creation of one and the possible deletion of several other edges, are not allowed. All decisions are taken selfishly by only considering the cost revenue of the network immediately after the decision of time t. In particular no vertex has any kind of long time strategy. The process terminates if a stable network is reached. For this model the following holds:

Theorem 13.3.5. *In the above setting we have:*
For $(\delta - c) > \delta^2 > 0$ the process terminates in a complete graph
in finite time.
For $(\delta - c) < 0$ the empty set is stable.
For $\delta^2 > (\delta - c) > 0$ $P_{star} :=$
$\Pr[Process\ terminates\ in\ finite\ time\ in\ a\ star] > 0,$
but $P_{star} \to 0$ for $n \to \infty$.

The first result is obvious as any new edge pays. The second just reformulates the stability Theorem 13.3.3. For the third part, note that a star can no longer emerge as soon as two disjoint pairs of vertices form their edges.

The model and the results, though remarkable, still leave lots of room for further refinement, generalization, and variation. For example, if a star has positive probability that tends to zero, then this could mean that one can expect networks in which some vertices will have high degree, but most vertices will show very low degree. This is a first indication of the much discussed structure of a power law.

13.3.2   Deterministic Impact of the Euclidean Distance

For the following geometric model we want to denote the Euclidean distance between a vertex i and a vertex j by $d(i, j)$. The idea of this model by Fabrikant, Koutsoupias, and Papadimitriou [196] is to iteratively construct a tree. In a first step a sequence $p_0, p_1, \ldots, p_n$ of vertices is distributed within a unit square or unit sphere. In the next step we insert edges successively. Here we want to distinguish between two opposite goals. On the one hand, we are interested in connecting vertices to their geometrically nearest neighbor. On the other hand, we are interested in a high degree of centrality for each vertex. In order to deal with this trade-off between the 'last mile' costs and the operation costs due to communication delays, we connect the vertices with edges in the following way. Vertex i becomes connected to the vertex j that fulfills $\min_{j<i} \alpha \cdot d(i,j) + h_j$, where h_j denotes the centrality measure and α the relative importance of both goals. Here centrality measures can be the average number of hops to other vertices, the maximum number of hops to another vertex, or the number of hops to a given center - a fixed vertex $v \in V$ (for more details on centrality measures, see Chapter 3).

The behavior of this model is of course highly dependent on the value α and, to a lesser extent, on the shape used to place the vertices. Let T denote the constructed tree in a unit square. And let us define h_j to be the number of hops from p_i to p_0 in the tree T. Then we can state the following properties of T for different values of α.

Theorem 13.3.6. *(Theorem 2.1. in [196])*
If T is generated as above then:

1. *If $\alpha < 1/\sqrt{2}$, then T is a star with vertex p_0 as its center.*
2. *If $\alpha = \Omega(\sqrt{n})$, then the degree distribution of T is exponential, that is, the expected number of vertices that have degree at least k is at most $n^2 \exp(-ck)$ for some constant c:*
   $E[|\{i : \text{ degree of } i \geq k\}|] < n^2 \exp(-ck).$
3. *If $\alpha \geq 4$ and $\alpha = o(\sqrt{n})$, then the degree distribution of T is a power law; specifically, the expected number of vertices with degree at least k is greater than $c \cdot (k/n)^{-\beta}$ for some constants c and β (that may depend on α though):*
   $E[|\{i : \text{ degree of } i \geq k\}|] > c(k/n)^{-\beta}.$ *Specifically, for $\alpha = o(\sqrt{3n})$ the constants are: $\beta \geq 1/6$ and $c = \mathcal{O}(\alpha^{-1/2})$.*

This theorem gives the impression that networks constructed by this algorithm have a degree sequence following a power law. But there are some points to add. The power law given in this theorem does not resemble the definition of power law given in this chapter. Here the authors analyzed the behavior of the degree distribution where only vertices with degree at least k come into consideration. Therefore, one has to take care when comparing results. A second point is that the results only hold for a very small number of vertices in the network. For a majority of vertices (all but $\mathcal{O}(n^{1/6})$) there is no statement made in the work by Fabrikant et al. Subsequently, Berger et al. [58] prove the real behavior of the degree distribution obtained by this model and show that "there is a very large number of vertices with almost maximum degree".

13.3.3 Probabilistic Impact of the Euclidean Distance

The model of Waxman [574] uses the Euclidean distance, henceforth denoted by $d(\cdot,\cdot)$, to determine the probability distribution used to generate a graph of the model. In a first step n points on a finite 2-dimensional lattice are chosen equiprobably to form the vertex set $V(G)$ of the graph G. Then each edge $\{i,j\}$, $i,j \in V$, of the complete graph on these vertices is chosen to be part of the edge set $E(G)$ with probability $\Pr(\{i,j\}) = \beta \exp \frac{-d(i,j)}{L\alpha}$. Thereby L denotes the maximum Euclidean distance of two lattice points, i.e., the diagonal of the lattice.

Increasing $\alpha \in (0, 1]$ will decrease the expected length of an edge, whereas increasing $\beta \in (0, 1]$ will result in a higher number of edges in expectation.

In a variant of the model, the function $d(\cdot, \cdot)$ is defined by random for each pair of chosen vertices. Thus, it will in general not even fulfill the triangle inequality.

13.4   Internet Topology

The Internet consists of two main levels, the router level and the Autonomous System level. Both are systems with certain properties, like a power law with specified exponent, a certain connectivity, and so on. These properties are analyzed by Faloutsos et al. [197] in detail. One goal is now to construct synthetic networks that resemble the Internet very much and, further, that can generate a prediction of the future Internet topology. There are two types of generators: The first type are model-oriented generators that implement only a specified set of models, as for example given in the previous sections. A universal topology generator, on the other hand, should further have the property of being extensible to new models that can be added in an easy way.

To have such a universal generator is interesting for researchers who need good synthetic topologies to simulate their Internet protocols and algorithms. Therefore very good generation tools are needed. Those tools should have at least the following characteristics to be usable for a wide range of researchers and their different applications, not only for Internet topologies (see also [417]).

1. *Representativeness:* The tool should generate accurate synthetic topologies where as many aspects of the target network as possible are reflected.
2. *Inclusiveness:* A single tool should combine the strengths of as many models as possible.
3. *Flexibility:* The tool should be able to generate networks of arbitrary size.
4. *Efficiency:* Even large topologies should be generated in reasonable CPU time and memory.
5. *Extensibility:* The generator should be easily extendible by new models by the user.
6. *User-friendliness:* There should be an easy to learn interface and mechanics of use.
7. *Interoperability:* There should be interfaces to the main simulation and visualization applications.
8. *Robustness* The tool should be robust in the sense of resilience to random failures and, moreover, have the capability to detect errors easily.

These are the desired characteristics of a generator tool. In order to reach the characteristics made above, there are some challenges that have not been solved yet in an acceptable way. Two main challenges in the field of topology generation are (quoted from [417]):

1. How do we develop an adapting and evolving generation tool that constitutes an interface between general Internet research and pure topology generation research? Through this interface, representative topologies, developed by the topology generation research community, can be made readily available to the Internet research community at large.

2. How do we design a tool that also achieves the goal of facilitating pure topology generation research? A researcher that devises a generation model should be able to test it readily without having to develop a topology generator from scratch.

The topology generators available today or, better, their underlying models can be classified as follows (after [417]). On the one hand, there are ad-hoc models that are based on educated guesses, like the model of Waxman [574] and further models [109, 157]. On the other hand there are measurement based models where measures can be, for example, a power law. We can divide this class into causality-oblivious and causality-aware models. By causality we think of some possible fundamental or physical causes, whereas causality-oblivious models orient themselves towards such abstract features as power laws. The INET model and generator, described in Section 13.4.2, and the PLRG model by Aiello et al. [9] belong to the first of these subclasses. The preferential attachment model and the topology generator BRITE belong to the causality-aware models.

13.4.1 Properties of the Internet's Topology

Faloutsos, Faloutsos and Faloutsos [197] analyzed the structure of the Internet topology at three different points in time, and especially analyzed the growing of special metrics. Some of the very obvious metrics are, for example, the rank of a vertex, i.e., the position of the vertex in a sorted list of vertex degrees in decreasing order, and the frequency of a vertex degree, i.e., how often a degree k occurs among all the vertices. Using the minimal distance between two vertices, i.e., the minimal number of edges on a path between the two vertices, one can determine the number of pairs of vertices $P(h)$ that are separated by a distance of no more than h. By taking this definition we obviously have the property that self-pairs are included in $P(h)$ and all other pairs are counted twice. A resultant metric is the average number of vertices $N(h)$ that lie in a distance of at most h hops.

In the Internet there are two levels worth evaluating (for details see [259]). In [197] the data collected by the Border Gateway Protocols (BGP) that stores all inter-domain connections is evaluated. There the authors looked at three special points in time. The first graph is from November 1997, the second from April 1998, and the third from December 1998.

By evaluating this information, and looking for power laws for the above mentioned metrics, they draw the following conclusions. The first conclusion is that the degree $d(v)$ of a vertex v is proportional to the rank r_v of the vertex, raised to the power of a constant, R. This yields the power law of the form

$d(v) \propto r_v^R$. R is defined as the slope in the graph of the function that maps $d(v)$ on the rank of v (denoted as $(d(v); \text{rank of v})$) in log-log plot. A second observation is about the frequency f_k of a vertex degree k: $f_k \propto k^O$ with O a constant. The constant O can be determined by determining the slope of the $(f_k; k)$ plot with a log-log scale. For the total number of pairs of vertices $P(h)$ they can only approximate the power law to the form $P(h) \propto h^H$, where H is the so called hop-plot exponent and is constant. In this case the constant H is defined by the slope of a plot in log-log scale of the $(P(h); h)$ graph.

13.4.2   INET – The InterNEt Topology Generator

This model-oriented topology generator (more details in [332]) tries to implement more than just the analyzed power law in the degree distribution. Several of the analyses of Faloutsos et al. result in exponential laws. The first exponential law they observed and determined an exact form for is the frequency of certain degrees.

$$f_k = \exp(at + b)k^O, \qquad (13.18)$$

where f_k is the frequency of a degree k. a, b, O are known constants and t is the time in months since November 1997. Having this equation, we can also predict the frequency of a degree k for t a month in the future.

A second exponential law they found was the degree growth.

$$k = \exp(pt + q)r^R \qquad (13.19)$$

The degree k at a given rank r also grows exponentially over time. Here p, q, R are known constants and t is again the number of months since November 1997. This law tells us that the value of the ith largest degree of the Internet grows exponentially. This does not necessarily mean that every AS's degree grows exponentially with time because the rank of a particular AS can change as the number of AS's increases.

Two further exponential laws are the pair size growth and the resultant neighborhood size growth.

$$P_t(h) = \exp(s_h t)P_0(h) \qquad (13.20)$$

The pair size within h hops, $P(h)$, grows exponentially with the factor $P_0(h)$, that is the pair size within h hops at time 0 (=November 1997). The neighborhood size within h hops, $A(h)$, grows exponentially as follows.

$$A_t(h) = \frac{P_t(h)}{P_0(h)} = \exp((\log P_0(h) - \log P_0(0)) + (s_h - s_0)t)$$
$$= A_0(h)\exp((s_h - s_0)t) \qquad (13.21)$$

Here $A_0(h)$ is the neighborhood size at time 0 (= November 1997). The value t is, as always, the number of months since time 0.

The INET topology generator now uses the observed and analyzed exponential laws to construct a network that strongly resembles the real Internet network evaluated at time t. In a first step, the user has to input the number of vertices and the fraction p of the number of vertices that have degree one. By assuming exponential growth of the number of AS's in the Internet, the generator computes the value of t – the number of months since November 1997. Then it is easy to also compute the distributions of degree frequency and rank. Since the second power law only holds for 98% of the vertices we have to assign degrees to the top 2% of the vertices using the rank distribution (13.19). p percent of the vertices are assigned degree one. The remaining vertices are assigned degrees following the frequency degree distribution. The edges are inserted into the initial graph G to be generated according to the following rules. First, a spanning tree is built among vertices with degree strictly larger than one. This is done by successively choosing uniformly at random a vertex with degree strictly larger than one that is not in the current tree G, and connecting it to a vertex in G with probability proportional to $\frac{k}{K}$. Here k is the degree of the vertex in G, and K is the sum of degrees of all vertices already in G that still have at least one unfilled degree. In a next step, $p|V|$ vertices with degree one are connected to vertices in G with proportional probability as above. In a final step, the remaining degrees in G, starting with the vertex with largest assigned degree, are connected to vertices with free degrees randomly picked, again using proportional probability. The connectivity of the graph is first tested by a feasibility test before actually inserting the edges.

Other Generators. There are several more topology generators available. The GT-ITM generator [109] is able to construct different topologies. One of them is a transit-stub network that has a well-defined hierarchical structure. The Tiers generator [157] is designed to provide a three level hierarchy that represents Wide Area Networks (WAN), Metropolitan Area Networks (MAN), and Local Area Networks (LAN). The generator BRITE [417] also contains several mechanisms to construct topologies. It not only includes the well-known basic model of Barabási and Albert on router and on AS level but also the Waxman model for both types. It can also illustrate and evaluate the networks made by the INET and GT-ITM generators, and the data obtained from the National Laboratory for Applied Network Research routing data [452].

By using the mentioned power laws observed in the Internet it is now easy to determine the representativeness of such a generator. Medina et al. [418] used the above mentioned topology generators to generate different kinds of topologies, and then evaluated them according to the existence of power laws. As a result they can say that the degree versus rank and the number of vertices versus degree power laws were not observed in all of the topologies. In this way the existence can be used to validate the accuracy of a generator. Power laws concerning the neighborhood size and the eigenvalues were found in all the generated topologies, but with different values of the exponent.

13.5   Chapter Notes

In 1959, Gilbert [244] introduced the model $(G_{n,p})$ in the sense of our second definition. In the same year Erdős and Rényi [181] presented a model parameterized by the number of vertices and the number of edges, n and m, which corresponds to the first definition we give, except that we fix the average vertex degree, $z = \frac{2m}{n}$. For a good introduction to this research area we refer to Bollobás [67, Chapter 7]. There Theorem 9 corresponds to our Theorem 13.1.1.

Our discussion about Local Search in Small Worlds is based on Kleinbergs pertinent work [360].

Further details on the exponential cutoff, and an evolutionary model that regards the exponential cutoff, are given in the very recent paper by Fenner, Levene, and Loizou [205].

In Bollobás et al. [68] further interesting behavior caused by certain initial choices of vertices and edges for the preferential attachment model by Barabási and Albert is given, and some more imprecisions are pointed out. For more details on the equivalence of (G_m^t) and (G_1^{tm}) see [68], too. Also, more explanations of other power law models and some mathematical background are given there.

Simple and efficient generators for standard random graphs, small worlds, and preferential attachment graphs are described in [43].

Generating functions are a concept for dealing with counting problems that is far more general than we present it here. Most books on combinatorics include a thorough discussion of generating functions (see for example [10]). A particular reference for generating functions only is [586], which can be downloaded at www.cis.upenn.edu/~wilf/. We mainly mathematically clarify the assertions found in [448]. There also further details on generating functions for bipartite graphs can be found.

The results in Section 13.2.3 are basically taken from Mihail et al. [420]. A proof for Theorem 13.2.4 is contained in [57, Chapter 6] (the original paper of Erdős and Gallai [180] is in Hungarian). The more precise Theorem 13.2.5 was firstly given and proven in Havel and Hakimi [288, 270], and it is stated as found in Aigner and Triesch [11]. For references on Markov-chain-processes see [420]. An overview of the algorithms for d-regular graphs can be found in Steger and Wormald [532]. They also construct a polynomial algorithm that works for all d and give an idea of how to implement that algorithm to obtain an expected running time in $\mathcal{O}(nd^2 + d^4)$.

We present the results given in Section 13.3.1 following A. Watts [572], though earlier work by Jackson and Wolinsky [323] prepared the ground.

The p^*-model is introduced in Section 10.2.5, and therefore it is omitted in this chapter.

14 Spectral Analysis

Andreas Baltz and Lasse Kliemann

A graph can be associated with several matrices, whose eigenvalues reflect structural properties of the graph. The adjacency matrix, the Laplacian, and the normalized Laplacian are in the main focus of spectral studies. How can the spectrum be used to analyze a graph? In particular, the following questions will be of interest:

- What can the spectrum tell us about subgraphs? Can we prove the existence or nonexistence of certain subgraphs by looking at the spectrum?
- Can certain eigenvalues be related to other global statistics (also called graph parameters), such as diameter, isoperimetric number or chromatic number?
- How can spectra help in classifying graphs?

   This chapter is organized as follows: In the first section, we review facts from linear algebra and point out some fundamental properties of graph spectra. In the next section, we summarize what methods we know to compute the spectrum. In the sections that follow, we give some answers and ideas regarding the above questions.

14.1 Fundamental Properties

We define the different spectra and point out some fundamental properties. We show why it makes sense to consider more than the adjacency spectrum and list the three spectra for some basic graph classes (Table 14.1).

14.1.1 Basics from Linear Algebra

Let $M = (m_{i,j}) \in \mathbb{C}^{n \times n}$ be an $n \times n$ matrix with complex numbers as entries. A non-zero vector $x \in \mathbb{C}^n$ is an *eigenvector* of M with corresponding *eigenvalue* $\lambda \in \mathbb{C}$ if x and λ satisfy the equation

$$Mx = \lambda x. \tag{14.1}$$

The vector $\mathbf{0}_n$ is excluded from the set of possible eigenvectors since *every* $\lambda \in \mathbb{C}$ is a solution to $M\mathbf{0}_n = \lambda \mathbf{0}_n$. Equation (14.1) has a non-zero solution if and only if $\text{rank}(M - \lambda I_n) < n$, which is equivalent to $\det(M - \lambda I_n) = 0$. Hence we can characterize the eigenvalues of M as roots of the polynomial $p_M(\lambda) := \det(M -$

U. Brandes and T. Erlebach (Eds.): Network Analysis, LNCS 3418, pp. 373–417, 2005.
© Springer-Verlag Berlin Heidelberg 2005

λI_n). This *characteristic polynomial* p_M does not change if M is replaced by $Q^{-1}MQ$ for some arbitrary non-singular matrix Q. So M and $Q^{-1}MQ$ have the same eigenvalues. The *spectrum* of M is defined as the multiset of all eigenvalues of M, where the multiplicity of an eigenvalue λ is its (algebraic) multiplicity as a root of p_M.

From now on we shall assume that M is a symmetric matrix, i.e. $M = M^\top$, with real valued entries. One can show that there exists a non-singular matrix Q such that $M' := Q^{-1}MQ$ has diagonal form and $Q^{-1} = Q^\top$. Clearly, each vector e_i from the standard basis of $\mathbb{R}^n$ is an eigenvector of M' with $\lambda_i := m'_{i,i}$, the ith entry on the diagonal of M', as its corresponding eigenvalue. Since the characteristic polynomial is unchanged,

$$\det(M - \lambda I_n) = \det(M' - \lambda I_n) = \prod_{i=1}^{n}(\lambda_i - \lambda). \qquad (14.2)$$

So the spectrum of M consists of n (not necessarily distinct) eigenvalues $\lambda_1, \ldots, \lambda_n \in \mathbb{R}$, and we have $\det(M) = \prod_{i=1}^{n} \lambda_i$. From the definition of the determinant as

$$\det M = \sum_{\pi \in S_n} \text{sign}(\pi) \cdot m_{1,\pi(1)} \cdots \cdot m_{n,\pi(n)}$$

we see that $\det(M - \lambda I_n) = \prod_{i=1}^{n}(m_{i,i} - \lambda) + R(\lambda)$, where $R(\lambda)$ is the sum corresponding to permutations from $S_n \setminus \{\text{id}\}$ and thus has degree $\leq n - 2$. Hence the coefficient of λ^{n-1} in the characteristic polynomial is $\sum_{i=1}^{n} m_{i,i}$, the *trace* of M, and by comparison with (14.2) we infer that $\text{trace}(M) = \sum_{i=1}^{n} \lambda_i$. Defining $v_i := Qe_i$ for all $i \in \{1, \ldots, n\}$, we have

$$Mv_i = MQe_i = QQ^{-1}MQe_i = Q\lambda_i e_i = \lambda_i v_i$$
$$v_i^\top v_j = (Qe_i)^\top Qe_j = e_i^\top Q^\top Qe_j = e_i^\top e_j,$$

i.e., $v_1, \ldots, v_n$ are orthonormal eigenvectors of M. Consequently, $\{v_1, \ldots, v_n\}$ is a basis of $\mathbb{R}^n$, and the multiplicity of λ_i as an eigenvalue equals the maximum cardinality of a corresponding set of linearly independent eigenvectors.

In sum we have observed the following facts.

Theorem 14.1.1 ([135, 247, 576]). *Let $M \in \mathbb{R}^{n \times n}$ be a matrix with $M = M^\top$, then*

1. *M has real eigenvalues $\lambda_1 \leq \ldots \leq \lambda_n$ and n orthonormal eigenvectors forming a basis of $\mathbb{R}^n$ [Spectral Theorem],*
2. *multiplicity of λ_i as an eigenvalue := multiplicity of λ_i as a root of the characteristic polynomial $\det(M - \lambda I_n)$ = cardinality of a maximum linearly independent set of eigenvectors corresponding to λ_i,*
3. *there is a matrix Q with $Q^\top = Q^{-1}$ such that $Q^\top MQ = \begin{pmatrix} \lambda_1 & & 0 \\ & \ddots & \\ 0 & & \lambda_n \end{pmatrix}$,*
4. *$\det(M) = \prod_{i=1}^{n} \lambda_i$, $\text{trace}(M) = \sum_{i=1}^{n} \lambda_i$.*

14.1.2   The Spectrum of a Graph

Given a multi-digraph $G = (V, E)$ with labeled vertices $v_1, \ldots, v_n$, we define the *adjacency matrix* $A = (a_{i,j})$ by

$$a_{i,j} := \text{multiplicity of edge } (v_i, v_j) \text{ in } E,$$

i.e., $a_{i,j}$ is equal to the number of arcs starting at vertex v_i and terminating at vertex v_j (for loops $(v_i, v_i) \in E$ it is sometimes useful to define $a_{i,i}$ as 2 instead of 1). The spectrum of G is the spectrum of the adjacency matrix of G (see Figure 14.1 for an example).

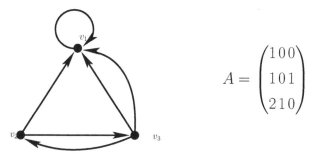

$$A = \begin{pmatrix} 1 & 0 & 0 \\ 1 & 0 & 1 \\ 2 & 1 & 0 \end{pmatrix}$$

Fig. 14.1. Example of an adjacency matrix

Note that though A depends on the order of labeling, the spectrum does not, since exchanging the labels of vertices v_i and v_j corresponds to exchanging row i with row j and column i with column j in A, which does not affect $\det(A)$ nor $\det(A - \lambda I_n)$. For the rest of this chapter we will identify V with the set of labels, i.e., we will assume that $V = \{1, \ldots, n\}$. Moreover we shall mainly focus on simple undirected graphs without loops. So, any graph $G = (V, E)$ will be assumed as simple, loopless, and undirected, unless explicitly stated otherwise. Hence the adjacency matrix A will (almost always) be a symmetric 0/1 matrix with a real spectrum of n eigenvalues λ_i, where we assume for convenience that $\lambda_1 \leq \lambda_2 \leq \ldots \leq \lambda_n$.

Notation. We will use both, spectrum(A) and spectrum(G) to denote the eigenvalues of an adjacency matrix A corresponding to a graph G. Moreover, when speaking of the spectrum of a graph, we will always have the adjacency spectrum in mind (unless stated otherwise).

Let $w \in \mathbb{C}^n$ be an arbitrary vector and let $\omega : V \to \mathbb{C}$ map each $i \in V$ on w_i. Since A represents a graph, the *i*th component of Aw, $\sum_{j=1}^{n} a_{i,j} w_j$, can be written as $\sum_{j \in N(i)} \omega(j)$. Now the equation $Ax = \lambda x$ has the following useful interpretation.

Remark 14.1.2.   1. A has eigenvalue λ if and only if there exists a non-zero weight function $\omega : V \to \mathbb{C}$ such that for all $i \in V$, $\lambda \omega(i) = \sum_{j \in N(i)} \omega(j)$.

2. The Spectral Theorem 14.1.1.1. ensures that we can restrict ourselves to considering real-valued weight functions. Moreover, we can assume the maximum weight to be non-negative (if $\max\{\omega(i);\ i \in V\} < 0$ then $\omega(i) < 0$ for all $i \in V$ and we can consider $-\omega$ instead of ω).

Consider an assignment of weights to the vertices of a triangle as depicted in Figure 14.2. From

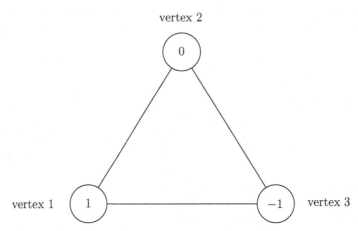

Fig. 14.2. A triangle with eigenvector components as weights

$$\omega(1) = 1 = -(0 + (-1)) = -(\omega(2) + \omega(3)),$$
$$\omega(2) = 0 = -(1 + (-1)) = -(\omega(1) + \omega(3)),$$
$$\omega(3) = -1 = -(1 + 0) = -(\omega(1) + \omega(2)),$$

we can conclude that -1 is an eigenvalue. Similarly, by assigning a weight of 1 to all vertices we can check that 2 is in the spectrum of the triangle. For another example, have a look at Figure 14.3, which proves that 2 is an eigenvalue of a star on 5 vertices.

Remark 14.1.2 enables us to prove the following claims.

Lemma 14.1.3 ([576]). *Let $G = (V, E)$ be a graph on n vertices with adjacency matrix A and eigenvalues $\lambda_1 \leq \ldots \leq \lambda_n$. Let Δ be the maximum vertex degree of G.*

1. *$\lambda_n \leq \Delta$.*
2. *If $G = G_1 \dot\cup G_2$ is the union of two disjoint graphs G_1 and G_2 then spectrum$(G) = $ spectrum$(G_1) \cup $ spectrum(G_2).*
3. *If G is bipartite then $\lambda \in$ spectrum$(G) \Leftrightarrow -\lambda \in$ spectrum(G).*
4. *If G is a simple cycle then spectrum$(G) = \left\{ 2\cos\left(\frac{2\pi k}{n}\right);\ k \in \{1, \ldots, n\} \right\}$.*

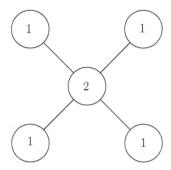

Fig. 14.3. A star with eigenvector components as weights

Proof. 1. Let ω be a non-zero weight function on the vertices such that $\lambda_n\omega(i) = \sum_{j\in N(i)}\omega(j)$ for all $i \in V$ and let i_0 be a vertex of maximum weight. Then $\lambda_n\omega(i_0) = \sum_{j\in N(i_0)}\omega(j) \le \Delta\omega(i_0)$ implies $\lambda_n \le \Delta$.

2. Let ω be a non-zero weight function for $\lambda_i \in \mathrm{spectrum}(G)$. Since ω is not identically zero, either $\omega|_{V(G_1)}$ or $\omega|_{V(G_2)}$ must not be identically zero and hence is a weight function for λ_i on G_1 or G_2. On the other hand, if ω is a weight function for $\lambda \in \mathrm{spectrum}(G_j)$ for $j = 1$ (or $j = 2$) then extending ω by defining $\omega(i) := 0$ for all $i \in V \setminus V(G_j)$, yields a non-zero weight function for λ on G.

3. Let ω be a weight function for λ on G. Let V_1, V_2 denote the partition classes of V. Define $\omega' : V \to \mathbb{R}$ by

$$i \mapsto \begin{cases} \omega(i), & \text{if } i \in V_1 \\ -\omega(i), & \text{if } i \in V_2. \end{cases}$$

Then for all $i \in V_1$,

$$-\lambda\omega'(i) = -\lambda\omega(i) = -\sum_{j\in N(i)}\omega(j) = \sum_{j\in N(i)}\omega'(j),$$

and for all $i \in V_2$,

$$-\lambda\omega'(i) = \lambda\omega(i) = \sum_{j\in N(i)}\omega(j) = \sum_{j\in N(i)}\omega'(j).$$

4. For the proof of this claim we will use complex weights. Assume that the edges of the cycle are $\{1, n\}$ and $\{i, i+1\}$, for all $i \in \{1, \ldots, n-1\}$. For each $k \in \{1, \ldots, n\}$ let $\tau_k := \exp(2\pi\mathbf{i}k/n)$ be an nth root of unity and put $\omega(j) := \tau_k^{j-1}$. (Here, $\mathbf{i}$ denotes the complex number $\mathbf{i}$ with $\mathbf{i}^2 = -1$ and *not* a vertex.) Then for all $j \in V$,

$$\sum_{l\in N(j)}\omega(l) = (\tau_k^{-1} + \tau_k) \cdot \tau_k^{j-1},$$

and thus $\tau_k^{-1} + \tau_k = \exp(-2\pi\mathbf{i}k/n) + \exp(2\pi\mathbf{i}k/n) = 2\cos\left(\frac{2\pi k}{n}\right)$ is an eigenvalue of G. $\qquad\square$

We mention two further results on the adjacency spectrum (see [135] for a proof): it can be shown that $\lambda_n = \Delta$ if and only if G has a Δ-regular component. For the smallest eigenvalue of G one can prove the lower bound $\lambda_1 \geq -\lambda_n$, where equality holds if and only if G has a bipartite component whose largest eigenvalue is equal to λ_n.

Let us determine the spectra of the complete bipartite graph K_{n_1,n_2} and the complete graph K_n.

Lemma 14.1.4 ([576]). *Let n_1, n_2, and n be positive integers.*

1. *For $G = K_{n_1,n_2}$, $\lambda_1 = -\sqrt{n_1 n_2}$, $\lambda_2 = \cdots = \lambda_{n-1} = 0$, and $\lambda_n = \sqrt{n_1 n_2}$.*
2. *For $G = K_n$, $\lambda_1 = \cdots = \lambda_{n-1} = -1$, $\lambda_n = n - 1$.*

Proof. 1. Since A is diagonalizable with eigenvalues as diagonal entries, the rank of A is equal to the number of non-zero eigenvalues. For K_{n_1,n_2}, the rank is 2, so A has two non-zero eigenvalues λ_i and λ_j. Note that the trace of A is both the sum of the eigenvalues and the number of loops in G. Hence, $\lambda_i + \lambda_j = 0$, and we conclude that the spectrum of G is $\lambda_1 = -c, \lambda_2 = \cdots = \lambda_{n-1} = 0, \lambda_n = c$ for some $c \in \mathbb{R}_{>0}$. Let us look at the characteristic polynomial, $\det(A - \lambda I_n) = (-c - \lambda)\lambda^{n-2}(c - \lambda) = \lambda^n - c^2\lambda^{n-2}$. Since λ appears only on the diagonal of $A - \lambda I_n$, terms in the permutation expansion that contribute to λ^{n-2} arise from those permutations that select $n - 2$ diagonal elements and 2 non-diagonal elements, $a_{i,j} = a_{j,i} = 1$. Choosing i and j completely determines the permutation, so there are exactly $n_1 \cdot n_2$ permutations contributing to λ^{n-2}, each with negative sign. Consequently, $c^2 = n_1 n_2$ and thus $\lambda_1 = -\sqrt{n_1 n_2}$, $\lambda_n = \sqrt{n_1 n_2}$.

2. For K_n, the adjacency matrix is $J - I_n$, where J is the matrix of all ones. Subtracting c from the diagonal of a matrix M shifts its eigenvalues by $-c$, since $Mx = \lambda x$ is equivalent to $(M - cI_n)x = (\lambda - c)x$. By induction on n it can be shown that the spectrum of J consists of a single eigenvalue being n and $n - 1$ eigenvalues equal to zero. Thus the spectrum of K_n is $\lambda_1 = \cdots = \lambda_{n-1} = -1$, $\lambda_n = n - 1$. $\quad\square$

The adjacency matrix is also useful for counting paths of length k in a graph.

Lemma 14.1.5 ([576]). *Let G be a multi-digraph possibly with loops. The (i,j)th entry of A^k counts the $i \to j$-paths of length k. The eigenvalues of A^k are λ_i^k.*

Proof. The first claim can be shown by induction on k. For the second claim, note that for every eigenvector x with corresponding eigenvalue λ, $A^k x = A^{k-1}(Ax) = \lambda A^{k-1}x = \cdots = \lambda^k x$. $\quad\square$

Corollary 14.1.6.   1. $\sum_{i=1}^n \lambda_i = $ *number of loops in G.*
2. $\sum_{i=1}^n \lambda_i^2 = 2 \cdot |E|$.
3. $\sum_{i=1}^n \lambda_i^3 = 6 \cdot$ *number of triangles in G.*

Now we can prove that the converse of the third claim in Lemma 14.1.3 is also true.

Lemma 14.1.7 ([576]). *G is bipartite if and only if the eigenvalues of G occur in pairs λ, λ' such that $\lambda = -\lambda'$.*

Proof. One direction of the claim has been shown in the proof of Lemma 14.1.3. For the other direction, note that $\lambda_i = -\lambda_j$ implies $\lambda_i^k = -\lambda_j^k$ for every odd k. Since $\operatorname{trace}(A^k) = \sum_{i=1}^{k} \lambda_i^k = 0$ counts the number of cycles of length k in G, we infer that in particular there are no odd simple cycles which means that G is bipartite. $\square$

We have seen that it is possible to retrieve certain structural properties of a graph (e.g. number of edges, number of triangles, whether or not it is bipartite) from its spectrum. However, the spectrum does not reflect all of the graph's structure. Take for example the graphs in Figure 14.4.

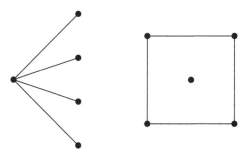

Fig. 14.4. Two non-isomorphic graphs with identical adjacency spectrum

Both have eigenvalues $\lambda_1 = -2$, $\lambda_2 = \lambda_3 = \lambda_4 = 0$, and $\lambda_5 = 2$, but they are not isomorphic. Such graphs are called *cospectral*. Obviously, we can not even determine from the spectrum if a graph is connected. Nevertheless, this can be achieved by looking at eigenvalues of another graph matrix, the Laplacian.

14.1.3 The Laplacian Spectrum

Let $G = (V, E)$ be an undirected multigraph (possibly with loops) with adjacency matrix A. Let $D = \operatorname{diag}(d(1), \ldots, d(n))$ be the diagonal matrix of vertex degrees. The *Laplacian* matrix $L = (l_{i,j})$ is defined as $L := D - A$, so if G is a simple undirected graph, then

$$l_{i,j} = \begin{cases} -1, & \text{if } \{i, j\} \in E \\ d(i), & \text{if } i = j \\ 0 & \text{otherwise.} \end{cases}$$

Another way of defining the Laplacian of an undirected simple graph is the following. Consider an arbitrary *orientation* of G, i.e. a mapping assigning each edge $e = \{i, j\}$ a direction by indicating whether i or j is to be viewed as the head of e. The *incidence matrix* $B = (b_{i,e})$ of the oriented graph (G, σ) is a $\{0, 1, -1\}$ matrix with rows and columns indexed by the vertices and edges of G, respectively, such that

$$b_{i,e} := \begin{cases} 1, & \text{if } i \text{ is the head of } e \\ -1, & \text{if } i \text{ is the tail of } e \\ 0 & \text{otherwise.} \end{cases}$$

It can be shown that independently of the choice of σ, $L = BB^\top$. Consequently, we obtain the following result.

Lemma 14.1.8. *For each $x \in \mathbb{C}^n$, $x^\top L x = x^\top B B^\top x = \sum_{\{i,j\} \in E} (x_i - x_j)^2$.*

Since A is real and symmetric, $L = D - A$ is also symmetric, and hence the Laplacian spectrum consists of n real eigenvalues $\lambda_1(L) \leq \cdots \leq \lambda_n(L)$. Again, we may interpret an eigenvector $x \in \mathbb{R}^n$ as an assignment of weights $\omega : V \to \mathbb{R}$, $i \mapsto x_i$. From this point of view, λ is an eigenvalue of L if there exists a non-zero (and not completely negative) weight function $\omega : V \to \mathbb{R}$ such that

$$\lambda \omega(i) = \sum_{j \in N(i)} (\omega(i) - \omega(j))$$

for all $i \in V$. Considering this equation for $i \in V$ with maximum weight, we see that $\lambda \omega(i) = \sum_{j \in N(i)} (\omega(i) - \omega(j)) \geq 0$, so all eigenvalues are non-negative. For $\omega \equiv 1$ we have $\lambda = \lambda \omega(i) = \sum_{j \in N(i)} (\omega(i) - \omega(j)) = 0$, hence the vector $\mathbf{1}_n$ is an eigenvector of L with eigenvalue 0.

Lemma 14.1.9 ([247]). *A graph G consists of k connected components if and only if $\lambda_1(L) = \cdots = \lambda_k(L) = 0$ and $\lambda_{k+1}(L) > 0$.*

Proof. Let B be the incidence matrix of an arbitrary orientation of G. For each component C of G define $z(C) \in \mathbb{R}^n$ by

$$z(C)_i := \begin{cases} 1, & \text{if } i \in V(C) \\ 0 & \text{otherwise.} \end{cases}$$

Then, $Z := \{z(C); C \text{ component of } G\}$ is linearly independent and $Lz(C) = BB^\top z(C) = \mathbf{0}_n$. Hence the connected component can be injectively mapped into a linearly independent set of eigenvectors with eigenvalue 0. On the other hand, if $z \in \mathbb{R}^n$ is a vector such that $Lz = BB^\top z = \mathbf{0}_n$, then $z^\top BB^\top z = 0$ implies $B^\top z = \mathbf{0}_n$, meaning that z must be constant on each connected component. Thus, z is a linear combination of elements from Z, and consequently we have exactly as many components as there are linearly independent eigenvectors corresponding to eigenvalue 0. $\qquad \square$

The Laplacian is useful for counting the number of spanning trees of a graph.

Theorem 14.1.10 (Matrix-Tree Theorem [135, 247, 427]). *The Laplacian matrix L of a graph G is related to spanning trees in the following way.*

1. *For every $i \in \{1, \ldots, n\}$ the number of spanning trees in G is equal to $|\det(L_i)|$, where L_i is obtained from the Laplacian L by deleting row i and column i.*

2. *Moreover, the number of spanning trees is equal to $\frac{1}{n} \prod_{i \geq 2} \lambda_i(L)$.*

While the Laplacian spectrum has the advantage over the adjacency spectrum of indicating the number of connected components of a graph, it fails to identify bipartite structures, as can be seen from the graphs in Figure 14.5 which are cospectral with respect to L.

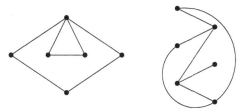

Fig. 14.5. Two cospectral graph with respect to the Laplacian

14.1.4 The Normalized Laplacian

A matrix whose spectrum enables us to recognize both, bipartite structure and connected components, can be obtained by multiplying L from left and right with the diagonal matrix $D^{-1/2}$, where the ith entry in the diagonal is $d(i)^{-1/2}$ if $d(i) > 0$ and 0 otherwise. This matrix is called the *normalized Laplacian* $\mathcal{L} = D^{-1/2} L D^{-1/2}$. For simple graphs, $\mathcal{L} = (\bar{l}_{i,j})$ satisfies

$$
\bar{l}_{i,j} = \begin{cases} 1, & \text{if } i = j \text{ and } d(i) > 0 \\ -\frac{1}{\sqrt{d(i)d(j)}}, & \text{if } \{i, j\} \in E \\ 0 & \text{otherwise.} \end{cases}
$$

λ is an eigenvalue of $\mathcal{L}$ if there is a non-zero weight function $\omega : V \to \mathbb{C}$, such that

$$
\lambda \omega(i) = \frac{1}{\sqrt{d(i)}} \sum_{j \in N(i)} \left(\frac{\omega(i)}{\sqrt{d(i)}} - \frac{\omega(j)}{\sqrt{d(j)}} \right).
$$

Again, $\mathcal{L}$ is symmetric with real valued entries, and we can order its n eigenvalues in the sequence $\lambda_1(\mathcal{L}) \leq \cdots \leq \lambda_n(\mathcal{L})$.

The following claims are proved in [125].

Lemma 14.1.11 ([125]). *Let G be a graph with normalized Laplacian matrix $\mathcal{L}$.*

1. $\lambda_1(\mathcal{L}) = 0$, $\lambda_n(\mathcal{L}) \leq 2$.
2. G is bipartite if and only if for each $\lambda(\mathcal{L})$, the value $2 - \lambda(\mathcal{L})$ is also an eigenvalue of $\mathcal{L}$.
3. If $\lambda_1(\mathcal{L}) = \cdots = \lambda_i(\mathcal{L}) = 0$ and $\lambda_{i+1}(\mathcal{L}) \neq 0$, then G has exactly i connected components.

14.1.5   Comparison of Spectra

If G is graph where each vertex has exactly d neighbors, then $L = dI_n - A$ and $\mathcal{L} = I - D^{-1/2}AD^{-1/2}$. This implies that for d-regular graphs, the three spectra are equivalent. In particular, if $d > 0$ and

$$\text{spectrum}(A) = (\lambda_1, \dots\dots\dots, \lambda_n) \quad \text{then}$$
$$\text{spectrum}(L) = (d - \lambda_n, \dots, d - \lambda_1) \quad \text{and}$$
$$\text{spectrum}(\mathcal{L}) = \left(1 - \frac{\lambda_n}{d}, \dots, 1 - \frac{\lambda_1}{d}\right).$$

In general, there is no simple relationship between the three spectra. Nevertheless, we can bound the eigenvalues of the adjacency matrix in terms of the Laplacian eigenvalues and the maximum and minimum vertex degrees.

Lemma 14.1.12 ([425]). *Let G be a graph with adjacency matrix A and Laplacian matrix L. If Δ and δ are the maximum and the minimum vertex degrees of G, respectively, then the kth smallest eigenvalue $\lambda_k(A)$ of A and the kth largest eigenvalue $\lambda_{n+1-k}(L)$ of L are related by*

$$\delta - \lambda_k(A) \leq \lambda_{n+1-k}(L) \leq \Delta - \lambda_k(A).$$

We will show this claim with the help of Courant-Fischer's characterization of the eigenvalues. This is a well-known theorem from linear algebra, however we include a proof for completeness.

Theorem 14.1.13 ([587]). *Let $M \in \mathbb{R}^{n \times n}$ be a real symmetric matrix with eigenvalues $\lambda_1 \leq \dots \leq \lambda_n$. Then, for all $k \in \{1, \dots, n\}$,*

$$\lambda_k = \min_{\substack{U \leq \mathbb{R}^n \\ \dim(U) = k}} \max_{\substack{x \in U \\ x \neq \mathbf{0}_n}} \frac{x^\top M x}{x^\top x}.$$

Proof. Let $\{v_1, \dots, v_n\}$ be an orthonormal basis of eigenvectors of M for $\mathbb{R}^n$ such that $Mv_i = \lambda_i v_i$ for all $i \in \{1, \dots, n\}$. For each $x \in \mathbb{R}^n$, $x \neq \mathbf{0}_n$, let $i_0(x) \in \{1, \dots, n\}$ be maximal subject to the condition

$$x \perp v_i \text{ for all } i < i_0(x).$$

In other words, $i_0(x)$ is the first index for which $x \not\perp v_{i_0(x)}$. (x cannot be orthogonal to all vectors from the basis, because $x \neq \mathbf{0}_n$.) Therefore, there exist scalars $\mu_{i_0(x)}, \dots, \mu_n \in \mathbb{R}$ such that

$$x = \sum_{i=i_0(x)}^{n} \mu_i v_i.$$

Consequently,

$$x^\top M x = x^\top M \sum_{i=i_0(x)}^{n} \mu_i v_i = x^\top \sum_{i=i_0(x)}^{n} \mu_i M v_i$$

$$= x^\top \sum_{i=i_0(x)}^{n} \mu_i \lambda_i v_i = \sum_{i=i_0(x)}^{n} \mu_i \lambda_i \underbrace{x^\top v_i}_{=\mu_i} \qquad (14.3)$$

$$\geq \lambda_{i_0(x)} \sum_{i=i_0(x)}^{n} \mu_i^2 = \lambda_{i_0(x)} x^\top x.$$

Obviously, equality holds for $x = v_k$, i.e., for all $k \in \{1, \ldots, n\}$ we have,

$$v_k{}^\top M v_k = \lambda_k v_k{}^\top v_k. \qquad (14.4)$$

We now make the following claim:

For every k-dimensional subspace $U \leq \mathbb{R}^n$ there exists $x \in U$, $x \neq \mathbf{0}_n$, such that $i_0(x) \geq k$. $\qquad (14.5)$

First, we explain how to prove the theorem from this claim. From (14.3) it follows that for every k-dimensional subspace U,

$$\max_{\substack{x \in U \\ x \neq \mathbf{0}_n}} \frac{x^\top M x}{x^\top x} \geq \lambda_k$$

On the other hand, because $\langle v_1, \ldots, v_k \rangle$ *is a k-dimensional subspace*, it follows from (14.4) that

$$\min_{\substack{U \leq \mathbb{R}^n \\ \dim(U)=k}} \max_{\substack{x \in U \\ x \neq \mathbf{0}_n}} \frac{x^\top M x}{x^\top x} = \lambda_k.$$

We now have to prove (14.5). For a contradiction, assume that for all $x \in U$, we have $i_0(x) < k$. By the definition of i_0 this means that for every $x \in U$ there exists an $i < k$ such that $x \not\perp v_i$ or equivalently, [1]

$$\langle v_1, \ldots, v_{k-1} \rangle^\perp \cap U = \mathbf{0}_n.$$

On the other hand

$$\dim \langle v_1, \ldots, v_{k-1} \rangle^\perp + \dim U = n - (k-1) + k = n + 1,$$

a contradiction. □

---

[1] Here, $\langle W \rangle$ denotes the linear hull of a subset $W \subseteq \mathbb{R}^n$, and $W^\perp$ denotes the set of all vectors that are orthogonal to each vector from W.

For later use, we state a simple corollary from this theorem.

Corollary 14.1.14. *(a) The largest eigenvalue λ_n of a real symmetric matrix $M \in \mathbb{R}^{n \times n}$ satisfies*

$$\lambda_n = \max_{\substack{x \in \mathbb{R}^n \\ x \neq 0_n}} \frac{x^\top M x}{x^\top x}.$$

(b) The second smallest eigenvalue of the Laplacian matrix satisfies

$$\lambda_2(L) = \min_{x \perp 1_n} \frac{x^\top L x}{x^\top x}.$$

Later, we will also use another characterization of Laplacian eigenvalues, which we cite without a proof. Let us call a vector $x \in \mathbb{R}^n$ *constant*, if all its entries are the same, i.e., if it is a multiple of 1_n.

Theorem 14.1.15 (Fiedler [211]).

$$\lambda_2(L) = n \min \left\{ \frac{\sum_{\{i,j\} \in E} (x_i - x_j)^2}{\sum_{\{i,j\} \in \binom{V}{2}} (x_i - x_j)^2} ; \ x \in \mathbb{R}^n \ non\text{-}constant \right\} \qquad (14.6)$$

$$\lambda_n(L) = n \max \left\{ \frac{\sum_{\{i,j\} \in E} (x_i - x_j)^2}{\sum_{\{i,j\} \in \binom{V}{2}} (x_i - x_j)^2} ; \ x \in \mathbb{R}^n \ non\text{-}constant \right\} \qquad (14.7)$$

Now, we turn to the proof of Lemma 14.1.12.

Proof (of Lemma 14.1.12). Since $\lambda_{n+1-k}(L)$ is the kth largest eigenvalue of L, $\delta - \lambda_{n+1-k}(L)$ is the kth smallest eigenvalue of the matrix $\delta I_n - L = A - (D - \delta I_n)$ which differs from A only on the diagonal, where the non-negative values $d(i) - \delta$ are subtracted. We have $\frac{x^\top (A - (D - \delta I_n))x}{x^\top x} = x^\top A x - r(x)$ for $r(x) := \frac{x^\top (D - \delta I_n) x}{x^\top x} \in \mathbb{R}_{\geq 0}$, and Theorem 14.1.13 gives

$$\delta - \lambda_{n+1-k}(L) = \lambda_k(\delta I_n - L)$$

$$= \lambda_k(A - (D - \delta I_n))$$

$$= \min_{\substack{U \leq \mathbb{R}^n \\ \dim(U)=k}} \max_{\substack{x \in U \\ x \neq 0_n}} \frac{x^\top (A - (D - \delta I_n)) x}{x^\top x}$$

$$= \min_{\substack{U \leq \mathbb{R}^n \\ \dim(U)=k}} \max_{\substack{x \in U \\ x \neq 0_n}} \left(\frac{x^\top A x}{x^\top x} - r(x) \right)$$

$$\leq \min_{\substack{U \leq \mathbb{R}^n \\ \dim(U)=k}} \max_{\substack{x \in U \\ x \neq 0_n}} \frac{x^\top A x}{x^\top x}$$

$$= \lambda_k(A).$$

The other inequality is obtained in a similar way.   □

Figure 14.6 shows two non-isomorphic graphs that are cospectral with respect to all three matrices.

 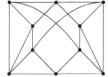

Fig. 14.6. Two graphs that are cospectral with respect to A, L, and $\mathcal{L}$

14.1.6 Examples

Table 14.1 lists the spectrum of the adjacency matrix A, the Laplacian L, and the normalized Laplacian $\mathcal{L}$ for some elementary graph classes. All graphs are assumed to have n vertices.

Table 14.1. Spectra of some elementary graph classes

graph class	spectrum(A)	spectrum(L)	spectrum($\mathcal{L}$)
simple path $G = P_n$	$2\cos\left(\frac{\pi k}{n+1}\right)$, $k \in \{1,\ldots,n\}$	$2 - 2\cos\left(\frac{\pi(k-1)}{n}\right)$, $k \in \{1,\ldots,n\}$	$1 - \cos\left(\frac{\pi(k-1)}{n-1}\right)$, $k \in \{1,\ldots,n\}$
simple cycle $G = C_n$	$2\cos\left(\frac{2\pi k}{n}\right)$, $k \in \{1,\ldots,n\}$	$2 - 2\cos\left(\frac{2\pi k}{n}\right)$, $k \in \{1,\ldots,n\}$	$1 - \cos\left(\frac{2\pi k}{n}\right)$, $k \in \{1,\ldots,n\}$
star $G = K_{1,n}$	$-\sqrt{n}, \sqrt{n}$, 0 ($n-2$ times)	$0, n$, 1 ($n-2$ times)	$0, 2$, 1 ($n-2$ times)
$G = K_{n_1,n_2}$	$-\sqrt{n_1 n_2}, \sqrt{n_1 n_2}$, $0(n-2$ times)	$0, n_1$ ($n_2 - 1$ times) n_2 ($n_1 - 1$ times), n	$0, 2$ 1 ($n-2$ times)
$G = K_n$	$1, -1$ ($n-1$ times)	$0, n$ ($n-1$ times)	$0, \frac{n}{n-1}$ ($n-1$ times)

14.2 Numerical Methods

To use the spectrum of a graph for analysis, we have to compute it (or parts of it) first. What methods are available? What are their running times and other characteristics?

14.2.1 Methods for Computing the Whole Spectrum of Small Dense Matrices

It is not at all obvious how to compute the spectrum of a matrix M efficiently, since already a straightforward evaluation of the characteristic polynomial as $\det(M - \lambda I_n)$ takes $\mathcal{O}(n!)$ steps. A better strategy is to utilize the fact that as a graph matrix, M is real and (in case of undirected graphs) symmetric and thus can be transformed into a diagonal matrix by means of a similarity transformation $M \to P^{-1}MP$. If we are interested only in eigenvalues, not eigenvectors, it is enough to transform the matrix M to be triangular, with all elements below (or above) the diagonal equal to zero. In this case the diagonal elements are already the eigenvalues.

There is a two-stage technique for implementing the diagonalization strategy. In the first stage we iteratively approximate P (and P^{-1}) as a product of certain 'atomic' transformations P_i designed for zeroing a particular off-diagonal element (Jacobi transformation [482]) or a whole particular row or column (Householder transformation [482], elimination method [482]). We stop after $\mathcal{O}(n^3)$ steps with a matrix $\tilde{P}$ such that $M_1 = (m_{i,j}) := \tilde{P}^{-1}M\tilde{P}$ has tridiagonal form (i.e., $m_{i,j} = 0$ whenever $|i - j| > 1$). Now the second stage starts, where we perform a QL- (or QR-) decomposition: the basic idea is that any real matrix M' can be decomposed in the form $M' = QL$ (or QR), such that Q is orthogonal (i.e., $Q^T = Q^{-1}$) and L (R) is lower (upper) triangular. Writing these factors in opposite order we get $M'' := LQ = Q^TQLQ = Q^TM'L$, where properties such as symmetry and tridiagonal form are preserved. The QL-algorithm consists of a sequence of transformations

$$M_i := Q_iL_i,$$
$$M_{i+1} := L_iQ_i = Q_i^TM_iQ_i,$$

and relies on the following theorem [535].

Theorem 14.2.1 ([535]).   *1. If M has eigenvalues of different absolute value $|\lambda_i|$ then M_s converts to lower triangular form as $s \to \infty$.*

2. *If M has an eigenvalue $|\lambda_i|$ of multiplicity p then M_s converts to lower triangular form as $s \to \infty$, except for a diagonal block matrix of order p, whose eigenvalues converge to λ_i.*

Note that if M has an eigenvalue with multiplicity greater than one, then the second part of Theorem 14.2.1 allows us to split the matrix into submatrices that can be diagonalized separately. For tridiagonal matrices, one iteration of the QL-algorithm can be performed in $\mathcal{O}(n)$ steps. With the technique of implicit shifts [482] a reasonably good convergence is achieved in $\mathcal{O}(n)$ steps, resulting in a total number of $\mathcal{O}(n^2)$ steps for the second stage. Thus the overall complexity of computing all eigenvalues with 'good precision' is $\mathcal{O}(n^3)$.

14.2.2 Methods for Computing Part of the Spectrum of Large Sparse Matrices

When our given matrix M is very large, the computational and storage costs required for the diagonalization strategy of the previous section become prohibitive. However, in many situations M is sparse and it suffices to determine a small subset of (extremal) eigenvalues. Suppose, we are interested in the largest eigenvalue of M. The eigenvalue equation $Mx = \lambda x$ is clearly equivalent to $\lambda = \frac{x^\top M x}{x^\top x}$. In fact, we have observed in Corollary 14.1.14 that

$$\lambda_n = \max \left\{ \frac{x^\top M x}{x^\top x};\ x \in \mathbb{R}^n \setminus \{0_n\} \right\}.$$

This suggests a simple algorithm: Choose an arbitrary starting vector x_1 with $x_1^\top x_1 = 1$, and then follow the direction of steepest ascend to successively obtain an improved approximation to λ_n. The *Lanczos algorithm* is a run-time and storage cost efficient algorithm that uses this idea to approximate extremal eigenvalues of M. It proceeds as follows: a given initial vector x_1 is normalized to length one. Then at each step i an orthonormal basis $(x_1, x_2, \ldots, x_i)$ for the space spanned by $(x_1, Mx_1, \ldots, M^{i-1}x_1)$ is constructed (this space is called 'Krylov space'). Let X_i denote the $n \times i$-matrix with $x_1, x_2, \ldots, x_i$ as column vectors. The matrix $T = X_i^\top M X_i$ has tridiagonal form. Its eigenvalues — which are easy to calculate (see [2]) — provide approximations to i eigenvalues of M. The method favors convergence to eigenvalues in the outermost part of the spectrum of A. The process is restarted every k steps for some fixed $k \ll n$ until sufficiently good convergence is achieved. We state the Lanczos method in its simplest form as Algorithm 29.

Algorithm 29: The Lanczos algorithm

---

1. Initialization: Choose the number of steps k, the desired number of eigenvalues r and an initial vector x_1; let $\beta_0 := x_1^\top x_1$, $x_1 := x_1/\beta_0$

2. Lanczos steps:
for $i = 1$ **to** k **do**
$\quad$ *(i)* $y := Mx_i$
$\quad$ *(ii)* $\alpha_i := x_i^\top y$
$\quad$ *(iii)* $x_{i+1} := y - \alpha_i x_i - \beta_{i-1} x_{i-1}$
$\quad$ *(iv)* $\beta_i := x_{i+1}^\top x_{i+1}$
$\quad$ *(v)* $x_{i+1} := x_{i+1}/\beta_i$;
Set $X_i := \mathrm{Mat}(x_1, \ldots, x_i)$

3. Eigenvalue computation: Compute the eigenvalues of $T := X_i^\top M X_i$.

4. Convergence test and restart: If the first r columns of T satisfy the convergence criteria then accept the corresponding eigenvalues and stop. Otherwise restart with a suitable new x_1.

By replacing M with $(M - \mu I_n)^{-1}$ in step 2(i) of the algorithm, this scheme is capable of approximating eigenvalues in the vicinity of any given value μ. In general, the minimum number of Lanczos iterations necessary to compute a desired set of eigenvalues is unknown. In practice, a fast convergence can be achieved via a restarting scheme (see [591]). Farkas et al. [198] were able to compute the spectrum of graphs with up to 40,000 vertices and 200,000 edges with the 'thick-restart' version of the Lanczos algorithm. The *Arnoldi Method* differs from the Lanczos algorithm in replacing steps (ii)–(iv) with

(ii)' $h_{i,j} := x_j^\top y$ for all $j \in \{1, \ldots, i\}$

(iii)' $x_{i+1} := y - \sum_{j+1}^i x_j h_{j,i}$

(iv)' $h_{i,i+1} := x_{i+1}^\top x_{i+1}$, $\beta_i := h_{i,i+1}$.

For symmetrical matrices, Arnoldi's and Lanczos' methods are mathematically equivalent, but the Lanczos algorithm uses fewer arithmetic operations per step by explicitly taking advantage of the symmetry of M. The advantage of Arnoldi's method is that it can be applied to treat asymmetric matrices. In that case, $H = X_i^\top M X_i$ is an upper Hessenberg matrix that can be reduced to block triangular form, allowing for an efficient computation of eigenvalues (see [509]).

14.3   Subgraphs and Operations on Graphs

What can the spectrum tell us about subgraphs in a graph? Do eigenvalues of subgraphs show up in the spectrum? Can we conclude that a certain graph is not a subgraph (or at least not an induced subgraph) by looking at the spectrum? What happens to the spectra of two graphs that are joined by taking the Cartesian product or the sum?

In this section, we will show some directions for answering these questions. We only consider the adjacency spectrum here, although many results also hold for the Laplacian spectra.

14.3.1   Interlacing Theorem

How do induced subgraphs manifest themselves in the spectrum? Obviously, not every induced subgraph inserts its eigenvalues or some of its eigenvalues into the spectrum of the whole graph. For example, the complete graph K_2 on two vertices has eigenvalues -1 and 1. But many graphs containing K_2 as an induced subgraph, i.e., an edge, do not have -1 or 1 in their spectrum. See Table 14.1 in Section 14.1.6 for examples.

However, there is an *interlacing* of the eigenvalues of induced subgraphs. Let $G = (V, E)$ be a graph on n vertices and let H be an induced subgraph of G with $n - 1$ vertices. (In other words: H is obtained from G be removing one vertex.) If, in our usual notation, $\lambda_1 \leq \ldots \leq \lambda_n$ are the eigenvalues of the adjacency matrix of G and $\mu_1 \leq \ldots \leq \mu_{n-1}$ are those of H, we have

$$\lambda_i \leq \mu_i \leq \lambda_{i+1} \quad \forall i \in \{1, \ldots, n-1\},$$

i.e., between two eigenvalues of G there lies exactly one eigenvalue of H. Consequently, if G has an eigenvalue with multiplicity k, then H has this eigenvalue with multiplicity $k - 1$.

For an induced subgraph on m vertices, by induction (over the number of vertices removed to obtain H from G) the interlacing generalizes to

$$\lambda_i \leq \mu_i \leq \lambda_{i+(n-m)} \quad \forall i \in \{1, \ldots, m\}. \tag{14.8}$$

We will now prove an even more general result than this, from which we will get (14.8) directly.

Theorem 14.3.1. *Let $n, m \in \mathbb{N}$ and $S \in \mathbb{R}^{n \times m}$ such that $S^\top S = \mathrm{Id}_m$. Let $A \in \mathbb{R}^{n \times n}$ be symmetric and $B := S^\top AS$. Then, for the eigenvalues $\lambda_1 \leq \ldots \leq \lambda_n$ of A and the eigenvalues $\mu_1 \leq \ldots \leq \mu_m$ of B we have the interlacing property (14.8), i.e.,*

$$\lambda_i \leq \mu_i \leq \lambda_{i+(n-m)} \quad \forall i \in \{1, \ldots, m\}. \tag{14.9}$$

The interlacing property of the adjacency spectrum follows from this. To see this, let A be the adjacency matrix of G. The adjacency matrix of an induced subgraph H is a principal submatrix of A, i.e., a matrix that is obtained by deleting the ith row and ith column of A for every vertex $i \in V \setminus V(H)$. Such a principal submatrix can be obtained from A in the same way as B is obtained from A in the above theorem. If $i_1, \ldots, i_k$ are the vertices removed from G to get H, we have to choose S to be the identity matrix with the i_jth row deleted, $j = 1, \ldots, k$.

Proof (of Theorem 14.3.1). The matrix S defines an injective mapping from $\mathbb{R}^m$ to $\mathbb{R}^n$. For a subset $U \subseteq \mathbb{R}^m$ we as usually denote by $S(U)$ the image of U under that mapping, i.e.,

$$S(U) := \{Su; \ u \in U\}.$$

If $U \leq \mathbb{R}^m$ is an i-dimensional subspace ($i \leq m$), then $S(U)$ is an i-dimensional subspace of $\mathbb{R}^n$, because S is injective.

Remember the characterization of eigenvalues given in Theorem 14.1.13. For every $i \in \{1, \ldots, m\}$ we have

$$\lambda_i = \min_{\substack{U \leq \mathbb{R}^n \\ \dim(U)=i}} \max_{\substack{x \in U \\ x \neq 0}} \frac{x^\top Ax}{x^\top x} \leq \min_{\substack{U \leq S(\mathbb{R}^m) \\ \dim(U)=i}} \max_{\substack{x \in U \\ x \neq 0}} \frac{x^\top Ax}{x^\top x}$$

$$= \min_{\substack{U \leq \mathbb{R}^m \\ \dim(U)=i}} \max_{\substack{x \in S(U) \\ x \neq 0}} \frac{x^\top Ax}{x^\top x} = \min_{\substack{U \leq \mathbb{R}^m \\ \dim(U)=i}} \max_{\substack{x \in U \\ x \neq 0}} \frac{(Sx)^\top A(Sx)}{(Sx)^\top (Sx)}$$

$$= \min_{\substack{U \leq \mathbb{R}^m \\ \dim(U)=i}} \max_{\substack{x \in U \\ x \neq 0}} \frac{x^\top (S^\top AS)x}{x^\top x} = \min_{\substack{U \leq \mathbb{R}^m \\ \dim(U)=i}} \max_{\substack{x \in U \\ x \neq 0}} \frac{x^\top Bx}{x^\top x}$$

$$= \mu_i.$$

This is the first inequality of (14.9).

Applying the same argument to $-A$ instead of A, we get for every $k \in \{0, \ldots, m-1\}$ that $-\lambda_{n-k} \leq -\mu_{m-k}$, which means

$$\lambda_{n-k} \geq \mu_{m-k}.$$

Setting $k := m - i$ yields the second inequality of (14.9). $\square$

From the Interlacing Theorem we immediately get the following corollary.

Corollary 14.3.2. *Let G and H be two graphs with eigenvalues $\lambda_1 \leq \ldots \leq \lambda_n$ and $\mu_1 \leq \ldots \leq \mu_m$ respectively. If $\mu_1 < \lambda_1$ or $\lambda_n < \mu_n$, then H does not occur as an induced subgraph of G.*

For example, let G be a graph with all eigenvalues smaller than 2.

The graphs C_j, $j \in \mathbb{N}$, all have 2 as their largest eigenvalue (see Table 14.1 in Section 14.1.6). Hence, G does not contain a cycle as an induced subgraph. Because in every cycle, we can find a simple cycle, in which in turn we can find an induced simple cycle, G even is cycle-free.

We also know that G has no vertices of degree larger than 3. Otherwise, G would have a $K_{1,j}$, $j \geq 4$, as an induced subgraph, which has $\sqrt{j}$ (as largest eigenvalue) in its spectrum.

14.3.2   Grafting

Let us now approach the question about the role of subgraphs from a different direction. Suppose you have a graph G and another graph H and would like to modify G in a way that gives G some of the eigenvalues of H.

For a disconnected graph the spectrum is the union of the spectra of its components. This is easy to see. It also gives us a straightforward method for accomplishing our goal in the given setup: Simply add H to G as a separate component. However this is not a suitable approach in case one has to preserve the connectivity of the graph.

Let λ be an eigenvalue of H which we would like to add to the spectrum of G. Let us first consider the special case that the eigenvector x corresponding to λ has an entry equal to zero, say, $x_{i_0} = 0$. Think of the graph G' as the union of G and H where the vertex i_0 of H has been identified with some arbitrary vertex of G. Formally, we define G' from $G = (V(G), E(G))$ and $H = (V(H), E(H))$ by picking a vertex $j_0 \in V(G)$ and setting

$$V(G') := V(G) \cup (V(H) \setminus \{i_0\}) \text{ and}$$
$$E(G') := E(G) \cup E(H - i_0) \cup \{\{j_0, i\}; \ i \in V(H), \{i_0, i\} \in E(H)\}.$$

We say that H is *grafted* onto G along i_0 and j_0.

To get an eigenvector for λ, assign 0 to all vertices of G in G'. To the vertices of $H - i_0$ in G' assign the values given by the eigenvector x. Using the combinatorial interpretation of eigenvectors (Remark 14.1.2), one easily verifies that this yields an eigenvector of G' for eigenvalue λ. See Figure 14.7 for an example.

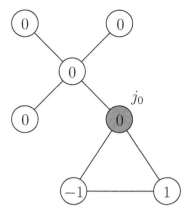

Fig. 14.7. A triangle grafted on a star of five vertices. The new graph has an eigenvalue -1, which the star alone did not have

But what if we do not have an eigenvector of H for λ with an entry equal to zero? We can still do a similar trick. We pick a vertex $i_0 \in V(H)$ and make two copies H^+ and H^- of H. The copies of a vertex $i \in V(H)$ are called i^+ and i^- respectively. We then take a new vertex i_1 and connect H^+ with H^- via two new edges $\{i_0^+, i_1\}$ and $\{i_1, i_0^-\}$. Call the resulting graph $\tilde{H}$.

Let λ be an eigenvalue of H and x a corresponding eigenvector. Then the following vector $\tilde{x}$ is an eigenvector of $\tilde{H}$ with the same eigenvalue λ.

$$\tilde{x}_{i^+} := x_i \text{ and } \tilde{x}_{i^-} := -x_i \quad \forall i \in V(H)$$

and

$$\tilde{x}_{i_1} := 0.$$

Now, $\tilde{H}$ can be grafted onto G along i_1 and an arbitrary vertex $j_0 \in V(G)$. We call such a construction a *symmetric graft*. Note, that if H is a tree, $\tilde{H}$ will be a tree as well. Symmetric grafts of trees play an important role in the analysis of spectra of random graphs [49]; see also the notes about sparse random graphs in 14.5.2.

As an example, consider the path on three vertices, P_3. This graph has an eigenvalue $\sqrt{2}$ with eigenvector $(1/\sqrt{2}, 1, 1/\sqrt{2})^{\top}$. Take the middle vertex as i_0. Figure 14.8 shows the symmetric construction ready to be grafted along i_1.

Finally we remark that grafting along one vertex can be extended in a natural way to grafting along several vertices.

14.3.3 Operations on Graphs

Consider two graphs $G_1 = (V_1, E_1)$ and $G_2 = (V_2, E_2)$. The *sum* $G_1 + G_2$ is a graph on $V_1 \times V_2$ where two vertices $(i_1, i_2), (j_1, j_2) \in V_1 \times V_2$ are connected if either (but not both!) $\{i_1, j_1\} \in E_1$ or $\{i_2, j_2\} \in E_2$. On the other hand, letting

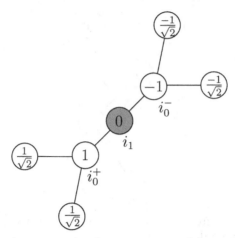

Fig. 14.8. A symmetric construction from two copies of P_3, ready to be grafted

(i_1, i_2) and (j_1, j_2) share an edge if and only if $\{i_1, j_1\} \in E_1$ and $\{i_2, j_2\} \in E_2$ defines the *Cartesian product* graph $G_1 \times G_2$. Figure 14.9 depicts the sum and the product of a simple path on four vertices with itself.

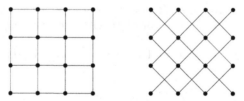

Fig. 14.9. The sum and the Cartesian product of a 4-vertex simple path with itself

It is easy to compute the eigenvalues of $G_1 + G_2$ and $G_1 \times G_2$ from the spectra of G_1 and G_2.

Lemma 14.3.3. *(a) spectrum$(G + H) = $ spectrum$(G) + $ spectrum(H).*
(b) spectrum$(G \times H) = $ spectrum$(G) \cdot $ spectrum(H).

More general, the spectrum of any *non-complete extended p-sum* of graphs is determined by the individual spectra.

Definition 14.3.4 ([135]). *Let $p \in \mathbb{N}_{\geq 2}$ and let $\mathcal{B} \subseteq \{0,1\}^p \setminus \mathbf{0}_p$. The non-complete extended p-sum of graphs $G_1 = (V_1, E_1), \ldots, G_p = (V_p, E_p)$ with basis $\mathcal{B}$ is the graph $NEpS(G_k)_{k=1}^p$ with vertex set $V_1 \times \cdots \times V_p$, in which two vertices $(i_1, \ldots, i_p), (j_1, \ldots, j_p)$ are adjacent if and only if there exists $\beta \in \mathcal{B}$ such that $i_k = j_k$ whenever $\beta_k = 0$ and $\{i_k, j_k\} \in E_k$ whenever $\beta_k = 1$.*

Theorem 14.3.5 ([135]). *Let $p \in \mathbb{N}_{\geq 2}$. For each $k \in \{1, \ldots, p\}$ let G_k be a graph on n_k vertices with spectrum $\lambda_{k1}, \ldots, \lambda_{kn_k}$. Then*

$$spectrum(NEpS(G_k)_{k=1}^p) = \left\{ \sum_{\beta \in \mathcal{B}} \lambda_{1i_1}^{\beta_1} \cdot \ldots \cdot \lambda_{pi_p}^{\beta_p}; (i_1, \ldots, i_k) \in I \right\},$$

where

$$I := \{1, \ldots, n_1\} \times \ldots \times \{1, \ldots, n_k\}.$$

Note that the sum and the Cartesian product of graphs are NEpS with $p = 2$ and basis $\mathcal{B} = \{(1,0), (0,1)\}$ resp. $\mathcal{B} = \{(1,1)\}$.

14.4   Bounds on Global Statistics

Certain eigenvalues, especially the extreme eigenvalues, give bounds for global statistics. Recall from Section 11.7 in Chapter 11 that a global statistic assigns to each graph (from some class of graphs) a single value. These statistics are sometimes also called *graph parameters*.

We will study a selection of eigenvalue bounds on graph parameters.

Average Degree

Recall that the degree $d(i)$ of a vertex i is the number of edges incident with i. Denote the *average degree* by $\bar{d}$, i.e.,

$$\bar{d} := \frac{1}{n} \sum_{i \in V} d(i). \tag{14.10}$$

The average degree is related to the largest eigenvalue of the adjacency matrix, as seen in the following observation.

Lemma 14.4.1. *Let G be a graph and $\bar{d}$ its average degree. Then*

$$\bar{d} \leq \lambda_n.$$

Proof. Let $x := \mathbf{1}_n$ be the n-dimensional vector with all entries equal to 1. Then by Corollary 14.1.14 we have

$$\lambda_n \geq \frac{x^\top A x}{x^\top x} = \frac{\sum_{i \in V} d(i)}{n} = \bar{d}.$$

$\square$

Diameter and Mean Distance

We only cite some results in this section. Recall that the diameter is the largest distance between two vertices. Eigenvalues of the Laplacian provide bounds on the diameter of a connected graph.

Theorem 14.4.2 ([126, 426]). *Let $\alpha > 1$. Then*

$$\mathrm{diam}(G) \leq 2 \left\lfloor \frac{\cosh^{-1}(n-1)}{\cosh^{-1}\left(\frac{\lambda_n(L)+\lambda_2(L)}{\lambda_n(L)-\lambda_2(L)}\right)} \right\rfloor + 1$$

and

$$\mathrm{diam}(G) \geq \left\lceil \frac{4}{n\lambda_2(L)} \right\rceil.$$

Recall that the mean distance $\bar{\rho}$ is the average over all distances between distinct vertices.

Theorem 14.4.3 ([426]). *The mean distance is bounded by Laplacian eigenvalues in the following way:*

$$\frac{1}{n-1}\left(\frac{2}{\lambda_2(L)} + \frac{n-2}{2}\right) \leq \bar{\rho}(G) \leq \frac{n}{n-1}\left\lceil \frac{\Delta+\lambda_2(L)}{4\lambda_2(L)} \ln(n-1) \right\rceil.$$

Connectivity

We already know that $\lambda_2(L)$ is non-zero if and only if the graph is connected. Fiedler [210] noticed more relations between $\lambda_2(L)$ and connectivity properties. $\lambda_2(L)$ is therefore also called the *algebraic connectivity*. We only cite his results.

Theorem 14.4.4 ([210]). *Let G be a graph and $\omega = \frac{\pi}{n}$. Let $\kappa(G)$ and $\eta(G)$ denote the minimum number of nodes and edges, respectively that have to be removed in order to make G disconnected. Then,*

1. $\lambda_2(L) \leq \kappa(G) \leq \eta(G)$,
2. $\lambda_2(L) \geq 2\eta(G)(1 - \cos\omega)$, and
3. $\lambda_2(L) \geq 2(\cos\omega - \cos 2\omega)\eta(G) - 2\cos\omega(1 - \cos\omega)\Delta(G)$.

Isoperimetric Number

The isoperimetric number is defined as follows:

$$i(G) := \min\left\{\frac{|E(X,Y)|}{\min\{|X|,|Y|\}};\ \emptyset \neq X \subsetneq V,\ Y = V \setminus X\right\},$$

where $E(X,Y)$ denotes the set of edges connecting X with Y. The definition characterizes $i(G)$ as the size of a smallest possible edge-cut separating as large a subset X as possible from the remaining larger part Y (we assume $|X| \leq |Y|$

w.l.o.g.). Hence, $i(G)$ is a measure of how many edges we have to remove from a network to isolate a large portion of nodes. This relates $i(G)$ to both, network connectivity and the min-bisection problem, which has important applications in VLSI design.

Obviously, by choosing $X = \{v\}$ with $d(v) = \delta(G)$, we see that

$$i(G) \leq \delta(G). \tag{14.11}$$

If G is disconnected, we have $i(G) = 0$.

A well-known relationship between the isoperimetric number and $\lambda_2(L)$ states that $i(G) \geq \frac{\lambda_2(L)}{2}$. For the case of $\lambda_2(L) \leq 2$, this bound is outperformed by the following result.

Theorem 14.4.5 ([176]). *The isoperimetric number is bounded from below by Laplacian eigenvalues in the following way:*

$$i(G) \geq \min\left\{1, \frac{\lambda_2(L)\lambda_n(L)}{2(\lambda_n(L) + \lambda_2(L) - 2)}\right\}.$$

We will now look at an upper bound following [425].

Proposition 14.4.6. *Let G be a graph with maximum degree Δ that is not a complete graph. Then,*

$$\lambda_2(L) \leq \Delta.$$

Proof. If G is disconnected, then $\lambda_2(L) = 0 \leq \Delta$. So let G be connected. Then G contains a path on three vertices P_3 as an induced subgraph. (Otherwise, G would be complete: Take two vertices i and j and by induction on the length of the shortest i-j path show that there exists an edge $\{i, j\}$.) By the examples in Table 14.1 in Section 14.1.6 we know that $\lambda_2(A(P_3)) = 0$. Hence, by the Interlacing Theorem, see (14.8), we get

$$0 = \lambda_2(A(P_3)) \leq \lambda_{n-1}(A).$$

We also know from Lemma 14.1.12 that

$$\lambda_2(L) \leq \Delta - \lambda_{n-1}(A) \leq \Delta.$$

$\square$

With the help of the preceding proposition, we can prove:

Theorem 14.4.7. *Let $G = (V, E)$ be a graph not equal to K_1, K_2 or K_3 with n vertices and m edges. Then*

$$i(G) \leq \sqrt{\lambda_2(L)(2\Delta - \lambda_2(L))}. \tag{14.12}$$

The proof takes up several pages and includes some more technical details. Nevertheless it is a nice proof and an interesting example, how we can obtain non-trivial bounds from Laplacian eigenvalues. It is therefore presented in an extra section 14.4.1 at the end of this section.

Expansion

Many times a good vertex expansion is a desired property of a network. One common definition capturing this property that 'all small sets of nodes have large neighborhoods' is the following.

$$c_V := \min\left\{ \frac{|N(S) \setminus S|}{|S|}; \ S \subseteq V, \ |S| \leq \frac{n}{2} \right\}.$$

A large vertex expansion c_V is crucial, e.g. for the construction of parallel sorting networks, superconcentrators, fault-tolerant networks, and networks for simulating random generators, the latter being an important means for derandomization via small sample spaces. Using probabilistic methods, one can generate a random network that will have good expansion properties almost surely, but it is hard to measure the expansion properties from the definition. Therefore the following bounds are very useful.

Theorem 14.4.8 ([19]). $\frac{\lambda_2(L)}{\frac{d}{2}+\lambda_2(L)} \leq c_V = \mathcal{O}(\sqrt{\lambda_2(L)}).$

A graph with expansion at least α is called an α-*magnifier*.

Let us prove a weak version of Theorem 14.4.8 for regular graphs.

Theorem 14.4.9 ([576]). *A d-regular graph is a* $\frac{\lambda_2(L)}{2d}$-*magnifier.*

Proof. Let $G = (V, E)$ be a d-regular graph on $V = \{1, \ldots, n\}$. Take a subset S of V of cardinality $s \leq n/2$. Define a vector $x \in \{s - n, s\}^n$ by

$$x_i := \begin{cases} s - n, & i \in S \\ s, & i \in V \setminus S. \end{cases}$$

By definition of L, $x^\top L x = x^\top (D - A) x = d \sum_{i=1}^n x_i^2 - \sum_{i=1}^n x_i \sum_{\{i,j\} \in E} x_j = \sum_{\{i,j\} \in E} (x_i - x_j)^2 = n^2 \cdot |E(S, V \setminus S)|$, where $E(S, V \setminus S)$ is the set of edges with exactly one end point in S. Since $\sum_i x_i = s(s - n) + (n - s)s = 0$, we see that x is perpendicular to the eigenvector $\mathbf{1}_n$ of L which corresponds to the eigenvalue 0. From Corollary 14.1.14 we conclude that $n^2 |(S, \bar{S})| = x^\top L x \geq \lambda_2(L) x^\top x = \lambda_2(L) s n(n - s)$, and consequently $|N(S)| \geq \frac{|(S,\bar{S})|}{d} \geq \frac{\lambda_2(L)s(n-s)}{dn} \geq \frac{\lambda_2(L)|S|}{2d}$ as claimed. $\qquad \square$

For results on the closely related Cheeger constants see [125].

Routing Number

Consider a set of (distinguishable) 'pebbles'. Initially, each pebble is located on a distinct vertex of the connected graph $G = (V, E)$, $|V| = n$. We are given a permutation π on V. The goal is to move each pebble which at the start is on vertex $i \in V$ to vertex $\pi(i)$. This has to be done in a number of steps of the following form. At each step, choose a set of disjoint edges $E_0 \subseteq E$ and interchange the pebbles at the endpoints of each edge in E_0.

The minimum number of steps required to accomplish the goal is denoted by $rt(G, \pi)$. The *routing number* $rt(G)$ of G is

$$rt(G) := \max_{\pi \in S_n} rt(G, \pi).$$

Now let G be connected and d-regular. We can upper bound the routing number in terms of λ_{n-1}. We only cite the following result.

Theorem 14.4.10 ([21]). *Let G be a d-regular connected graph. Then we have $\lambda_{n-1} < d$ and*

$$rt(G) = \mathcal{O}\left(\frac{d^2}{(d - \lambda_{n-1})^2} \log^2 n\right).$$

Chromatic Number

A *coloring* of a graph is an assignment of colors, e.g., natural numbers, to the vertices such that adjacent vertices have different colors. We speak of a *k-coloring* if the graph is assigned k different colors. The *chromatic number* of a graph is the minimum number of colors required for a coloring. We denote the chromatic number by $\chi(G)$.

It is well-known that computing the chromatic number is $\mathcal{NP}$-hard. However, eigenvalues provide us with lower and upper bounds. We consider only eigenvalues of the adjacency matrix.

Theorem 14.4.11 ([428]). *Let G be a graph. Then*

$$\chi(G) \le 1 + \lambda_n.$$

Proof. Let H be a subgraph of G without isolated vertices and such that

$$\chi(H) = \chi(G) \text{ and for every edge } e \text{ in } H \text{ we have } \chi(H - e) < \chi(H). \quad (14.13)$$

It is easy to show that such a subgraph always exists and that in fact $\chi(H - e) = \chi(H) - 1$. We have $\chi(H) \le \delta(H) + 1$. To see this, assume $\chi(H) > \delta(H) + 1$ and let $i \in V(H)$ be a vertex with $d_H(i) = \delta(H)$. Let $j \in N(i)$ and $e = \{i, j\}$. Then $k := \chi(H - e) = \chi(H) - 1 > \delta(H)$. Because $d_H(i) = \delta(H)$, we can construct a k-coloring of H from a k-coloring of $H - e$. This is a contradiction to (14.13).

From Lemma 14.4.1 and the Interlacing Theorem (14.8), we derive

$$\chi(G) - 1 = \chi(H) - 1 \le \delta(H) \le \lambda_n(H) \le \lambda_n(G).$$

$\square$

It is possible to actually find a $1 + \lambda_n$–coloring in polynomial time; see [428] for details.

Theorem 14.4.12 ([246, 268]). *Let G be a graph. Then*

$$1 - \frac{\lambda_n}{\lambda_1} \le \chi(G).$$

Theorem 14.4.13 ([135]). *Let G be a graph. Then*

$$\frac{n}{n - \lambda_n} \le \chi(G).$$

Independence Number

An *independent set* is a set of vertices such that none of them is adjacent to another vertex from that set. Independent sets are also called *stable sets*. For a graph G, the *independence number* $\alpha(G)$ is the cardinality of an independent set of maximal cardinality among all independent sets of G.

We will follow [427] in extending the following result due to Hoffman and Lovász [397].

Theorem 14.4.14. *Let G be a d-regular graph. Then*

$$\alpha(G) \leq n \left(1 - \frac{d}{\lambda_n(L)} \right).$$

Now, let $G = (V, E)$ be a graph with n vertices and degrees

$$d_1 \leq d_2 \ldots \leq d_n.$$

Set

$$\bar{d}_s := \frac{1}{s} \sum_{i \in \{1,\ldots,s\}} d_i \text{ for all } s \in \{1,\ldots,n\}.$$

Then the sequence $\bar{d}_1, \bar{d}_2, \ldots, \bar{d}_n$ is non-decreasing and for a d-regular graph $\bar{d}_s = d$ for all $s \in \{1,\ldots,n\}$.

Theorem 14.4.15. *Let s_0 be the smallest integer such that*

$$\bar{d}_{s_0} > \frac{\lambda_n(L)(n - s_0)}{n}.$$

Then

$$\alpha(G) \leq s_0 - 1,$$

and consequently

$$\alpha(G) \leq n \left(1 - \frac{\bar{d}_{s_0 - 1}}{\lambda_n(L)} \right).$$

Proof. We will show that

$$\frac{\lambda_n(L)(n - s)}{n} \geq \bar{d}_s \tag{14.14}$$

whenever there is a stable set of size $s > 1$ in G. To this end, let $S \subseteq V$ be a stable set of size $s := |S| > 1$. Equation (14.14) is true for $s = n$, since then $\bar{d}_i = 0$ for all $i \in \{1,\ldots,n\}$. So let $s < n$. Define $x \in \mathbb{R}^n$ by

$$x_i := \begin{cases} 0 & \text{if } i \in S \\ 1 & \text{otherwise} \end{cases}.$$

Then x is non-constant, i.e., not a multiple of $\mathbf{1}$. By Fiedler's characterization, Theorem 14.1.15, equation (14.7), we have

$$\lambda_n(L) \geq n \frac{\sum_{\{i,j\} \in E} (x_i - x_j)^2}{\sum_{\{i,j\} \in \binom{V}{2}} (x_i - x_j)^2} = n \frac{|E(S, V \setminus S)|}{s(n-s)}.$$

Because S contains no edges, we have $|E(S, V \setminus S)| \geq s\bar{d}_s$, and so

$$\lambda_n(L) \geq n \frac{s\bar{d}_s}{s(n-s)} = n \frac{\bar{d}_s}{n-s}.$$

This yields (14.14).

Now let s_0 be the first integer that violates (14.14). Then, there is no stable set of cardinality s_0. Because $(\bar{d}_s)_{s=1,\dots,n}$ is non-decreasing, all $s \geq s_0$ violate (14.14) as well, and so there cannot be a stable set larger than $s_0 - 1$.   □

Bisection Width

Given a graph with an even number of vertices, the Minimum Bisection problem aims at partitioning the vertices into two classes of equal size that are connected by as few edges as possible. The minimum number of edges between the two classes is called the *bisection width* of the graph. The decision version of the Minimum Bisection problem is $\mathcal{NP}$-complete [241], and the currently best polynomial approximation algorithm is guaranteed only to stay within a multiplicative error of $\mathcal{O}(\log^2 n)$ [203]. The following bound on the bisection width is a special case of a result of Alon and Milman [23].

Lemma 14.4.16. *Let $G = (V, E)$ be a graph on $\{1, \dots, n\}$, where n is an even positive integer, then*

$$bw(G) \geq \frac{n}{4} \lambda_2(L).$$

Proof. Let S be an arbitrary subset of V of cardinality $\frac{n}{2}$ and define

$$x_i := \begin{cases} 1, & i \in S \\ -1, & i \notin S \end{cases}$$

for all $i \in V$. Then $\sum_{i \in V} x_i = 0$. Hence $x \perp \mathbf{1}_n$ and by Corollary 14.1.14 and Lemma 14.1.8 we have $n\lambda_2(L) = x^\top x \lambda_2(L) \leq x^\top Lx = \sum_{\{i,j\} \in E} (x_i - x_j)^2 = \sum_{\{i,j\} \in E(S, V \setminus S)} (x_i - x_j)^2 = 4 \cdot |E(S, V \setminus S)|$. Choosing S as one class of a minimum bisection yields the claim.   □

Bezrukow et al. [60] have shown that this bound is tight if and only if all vertices are incident with exactly $\frac{\lambda_2(L)}{2}$ cut edges, which is true, e.g., for complete graphs, complete bipartite graphs, hypercubes, and the Petersen graph. However, for the $\sqrt{n} \times \sqrt{n}$ grid graph, the bisection width is $\sqrt{n}$ while $\lambda_2(L) = 2 - 2\cos(\pi/\sqrt{n}) \approx \pi^2/n$ [60] and hence $\frac{n}{4} \cdot \lambda_2(L) \approx \frac{\pi^2}{4}$. So the gap between the optimal bisection width and the bound of Lemma 14.4.16 can be large.

As mentioned earlier, the bisection width is closely related to the isoperimetric number: directly from the definition of $i(G)$ and $bw(G)$ we obtain $i(G) \leq \frac{2bw(G)}{n}$. Hence, lower bounds on $i(G)$ yield lower bounds on $bw(G)$.

14.4.1   Proof of Theorem 14.4.7

Recall that $G = (V, E)$ is a graph not equal to K_1, K_2 or K_3 with n vertices and m edges. We have to prove

$$i(G) \leq \sqrt{\lambda_2(L)(2\Delta - \lambda_2(L))}. \tag{14.15}$$

For the proof, let us write $\lambda := \lambda_2(L)$, $\delta := \delta(G)$, and $\Delta := \Delta(G)$.

If $\lambda = 0$, then G is disconnected and so $i(G) = 0$, and we are done. The case that G is a complete graph on $n \geq 4$ vertices or more can be dealt with easily by using $\lambda = n$, see 14.1.6.

Hence we may assume that G is not a complete graph. Then by Proposition 14.4.6, we have $\lambda \leq \Delta$. If $\delta < \lambda$, we have

$$\lambda(2\Delta - \lambda) > \delta(2\Delta - \lambda) \geq \delta(2\Delta - \Delta) = \delta\Delta \geq \delta^2 \underset{(14.11)}{\geq} i(G)^2.$$

This is (14.15). We now may assume that

$$\lambda \leq \delta. \tag{14.16}$$

Let $y \in \mathbb{R}^n$ be an eigenvector for λ and set $W := \{i \in V;\ y_i > 0\}$. Then, perhaps after switching from y to $-y$, we have $|W| = \min\{|W|, |V \setminus W|\}$. Define

$$g_i := \begin{cases} y_i & \text{if } i \in W \\ 0 & \text{otherwise} \end{cases}.$$

Let $E(W) \subseteq E$ be the edges between vertices from W. We have

$$\sum_{i \in W} \left(d(i)y_i \pm \sum_{j:\, \{i,j\} \in E} y_j \right) y_i$$

$$= \sum_{i \in W} \sum_{j:\, \{i,j\} \in E} (y_i \pm y_j)y_i$$

$$= \sum_{\{i,j\} \in E(W)} ((y_i \pm y_j)y_i + (y_j \pm y_i)y_j) + \sum_{\{i,j\} \in E(W, V \setminus W)} (y_i \pm y_j)y_i$$

$$= \sum_{\{i,j\} \in E(W)} (y_i \pm y_j)^2 + \sum_{\{i,j\} \in E(W, V \setminus W)} (y_i \pm y_j)y_i \tag{14.17}$$

$$= \sum_{\{i,j\} \in E(W)} (y_i \pm y_j)^2 + \sum_{i \in W} d(i)y_i^2 \pm \sum_{\{i,j\} \in E(W, V \setminus W)} y_j y_i$$

$$= \sum_{\{i,j\} \in E} (g_i \pm g_j)^2 - \sum_{i \in W} d(i)g_i^2 + \sum_{i \in W} d(i)y_i^2 \pm \sum_{\{i,j\} \in E(W, V \setminus W)} y_j y_i$$

$$= \sum_{\{i,j\} \in E} (g_i \pm g_j)^2 \pm \sum_{\{i,j\} \in E(W, V \setminus W)} y_j y_i.$$

Keep in mind that when summing over all edges $\{i, j\} \in E$, the terms must not depend on which end of the edge actually is i and which is j. Observe that this is always the case here, e.g., in the preceding calculation because $(g_i \pm g_j)^2 = (g_j \pm g_i)^2$ for all $i, j \in V$.

Using the eigenvalue property of y we get

$$\lambda y_i = d(i)y_i - \sum_{j:\, \{i,j\}\in E} y_j \quad \forall i \in V. \tag{14.18}$$

This, together with the '$-$' version of (14.17), yields

$$\lambda \sum_{i\in W} y_i^2 = \sum_{i\in W} \left(d(i)y_i - \sum_{j:\, \{i,j\}\in E} y_j \right) y_i$$

$$= \sum_{(14.17)\ \{i,j\}\in E} (g_i - g_j)^2 - \sum_{\{i,j\}\in E(W,V\setminus W)} y_j y_i, \tag{14.19}$$

and using the '$+$' version of (14.17)

$$(2\Delta - \lambda) \sum_{i\in W} y_i^2$$

$$= 2\Delta \sum_{i\in W} y_i^2 - \sum_{i\in W} d(i)y_i^2 + \sum_{i\in W} \sum_{j:\, \{i,j\}\in E} y_j y_i$$

$$\geq \sum_{i\in W} d(i)y_i^2 + \sum_{i\in W} \sum_{j:\, \{i,j\}\in E} y_j y_i \tag{14.20}$$

$$= \sum_{(14.17)\ \{i,j\}\in E} (g_i + g_j)^2 + \sum_{\{i,j\}\in E(W,V\setminus W)} y_j y_i.$$

Set $\alpha := \sum_{\{i,j\}\in E(W,V\setminus W)} y_i y_j$. Because the left and right hand sides of (14.19) are not negative, using (14.20) we derive

$$\lambda(2\Delta - \lambda) \left(\sum_{i\in W} y_i^2 \right)^2$$

$$\geq \sum_{\{i,j\}\in E} (g_i + g_j)^2 \sum_{\{i,j\}\in E} (g_i - g_j)^2$$

$$+ \alpha \left(\sum_{\{i,j\}\in E} (g_i - g_j)^2 - \sum_{\{i,j\}\in E} (g_i + g_j)^2 \right) - \alpha^2 \tag{14.21}$$

$$= \sum_{\{i,j\}\in E} (g_i + g_j)^2 \sum_{\{i,j\}\in E} (g_i - g_j)^2 - \alpha \left(4 \sum_{\{i,j\}\in E(W)} y_i y_j + \alpha \right).$$

We would like to drop the 'α' term completely in this equation. To this end, observe that by the definition of W we have $\alpha \leq 0$. Furthermore, using again the eigenvalue property of λ (see also (14.18)), we have

$$4 \sum_{\{i,j\}\in E(W)} y_i y_j + \alpha$$

$$= 2 \sum_{\{i,j\}\in E(W)} y_i y_j + 2 \sum_{\{i,j\}\in E(W)} y_i y_j + \sum_{\{i,j\}\in E(W,V\setminus W)} y_i y_j$$

$$= 2 \sum_{\{i,j\}\in E(W)} y_i y_j + \sum_{i\in W} y_i \sum_{j:\{i,j\}\in E} y_j$$

$$= 2 \underbrace{\sum_{\{i,j\}\in E(W)} y_i y_j}_{\geq 0} + \sum_{i\in W} \underbrace{(d(i) - \lambda) y_i^2}_{\geq 0}$$

$$\geq 0,$$

because of the definition of W and (14.16). We thus have by (14.21)

$$\lambda(2\Delta - \lambda) \left(\sum_{i\in W} y_i^2 \right)^2 \geq \sum_{\{i,j\}\in E} (g_i + g_j)^2 \sum_{\{i,j\}\in E} (g_i - g_j)^2. \tag{14.22}$$

Define $v, w \in \mathbb{R}^m$ by $v_{\{i,j\}} := g_i + g_j$ and $w_{\{i,j\}} := |g_i - g_j|$. We apply the Cauchy-Schwartz inequality to v and w, getting

$$\left(\sum_{\{i,j\}\in E} |g_i^2 - g_j^2| \right)^2$$

$$= \left(\sum_{\{i,j\}\in E} (g_i + g_j)|g_i - g_j| \right)^2$$

$$= \langle v, w \rangle^2$$

$$\leq \|v\|^2 \|w\|^2 \tag{14.23}$$

$$= \sum_{\{i,j\}\in E} (g_i + g_j)^2 \sum_{\{i,j\}\in E} (g_i - g_j)^2$$

$$\underset{(14.22)}{\leq} \lambda(2\Delta - \lambda) \left(\sum_{i\in W} y_i^2 \right)^2.$$

We will now bound this from below. Let $0 = t_0 < t_1 < \ldots < t_N$ be all the different values of the components of g. Define $V_k := \{i \in V; g_i \geq t_k\}$ for $k \in \{0, \ldots, N\}$ and for convenience $V_{N+1} := \emptyset$. Then, for $k \in \{1, \ldots, N+1\}$ we have $V_k \subseteq W$ and therefore $|V_k| \leq |W|$, hence $|V_k| = \min\{|V_k|, |V \setminus V_k|\}$. It also holds that $V_N \subseteq V_{N-1} \subseteq \ldots V_1 = W \subseteq V_0 = V$ and that $|V_k| - |V_{k+1}|$ is the number of entries in g equal to t_k for all $k \in \{0, \ldots, N\}$.

We will later show that we can express the sum $\sum_{\{i,j\}\in E} |g_i^2 - g_j^2|$ in a convenient way:

$$\sum_{\{i,j\}\in E} \left|g_i^2 - g_j^2\right| = \sum_{k=1}^{N} \sum_{\substack{\{i,j\}\in E \\ g_i < g_j = t_k}} (g_j^2 - g_i^2)$$

$$\underset{\text{see below}}{=} \sum_{k=1}^{N} \sum_{\{i,j\}\in E(V_k,V\setminus V_k)} (t_k^2 - t_{k-1}^2) = \sum_{k=1}^{N} |E(V_k, V\setminus V_k)|\,(t_k^2 - t_{k-1}^2)$$

$$\geq i(G) \sum_{k=1}^{N} |V_k|\,(t_k^2 - t_{k-1}^2) = i(G) \sum_{k=0}^{N} t_k^2(|V_k| - |V_{k+1}|),$$

since $V_{N+1} = \emptyset$ and $t_0 = 0$. Now we can conclude that $\sum_{\{i,j\}\in E}\left|g_i^2 - g_j^2\right| \geq i(G)\sum_{i\in V} g_i^2 = i(G)\sum_{i\in W} y_i^2$. This together with (14.23) yields the claim of the theorem.

We now only have left to prove the validity of the transformation of the sum, i.e.,

$$\sum_{k=1}^{N} \sum_{\substack{\{i,j\}\in E \\ g_i < g_j = t_k}} (g_j^2 - g_i^2) = \sum_{k=1}^{N} \sum_{\{i,j\}\in E(V_k,V\setminus V_k)} (t_k^2 - t_{k-1}^2). \tag{14.24}$$

This will be done by induction on N. The case $N = 1$ is clear. So let $N > 1$ and assume that (14.24) has already been proven for instances with $N - 1$ instead of N, i.e., instances where we have a vector $\tilde{g}$ on a graph $\tilde{G} = (\tilde{V}, \tilde{E})$ assuming only N different values $0 = \tilde{t}_0 < \ldots < \tilde{t}_{N-1}$ on its components and where subsets $\tilde{V}_{N-1} \subseteq \tilde{V}_{N-2} \subseteq \ldots \tilde{V}_1 = \tilde{W} \subseteq \tilde{V}_0 = \tilde{V}$ are defined accordingly.

We will make use of this for the following instance. Define $\tilde{G} := G - V_N$ (the vertices and edges of $\tilde{G}$ are $\tilde{V}$ and $\tilde{E}$, respectively) and let $\tilde{g}$ be the restriction of g on $\tilde{V}$. We then have $\tilde{t}_k = t_k$ for all $k \in \{0, \ldots, N-1\}$. If we then define the sets $\tilde{V}_k$ accordingly, we also have $\tilde{V}_k = V_k \setminus V_N$ for all $k \in \{0, \ldots, N-1\}$. Note that $V_N \subseteq V_k$ for all $k \in \{0, \ldots, N-1\}$, so the sets $\tilde{V}_k$ differ from the sets V_k exactly by the vertices in V_N.

By induction, we have

$$\sum_{k=1}^{N} \sum_{\substack{\{i,j\}\in E \\ g_i<g_j=t_k}} (g_j^2 - g_i^2)$$

$$= \sum_{k=1}^{N-1} \sum_{\substack{\{i,j\}\in E \\ g_i<g_j=t_k}} (g_j^2 - g_i^2) + \sum_{\substack{\{i,j\}\in E \\ g_i<g_j=t_N}} (g_j^2 - g_i^2)$$

$$= \sum_{k=1}^{N-1} \sum_{\substack{\{i,j\}\in E \\ \tilde{g}_i<\tilde{g}_j=t_k}} (\tilde{g}_j^2 - \tilde{g}_i^2) + \sum_{\substack{\{i,j\}\in E \\ g_i<g_j=t_N}} (g_j^2 - g_i^2) \qquad (14.25)$$

$$\underset{\text{induction}}{=} \sum_{k=1}^{N-1} \sum_{\{i,j\}\in E_{\tilde{G}}(\tilde{V}_k,\tilde{V}\setminus\tilde{V}_k)} (\tilde{t}_k^2 - \tilde{t}_{k-1}^2) + \sum_{\substack{\{i,j\}\in E \\ g_i<g_j=t_N}} (g_j^2 - g_i^2)$$

$$= \underbrace{\sum_{k=1}^{N-1} \sum_{\{i,j\}\in E_{\tilde{G}}(\tilde{V}_k,\tilde{V}\setminus\tilde{V}_k)} (t_k^2 - t_{k-1}^2)}_{(*)} + \sum_{\substack{\{i,j\}\in E \\ g_i<g_j=t_N}} (g_j^2 - g_i^2).$$

Observe that the cuts $E_{\tilde{G}}(\tilde{V}_k, \tilde{V} \setminus \tilde{V}_k)$ only consist of edges in $\tilde{E}$. If we switch to cuts in G, we have to subtract some edges afterwards, namely those with one end in V_N. This way, we get for the sum $(*)$ the following:

$$\sum_{k=1}^{N-1} \sum_{\{i,j\}\in E_{\tilde{G}}(\tilde{V}_k,\tilde{V}\setminus\tilde{V}_k)} (t_k^2 - t_{k-1}^2)$$

$$= \sum_{k=1}^{N-1} \sum_{\{i,j\}\in E(V_k,V\setminus V_k)} (t_k^2 - t_{k-1}^2) - \underbrace{\sum_{k=1}^{N-1} \sum_{\substack{\{i,j\}\in E(V_k,V\setminus V_k) \\ j\in V_N}} (t_k^2 - t_{k-1}^2)}_{(+)}. \qquad (14.26)$$

We will inspect the 'corrective' term $(+)$ closer. To this end, for each $i \in V$ let $k(i)$ be the smallest index such that $i \in V \setminus V_{k(i)}$. We then have $g_i = t_{k(i)-1}$ for all $i \in V$ and

$$\sum_{\substack{k=1}}^{N-1} \sum_{\substack{\{i,j\}\in E(V_k,V\setminus V_k) \\ j\in V_N}} (t_k^2 - t_{k-1}^2)$$

$$= \sum_{\substack{\{i,j\}\in E(V_N,V\setminus V_N) \\ j\in V_N}} \sum_{\substack{k\in\{1,\dots,N-1\} \\ i\in V\setminus V_k}} (t_k^2 - t_{k-1}^2)$$

$$= \sum_{\substack{\{i,j\}\in E(V_N,V\setminus V_N) \\ j\in V_N}} \sum_{k=k(i)}^{N-1} (t_k^2 - t_{k-1}^2) \tag{14.27}$$

$$\underset{\text{telescope}}{=} \sum_{\substack{\{i,j\}\in E(V_N,V\setminus V_N) \\ j\in V_N}} (t_{N-1}^2 - t_{k(i)-1}^2)$$

$$= \sum_{\substack{\{i,j\}\in E(V_N,V\setminus V_N) \\ j\in V_N}} (t_{N-1}^2 - g_i^2).$$

We can now continue our work on (14.25). Note that

$$\{(i,j);\ \{i,j\}\in E(V_N, V\setminus V_N), j\in V_N\}$$
$$= \{(i,j);\ \{i,j\}\in E, g_i < g_j = t_N\}. \tag{14.28}$$

This will later allow us to combine the last sum from (14.25) and that from (14.27).

Putting everything together, we see:

$$\sum_{\substack{k=1}}^{N} \sum_{\substack{\{i,j\}\in E \\ g_i<g_j=t_k}} (g_j^2 - g_i^2)$$

$$\underset{(14.25)}{=} \underbrace{\sum_{k=1}^{N-1} \sum_{\{i,j\}\in E_{\bar{G}}(\tilde{V}_k,\tilde{V}\setminus\tilde{V}_k)} (t_k^2 - t_{k-1}^2)}_{(*)} + \sum_{\substack{\{i,j\}\in E \\ g_i<g_j=t_N}} (g_j^2 - g_i^2)$$

$$\underset{(14.26)}{=} \sum_{k=1}^{N-1} \sum_{\{i,j\}\in E(V_k,V\setminus V_k)} (t_k^2 - t_{k-1}^2) - \underbrace{\sum_{k=1}^{N-1} \sum_{\substack{\{i,j\}\in E(V_k,V\setminus V_k) \\ i\in V_N}} (t_k^2 - t_{k-1}^2)}_{(+)}$$

$$+ \sum_{\substack{\{i,j\}\in E \\ g_i<g_j=t_N}} (g_j^2 - g_i^2).$$

Apply (14.27) to the $(+)$ term and get

$$\ldots = \sum_{k=1}^{N-1} \sum_{\{i,j\} \in E(V_k, V \setminus V_k)} (t_k^2 - t_{k-1}^2) - \sum_{\substack{\{i,j\} \in E(V_N, V \setminus V_N) \\ j \in V_N}} (t_{N-1}^2 - g_i^2)$$

$$+ \sum_{\substack{\{i,j\} \in E \\ g_i < g_j = t_N}} (\underbrace{g_j^2}_{=t_N^2} - g_i^2)$$

$$\underset{(14.28)}{=} \sum_{k=1}^{N-1} \sum_{\{i,j\} \in E(V_k, V \setminus V_k)} (t_k^2 - t_{k-1}^2)$$

$$+ \sum_{\substack{\{i,j\} \in E(V_N, V \setminus V_N) \\ j \in V_N}} (t_N^2 - g_i^2 - t_{N-1}^2 + g_i^2)$$

$$= \sum_{k=1}^{N-1} \sum_{\{i,j\} \in E(V_k, V \setminus V_k)} (t_k^2 - t_{k-1}^2) + \sum_{\substack{\{i,j\} \in E(V_N, V \setminus V_N) \\ j \in V_N}} (t_N^2 - t_{N-1}^2)$$

$$= \sum_{k=1}^{N} \sum_{\{i,j\} \in E(V_k, V \setminus V_k)} (t_k^2 - t_{k-1}^2).$$

This is (14.24). □

14.5   Heuristics for Graph Identification

We know several models for random graphs from Chapter 13. Given some graph, which one of these models describes it best? In this chapter, we will point out some ideas for using spectral methods to recognize graphs from different random graph models.

We will only consider the adjacency spectrum in this chapter. In the first section we will define some new graph parameters based on this spectrum. In the subsequent sections we will look at histogram plots of the spectrum and at the behavior of these new parameters for different models of random graphs.

In a final section, we briefly review some analytical results concerning the adjacency spectrum of random power law graphs.

14.5.1   New Graph Statistics

The spectrum of a graph as well as the set of its eigenvectors are graph statistics, more precisely they are global distributions. See Section 11.7 in Chapter 11 for a general discussion of graph statistics.

A multigraph is — up to isomorphy — fully determined by its spectrum plus eigenvectors [135, Theorem 1.8, p. 44]. It seems reasonable to define new statistics which store only selected parts of that information in order to retrieve relevant characteristics.

Inverse Participation Ratio. Recall that we can interpret an eigenvector as an assignment of weights to the vertices; see Remark 14.1.2. We will ask, how these weights are distributed. Are there a few vertices that have relatively large weights compared to the rest (we also speak of a high localization in this case), or are the weights very similar for all vertices?

Let G be a graph and let $w_1, \ldots, w_n$ be the *normalized* eigenvectors of its adjacency matrix. We define the *inverse participation ratio* [198] of the jth eigenvector as

$$I_j(G) := \sum_{k=1}^n \left((w_j)_k\right)^4 . \qquad (14.29)$$

We also write I_j when dealing with only one graph at a time.

Because the eigenvectors are normalized, we have $I_j(G) \in [\frac{1}{n}, 1]$ for all $j \in \{1, \ldots, n\}$. The inverse participation ratio measures an eigenvector's extent of localization. This becomes clear by considering extreme cases. First let w_j define perfectly evenly distributed weights on the graph, i.e., $(w_j)_k := \frac{1}{\sqrt{n}}$ for all $k \in \{1, \ldots, n\}$. Then, $I_j = \frac{1}{n}$, i.e., the inverse participation ration attains its minimal value.

On the other hand, if for an arbitrary vertex $k_0 \in V$, we have

$$(w_j)_k = \begin{cases} 1 & \text{if } k = k_0 \\ 0 & \text{otherwise} \end{cases},$$

then $I_j = 1$, i.e., the inverse participation ration attains its maximum value.

Offset of Largest Eigenvalue. In certain classes of graphs, the largest eigenvalue tends to 'escape' from the rest of the spectrum. For example, this is the case in certain models of random graphs. It therefore is interesting to study the quantity

$$R(G) := \frac{\lambda_n - \lambda_{n-1}}{\lambda_{n-1} - \lambda_1}. \qquad (14.30)$$

$R(G)$ captures the offset of λ_n from the second largest eigenvalue λ_{n-1} normalized with respect to the range of the rest of the spectrum.

There is a correlation between $R(G)$ and the chromatic number $\chi(G)$, as was noticed in [198]. Define ε by the equation

$$-\lambda_1 = \lambda_{n-1} + \varepsilon.$$

For $|\varepsilon|$ small, the smallest interval containing all eigenvalues except λ_n is almost centered around 0. (For certain random graphs, $|\varepsilon|$ is in fact small.) Recall from 14.4 that there are bounds on the chromatic number in terms of extremal eigenvalues. In particular, we have by Theorem 14.4.12,

$$\frac{\chi(G)}{2} - 1 \geq \frac{-\lambda_n}{2\lambda_1} - \frac{1}{2} = \frac{-\lambda_n - \lambda_1}{2\lambda_1}$$

$$= \frac{-\lambda_n + \lambda_{n-1} + \varepsilon}{\lambda_1 - \lambda_{n-1} - \varepsilon} = \frac{\lambda_n - \lambda_{n-1} - \varepsilon}{\lambda_{n-1} - \lambda_1 + \varepsilon}.$$

So, for graphs with $\varepsilon = 0$ the chromatic number is lower bounded in terms of $R(G)$ as

$$\chi(G) \geq 2R(G) + 2.$$

For small $|\varepsilon|$, this equation obviously still holds in an approximate sense.

14.5.2 Spectral Properties of Random Graphs

Can random graphs from different models be distinguished by looking at spectral properties? One way of approaching this problem is to study the outcome of a large number of random experiments. We will present some results of some such experiments and discuss possible interpretations. The three random models under consideration are the model by Erdős and Rényi, $\mathcal{G}(n, p)$, the scale-free graphs by Barabási and Albert — see 13.1.4 — and the small-world graphs by Watts and Strogatz — see 13.1.2.

First, we investigate how these models differ in their spectra by looking at histogram plots of the eigenvalues of a number of randomly generated graphs. Then, we compare the inverse participation ratios, defined in (14.29). Finally we also look at the offsets of the largest eigenvalues, defined in (14.30). Our focus will be on $\mathcal{G}(n, p)$ and Barabási-Albert-like graphs. For details on small-world graphs, see [198].

For our numerical experiments, we wrote a small C program (about 600 lines of code). To compute the eigenvalues and later (for the inverse participation ratios) also the eigenvectors, we use the function `ssyev` from the Sun Performance Library [537]. The actual plots were made using `gnuplot` [588].

Spectra. Let us start with $\mathcal{G}(n, p)$. We generated 100 graphs randomly according to $\mathcal{G}(2000, \frac{1}{2})$ and computed their spectra. A first observation is that the largest eigenvalue is significantly set off from the rest of the spectrum in all of these experiments. Because this offset will be explicitly regarded later, we for now exclude the largest eigenvalue from our considerations. So, when talking of 'spectrum', for the rest of this section, we mean 'spectrum without the largest eigenvalue'.

To get a visualization of all the spectra, we computed a histogram of the eigenvalues of all 100 random graphs. This technique will be used for the other random graph models as well. The quantity approximated by those histograms is also known as the *spectral density*. Although we do not introduce this notion formally [198], we will in the following occasionally use that term when referring to the way eigenvalues are distributed.

All histograms, after being normalized to comprise an area of 1, were scaled in the following way. Let $\lambda_1, \ldots, \lambda_N$ be the eigenvalues. (In our case $N = 100 \cdot (2000 - 1)$, for we have 100 graphs with 2000 eigenvalues each, and we exclude the largest of them for each graph). Define $\bar{\lambda}$ as the mean of all these values

$$\bar{\lambda} = \frac{1}{N} \sum_{i=1}^{N} \lambda_i.$$

Then, we compute the standard deviation

$$\sigma = \sqrt{\sum_{i=1}^{N} (\lambda_i - \bar{\lambda})^2}.$$

The x-axis of our plots is scaled by $\frac{1}{\sigma}$ and the y-axis is scaled by σ. This makes it easy to compare spectra of graphs of different sizes.

Figure 14.10 shows the scaled histogram. The semicircle form actually is no surprise. It follows from a classical result from random matrix theory known as the *semicircle law*. It originally is due to Wigner [584, 585] and has later been refined by a number of researchers [34, 233, 337].

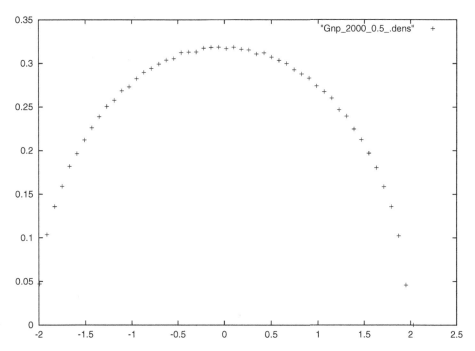

Fig. 14.10. A histogram of the union of the spectra of 100 random graphs from $\mathcal{G}(2000, \frac{1}{2})$

Now we look at graphs that originate from an evolutionary process with preferential attachment, like the Barabási-Albert graphs. We implemented a very simple algorithm for creating graphs with preferential attachment. There is only one parameter m (apart from the number of vertices n). The process starts with m unconnected vertices. Whenever a new vertex arrives, it is connected with m edges to older vertices. To this end, the algorithm m times chooses a vertex out of the set of old vertices at random. The probability of an old vertex being

chosen is proportional to its degree; at the very first step, when all vertices have
degree 0, each of them is equally likely to be chosen.

This way of generating preferential attachment graphs is slightly different
from what usually is considered in that multiple edges are allowed. However,
the resulting histograms of the spectra look very similar to those of [198], where
multiple edges do not occur. See Figure 14.11 for our results. For comparison,
Figure 14.12 shows plots for our preferential attachment graphs and for $\mathcal{G}(n, p)$
graphs in one figure.

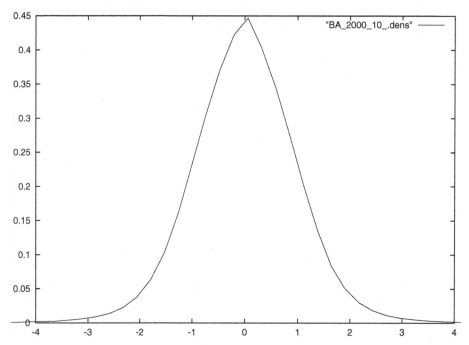

Fig. 14.11. A smoothed histogram of the union of the spectra of 100 random graphs
with preferential attachment on 2000 vertices and with $m = 10$

The preferential attachment graph obviously has significantly more small
eigenvalues than the $\mathcal{G}(n, p)$ graph. Since it essentially is connected[2] these small
eigenvalues cannot originate from small connected components. In [198] it is
suggested that these eigenvalues belong to eigenvectors that are highly localized.
This is supported by high inverse participation ratios, which we will observe later.

An interesting case for our studies of $\mathcal{G}(n, p)$ are *sparse random graphs*. We
used to look at large values of n, for which according to the semicircle law, our
histograms should more and more (with increasing n) resemble a semicircle. In

---

[2] There might be a number of less than $m = 10$ isolated vertices.

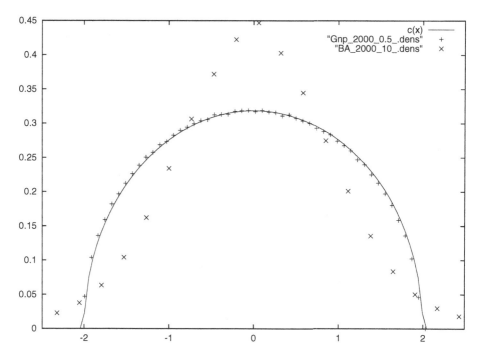

Fig. 14.12. Histograms of spectra of 100 random graphs with preferential attachment ($n = 2000$, $m = 10$) and from $\mathcal{G}(2000, \frac{1}{2})$ each. The solid line marks an ideal semicircle

the case displayed in Figure 14.10 we have an expected degree of approximately[3] $pn = \frac{1}{2}2000 = 1000$. This quantity pn is also referred to as the *connectivity* of the graph, although we will avoid this term, because it has already been extensively used in another context (see Chapter 7).

What if we look at graphs with fewer edges than in Figure 14.10, say, graphs in $\mathcal{G}(n, p)$ where $pn = 5$? Figure 14.13 shows that the spectral densities of such graphs rise above the semicircle in the vicinity of 0. Reasons for this have been discussed in [198].

One can also notice small peaks at larger eigenvalues. They become more obvious in plots of graphs with even lower expected degree. To study them in more detail, it is helpful to look at the *cumulative distribution* of eigenvalues, i.e., at the function

$$x \mapsto |\{i;\ \lambda_i \leq x\}|$$

Figure 14.14 shows an examplary plot.

---

[3] The expected degree is in fact $p(n-1)$. To see this, fix a vertex $i \in V$. This vertex has $n-1$ potential neighbors. Let $X_1, \ldots, X_{n-1}$ be 0/1 random variables, where X_j indicates whether or not an edge is drawn from i to j. If we put $X := \sum_{j=1}^{n-1} X_i$, then $\mathbb{E}[X]$ is the expected degree of i. The claim follows from linearity of expectation.

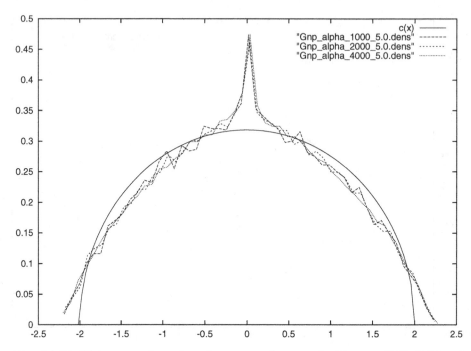

Fig. 14.13. Histograms of spectra of sparse random graphs. For $n = 1000$, 2000 and 4000 each we created 10 random graphs from $\mathcal{G}(n, p)$ with p set to satisfy $pn = 5$. The solid line marks an ideal semicircle

In [49], the reasons for the peaks are investigated. It is argued that they originate from small connected components and, for $pn > 1$, also from small trees grafted on the giant component.

Inverse Participation Ratios. Looking at Figure 14.12 has already lead us to conjecture that our graphs with preferential attachment have highly localized eigenvectors. To investigate this further, we created random graphs in all three models and plotted the inverse participation ratios of their eigenvectors. Figure 14.15 shows all three plots. Each circle represents an eigenvector. The x position of the circle corresponds to the eigenvalue and the y position to the inverse participation ratio. Note that in the second plot (which shows the results for the Barabási-Albert-like graph), there are even eigenvectors with inverse participation ratio near to 1.

It is also interesting that we obtain very distinguishing plots even for small numbers of vertices; all considered graphs only had 100 vertices each. We randomly created more such graphs and always could observe the same characteristics. Obviously one can recognize the $\mathcal{G}(n, p)$ by the fact that all eigenvectors have their inverse participation ratios rather evenly distributed in a small band at the bottom of the diagram. The Barabási-Albert-like graph exhibits as a salient

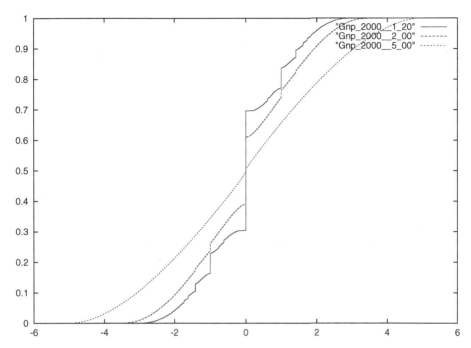

Fig. 14.14. The cumulative distribution of eigenvalues for sparse random graphs. For $n = 2000$ and $pn = 1.2$, 2.00 and 5.00 each we created 10 random graphs from $\mathcal{G}(n, p)$

feature eigenvectors with high inverse participation ratio. The small-world graph shows a very asymmetric structure.

We were not able to produce likewise distinguishing plots of the spectral density for such small graphs.

Offset of Largest Eigenvalue. We already mentioned that in the $\mathcal{G}(n, p)$ model as well as in the graphs with preferential attachment, the largest eigenvalue is significantly set off from the rest of the spectrum. The offset R, see (14.30), in these two models differs from that in the small-world graphs by several orders of magnitude. But there is also a difference between the R values of sparse $\mathcal{G}(n, p)$ graphs and graphs with preferential attachment. For the former, with increasing number of vertices and constant average degree, the R values seem to stay constant, while for the latter they decrease. See [198] for numerical results.

14.5.3 Random Power Law Graphs

In [127, 128], first the adjacency spectrum of a very general class of random graphs is examined. Given a sequence $\boldsymbol{w} = (w_1, w_2, \ldots, w_n)$ of non-negative reals satisfying

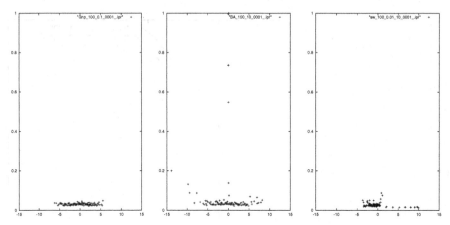

Fig. 14.15. Inverse participation ratios of three random graphs. The models are: $\mathcal{G}(n,p)$ (`Gnp_100_0.1_0001_ipr`), preferential attachment (`BA_100_10_0001_ipr`), small-world (`sw_100_0.01_10_0001_ipr`). All graphs have an expected degree of 10

$$\max_{i\in\{1,\ldots,n\}} w_i^2 < \sum_{i=1}^{n} w_i,$$

a random graph $G(\boldsymbol{w})$ is constructed by inserting an edge between vertices i and j with probability

$$\frac{w_i w_j}{\sum_{k=1}^{n} w_k}.$$

It is easy to verify that in such a graph, vertex i has expected degree w_i.

A useful notation is the *second order average degree*, defined by

$$\tilde{d} := \frac{\sum_{i=1}^{n} w_i^2}{\sum_{i=1}^{n} w_i}.$$

Furthermore, we denote the largest expected degree by m and the average expected degree by d.

We can prove results on the largest eigenvalue of the adjacency spectrum of a random graph from $G(\boldsymbol{w})$, which hold almost surely[4] and under certain conditions on $\tilde{d}$ and m. We can also make such statements on the k largest eigenvalues, provided that $\tilde{d}$, m, and the k largest expected degrees behave appropriately.

An interesting application of these results concerns random power law graphs: we can choose the sequence $\boldsymbol{w}$ suitably, so that the expected number of vertices of degree k is proportional to $k^{-\beta}$, for some given β. Under consideration were values of $\beta > 2$.

Theorem 14.5.1 ([127, 128]). *Let G be a random power law graph with exponent β and adjacency spectrum $\lambda_1, \ldots, \lambda_n$.*

---

[4] I.e., with probability tending to 1 as n tends to ∞.

1. *For $\beta \geq 3$ and*
$$m > d^2 \log^3 n, \tag{14.31}$$
   we have almost surely
$$\lambda_n = (1 + o(1))\sqrt{m}.$$

2. *For $2.5 < \beta < 3$ and*
$$m > d^{\frac{\beta-2}{\beta-2.5}} \log^{\frac{3}{\beta-2.5}} n, \tag{14.32}$$
   we have almost surely
$$\lambda_n = (1 + o(1))\sqrt{m}.$$

3. *For $2 < \beta < 2.5$ and*
$$m > \log^{\frac{3}{2.5-\beta}} n,$$
   we have almost surely
$$\lambda_n = (1 + o(1))\tilde{d}.$$

4. *For $2.5 < \beta$ and $k < n \left(\dfrac{d}{m \log n}\right)^{\beta-1}$, almost surely the k largest eigenvalues of G have power law distribution with exponent $2\beta - 1$, provided m is large enough (satisfying (14.31) and (14.32)).*

We remark that the second order average degree $\tilde{d}$ can actually be computed in these cases. For details see [127, 128].

14.6   Chapter Notes

Fundamental Properties

The adjacency spectrum is discussed in a vast amount of papers and textbooks. The results presented in this chapter are taken from [135, 247, 576]. More on the Laplacian can be found in [135] (under the term 'admittance matrix') and in [427]. The normalized Laplacian is extensively studied in [125]. While the eigenvalues alone do not generally determine the structure of a graph (as shown by cospectral graphs), eigenvalues *plus* eigenvectors do: if $u_1, u_2, \ldots, u_n$ are linearly independent eigenvectors of A corresponding to $\lambda_1, \lambda_2, \ldots, \lambda_n$ respectively, then $A = UDU^{-1}$, where $U :=\text{Mat}(u_1, \ldots, u_n)$ is the matrix with u_i as column vectors and $D :=\text{diag}(\lambda_1, \ldots, \lambda_n)$ is the diagonal matrix with entries λ_i. Already the knowledge of some eigenvectors can be very useful to recognize important properties of a graph. This is elaborated on in [136]. The cited references contain also many results on the spectrum of regular graphs which we treated rather stepmotherly since in network analysis we are typically dealing with highly non-regular graphs.

Numerical Methods

The diagonalization strategy for small dense matrices is comprehensively treated in [482, 535]. For a discussion of QR-like algorithms including parallelizable versions see [571]. More on the Lanczos method can be found in [467, 591]. An Arnoldi code for real asymmetric matrices is discussed in [509].

Subgraphs and Operations on Graphs

The Interlacing Theorem plays an important role in many publications on spectral graph theory. In addition to the already mentioned literature, we point the reader to [561]. More on graph operations and resulting spectra can be found in [135, Chapter 2] and [427]. As already mentioned, symmetric grafting also plays a role in [49].

Bounds on Global Statistics

For more results on the connection between eigenvalues and graph parameters see [427, 428] and the references therein, such as [426, 425]. More on the role of the eigenvalues of the normalized Laplacian for graph parameters can be found in [125]. More on the connection between $\lambda_2(L)$ and expansion properties can be found in [19]. Better spectral lower bounds on the bisection width have been given by Donath, Hoffman [159], Boppana [78], and Rendl, Wolkowicz [491]. Boppana's technique also yields an efficient algorithm with good average case behavior. For a discussion of spectral methods for the multiway partition problem of finding a k-partition with prescribed part sizes see [27, 389].

Heuristics for Graph Identification

For further reading, see [198, 49] as well as the already mentioned [127, 128]. The behavior of certain extremal eigenvalues is also considered in [337].

15 Robustness and Resilience

Gunnar W. Klau and René Weiskircher

Intuitively, a complex network is *robust* if it keeps its basic functionality even under failure of some of its components. The study of robustness in networks is important because a thorough understanding of the behavior of certain classes of networks under failures and attacks may help to protect, for instance, communication networks like the Internet against assaults or to exploit weaknesses of metabolic networks in drug design.

Often, we distinguish between random failure and intentional attacks. Examples for random and intentional component failures in real-world complex networks are, for instance, mutations in a cell, pharmaceutical or environmental stress on metabolic networks, router failures in the Internet, or intentional attacks on airline or highway networks. We will see that some networks like the Internet are very robust against random drop-outs of routers but may suffer heavily from targeted attacks against well-chosen central routers.

This chapter is dedicated to network statistics that are of interest with respect to a network's robustness or its resilience against repeated component failure. We give an overview of a variety of statistics and discuss their applicability in practice in terms of usefulness and computational complexity. Often, research on robustness focuses on how these statistics change, by analyzing or measuring the effects if a network undergoes a sequence of component failures. Wherever possible we try to relate the different statistics and discuss their advantages and disadvantages. In many cases, we use examples to illustrate the definitions.

We chose to organize this chapter as follows: We distinguish between worst case, average, and probabilistic statistics. Sections 15.1 and 15.2 cover worst case connectivity and distance measures. Average robustness statistics (Section 15.3) allow a more global perspective on robustness properties whereas probabilistic statistics (Section 15.4) consider the failure probabilities implicitly. While, roughly speaking, the statistics become more and more meaningful the more they are located towards the end of this chapter, they are also more difficult to compute. We conclude this chapter in Section 15.5 with final remarks and list some open problems.

15.1 Worst-Case Connectivity Statistics

This section deals with statistics that answer questions of the form "What is the minimum number of edges or vertices that have to be deleted from the network

U. Brandes and T. Erlebach (Eds.): Network Analysis, LNCS 3418, pp. 417–437, 2005.
© Springer-Verlag Berlin Heidelberg 2005

such that the resulting network is disconnected and has property P ?". These are worst case statistics because the deletion of an arbitrary set of vertices or edges of the same size may not cause the same effect. So we implicitly assume that the vertex or edge failures are not random but targeted for maximum effect.

15.1.1 Classical Connectivity

Classical connectivity is the basis of many robustness statistics. A network is called *connected*, if there exists a path between every pair of vertices in the network. In many applications, connectedness is a necessary condition for a network to fulfill its purpose. Therefore, one measure of robustness of a network is the number of vertices or edges that have to be removed to make the network unconnected. These are called the *vertex-connectivity* and *edge-connectivity* of the network, respectively. They are treated in depth in Chapter 7. Here we only look at connectivity as a measure of the robustness of a network.

If a network loses its functionality completely as soon as it is not connected anymore, connectivity is indeed a good measure for its robustness. But if we are concerned with the case where the usefulness of a network is not seriously affected by disconnecting a small set of vertices from the network, connectivity is not a meaningful measure. Consider the Internet as an example. A desktop computer is only connected to the net via one link to a provider or server. Cutting this link disconnects the net but has only a negligible influence on the functionality of the whole Internet. Yet the edge-connectivity of the net is only one. Similarly, the failure of a small router will only disconnect a handful of clients from the net but proves that the Internet has vertex connectivity one.

15.1.2 Cohesiveness

The notion of cohesiveness was introduced by Akiyama et al. in [13] and defines for each vertex of the network to what extent it contributes to the connectivity.

Definition 15.1.1. *Let $\kappa(G)$ be the vertex-connectivity of G (see the definition in Section 2.2.4). Let $G - v$ be the network obtained from G by removing vertex v. For any vertex v of G, the cohesiveness $c(v)$ is defined as follows:*

$$c(v) = \kappa(G) - \kappa(G - v)$$

Vertex 7 in Figure 15.1(a) has a cohesiveness of -2, because the network has vertex-connectivity 1 if vertex 7 is present and vertex connectivity 3 if we delete it. On the other hand, vertex 6 in Figure 15.1(b) has cohesiveness 1 because if we remove it from the network, the vertex-connectivity drops from 3 to 2.

It follows from the definition that the cohesiveness of a vertex cannot be greater than 1. Intuitively, a vertex with negative cohesiveness is an outlier of the network while a vertex with cohesiveness 1 is central. It can be shown that a network can have at most one vertex with negative cohesiveness and that the neighborhood of this *negative vertex* contains the only set of vertices of

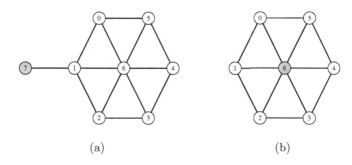

(a)   (b)

Fig. 15.1. Example graphs for the cohesiveness of a vertex. Vertex 7 in Figure 15.1(a) has cohesiveness -2 and vertex 6 in Figure 15.1(b) cohesiveness 1

size $\kappa(G)$ whose removal disconnects the network. Consider as an example the network shown in Figure 15.1(a), where vertex 7 is the only vertex with negative cohesiveness. The only neighbor of vertex 7 is vertex 1 and this is the only vertex whose deletion splits the network.

Even though a network can have at most one negative vertex, we can compute a set of loosely connected vertices by removing the negative vertex and then looking for the next negative vertex. This algorithm could be used to find loosely connected vertices in a network because a negative vertex is at the periphery of the graph. A drawback of this approach is that this algorithm may stop after a few vertices even for big networks because there are no more vertices with negative cohesiveness.

The cohesiveness of a vertex can be computed using standard connectivity algorithms (see Chapter 7). To compute the cohesiveness of every vertex, the connectivity algorithm has to be called n times where n is the number of vertices in the network.

15.1.3 Minimum m-Degree

The statistics we have mentioned so far make statements about the connectivity of a network. The m-degree was introduced in [65] by Boesch and Thomas. It is concerned with the state of the network after disconnection.

Definition 15.1.2. *The* minimum m-degree $\xi(m)$ *of a network is the smallest number of edges that must be removed to disconnect the network into two connected components G_1 and G_2 where G_1 contains exactly m vertices.*

Table 15.1 shows the m-degrees for the network in Figure 15.2.

Let $G = (V, E)$ be a network with $|V| = n$. Boesch and Thomas showed in [65] the following properties of the minimum m-degree:

- $\xi(m) = \xi(n - m)$.
- $\xi(m) \geq m(\delta(G) - m + 1)$ where $\delta(G)$ is the minimum degree of any vertex in G.

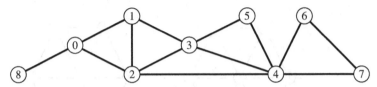

Fig. 15.2. Example network for the minimum m-degree

Table 15.1. The m-degrees for the network in Figure 15.2

1-degree	2-degree	3-degree	4-degree	5-degree
1	2	3	3	3

– Let G be a regular network with degree $r \le n/2$, $n > 2$ and $m \ge l$. Then

$$r \ge \lfloor \xi(m)/m \rfloor + \lceil \xi(l)/l \rceil \ .$$

There is no asymptotically faster algorithm known for computing the minimum m-degree than trying all sets of vertices of size m and check if the graphs induced by the set and by its complement are connected. If this is the case, we count the number of edges connecting vertices in the set with vertices outside. The minimum over all sets is the m-degree. This results in a running time of $\mathcal{O}(\binom{n}{m}|E|)$.

The main problem of this statistics is that the splitting of the graph has to result in two *connected* components, so it does not express an intuitive concept of robustness. The network in Figure 15.3 has 3-degree 3 while the deletion of the two thick edges is enough to split a component with three vertices from the network.

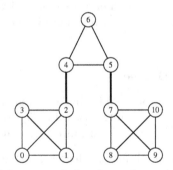

Fig. 15.3. A counter-intuitive example for the m-degree statistics

15.1.4   Toughness

The toughness of a network was introduced by Chvátal [129]. It measures the number of internally connected components that the graph can be broken into by the failure of a certain number of vertices.

Definition 15.1.3. *Let S be a subset of the vertices of G and let $K(G - S)$ be the number of internally connected components that G is split into by the removal of S. The* toughness *of G is defined as follows:*

$$t(G) = \min_{S \subseteq V, K(G-S) > 1} \left\{ \frac{|S|}{K(G - S)} \right\}$$

The edge-toughness *of a network is defined analogously for edges.*

Intuitively, the toughness of a network is high if even the removal of a large number of vertices splits the network only into few components. Conversely, if a network can be split into many components by removing a small number of vertices, its toughness is small.

The toughness of a complete network is defined as infinite. The network with the smallest toughness is a star. Removing the central vertex splits the network into components of size one and so the toughness of a star with n vertices is $\frac{1}{n-1}$. Note that the central vertex is also the only one whose removal splits the graph.

It is $\mathcal{NP}$-hard to decide for a general graph if it has toughness at least t [48]. If the network is a tree, the toughness is $\frac{1}{\Delta(G)}$ where $\Delta(G)$ is the maximum degree of any vertex. The toughness of the complete bipartite network $K_{m,n}$ with $m \leq n$ and $n \geq 2$ is $\frac{m}{n}$.

The toughness of a circle is one and it follows that the toughness of a Hamiltonian graph is at least one. In [129], Chvátal also showed a connection between the *independence number* of a network and the toughness. The independence number β_0 is the size of the largest subset S of the vertices with the property that there is no edge in the network connecting two vertices in S. The toughness of G is lower-bounded by $\kappa(G)/\beta_0(G)$ and upper bounded by $(n - \beta_0(G))/\beta_0$.

15.1.5   Conditional Connectivity

Conditional connectivity was introduced by Harary in [276] and is a generalization of the minimum m-degree. The measure is parameterized with a property P that has to hold for all the components created by deleting vertices from the network.

Definition 15.1.4. *The* P-connectivity $\kappa(G : P)$ *of network G is the smallest number of vertices that have to be deleted from the network such that the remaining network G' has the following properties:*

1. *G' is not connected.*
2. *Every connected component of G' has property P.*

Conditional edge-connectivity is defined analogously for the deletion of edges. Conditional connectivity is potentially very useful in practice because the property P can be chosen according to the characteristics of the task that the network should accomplish. An example could be defining P as: "The component has at most k vertices". The conditional connectivity would then correspond to the size of the smallest subset of vertices we have to delete to split the network into components of at most k vertices each. Classical connectivity is a special case of conditional connectivity where $P = \emptyset$.

If we define a sequence $S = (P_1, \ldots, P_k)$ of properties according to our application such that P_{i+1} implies P_i for $1 \leq i \leq k - 1$, we obtain a vector of conditional connectivity

$$(\kappa(G : P_1), \ldots, \kappa(G : P_k)) \ .$$

If the properties are defined to model increasing degradation of the network with respect to the application, this vector gives upper bounds for the usefulness of the system with respect to the number of failed vertices.

A similar measure is *general connectivity*, also introduced by Harary [277]. If G is a network with property P and Y is a subset of the vertices (edges) of G, then $\kappa(G, Y : P)$ is the smallest set $X \subset Y$ of vertices (edges) in G whose removal results in a network G' that does not have property P. Conditional connectivity is a special case of general connectivity.

The main drawback of these statistics is that there is no efficient algorithm known that computes them for a general graph.

15.2   Worst-Case Distance Statistics

The statistics in this section make statements about the increase of distances in the network caused by the deletion of vertices or edges. These are again worst-case statistics because they give the smallest number of vertices or edges that have to be deleted in order to increase the distances. All the statistics we present in this section are only defined until the network becomes disconnected by the removal of vertices and edges.

15.2.1   Persistence

The *persistence* of a network is the minimum number of vertices that have to be deleted in order to increase the diameter (the longest distance between a pair of vertices in the network). Again, an analogous notion is defined for the deletion of edges (*edge persistence*). Persistence was introduced by Boesch, Harary and Kabell in [64] where they also present the following properties of the persistence of a network:

- The persistence of a network with diameter $2 \leq d \leq 4$ is equal to the minimum over all pairs of non-adjacent vertices i and j of the maximum number of vertex-disjoint i, j-paths of length no more than d.
- The *edge-persistence* of a network with diameter $d \in \{2, 3\}$ is the minimum over all pairs of vertices i, j of the maximum number of edge-disjoint i, j-paths of length no more than d.

There are many theoretic results on persistence that mainly establish connections between connectivity and persistence, see for example [74, 475]. The *persistence vector* is an extension of the persistence concept. The i-th component of $P(G) = (p_1, \ldots, p_n)$ is the worst-case diameter of G if i vertices are removed. This is the same concept as the vertex-deleted diameter sequence we introduce in Section 15.2.2.

The main drawback of persistence is that there is no efficient algorithm known to compute it.

15.2.2   Incremental Distance and Diameter Sequences

Krishnamoorthy, Thulasiraman, and Swamy have studied the increase of distances in a network caused by the deletion of vertices and edges [371]. They introduce for a network G four sequences A, B, D, and T defined as follows:

Definition 15.2.1. *Let $d(u, v) = d_G(u, v)$ be the distance of the two vertices u and v in G. Let $d(G)$ be the diameter of G. Let l be the vertex connectivity of G and m the edge-connectivity. Then the sequences A, B, D and T are defined as follows:*

$$
\begin{aligned}
a_i &= \max_{|V_i|=i}\{d_{G-V_i}(u,v) - d(u,v) \mid u, v \in V - V_i\} \ \textit{for } 1 \le i \le l-1 \\
b_i &= \max_{|E_i|=i}\{d_{G-E_i}(u,v) - d(u,v)\} && \textit{for } 1 \le i \le m-1 \\
d_i &= \max_{|V_i|=i}\{d(G - V_i)\} && \textit{for } 1 \le i \le l-1 \\
t_i &= \max_{|E_i|=i}\{d(G - E_i)\} && \textit{for } 1 \le i \le m-1 \ .
\end{aligned}
$$

Sequence A is called the *vertex-deleted incremental distance sequence*, B the *edge-deleted incremental distance sequence*, D the *vertex-deleted diameter sequence* and T the *edge-deleted diameter sequence*.

Entry i in sequence A is the maximum increase of the distance between a pair of vertices caused by the deletion of i vertices from G. The sequence B contains the maximum increase in distance for the deletion of edges. Entry i in sequence D is the maximum diameter of the graph caused by deleting i vertices, and sequence T is the analogous sequence for the deletion of edges. Table 15.2 contains the four sequences for the network shown in Figure 15.4.

Table 15.2. The vertex- and edge-deletion-sequences for the network of Figure 15.4

A	(1,2)
B	(3,3)
D	(3,4)
T	(4,4)

It is easy to see that the A, B and T sequences are always monotonically nondecreasing. The entries of the A sequence are non-negative and the entries in the B sequence at least 1. If G is complete the four sequences are as follows:

– $A = (0, \ldots, 0)$

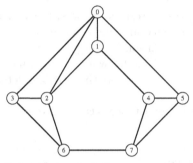

Fig. 15.4. Example graph for incremental distance sequences

- $B = (1, \ldots, 1)$
- $D = (1, \ldots, 1)$
- $T = (2, \ldots, 2)$

Krishnamoorthy, Thulasiraman and Swamy show that the largest increase in the distance between any pair of vertices caused by the deletion of i vertices or edges can always be found among the neighbors of the deleted objects. This speeds up the computation of the sequences significantly and also simplifies the definitions of A and B. These sequences can also be defined as follows (note that $N(V_i)$ is the set of vertices adjacent to vertices in the set V_i and $N(E_i)$ is the set of vertices incident to edges in E_i):

$$a_i = \max_{|V_i|=i} \{d_{G-V_i}(u,v) - d(u,v) \mid u,v \in N(V_i)\} \text{ for } 1 \le i \le l-1$$
$$b_i = \max_{|E_i|=i} \{d_{G-E_i}(u,v) - d(u,v) \mid u,v \in N(E_i)\} \text{ for } 1 \le i \le m-1$$

The vertex- and edge-deletion sequences are a worst case measure for the increase in distance caused by the failure of vertices or edges and they do not make any statements about the state of the graph after disconnection occurred. So these measures are only suited for applications where distance is crucial and disconnection makes the whole network unusable. Even with the improvement mentioned above, computing the sequences is still only possible for graphs with low connectivity.

15.3 Average Robustness Statistics

The statistics in this section make statements about the average number of vertices or edges that have to fail in order for the network to have a certain property or build an average of local properties in order to cover global aspects of the network.

15.3.1   Mean Connectivity

All of the measures introduced so far are worst-case measures. The *mean connectivity* introduced by Tainiter [538, 539] tries to make statements about the probability that a network is disconnected by the random deletion of edges.

Definition 15.3.1. *Let $G = (V, E)$ be a connected network with n vertices and m edges. Let $S(G)$ be the set of all $m!$ orderings of the edges and $G_0 = (V, \emptyset)$. For each ordering $s \in S(G)$ we define the number $\xi(s)$ as follows: We insert the edges of G into G_0 in the sequence given by s. We define $\xi(s)$ as the index of the edge that transforms the network from disconnected to connected. The mean connectivity of G is then defined as follows:*

$$\mathcal{M}(G) = m - \frac{1}{m!} \sum_{s \in S(G)} \xi(s)$$

Figure 15.5 shows a graph with mean connectivity $3/4$. This can be seen as follows: For every edge-sequence where the edge $(2, 3)$ does not come last, we have $\xi(s) = 3$. For all other sequences, we have $\xi(s) = 4$. Since there are six sequences where edge $(2, 3)$ is last and 24 sequences in total, the mean connectivity of the graph is $3/4$.

Note that $\mathcal{M}(G)$ is not the same as the mean number of edges we have to delete to disconnect G. If we look at all sequences of deleting edges and compute the mean index where the graph becomes disconnected, we obtain the value $7/4$ for the graph in Figure 15.5.

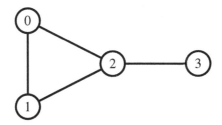

Fig. 15.5. A graph with mean connectivity $3/4$

Tainiter has shown the following properties of this measure:

- If $G = (V, E')$ with $E' \subseteq E$ is a connected sub-network of $G = (V, E)$ then $\mathcal{M}(G') \leq \mathcal{M}(G)$
- Let G be a network with n vertices and m edges. We construct a new network G' by adding one new vertex and h edges that connect it to vertices in G. Let $\mathcal{M}(G, k)$ be the number of edge-sequences for G with $\xi(s) = k$. Then the following inequality is satisfied:

$$\mathcal{M}(G') - \mathcal{M}(G) \geq \frac{\mathcal{M}(G) + 1}{m + 1} - \frac{1}{h + 1} \sum_{k=n-1}^{m} \mathcal{M}(G, k) \frac{(h + m - k + 1)!}{(m - k)!(m + h)!}$$

– The following bounds are tight:

$$\lambda(G) - 1 \leq \mathcal{M}(G) \leq m - n + 1$$

where $\lambda(G)$ is the edge-connectivity of G. An example where both bounds are tight is a circle where we have $\lambda(G) = 2$ and $\mathcal{M}(G) = 1$.

If the difference between the mean connectivity and the classical edge-connectivity is large, then there must be connectivity bottlenecks in the network. It follows that the connectivity of the network can be strengthened by inserting only a few edges to bridge the bottleneck. An example would be a complete graph with one 'dangling' vertex connected to the rest of the graph by a single edge. With each edge we add to the dangling vertex, we can increase the connectivity of the graph by one. The principal drawback of the measure is again the fact that there is no efficient algorithm known for computing it. Also, it is useful only in the case of random edge failures.

15.3.2   Average Connected Distance and Fragmentation

In 1999, the article [17] received a lot of attention in the scientific world. Albert, Jeong, and Barabási simulate random vertex failures and intentional attacks at the highest-degree vertices in random and scale-free networks. They measure the effects on two parameters of the network, namely on the *average connected distance* and on the *fragmentation*.

The average connected distance $\bar{d}$ is the average length of the shortest paths between connected pairs of nodes in the network as defined in Section 11.2.[1]

Fragmentation measures the decay of a network in terms of the size of its connected components.

Definition 15.3.2 (Fragmentation). *Let G be a network with k connected components $S_1, \ldots, S_k$. The* fragmentation $\mathrm{frag}(G) = (\mathrm{frag}_1(G), \mathrm{frag}_2(G))$ *is defined by two parameters: The relative size of the largest component*

$$\mathrm{frag}_1 = \frac{\max_{i=1}^{k} |S_k|}{\sum_{i=1}^{k} |S_k|}$$

and the average size of an isolated component

$$\mathrm{frag}_2 = \frac{\sum_{i=1}^{k} |S_k| - \max_{i=1}^{k} |S_k|}{k - 1} \quad ,$$

where $|S_k|$ denotes the number of vertices in the kth component.

---

[1] In [17], the authors use the term *interconnectedness* which corresponds to the classical average distance. In their experiments, however, they measure the average connected distance. The classical average distance becomes ∞ as soon as the graph becomes disconnected.

Figure 15.6 shows the effect of vertex failures and attacks on the average connected distance $\bar{d}$ for randomly generated networks whose degree distributions follow a Poisson distribution and a power-law distribution, respectively. The Poisson networks suffer equally from random and targeted failures. Every vertex plays more or less the same role, and deleting one of them affects the average connected distance, on average, only slightly if at all. The scale-free network, in contrast, is very robust to failures in terms of average connected distance. The probability that a high-degree vertex is deleted is quite small and since those vertices are responsible for the short average distance in scale-free networks, the distances almost do not increase at all when deleting vertices randomly. If, however, those vertices are the aim of an attack, the average connected distance increases quickly. Simulations on small fragments of the Internet router graph and the WWW graph show a similar behavior as the random scale-free network, see [17].

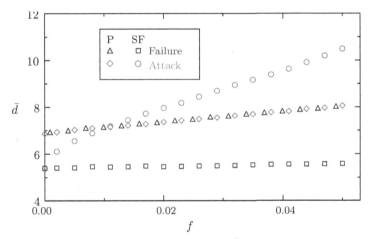

Fig. 15.6. Changes in average connected distance $\bar{d}$ of randomly generated networks ($|V| = 10,000, |E| = 20,000$) with Poisson (P) and scale-free (SF) degree distribution after randomly removing $f|V|$ vertices (source: [17])

The increase in average connected distance alone does not say much about the connectivity status of the network in terms of fragmentation. It is possible to create networks with small average connected distance that consist of many disconnected components (imagine a large number of disconnected triangles: their average connected distance is 1). Therefore, Albert et al. also measure the fragmentation process under failure and attack.

Figure 15.7 shows the results of the experimental study on fragmentation. The Poisson network shows a threshold-like behavior for $f > f_c \approx 0.28$ when frag_1, the relative size of the largest component, becomes almost zero. Together with the behavior of frag_2, the average size of the disconnected components, that reaches a peak of 2 at this point, this indicates the breakdown scenario as shown also in Figure 15.8: Removing few vertices disconnects only single vertices. The components become larger as f reaches the percolation threshold f_c. After that, the system falls apart. As in Figure 15.6, the results are the same for random and targeted failures in networks with Poisson degree distribution.

The process looks different for scale-free networks (again, the data for the router and WWW graphs look similar as for the randomly generated scale-free networks). For random deletion of vertices no percolation threshold can be observed: the system shows a behavior known as *graceful degradation*. In case of attacks, we see the same breakdown scenario as for the Poisson network, with an earlier percolation threshold $f_c \approx 0.18$.

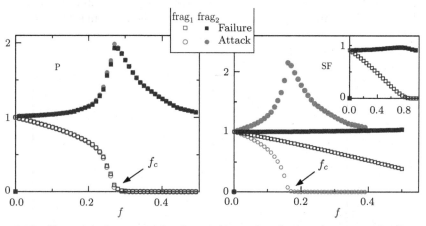

Fig. 15.7. Changes in fragmentation frag $= (\text{frag}_1, \text{frag}_2)$ of random networks (Poisson degree distribution: P, scale-free degree distribution: SF) after randomly removing $f|V|$ vertices. The inset in the upper right corner shows the scenario for the full range of deletions in scale-free networks (source: [17])

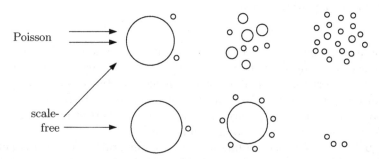

Fig. 15.8. Breakdown scenarios of networks with Poisson degree and scale-free distribution (source: [17])

In summary the experimental study shows that scale-free networks are tolerant against random failures but highly sensitive to targeted attacks. Since the Internet is believed to have a scale-free structure, the findings confirm the vulnerability of this network which is often paraphrased as the 'Achilles heel of the Internet'.

Broder et al. study the structure of the web more thoroughly and come to the conclusion that the web has a 'bow tie structure' as depicted in Figure 4.1 on page 77 in Chapter 3 [102]. Their experimental results on the web graph W reveal that the world wide web is robust against attacks. Deleting all vertices $\{v \in V(W) \mid d^-(v) \geq 5\}$ does not decrease the size of the largest component dramatically, it still contains approximately 30% of the vertices. This apparent contradiction to the results of Albert et al. can be explained by the fact that

$$\frac{|\{v \in V(W) \mid d^-(v) \geq 5\}|}{|V(W)|}$$

is still below the percolation threshold and is thus just another way to look at the same data: while 'deleting all vertices with high degree' sounds drastic this is still a set of small cardinality.

A number of application-oriented papers use the average connected distance and fragmentation as the measures of choice in order to show the robustness properties of the corresponding network. For example, Jeong et al. study the protein interaction network of the yeast proteome (*S. cervisiae*) and show that it is robust against random mutations of proteins but susceptible to the destruction of the highest degree proteins [327]. Using average connected distance and fragmentation to study epidemic propagation networks leads to the advice to take care of the hubs first, when it comes to deciding a vaccination strategy (see, e.g., [469]).

Holme et al. [305] study slightly more complex attacks on networks. Besides attacks on vertices they also consider deleting edges and choose betweenness centrality as an alternative selection criterion for deletion. In addition, they investigate in how far recalculating the selection criteria after each deletion alters the results. They show empirically that attacks based on recalculated values are more effective.

On the theoretical side Cohen et al. [130] and, independently, Callaway et al. [108] study the fragmentation process on scale-free networks analytically. While the first team of authors uses percolation theory, Callaway and his colleagues obtain more general results for arbitrary degree distributions using generating functions (see Section 13.2.2 in Chapter 13). The theoretical analyses confirm the results of the empirical studies and yield the same percolation thresholds as shown in the figures above.

15.3.3   Balanced-Cut Resilience

Among other statistics, Tangmunarunkit et al. use a new measure of robustness to link failures in their experimental study [541]. The aim of their experiments is to evaluate generators that supposedly simulate the Internet topology. Besides *expansion* and *distortion* (see Chapter 11), the authors measure the similarity of generated and real networks with respect to the size of a balanced cut through the network. In terms of the new statistics, a network is resilient to component failure if the average size of a balanced cut within an h-neighborhood around each vertex is large. We give a more formal definition:

Definition 15.3.3 (Balanced-cut resilience). *Let $G = (V, E)$ be a network with n vertices, and let the capacity of each edge in G be equal to one. The minimum balanced cut of G is the capacity of a minimum cut such that the two resulting vertex sets contain approximately the same number, namely $\lfloor \frac{n}{2} \rfloor$ and $\lceil \frac{n}{2} \rceil$, of vertices. The balanced-cut resilience $R(N(v, h))$ is the average size of a minimum balanced cut within the h-neighborhood $\text{Neigh}_h(v)$ around each vertex v, that is,*

$$R(N(v, h)) = \frac{1}{n} \left(\sum_{v \in V} \text{min. balanced cut in } \text{Neigh}_h(v) \right) .$$

The h-neighborhood of a vertex v contains all vertices with distance less than or equal to h from v, see also the definition on page 296 in Chapter 11. The balanced-cut resilience is a function of the number of nodes $N(v, h)$ in the h-neighborhood of a vertex v, not the radius h itself, to factor out the fact that networks with high expansion have more nodes in neighborhoods of the same radius. Clearly, we have $R(h) = 1$ for paths and trees. The resilience of random graphs in the Erdős-Rényi model with average degree k is proportional to kn, whereas it is proportional to n for complete graphs, see [541]. For regular grid graphs, the balanced-cut resilience grows with $\sqrt{n}$. See Figure 15.9 for an illustrative example.

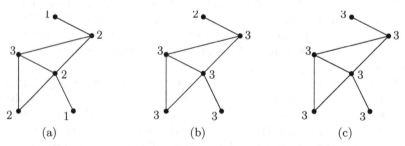

Fig. 15.9. Balanced-cut resilience for an example graph. Balanced cut shown for each vertex for (a) 1-neighborhoods, (b) 2-neighborhoods, and (c) 3-neighborhoods

Computing a minimum balanced cut is $\mathcal{NP}$-hard [240] and thus the drawback of this statistics is certainly its computational complexity which makes it impractical for large networks. There are, however, a number of heuristics that yield reasonably good values so that the balanced-cut resilience can at least be estimated. Karypis and Kumar [348], for instance, propose a multilevel partitioning heuristics that runs in time $\mathcal{O}(m)$ where m is the number of edges in the network.

15.3.4   Effective Diameter

Palmer et al. introduce in [462] the *effective eccentricity* and the *effective diameter* as measures of resilience against vertex and edge failures. These statistics are based on the hop-plot and we recall their definitions (see also Sections 11.2.4 and 11.2.3 on neighborhoods and eccentricity in Chapter 11):

Definition 15.3.4 (Effective eccentricity, effective diameter). *The effective eccentricity $\varepsilon_{\text{eff}}(v, r)$, $0 \leq r \leq 1$, of a vertex v is the smallest h such that the number of vertices $N(v, h)$ within a h-neighborhood of v is at least r times the total number of vertices, that is,*

$$\varepsilon_{\text{eff}}(v, r) = \min\{h \in \mathbb{N} \mid N(v, h) \geq rn\} \ .$$

The effective diameter $\text{diam}_{\text{eff}}(r)$ *of a network is the smallest h such that the number of pairs within a h-neighborhood is at least r times the total number of reachable pairs:*

$$\text{diam}_{\text{eff}}(r) = \min\{h \in \mathbb{N} \mid P(h) \geq rP(\infty)\} \ ,$$

where P denotes the number of pairs within a certain neighborhood (hop-plot), that is,

$$P(h) := \left|\{(u, v) \in V^2 \mid d(u, v) \leq h\}\right| = \sum_{v \in V} N(v, h) \ ,$$

see also Chapter 11. In the case that this distribution follows the power law $P(h) = (n + 2m)h^{\mathcal{H}}$, the value $\mathcal{H}$ is also referred to as the hop-plot exponent.

The authors perform experiments on the network of approximately 285,000 routers in the Internet to investigate in how far and under which circumstances the effective diameter of the router network changes. The experiments consist of deleting either edges or vertices of the network and recomputing the effective diameter diam_{eff} after each deletion, using a value of 0.9 for the parameter r. Since an exact calculation of this statistics would take days, they exploit the approximate neighborhood function described in Section 11.2.6 of Chapter 11. Using these estimated values leads to a speed-up factor of 400.

Figures 15.10 and 15.11 show the effect of link and router failures on the Internet graph. Confirming previous studies, the plots show that the Internet is very robust against random failures but highly sensitive to failure of high degree vertices. Also, deleting vertices with low effective eccentricity first rapidly decreases the connectivity.

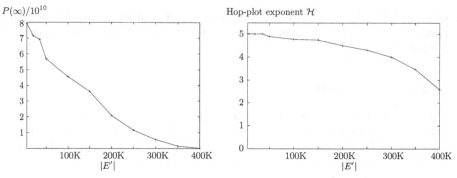

Fig. 15.10. Effect of edge deletions (link failures) on the network of 285,000 routers (source: [462]). The set E' denotes the deleted edges

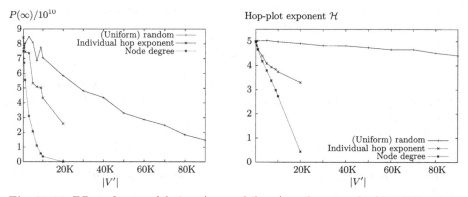

Fig. 15.11. Effect of vertex deletions (router failures) on the network of 285,000 routers (source: [462]). The set V' denotes the deleted edges

15.4   Probabilistic Robustness Statistics

This section describes robustness statistics that explicitly consider the failure probabilities of network components and are thus more appropriate to describe untargeted component failure. We present two different approaches to determine the probability of network disconnection given the failure probability: the *reliability polynomial* and *probabilistic resilience*.

We chose not to cover purely theoretical approaches such as the symbolic approach to robustness by Flajolet et al. [214], in which the authors define a measure of robustness by determining the expected number of edge-disjoint paths to get from a start vertex s to a target vertex t in a graph.

15.4.1   Reliability Polynomial

The *reliability polynomial* was already used in 1977 by Boorstyn and Frank [75].

Definition 15.4.1. *Let G be a connected network with n vertices and m edges. We assume that the edges of G fail independently with probability $1-p$ where $0 \leq p \leq 1$. The reliability polynomial $R(G,p)$ is the probability that G is connected.*

Obvious properties of the reliability polynomial $R(G,p)$ are:

1. $R(G,0) = 0$, $R(G,1) = 1$.
2. $p_1 < p_2$ implies $R(G, p_1) < R(G, p_2)$.
3. Let G be a connected graph and G_{-e} be the graph obtained from G by removing e. Let G_e be the graph obtained from G by contracting e. Then the following equality holds:

$$R(G,p) = (1-p)R(G_{-e},p) + pR(G_e,p) \ .$$

4. If G is a tree with m edges, than we have $R(G,p) = p^m$.

In his doctoral thesis [497], Rosenthal showed that it is $\mathcal{NP}$-hard to decide for a given edge failure probability if the probability that the network is connected is at least a certain value q. The same is true if we are given a failure probability for vertices and edges. In [480], Pönitz and Tittmann have shown that the problem can be solved in time $\mathcal{O}((2n+m)B(k))$ for graphs with pathwidth k where $B(k)$ is the Bell number of k. The bell number of k is the number of ways the set of natural numbers from 1 to k can be partitioned into nonempty subsets. It follows that the problem is polynomially solvable for graphs with bounded pathwidth. Figure 15.12 shows a graph with pathwidth two from [480] together with a plot of its reliability polynomial. The polynomial has the following formula:

$$R(G,p) = 55p^5 - 155p^6 + 169p^7 - 84p^8 + 16p^9$$

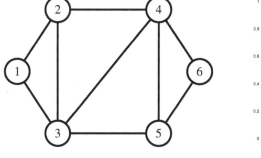
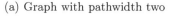
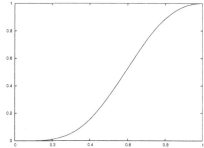

(a) Graph with pathwidth two   (b) The reliability polynomial

Fig. 15.12. A graph and a plot of its reliability polynomial

There is no polynomial time algorithm known to compute the reliability polynomial for general graphs.

15.4.2   Probabilistic Resilience

In contrast to the deterministic probability measures presented in Section 15.1 on worst-case connectivity statistics, Najjar and Gaudiot study a probabilistic variant of connectivity [438]. The authors consider a class of regular networks and examine the probability of disconnection through random vertex failures.

They define the *disconnection probability* of a network G as

$$P(G, i) = \Pr[G \text{ disconnected exactly after } i\text{th failure}]$$

Motivated by the architectures of large-scale computer clusters the authors study a family $\mathcal{F}$ of k-regular graphs that includes, for example, tori and hypercubes. They show that for networks in $\mathcal{F}$ the disconnection probability $P(G, i)$ can be approximated by the term

$$P_1(G, i) = \Pr[G \text{ disconnected exactly after } i\text{th failure}$$
$$\text{and one component contains exactly one vertex}] ,$$

that is, the disconnection probability can be estimated by the probability of disconnecting only one vertex from the network. For networks in the family $\mathcal{F}$, $P_1(G, i)$ and thus an estimation of $P(G, i)$ can be derived analytically.

The function $P(G, i)$ is a bell-shaped curve whose height increases with n, the number of vertices in the network, whereas the x-coordinate of the maximum depends on k, the degree of the vertices (see Figure 15.13). The larger the connectivity of a regular network in terms of k the more failures are needed until disconnection occurs. The authors confirm their theoretical predictions by running Monte-Carlo experiments on a large number of graphs from $\mathcal{F}$.

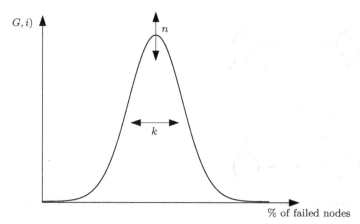

Fig. 15.13. The probability $P(G, i)$ for members of $\mathcal{F}$. The number of vertices in the network, n, determines the height of the curve. Their vertex degree, k, determines the offset on the abscissa

The concept of disconnection probability enables us to define a probabilistic version of connectivity: *probabilistic resilience*. Intuitively, a resilient network should sustain a large number of vertex failures until it becomes disconnected.

Definition 15.4.2 (Probabilistic resilience). *Let G be a network with n vertices. The probabilistic resilience[2] $\mathrm{res}_{prob}(G, p)$ is the largest number of vertex failures such that G is still connected with probability $1 - p$, that is,*

$$\mathrm{res}_{prob}(G, p) = \max\{I \mid \sum_{i=1}^{I} P(G, i) \leq p\} \ .$$

The relative probabilistic resilience relates $\mathrm{res}_{prob}(G, p)$ to the size of G:

$$\overline{\mathrm{res}}_{prob}(G, p) = \frac{\mathrm{res}_{prob}(G, p)}{n} \ .$$

Clearly, this probabilistic measure is related to classical connectivity, and the identity $\mathrm{res}_{prob}(G, 0) = \kappa(G) - 1$ holds.

Analyzing $P(G, i)$ for regular graphs shows that the probabilistic resilience $\mathrm{res}_{prob}(G, p)$ grows with the size of G. The relative probabilistic resilience $\overline{\mathrm{res}}_{prob}(G, p)$, however, decreases with the size if the degree of the network remains constant. Therefore, the relative resilience increases for hypercubes and decreases for tori with increasing network size.

It is quite difficult to compute the probabilistic resilience for more complicated families of networks than $\mathcal{F}$. Even in this case, $P(G, i)$ can only be estimated. Nevertheless, the probabilistic variant of connectedness seems well-suited to describe system degradation under random component failure. Due to its analytical complexity, however, it will most likely be used only in empirical evaluations.

15.5   Chapter Notes

Many different statistics have been studied in order to describe how networks change under component failures or intentional attacks. In this chapter we have given an overview of analyses and experimental results that aim at describing robustness and resilience properties of complex networks.

We first looked at worst case connectivity statistics that implicitly assume optimal attacks. Apart from classical connectivity, we also considered cohesiveness, the minimum m-degree, toughness and conditional connectivity. Only the first two measures can be computed in polynomial time. For a fixed parameter m, the minimum m-degree is also computable in polynomial time. Toughness is known to be $\mathcal{NP}$-hard and the complexity of conditional connectivity depends on the chosen property.

In an application, the function of a network might not only depend on its connectivity, but also on the length of the shortest paths. In Section 15.2, we

---

[2] In the original paper [438], Najjar and Gaudiot use the term *network resilience*.

looked at two worst case distance statistics, namely the persistence and incremental distance sequences. The second concept is more general than the first but for neither of them a polynomial time algorithm is known.

The main drawback of all the worst case statistics is that they make no statements about the results of random edge- or vertex-failures. Therefore, we looked at average robustness statistics in Section 15.3. The two statistics in this section for which no polynomial algorithm is known (mean connectivity and balanced-cut resilience) make statements about the network when edges fail while the two other statistics (average distance/fragmentation and effective diameter) only characterize the current state of a network. Hence, they are useful to measure robustness properties of a network only if they are repeatedly evaluated after successive edge deletions—either in an experiment or analytically.

In Section 15.4, we presented two statistics that give the probability that the network under consideration is still connected after the random failure of edges or vertices. The reliability polynomial gives the probability that the graph is connected given a failure probability for the edges while the probabilistic resilience for a network and a number i is the probability that the network disconnects after exactly i failures. There is no polynomial time algorithm known to compute any of these two statistics for general graphs.

The ideal statistics for describing the robustness of a complex network depend on the application and the type of the failures that are expected. If a network ceases to be useful after it is disconnected, statistics that describe the connectivity of the graph are best suited. If distances between vertices must be small, diameter-based statistics are preferable.

For random failures, the average and probabilistic statistics are the most promising while the effects of deliberate attacks are best captured by worst case statistics. So the ideal measure for deliberate attacks seems to be generalized connectivity but this has the drawback that it is hard to compute. A probabilistic version of generalized connectivity would be ideal for random failures.

In practice, an experimental approach to robustness seems to be most useful. The simultaneous observation of changes in average connected distance and fragmentation is suitable in many cases. One of the central results regarding robustness is certainly that scale-free networks are on the one hand tolerant against random failure but on the other hand exposed to intentional attacks.

Robustness is already a very complex topic but there are still many features of real-world networks that we have not touched in this chapter. Examples include the bandwidth of edges or the importance of vertices in an application as well as routing protocols and delay on edges.

Another interesting area are networks where the failures of elements are not independent of each other. In power networks for example, the failure of a power line puts more stress on other lines and thus makes their failure more likely, which might cause a domino effect.

At the moment, there are no deterministic polynomial algorithms that can answer meaningful questions about the robustness of complex real-world networks. If there are no major theoretic breakthroughs the most useful tools in this field will be simulations and heuristics.

Acknowledgments. The authors thank the editors, the co-authors of this book, and the anonymous referee for valuable comments.

Bibliography

1. Serge Abiteboul, Mihai Preda, and Gregory Cobena. Adaptive on-line page importance computation. In *Proceedings of the 12th International World Wide Web Conference (WWW12)*, pages 280–290, Budapest, Hungary, 2003.
2. Forman S. Acton. *Numerical Methods that Work*. Mathematical Association of America, 1990.
3. Alan Agresti. *Categorical Data Analysis*. Wiley, 2nd edition, 2002.
4. Alfred V. Aho, John E. Hopcroft, and Jeffrey D. Ullman. *The Design and Analysis of Computer Algorithms*. Addison-Wesley, 1974.
5. Alfred V. Aho, John E. Hopcroft, and Jeffrey D. Ullman. *Data Structures and Algorithms*. Addison-Wesley, 1983.
6. Ravindra K. Ahuja, Thomas L. Magnanti, and James B. Orlin. *Network Flows: Theory, Algorithms, and Applications*. Prentice Hall, 1993.
7. Ravindra K. Ahuja and James B. Orlin. A fast and simple algorithm for the maximum flow problem. *Operations Research*, 37(5):748–759, September/October 1989.
8. Ravindra K. Ahuja and James B. Orlin. Distance-based augmenting path algorithms for the maximum flow and parametric maximum flow problems. *Naval Research Logistics Quarterly*, 38:413–430, 1991.
9. William Aiello, Fan R. K. Chung, and Linyuan Lu. A random graph model for massive graphs. In *Proceedings of the 32nd Annual ACM Symposium on the Theory of Computing (STOC'00)*, pages 171–180, May 2000.
10. Martin Aigner. *Combinatorial Theory*. Springer-Verlag, 1999.
11. Martin Aigner and Eberhard Triesch. Realizability and uniqueness in graphs. *Discrete Mathematics*, 136:3–20, 1994.
12. Donald Aingworth, Chandra Chekuri, and Rajeev Motwani. Fast estimation of diameter and shortest paths (without matrix multiplication). In *Proceedings of the 7th Annual ACM–SIAM Symposium on Discrete Algorithms (SODA'96)*, 1996.
13. Jin Akiyama, Francis T. Boesch, Hiroshi Era, Frank Harary, and Ralph Tindell. The cohesiveness of a point of a graph. *Networks*, 11(1):65–68, 1981.
14. Richard D. Alba. A graph theoretic definition of a sociometric clique. *Journal of Mathematical Sociology*, 3:113–126, 1973.
15. Réka Albert and Albert-László Barabási. Statistical mechanics of complex networks. *Reviews of Modern Physics*, 74(1):47–97, 2002.
16. Réka Albert, Hawoong Jeong, and Albert-László Barabási. Diameter of the world wide web. *Nature*, 401:130–131, September 1999.
17. Réka Albert, Hawoong Jeong, and Albert-László Barabási. Error and attack tolerance of complex networks. *Nature*, 406:378–382, July 2000.
18. Mark S. Aldenderfer and Roger K. Blashfield. *Cluster Analysis*. Sage, 1984.
19. Noga Alon. Eigenvalues and expanders. *Combinatorica*, 6(2):83–96, 1986.
20. Noga Alon. Generating pseudo-random permutations and maximum flow algorithms. *Information Processing Letters*, 35(4):201–204, 1990.

21. Noga Alon, Fan R. K. Chung, and Ronald L. Graham. Routing permutations on graphs via matchings. *SIAM Journal on Discrete Mathematics*, 7:513–530, 1994.

22. Noga Alon, Michael Krivelevich, and Benny Sudakov. Finding a large hidden clique in a random graph. *Randoms Structures and Algorithms*, 13(3–4):457–466, 1998.

23. Noga Alon and Vitali D. Milman. λ_1, isoperimetric inequalities for graphs, and superconcentrators. *Journal of Combinatorial Theory Series B*, 38:73–88, 1985.

24. Noga Alon and Joel Spencer. *The Probabilistic Method*. Wiley, 1992.

25. Noga Alon, Joel Spencer, and Paul Erdős. *The Probabilistic Method*. Wiley, 1992.

26. Noga Alon, Raphael Yuster, and Uri Zwick. Finding and counting given length cycles. *Algorithmica*, 17(3):209–223, 1997.

27. Charles J. Alpert and Andrew B. Kahng. Recent directions in netlist partitioning: A survey. *Integration: The VLSI Journal*, 19(1-2):1–81, 1995.

28. Ashok T. Amin and S. Louis Hakimi. Graphs with given connectivity and independence number or networks with given measures of vulnerability and survivability. *IEEE Transactions on Circuit Theory*, 20(1):2–10, 1973.

29. Carolyn J. Anderson, Stanley Wasserman, and Bradley Crouch. A p^* primer: Logit models for social networks. *Social Networks*, 21(1):37–66, January 1999.

30. Carolyn J. Anderson, Stanley Wasserman, and Katherine Faust. Building stochastic blockmodels. *Social Networks*, 14:137–161, 1992.

31. James G. Anderson and Stephen J. Jay. The diffusion of medical technology: Social network analysis and policy research. *The Sociological Quarterly*, 26:49–64, 1985.

32. Jacob M. Anthonisse. The rush in a directed graph. Technical Report BN 9/71, Stichting Mathematisch Centrum, 2e Boerhaavestraat 49 Amsterdam, October 1971.

33. Arvind Arasu, Jasmine Novak, Andrew S. Tomkins, and John Tomlin. PageRank computation and the structure of the web: experiments and algorithms. short version appeared in Proceedings of the 11th International World Wide Web Conference, Poster Track, November 2001.

34. Ludwig Arnold. On the asymptotic distribution of the eigenvalues of random matrices. *Journal of Mathematical Analysis and Applications*, 20:262–268, 1967.

35. Sanjeev Arora, David R. Karger, and Marek Karpinski. Polynomial time approximation schemes for dense instances of $\mathcal{NP}$-hard problems. *Journal of Computer and System Sciences*, 58(1):193–210, 1999.

36. Sanjeev Arora, Satish Rao, and Umesh Vazirani. Expander flows, geometric embeddings and graph partitioning. In *Proceedings of the 36th Annual ACM Symposium on the Theory of Computing (STOC'04)*, pages 222–231. ACM Press, 2004.

37. Yuichi Asahiro, Refael Hassin, and Kazuo Iwama. Complexity of finding dense subgraphs. *Discrete Applied Mathematics*, 121(1–3):15–26, 2002.

38. Yuichi Asahiro, Kazuo Iwama, Hisao Tamaki, and Takeshi Tokuyama. Greedily finding a dense subgraph. *Journal of Algorithms*, 34(2):203–221, 2000.

39. Giorgio Ausiello, Pierluigi Crescenzi, Giorgio Gambosi, Viggo Kann, and Alberto Marchetti-Spaccamela. *Complexity and Approximation - Combinatorial Optimization Problems and Their Approximability Properties*. Springer-Verlag, 2nd edition, 2002.

40. Albert-László Barabási and Réka Albert. Emergence of scaling in random networks. *Science*, 286:509–512, 1999.

41. Alain Barrat and Martin Weigt. On the properties of small-world network models. *The European Physical Journal B*, 13:547–560, 2000.

42. Vladimir Batagelj. Notes on blockmodeling. *Social Networks*, 19(2):143–155, April 1997.

43. Vladimir Batagelj and Ulrik Brandes. Efficient generation of large random networks. *Physical Review E*, 2005. To appear.
44. Vladimir Batagelj and Anuška Ferligoj. Clustering relational data. In Wolfgang Gaul, Otto Opitz, and Martin Schader, editors, *Data Analysis*, pages 3–15. Springer-Verlag, 2000.
45. Vladimir Batagelj, Anuška Ferligoj, and Patrick Doreian. Generalized blockmodeling. *Informatica: An International Journal of Computing and Informatics*, 23:501–506, 1999.
46. Vladimir Batagelj and Andrej Mrvar. Pajek – A program for large network analysis. *Connections*, 21(2):47–57, 1998.
47. Vladimir Batagelj and Matjaž Zaveršnik. An $\mathcal{O}(m)$ algorithm for cores decomposition of networks. Technical Report 798, IMFM Ljublana, Ljubljana, 2002.
48. Douglas Bauer, S. Louis Hakimi, and Edward F. Schmeichel. Recognizing tough graphs is NP-hard. *Discrete Applied Mathematics*, 28:191–195, 1990.
49. Michel Bauer and Olivier Golinelli. Random incidence matrices: moments of the spectral density. *Journal of Statistical Physics*, 103:301–307, 2001. arXiv cond-mat/0007127.
50. Alex Bavelas. A mathematical model for group structure. *Human Organizations*, 7:16–30, 1948.
51. Alex Bavelas. Communication patterns in task oriented groups. *Journal of the Acoustical Society of America*, 22:271–282, 1950.
52. Murray A. Beauchamp. An improved index of centrality. *Behavioral Science*, 10:161–163, 1965.
53. M. Becker, W. Degenhardt, Jürgen Doenhardt, Stefan Hertel, G. Kaninke, W. Keber, Kurt Mehlhorn, Stefan Näher, Hans Rohnert, and Thomas Winter. A probabilistic algorithm for vertex connectivity of graphs. *Information Processing Letters*, 15(3):135–136, October 1982.
54. Richard Beigel. Finding maximum independent sets in sparse and general graphs. In *Proceedings of the 10th Annual ACM–SIAM Symposium on Discrete Algorithms (SODA'99)*, pages 856–857. IEEE Computer Society Press, 1999.
55. Lowell W. Beineke and Frank Harary. The connectivity function of a graph. *Mathematika*, 14:197–202, 1967.
56. Lowell W. Beineke, Ortrud R. Oellermann, and Raymond E. Pippert. The average connectivity of a graph. *Discrete Mathematics*, 252(1):31–45, May 2002.
57. Claude Berge. *Graphs*. North-Holland, 3rd edition, 1991.
58. Noam Berger, Béla Bollobás, Christian Borgs, Jennifer Chayes, and Oliver M. Riordan. Degree distribution of the FKP network model. In *Proceedings of the 30th International Colloquium on Automata, Languages, and Programming (ICALP'03)*, pages 725–738, 2003.
59. Julian E. Besag. Spatial interaction and the statistical analysis of lattice systems (with discussion). *Journal of the Royal Statistical Society, Series B*, 36:196–236, 1974.
60. Sergej Bezrukov, Robert Elsässer, Burkhard Monien, Robert Preis, and Jean-Pierre Tillich. New spectral lower bounds on the bisection width of graphs. *Theoretical Computer Science*, 320:155–174, 2004.
61. Monica Bianchini, Marco Gori, and Franco Scarselli. Inside PageRank. *ACM Transactions on Internet Technology*, 2004. in press.
62. Robert E. Bixby. The minimum number of edges and vertices in a graph with edge connectivity n and m n-bonds. *Bulletin of the American Mathematical Society*, 80(4):700–704, 1974.
63. Robert E. Bixby. The minimum number of edges and vertices in a graph with edge connectivity n and m n-bonds. *Networks*, 5:253–298, 1981.

64. Francis T. Boesch, Frank Harary, and Jerald A. Kabell. Graphs as models of communication network vulnerability: Connectivity and persistence. *Networks*, 11:57–63, 1981.
65. Francis T. Boesch and R. Emerson Thomas. On graphs of invulnerable communication nets. *IEEE Transactions on Circuit Theory*, CT-17, 1970.
66. Béla Bollobás. *Extremal graph theory*. Academic Press, 1978.
67. Béla Bollobás. *Modern Graph Theory*, volume 184 of *Graduate Texts in Mathematics*. Springer-Verlag, 1998.
68. Béla Bollobás and Oliver M. Riordan. Mathematical results on scale-free random graphs. In Stefan Bornholdt and Heinz Georg Schuster, editors, *Handbook of Graphs and Networks: From the Genome to the Internet*, pages 1–34. Wiley-VCH, 2002.
69. Béla Bollobás, Oliver M. Riordan, Joel Spencer, and Gábor Tusnády. The degree sequence of a scale-free random graph process. *Randoms Structures and Algorithms*, 18:279–290, 2001.
70. Immanuel M. Bomze, Marco Budinich, Panos M. Pardalos, and Marcello Pelillo. The maximum clique problem. In Ding-Zhu Du and Panos M. Pardalos, editors, *Handbook of Combinatorial Optimization (Supplement Volume A)*, volume 4, pages 1–74. Kluwer Academic Publishers Group, 1999.
71. Phillip Bonacich. Factoring and weighting approaches to status scores and clique identification. *Journal of Mathematical Sociology*, 2:113–120, 1972.
72. Phillip Bonacich. Power and centrality: A family of measures. *American Journal of Sociology*, 92(5):1170–1182, 1987.
73. Phillip Bonacich. What is a homomorphism? In Linton Clarke Freeman, Douglas R. White, and A. Kimbal Romney, editors, *Research Methods in Social Network Analysis*, chapter 8, pages 255–293. George Mason University Press, 1989.
74. J. Bond and Claudine Peyrat. Diameter vulnerability in networks. In Yousef Alavi, Gary Chartrand, Linda Lesniak, Don R. Lick, and Curtiss E. Wall, editors, *Graph Theory with Applications to Algorithms and Computer Science*, pages 123–149. Wiley, 1985.
75. Robert R. Boorstyn and Howard Frank. Large scale network topological optimization. *IEEE Transaction on Communications*, Com-25:29–37, 1977.
76. Kellogg S. Booth. Problems polynomially equivalent to graph isomorphism. Technical report, CS-77-04, University of Ljublana, 1979.
77. Kellogg S. Booth and George S. Lueker. Linear algorithms to recognize interval graphs and test for consecutive ones property. *Proceedings of the 7th Annual ACM Symposium on the Theory of Computing (STOC'75)*, pages 255–265, 1975.
78. Ravi B. Boppana. Eigenvalues and graph bisection: an average case analysis. In *Proceedings of the 28th Annual IEEE Symposium on Foundations of Computer Science (FOCS'87)*, pages 280–285, October 1987.
79. Ravi B. Boppana and Magnús M. Halldórsson. Approximating maximum independent sets by excluding subgraphs. *BIT*, 32(2):180–196, 1992.
80. Ravi B. Boppana, Johan Håstad, and Stathis Zachos. Does co-$\mathcal{NP}$ have short interactive proofs? *Information Processing Letters*, 25:127–132, 1987.
81. Stephen P. Borgatti. Centrality and AIDS. *Connections*, 18(1):112–115, 1995.
82. Stephen P. Borgatti and Martin G. Everett. The class of all regular equivalences: Algebraic structure and computation. *Social Networks*, 11(1):65–88, 1989.
83. Stephen P. Borgatti and Martin G. Everett. Two algorithms for computing regular equivalence. *Social Networks*, 15(4):361–376, 1993.
84. Stephen P. Borgatti and Martin G. Everett. Models of core/periphery structures. *Social Networks*, 21(4):375–395, 1999.
85. Stephen P. Borgatti and Martin G. Everett. A graph-theoretic perspective on centrality. Unpublished manuscript, 2004.

86. Stephen P. Borgatti, Martin G. Everett, and Paul R. Shirey. LS sets, lambda sets and other cohesive subsets. *Social Networks*, 12(4):337–357, 1990.

87. Allan Borodin, Gareth O. Roberts, Jeffrey S. Rosenthal, and Panayiotis Tsaparas. Finding authorities and hubs from link structures on the world wide web. In *Proceedings of the 10th International World Wide Web Conference (WWW10)*, pages 415–429, Hong Kong, 2001.

88. Rodrigo A. Botagfogo, Ehud Rivlin, and Ben Shneiderman. Structural analysis of hypertexts: Identifying hierarchies and useful metrics. *ACM Transactions on Information Systems*, 10(2):142–180, 1992.

89. John P. Boyd. *Social Semigroups*. George Mason University Press, 1991.

90. John P. Boyd and Martin G. Everett. Relations, residuals, regular interiors, and relative regular equivalence. *Social Networks*, 21(2):147–165, April 1999.

91. Stephen Boyd and Lieven Vandenberghe. *Convex Optimization*. Cambridge University Press, 2004.

92. Ulrik Brandes. A faster algorithm for betweenness centrality. *Journal of Mathematical Sociology*, 25(2):163–177, 2001.

93. Ulrik Brandes and Sabine Cornelsen. Visual ranking of link structures. *Journal of Graph Algorithms and Applications*, 7(2):181–201, 2003.

94. Ulrik Brandes and Daniel Fleischer. Centrality measures based on current flow. In *Proceedings of the 22nd International Symposium on Theoretical Aspects of Computer Science (STACS'05)*, volume 3404 of *Lecture Notes in Computer Science*, 2005. To appear.

95. Ulrik Brandes, Marco Gaertler, and Dorothea Wagner. Experiments on graph clustering algorithms. In *Proceedings of the 11th Annual European Symposium on Algorithms (ESA'03)*, volume 2832 of *Lecture Notes in Computer Science*, pages 568–579, September 2003.

96. Ulrik Brandes, Patrick Kenis, and Dorothea Wagner. Communicating centrality in policy network drawings. *IEEE Transactions on Visualization and Computer Graphics*, 9(2):241–253, 2003.

97. Ulrik Brandes and Jürgen Lerner. Structural similarity in graphs. In *Proceedings of the 15th International Symposium on Algorithms and Computation (ISAAC'04)*, volume 3341 of *Lecture Notes in Computer Science*, pages 184–195, 2004.

98. Ronald L. Breiger. Toward an operational theory of community elite structures. *Quality and Quantity*, 13:21–57, 1979.

99. Ronald L. Breiger, Scott A. Boorman, and Phipps Arabie. An algorithm for clustering relational data with applications to social network analysis and comparison with multidimensional scaling. *Journal of Mathematical Psychology*, 12:328–383, 1975.

100. Ronald L. Breiger and James G. Ennis. Personae and social roles: The network structure of personality types in small groups. *The Sociological Quarterly*, 42:262–270, 1979.

101. Sergey Brin and Lawrence Page. The anatomy of a large-scale hypertextual Web search engine. *Computer Networks and ISDN Systems*, 30(1–7):107–117, 1998.

102. Andrei Broder, Ravi Kumar, Farzin Maghoul, Prabhakar Raghavan, Sridhar Rajagopalan, Raymie Stata, Andrew S. Tomkins, and Janet Wiener. Graph structure in the Web. *Computer Networks: The International Journal of Computer and Telecommunications Networking*, 33(1–6):309–320, 2000.

103. Coen Bron and Joep A. G. M. Kerbosch. Algorithm 457: Finding all cliques of an undirected graph. *Communications of the ACM*, 16(9):575–577, 1973.

104. Tian Bu and Don Towsley. On distinguishing between Internet power law topology generators. In *Proceedings of Infocom'02*, 2002.

105. Mihai Bădoiu. Approximation algorithm for embedding metrics into a two-dimensional space. In *Proceedings of the 14th Annual ACM–SIAM Symposium on Discrete Algorithms (SODA'03)*, pages 434–443, 2003.

106. Horst Bunke and Kim Shearer. A graph distance metric based on the maximal common subgraph. *Pattern Recognition Letters*, 19:255–259, 1998.

107. Ronald S. Burt. Positions in networks. *Social Forces*, 55:93–122, 1976.

108. Duncan S. Callaway, Mark E. J. Newman, Steven H. Strogatz, and Duncan J. Watts. Network robustness and fragility: Percolation on random graphs. *Physical Review Letters*, 25(85):5468–5471, December 2000.

109. Kenneth L. Calvert, Matthew B. Doar, and Ellen W. Zegura. Modeling Internet topology. *IEEE Communications Magazine*, 35:160–163, June 1997.

110. A. Cardon and Maxime Crochemore. Partitioning a graph in $\mathcal{O}(|a|\log_2|v|)$. *Theoretical Computer Science*, 19:85–98, 1982.

111. Tami Carpenter, George Karakostas, and David Shallcross. Pracical Issues and Algorithms for Analyzing Terrorist Networks. invited talk at WMC 2002, 2002.

112. Peter J. Carrington, Greg H. Heil, and Stephen D. Berkowitz. A goodness-of-fit index for blockmodels. *Social Networks*, 2:219–234, 1980.

113. Moses Charikar. Greedy approximation algorithms for finding dense components in a graph. In *Proceedings of the 3rd International Workshop on Approximatin Algorithms for Combinatorial Optimization (APPROX'00)*, volume 1931 of *Lecture Notes in Computer Science*, pages 84–95. Springer-Verlag, 2000.

114. Gary Chartrand. A graph-theoretic approach to a communications problem. *SIAM Journal on Applied Mathematics*, 14(5):778–781, July 1966.

115. Gary Chartrand, Gary L. Johns, Songlin Tian, and Steven J. Winters. Directed distance on digraphs: Centers and medians. *Journal of Graph Theory*, 17(4):509–521, 1993.

116. Gary Chartrand, Grzegorz Kubicki, and Michelle Schultz. Graph similarity and distance in graphs. *Aequationes Mathematicae*, 55(1-2):129–145, 1998.

117. Qian Chen, Hyunseok Chang, Ramesh Govindan, Sugih Jamin, Scott Shenker, and Walter Willinger. The origin of power laws in internet topologies revisited. In *Proceedings of Infocom'02*, 2002.

118. Joseph Cheriyan and Torben Hagerup. A randomized maximum-flow algorithm. *SIAM Journal on Computing*, 24(2):203–226, 1995.

119. Joseph Cheriyan, Torben Hagerup, and Kurt Mehlhorn. An $o(n^3)$-time maximum-flow algorithm. *SIAM Journal on Computing*, 25(6):144–1170, December 1996.

120. Joseph Cheriyan and John H. Reif. Directed s-t numberings, rubber bands, and testing digraph k-vertex connectivity. In *Proceedings of the 3rd Annual ACM–SIAM Symposium on Discrete Algorithms (SODA'92)*, pages 335–344, January 1992.

121. Joseph Cheriyan and Ramakrishna Thurimella. Fast algorithms for k-shredders and k-node connectivity augmentation. *Journal of Algorithms*, 33:15–50, 1999.

122. Boris V. Cherkassky. An algorithm for constructing a maximal flow through a network requiring $\mathcal{O}(n^2\sqrt{p})$ operations. *Mathematical Methods for Solving Economic Problems*, 7:117–126, 1977. (In Russian).

123. Boris V. Cherkassky. A fast algorithm for constructing a maximum flow through a network. In *Selected Topics in Discrete Mathematics: Proceedings of the Moscow Discrete Mathematics Seminar, 1972-1990*, volume 158 of *American Mathematical Society Translations – Series 2*, pages 23–30. AMS, 1994.

124. Steve Chien, Cynthia Dwork, Ravi Kumar, and D. Sivakumar. Towards exploiting link evolution. In *Workshop on Algorithms and Models for the Web Graph*, November 2002.

125. Fan R. K. Chung. *Spectral Graph Theory*. CBMS Regional Conference Series in Mathematics. American Mathematical Society, 1997.

126. Fan R. K. Chung, Vance Faber, and Thomas A. Manteuffel. An upper bound on the diameter of a graph from eigenvalues associated with its laplacian. *SIAM Journal on Discrete Mathematics*, 7(3):443–457, 1994.

127. Fan R. K. Chung, Linyuan Lu, and Van Vu. Eigenvalues of random power law graphs. *Annals of Combinatorics*, 7:21–33, 2003.

128. Fan R. K. Chung, Linyuan Lu, and Van Vu. The spectra of random graphs with given expected degree. *Proceedings of the National Academy of Science of the United States of America*, 100(11):6313–6318, May 2003.

129. Vašek Chvátal. Tough graphs and hamiltionian circuits. *Discrete Mathematics*, 5, 1973.

130. Reuven Cohen, Keren Erez, Daniel ben Avraham, and Shlomo Havlin. Resilience of the Internet to random breakdown. *Physical Review Letters*, 21(85):4626–4628, November 2000.

131. Colin Cooper and Alan M. Frieze. A general model of web graphs. *Randoms Structures and Algorithms*, 22:311–335, 2003.

132. Don Coppersmith and Shmuel Winograd. Matrix multiplication via arithmetic progressions. *Journal of Symbolic Computation*, 9(3):251–280, 1990.

133. Thomas H. Cormen, Charles E. Leiserson, Ronald L. Rivest, and Clifford Stein. *Introduction to Algorithms*. MIT Press, 2nd edition, 2001.

134. Trevor F. Cox and Michael A. A. Cox. *Multidimensional Scaling*. Monographs on Statistics and Applied Probability. Chapman & Hall/CRC, 2nd edition, 2001.

135. Dragoš M. Cvetković, Michael Doob, and Horst Sachs. *Spectra of Graphs*. Johann Ambrosius Barth Verlag, 1995.

136. Dragoš M. Cvetković, Peter Rowlinson, and Slobodan Simic. *Eigenspaces of Graphs*. Cambridge University Press, 1997.

137. Andrzej Czygrinow. Maximum dispersion problem in dense graphs. *Operations Research Letter*, 27(5):223–227, 2000.

138. Peter Dankelmann and Ortrud R. Oellermann. Bounds on the average connectivity of a graph. *Discrete Applied Mathematics*, 129:305–318, August 2003.

139. George B. Dantzig. Application of the simplex method to a transportation problem. In Tjalling C. Koopmans, editor, *Activity Analysis of Production and Allocation*, volume 13 of *Cowles Commission for Research in Economics*, pages 359–373. Wiley, 1951.

140. George B. Dantzig. Maximization of a linear function of variables subject to linear inequalities. In Tjalling C. Koopmans, editor, *Activity Analysis of Production and Allocation*, volume 13 of *Cowles Commission for Research in Economics*, pages 339–347. Wiley, 1951.

141. George B. Dantzig and Delbert R. Fulkerson. On the max-flow min-cut theorem of networks. In *Linear Inequalities and Related Systems*, volume 38 of *Annals of Mathematics Studies*, pages 215–221. Princeton University Press, 1956.

142. Camil Demetrescu and Giuseppe F. Italiano. A new approach to dynamic all pairs shortest paths. In *Proceedings of the 35th Annual ACM Symposium on the Theory of Computing (STOC'03)*, pages 159–166, June 2003.

143. Guiseppe Di Battista and Roberto Tamassia. Incremental planarity testing. In *Proceedings of the 30th Annual IEEE Symposium on Foundations of Computer Science (FOCS'89)*, pages 436–441, October/November 1989.

144. Guiseppe Di Battista and Roberto Tamassia. On-line maintenance of triconnected components with SPQR-trees. *Algorithmica*, 15:302–318, 1996.

145. Reinhard Diestel. *Graph Theory*. Graduate Texts in Mathematics. Springer-Verlag, 2nd edition, 2000.

146. Edsger W. Dijkstra. A note on two problems in connection with graphs. *Numerische Mathematik*, 1:269–271, 1959.

147. Stephen Dill, Ravi Kumar, Kevin S. McCurley, Sridhar Rajagopalan, D. Sivakumar, and Andrew S. Tomkins. Self-similarity in the web. *ACM Transactions on Internet Technology*, 2(3):205–223, August 2002.
148. Chris H. Q. Ding, Xiaofeng He, Parry Husbands, Hongyuan Zha, and Horst D. Simon. PageRank, HITS and a unified framework for link analysis. LBNL Tech Report 49372, NERSC Division, Lawrence Berkeley National Laboratory, University of California, Berkeley, CA, USA, November 2001. updated Sept. 2002 (LBNL-50007), presented in the poster session of the Third SIAM International Conference on Data Mining, San Francisco, CA, USA, 2003.
149. Chris H. Q. Ding, Hongyuan Zha, Xiaofeng He, Parry Husbands, and Horst D. Simon. Link analysis: Hubs and authorities on the world wide web. *SIAM Review*, 46(2), 2004. to appear, published electronically May, 3, 2004.
150. Yefim Dinitz. Algorithm for solution of a problem of maximum flow in a network with power estimation. *Soviet Mathematics-Doklady*, 11(5):1277–1280, 1970.
151. Yefim Dinitz. Bitwise residual decreasing method and transportation type problems. In A. A. Fridman, editor, *Studies in Discrete Mathematics*, pages 46–57. Nauka, 1973. (In Russian).
152. Yefim Dinitz. Finding shortest paths in a network. In Y. Popkov and B. Shmulyian, editors, *Transportation Modeling Systems*, pages 36–44. Institute for System Studies, Moscow, 1978.
153. Yefim Dinitz, Alexander V. Karzanov, and M. V. Lomonosov. On the structure of the system of minimum edge cuts in a graph. In A. A. Fridman, editor, *In Studies in Discrete Optimization*, pages 290–306. Nauka, 1976.
154. Yefim Dinitz and Ronit Nossenson. Incremental maintenance of the 5-edge-connectivity classes of a graph. In *Proceedings of the 7th Scandinavian Workshop on Algorithm Theory (SWAT'00)*, volume 1851 of *Lecture Notes in Mathematics*, pages 272–285. Springer-Verlag, July 2000.
155. Yefim Dinitz and Jeffery Westbrook. Maintaining the classes of 4-edge-connectivity in a graph on-line. *Algorithmica*, 20(3):242–276, March 1998.
156. Gabriel A. Dirac. Extensions of Turán's theorem on graphs. *Acta Mathematica Academiae Scientiarum Hungaricae*, 14:417–422, 1963.
157. Matthew B. Doar. A better model for generating test networks. In *IEEE GLOBECOM'96*, 1996.
158. Wolfgang Domschke and Andreas Drexl. *Location and Layout Planning: An International Bibliography*. Springer-Verlag, Berlin, 1985.
159. William E. Donath and Alan J. Hoffman. Lower bounds for the partitioning of graphs. *IBM Journal of Research and Development*, 17(5):420–425, 1973.
160. Patrick Doreian. Using multiple network analytic tools for a single social network. *Social Networks*, 10:287–312, 1988.
161. Patrick Doreian and Louis H. Albert. Partitioning political actor networks: Some quantitative tools for analyzing qualitative networks. *Journal of Quantitative Anthropology*, 1:279–291, 1989.
162. Patrick Doreian, Vladimir Batagelj, and Anuška Ferligoj. Symmetric-acyclic decompositions of networks. *Journal of Classification*, 17(1):3–28, 2000.
163. Patrick Doreian, Vladimir Batagelj, and Anuška Ferligoj. Generalized blockmodeling of two-mode network data. *Social Networks*, 26(1):29–53, 2004.
164. Sergey N. Dorogovtsev and Jose Ferreira F. Mendes. Evolution of networks. *Advances in Physics*, 51(4):1079–1187, June 2002.
165. Sergey N. Dorogovtsev and Jose Ferreira F. Mendes. *Evolution of Networks*. Oxford University Press, 2003.
166. Sergey N. Dorogovtsev, Jose Ferreira F. Mendes, and Alexander N. Samukhin. Structure of growing networks: Exact solution of the Barabási-Albert's model. http://xxx.sissa.it/ps/cond-mat/0004434, April 2000.

167. Rodney G. Downey and Michael R. Fellows. Fixed-parameter tractability and completeness II. On completeness for W[1]. *Theoretical Computer Science*, 141(1–2):109–131, 1995.

168. Rodney G. Downey and Michael R. Fellows. *Parameterized Complexity*. Monographs in Computer Science. Springer-Verlag, 1999.

169. Zvi Drezner and Horst W. Hamacher, editors. *Facility Location: Application and Theory*. Springer-Verlag, 2002.

170. Petros Drineas, Alan M. Frieze, Ravi Kannan, Santosh Vempala, and V. Vinay. Clustering large graphs via the singular value decomposition. *Machine Learning*, 56:9–33, 2004.

171. Jack Edmonds. Edge-disjoint branchings. In Randall Rustin, editor, *Courant Computer Science Symposium 9: Combinatorial Algorithms (1972)*, pages 91–96. Algorithmics Press, 1973.

172. Jack Edmonds and Richard M. Karp. Theoretical improvements in algorithmic efficiency for network flow problems. *Journal of the ACM*, 19(2):248–264, April 1972.

173. Eugene Egerváry. On combinatorial properties of matrices. *Mat. Lapok*, 38:16–28, 1931.

174. Friedrich Eisenbrand and Fabrizio Grandoni. On the complexity of fixed parameter clique and dominating set. *Theoretical Computer Science*, 326(1–3):57–67, 2004.

175. Peter Elias, Amiel Feinstein, and Claude E. Shannon. A note on the maximum flow through a network. *IRE Transactions on Information Theory*, 2(4):117–119, December 1956.

176. Robert Elsässer and Burkhard Monien. Load balancing of unit size tokens and expansion properties of graphs. In *Proceedings of the 15th Annual ACM Symposium on Parallel Algorithms and Architectures (SPAA'03)*, pages 266–273, 2003.

177. Lars Engebretsen and Jonas Holmerin. Towards optimal lower bounds for clique and chromatic number. *Theoretical Computer Science*, 299(1-3):537–584, 2003.

178. David Eppstein. Fast hierarchical clustering and other applications of dynamic closest pairs. *j-ea*, 5:1–23, 2000.

179. David Eppstein and Joseph Wang. Fast approximation of centrality. In *Proceedings of the 12th Annual ACM–SIAM Symposium on Discrete Algorithms (SODA'01)*, 2001.

180. Paul Erdős and Tibor Gallai. Graphs with prescribed degrees of vertices (in hungarian). *Matematikai Lapok*, 11:264–274, 1960.

181. Paul Erdős and Alfred Rényi. On random graphs I. *Publicationes Mathematicae Debrecen*, 6:290–297, 1959.

182. Abdol-Hossein Esfahanian. Lower-bounds on the connectivities of a graph. *Journal of Graph Theory*, 9(4):503–511, 1985.

183. Abdol-Hossein Esfahanian and S. Louis Hakimi. On computing the connectivities of graphs and digraphs. *Networks*, 14(2):355–366, 1984.

184. Stephen Eubank, V. S. Anil Kumar, Madhav V. Marathe, Aravind Srinivasan, and Nan Wang. Structural and algorithmic aspects of massive social networks. In *Proceedings of the 14th Annual ACM–SIAM Symposium on Discrete Algorithms (SODA'04)*, pages 718–727, 2004.

185. Shimon Even. *Algorithmic Combinatorics*. Macmillan, 1973.

186. Shimon Even. An algorithm for determining whether the connectivity of a graph is at least k. *SIAM Journal on Computing*, 4(3):393–396, September 1975.

187. Shimon Even. *Graph Algorithms*. Computer Science Press, 1979.

188. Shimon Even and Robert E. Tarjan. Network flow and testing graph connectivity. *SIAM Journal on Computing*, 4(4):507–518, December 1975.

189. Martin G. Everett. Graph theoretic blockings k-plexes and k-cutpoints. *Journal of Mathematical Sociology*, 9:75–84, 1982.

190. Martin G. Everett and Stephen P. Borgatti. Role colouring a graph. *Mathematical Social Sciences*, 21:183–188, 1991.

191. Martin G. Everett and Stephen P. Borgatti. Regular equivalence: General theory. *Journal of Mathematical Sociology*, 18(1):29–52, 1994.

192. Martin G. Everett and Stephen P. Borgatti. Analyzing clique overlap. *Connections*, 21(1):49–61, 1998.

193. Martin G. Everett and Stephen P. Borgatti. Peripheries of cohesive subsets. *Social Networks*, 21(4):397–407, 1999.

194. Martin G. Everett and Stephen P. Borgatti. Extending centrality. In Peter J. Carrington, John Scott, and Stanley Wasserman, editors, *Models and Methods in Social Network Analysis*. Cambridge University Press, 2005. To appear.

195. Martin G. Everett, Philip Sinclair, and Peter Dankelmann. Some centrality results new and old. Submitted, 2004.

196. Alex Fabrikant, Elias Koutsoupias, and Christos H. Papadimitriou. Heuristically optimized trade-offs: A new paradigm for power laws in the Internet. In *Proceedings of the 29th International Colloquium on Automata, Languages, and Programming (ICALP'02)*, volume 2380 of *Lecture Notes in Computer Science*, 2002.

197. Michalis Faloutsos, Petros Faloutsos, and Christos Faloutsos. On power-law relationships of the Internet topology. In *Proceedings of SIGCOMM'99*, 1999.

198. Illés Farkas, Imre Derényi, Albert-László Barabási, and Tamás Vicsek. Spectra of "real-world" graphs: Beyond the semicircle law. *Physical Review E*, 64, August 2001.

199. Katherine Faust. Comparison of methods for positional analysis: Structural and general equivalences. *Social Networks*, 10:313–341, 1988.

200. Katherine Faust and John Skvoretz. Logit models for affiliation networks. *Sociological Methodology*, 29(1):253–280, 1999.

201. Uriel Feige, Guy Kortsarz, and David Peleg. The dense k-subgraph problem. *Algorithmica*, 29(3):410–421, 2001.

202. Uriel Feige and Robert Krauthgamer. Finding and certifying a large hidden clique in a semirandom graph. *Randoms Structures and Algorithms*, 16(2):195–208, 2000.

203. Uriel Feige and Robert Krauthgamer. A polylogarithmic approximation of the minimum bisection. *SIAM Journal on Computing*, 31(4):1090–1118, 2002.

204. Uriel Feige and Michael A. Seltser. On the densest k-subgraph problem. Technical Report CS97-16, Department of Applied Mathematics and Computer Science, The Weizmann Institute of Science, Rehovot, Israel, 1997.

205. Trevor Fenner, Mark Levene, and George Loizou. A stochastic evolutionary model exhibiting power-law behaviour with an exponential cutoff. http://xxx.sissa.it/ps/cond-mat/0209463, June 2004.

206. Anuška Ferligoj, Patrick Doreian, and Vladimir Batagelj. Optimizational approach to blockmodeling. *Journal of Computing and Information Technology*, 4:63–90, 1996.

207. Jean-Claude Fernandez. An implementation of an efficient algorithm for bisimulation equivalence. *Science of Computer Programming*, 13(1):219–236, 1989.

208. William L. Ferrar. *Finite Matrices*. Oxford University Press, London, 1951.

209. Jiří Fiala and Daniël Paulusma. The computational complexity of the role assignment problem. In *Proceedings of the 30th International Colloquium on Automata, Languages, and Programming (ICALP'03)*, pages 817–828. Springer-Verlag, 2003.

210. Miroslav Fiedler. Algebraic connectivity of graphs. *Czechoslovak Mathematical Journal*, 23(98):289–305, 1973.

211. Miroslav Fiedler. A property of eigenvectors of nonnegative symmetric matrices and its application to graph theory. *Czechoslovak Mathematical Journal*, 1:619–633, 1975.

212. Stephen E. Fienberg and Stanley Wasserman. Categorical data analysis of a single sociometric relation. In Samuel Leinhardt, editor, *Sociological Methodology*, pages 156–192. Jossey Bass, 1981.

213. Stephen E. Fienberg and Stanley Wasserman. Comment on an exponential family of probability distributions. *Journal of the American Statistical Association*, 76(373):54–57, March 1981.

214. Philippe Flajolet, Kostas P. Hatzis, Sotiris Nikoletseas, and Paul Spirakis. On the robustness of interconnections in random graphs: A symbolic approach. *Theoretical Computer Science*, 287(2):515–534, September 2002.

215. Philippe Flajolet and G. Nigel Martin. Probabilistic counting algorithms for data base applications. *Journal of Computer and System Sciences*, 31(2):182–209, 1985.

216. Lisa Fleischer. Building chain and cactus representations of all minimum cuts from Hao-Orlin in the same asymptotic run time. *Journal of Algorithms*, 33(1):51–72, October 1999.

217. Robert W. Floyd. Algorithm 97: Shortest path. *Communications of the ACM*, 5(6):345, 1962.

218. Lester R. Ford, Jr. and Delbert R. Fulkerson. Maximal flow through a network. *Canadian Journal of Mathematics*, 8:399–404, 1956.

219. Lester R. Ford, Jr. and Delbert R. Fulkerson. A simple algorithm for finding maximal network flows and an application to the Hitchcock problem. *Canadian Journal of Mathematics*, 9:210–218, 1957.

220. Lester R. Ford, Jr. and Delbert R. Fulkerson. *Flows in Networks*. Princeton University Press, 1962.

221. Scott Fortin. The graph isomorphism problem. Technical Report 96-20, University of Alberta, Edmonton, Canada, 1996.

222. Ove Frank and David Strauss. Markov graphs. *Journal of the American Statistical Association*, 81:832–842, 1986.

223. Greg N. Frederickson. Ambivalent data structures for dynamic 2-edge-connectivity and k smallest spanning trees. In *Proceedings of the 32nd Annual IEEE Symposium on Foundations of Computer Science (FOCS'91)*, pages 632–641, October 1991.

224. Michael L. Fredman. New bounds on the complexity of the shortest path problem. *SIAM Journal on Computing*, 5:49–60, 1975.

225. Michael L. Fredman and Dan E. Willard. Trans-dichotomous algorithms for minimum spanning trees and shortest paths. *Journal of Computer and System Sciences*, 48(3):533–551, 1994.

226. Linton Clarke Freeman. A set of measures of centrality based upon betweeness. *Sociometry*, 40:35–41, 1977.

227. Linton Clarke Freeman. Centrality in social networks: Conceptual clarification I. *Social Networks*, 1:215–239, 1979.

228. Linton Clarke Freeman. *The Development of Social Network Analysis: A Study in the Sociology of Science*. Booksurge Publishing, 2004.

229. Linton Clarke Freeman, Stephen P. Borgatti, and Douglas R. White. Centrality in valued graphs: A measure of betweenness based on network flow. *Social Networks*, 13(2):141–154, 1991.

230. Noah E. Friedkin. Structural cohesion and equivalence explanations of social homogeneity. *Sociological Methods and Research*, 12:235–261, 1984.

231. Delbert R. Fulkerson and George B. Dantzig. Computation of maximal flows in networks. *Naval Research Logistics Quarterly*, 2:277–283, 1955.

232. Delbert R. Fulkerson and G. C. Harding. On edge-disjoint branchings. *Networks*, 6(2):97–104, 1976.

233. Zoltán Füredi and János Komlós. The eigenvalues of random symmetric matrices. *Combinatorica*, 1(3):233–241, 1981.

234. Harold N. Gabow. Scaling algorithms for network problems. *Journal of Computer and System Sciences*, 31(2):148–168, 1985.

235. Harold N. Gabow. Path-based depth-first search for strong and biconnected components. *Information Processing Letters*, 74:107–114, 2000.

236. Zvi Galil. An $\mathcal{O}(V^{5/3}E^{2/3})$ algorithm for the maximal flow problem. *Acta Informatica*, 14:221–242, 1980.

237. Zvi Galil and Giuseppe F. Italiano. Fully dynamic algorithms for edge connectivity problems. In *Proceedings of the 23rd Annual ACM Symposium on the Theory of Computing (STOC'91)*, pages 317–327, May 1991.

238. Zvi Galil and Amnon Naamad. An $\mathcal{O}(EV \log^2 V)$ algorithm for the maximal flow problem. *Journal of Computer and System Sciences*, 21(2):203–217, October 1980.

239. Giorgio Gallo, Michail D. Grigoriadis, and Robert E. Tarjan. A fast parametric maximum flow algorithm and applications. *SIAM Journal on Computing*, 18(1):30–55, 1989.

240. Michael R. Garey and David S. Johnson. *Computers and Intractability. A Guide to the Theory of $\mathcal{NP}$-Completeness*. W. H. Freeman and Company, 1979.

241. Michael R. Garey, David S. Johnson, and Larry J. Stockmeyer. Some simplified $\mathcal{NP}$-complete graph problems. *Theoretical Computer Science*, 1:237–267, 1976.

242. Christian Gawron. An iterative algorithm to determine the dynamic user equilibrium in a traffic simulation model. *International Journal of Modern Physics C*, 9(3):393–408, 1998.

243. Andrew Gelman, John B. Carlin, Hal S. Stern, and Donald B. Rubin. *Bayesian Data Analysis*. Chapman & Hall Texts in Statistical Science. Chapman & Hall/CRC, 2nd edition, June 1995.

244. Horst Gilbert. Random graphs. *The Annals of Mathematical Statistics*, 30(4):1141–1144, 1959.

245. Walter R. Gilks, Sylvia Richardson, and David J. Spiegelhalter. *Markov Chain Monte Carlo in Practice*. Interdisciplinary Statistics. Chapman & Hall/CRC, 1996.

246. Chris Godsil. Tools from linear algebra. Research report, University of Waterloo, 1989.

247. Chris Godsil and Gordon Royle. *Algebraic Graph Theory*. Graduate Texts in Mathematics. Springer-Verlag, 2001.

248. Andrew V. Goldberg. Finding a maximum density subgraph. Technical Report UCB/CSB/ 84/171, Department of Electrical Engineering and Computer Science, University of California, Berkeley, CA, 1984.

249. Andrew V. Goldberg. A new max-flow algorithm. Technical Memo MIT/LCS/TM-291, MIT Laboratory for Computer Science, November 1985.

250. Andrew V. Goldberg and Satish Rao. Beyond the flow decomposition barrier. *Journal of the ACM*, 45(5):783–797, 1998.

251. Andrew V. Goldberg and Satish Rao. Flows in undirected unit capacity networks. *SIAM Journal on Discrete Mathematics*, 12(1):1–5, 1999.

252. Andrew V. Goldberg and Robert E. Tarjan. A new approach to the maximum-flow problem. *Journal of the ACM*, 35(4):921–940, 1988.

253. Andrew V. Goldberg and Kostas Tsioutsiouliklis. Cut tree algorithms: An experimental study. *Journal of Algorithms*, 38(1):51–83, 2001.

254. Donald L. Goldsmith. On the second order edge connectivity of a graph. *Congressus Numerantium*, 29:479–484, 1980.

255. Donald L. Goldsmith. On the n-th order edge-connectivity of a graph. *Congressus Numerantium*, 32:375–381, 1981.

256. Gene H. Golub and Charles F. Van Loan. *Matrix Computations*. John Hopkins University Press, 3rd edition, 1996.

257. Ralph E. Gomory and T.C. Hu. Multi-terminal network flows. *Journal of SIAM*, 9(4):551–570, December 1961.
258. Ralph E. Gomory and T.C. Hu. Synthesis of a communication network. *Journal of SIAM*, 12(2):348–369, 1964.
259. Ramesh Govindan and Anoop Reddy. An analysis of Internet inter-domain topology and route stability. In *Proceedings of Infocom'97*, 1997.
260. Fabrizio Grandoni and Giuseppe F. Italiano. Decremental clique problem. In *Proceedings of the 30th International Workshop on Graph-Theoretical Conecpts in Computer Science (WG'04)*, Lecture Notes in Computer Science. Springer-Verlag, 2004. To appear.
261. George Grätzer. *General Lattice Theory*. Birkhäuser Verlag, 1998.
262. Jerrold R. Griggs, Miklós Simonovits, and George Rubin Thomas. Extremal graphs with bounded densities of small subgraphs. *Journal of Graph Theory*, 29(3):185–207, 1998.
263. Geoffrey Grimmett and Colin J. H. McDiarmid. On colouring random graphs. *Mathematical Proceedings of the Cambridge Philosophical Society*, 77:313–324, 1975.
264. Dan Gusfield. Connectivity and edge-disjoint spanning trees. *Information Processing Letters*, 16(2):87–89, 1983.
265. Dan Gusfield. Very simple methods for all pairs network flow analysis. *SIAM Journal on Computing*, 19(1):143–155, 1990.
266. Ronald J. Gutman. Reach-based routing: A new approach to shortest path algorithms optimized for road networks. In *Proceedings of the 6th Workshop on Algorithm Engineering and Experiments (ALENEX'04)*, Lecture Notes in Computer Science, pages 100–111. SIAM, 2004.
267. Carsten Gutwenger and Petra Mutzel. A linear time implementation of SPQR-trees. In *Proceedings of the 8th International Symposium on Graph Drawing (GD'00)*, volume 1984 of *Lecture Notes in Computer Science*, pages 70–90, January 2001.
268. Willem H. Haemers. Eigenvalue methods. In Alexander Schrijver, editor, *Packing and Covering in Combinatorics*, pages 15–38. Mathematisch Centrum, 1979.
269. Per Hage and Frank Harary. *Structural models in anthropology*. Cambridge University Press, 1st edition, 1983.
270. S. Louis Hakimi. On the realizability of a set of integers as degrees of the vertices of a linear graph. *SIAM Journal on Applied Mathematics*, 10:496–506, 1962.
271. S. Louis Hakimi. Optimum location of switching centers and the absolute centers and medians of a graph. *Operations Research*, 12:450–459, 1964.
272. Jianxiu Hao and James B. Orlin. A faster algorithm for finding the minimum cut in a graph. In *Proceedings of the 3rd Annual ACM–SIAM Symposium on Discrete Algorithms (SODA'92)*, pages 165–174, January 1992.
273. Frank Harary. Status and contrastatus. *Sociometry*, 22:23–43, 1959.
274. Frank Harary. The maximum connectivity of a graph. *Proceedings of the National Academy of Science of the United States of America*, 48(7):1142–1146, July 1962.
275. Frank Harary. A characterization of block-graphs. *Canadian Mathematical Bulletin*, 6(1):1–6, January 1963.
276. Frank Harary. Conditional connectivity. *Networks*, 13:347–357, 1983.
277. Frank Harary. General connectivity. In Khee Meng Koh and Hian-Poh Yap, editors, *Proceedings of the 1st Southeast Asian Graph Theory Colloquium*, volume 1073 of *Lecture Notes in Mathematics*, pages 83–92. Springer-Verlag, 1984.
278. Frank Harary and Per Hage. Eccentricity and centrality in networks. *Social Networks*, 17:57–63, 1995.
279. Frank Harary and Yukihiro Kodama. On the genus of an *n*-connected graph. *Fundamenta Mathematicae*, 54:7–13, 1964.

280. Frank Harary and Helene J. Kommel. Matrix measures for transitivity and balance. *Journal of Mathematical Sociology*, 6:199–210, 1979.

281. Frank Harary and Robert Z. Norman. The dissimilarity characteristic of Husimi trees. *Annals of Mathematics*, 58(2):134–141, 1953.

282. Frank Harary and Herbert H. Paper. Toward a general calculus of phonemic distribution. *Language: Journal of the Linguistic Society of America*, 33:143–169, 1957.

283. Frank Harary and Geert Prins. The block-cutpoint-tree of a graph. *Publicationes Mathematicae Debrecen*, 13:103–107, 1966.

284. Frank Harary and Ian C. Ross. A procedure for clique detection using the group matrix. *Sociometry*, 20:205–215, 1957.

285. David Harel and Yehuda Koren. On clustering using random walks. In *Proceedings of the 21st Conference on Foundations of Software Technology and Theoretical Computer Science (FSTTCS'01)*, volume 2245 of *Lecture Notes in Computer Science*, pages 18–41. Springer-Verlag, 2001.

286. Erez Hartuv and Ron Shamir. A clustering algorithm based on graph connectivity. *Information Processing Letters*, 76(4-6):175–181, 2000.

287. Johan Håstad. Clique is hard to approximate within $n^{1-\varepsilon}$. *Acta Mathematica*, 182:105–142, 1999.

288. Vaclav Havel. A remark on the existence of finite graphs (in czech). *Casopis Pest. Math.*, 80:477–480, 1955.

289. Taher H. Haveliwala. Topic-sensitive pagerank: A context-sensitive ranking algorithm for web search. *IEEE Transactions on Knowledge and Data Engineering*, 15(4):784–796, 2003.

290. Taher H. Haveliwala and Sepandar D. Kamvar. The second eigenvalue of the Google matrix. Technical report, Stanford University, March 2003.

291. Taher H. Haveliwala, Sepandar D. Kamvar, and Glen Jeh. An analytical comparison of approaches to personalized PageRank. Technical report, Stanford University, June 2003.

292. Taher H. Haveliwala, Sepandar D. Kamvar, Dan Klein, Christopher D. Manning, and Gene H. Golub. Computing PageRank using power extrapolation. Technical report, Stanford University, July 2003.

293. George R. T. Hendry. On graphs with a prescribed median. I. *Journal of Graph Theory*, 9:477–487, 1985.

294. Michael A. Henning and Ortrud R. Oellermann. The average connectivity of a digraph. *Discrete Applied Mathematics*, 140:143–153, May 2004.

295. Monika R. Henzinger and Michael L. Fredman. Lower bounds for fully dynamic connectivity problems in graphs. *Algorithmica*, 22(3):351–362, 1998.

296. Monika R. Henzinger and Valerie King. Fully dynamic 2-edge connectivity algorithm in polylogarithmic time per operation. SRC Technical Note 1997-004a, Digital Equipment Corporation, Systems Research Center, Palo Alto, California, June 1997.

297. Monika R. Henzinger and Johannes A. La Poutré. Certificates and fast algorithms for biconnectivity in fully-dynamic graphs. SRC Technical Note 1997-021, Digital Equipment Corporation, Systems Research Center, Palo Alto, California, September 1997.

298. Monika R. Henzinger, Satish Rao, and Harold N. Gabow. Computing vertex connectivity: New bounds from old techniques. In *Proceedings of the 37th Annual IEEE Symposium on Foundations of Computer Science (FOCS'96)*, pages 462–471, October 1996.

299. Wassily Hoeffding. Probability inequalities for sums of bounded random variables. *Journal of the American Statistical Association*, 58(301):713–721, 1963.

300. Karen S. Holbert. A note on graphs with distant center and median. In V. R. Kulli, editor, *Recent Sudies in Graph Theory*, pages 155–158, Gulbarza, India, 1989. Vishwa International Publications.

301. Paul W. Holland, Kathryn B. Laskey, and Samuel Leinhardt. Stochastic blockmodels: First steps. *Social Networks*, 5:109–137, 1983.

302. Paul W. Holland and Samuel Leinhardt. An exponential family of probability distributions for directed graphs. *Journal of the American Statistical Association*, 76(373):33–50, March 1981.

303. Jacob Holm, Kristian de Lichtenberg, and Mikkel Thorup. Poly-logarithmic deterministic fully-dynamic algorithms for connectivity, minimum spanning tree, 2-edge, and biconnectivity. *Journal of the ACM*, 48(4):723–760, 2001.

304. Petter Holme. Congestion and centrality in traffic flow on complex networks. *Advances in Complex Systems*, 6(2):163–176, 2003.

305. Petter Holme, Beom Jun Kim, Chang No Yoon, and Seung Kee Han. Attack vulnerability of complex networks. *Physical Review E*, 65(056109), 2002.

306. Klaus Holzapfel. *Density-based clustering in large-scale networks*. PhD thesis, Technische Universität München, 2004.

307. Klaus Holzapfel, Sven Kosub, Moritz G. Maaß, Alexander Offtermatt-Souza, and Hanjo Täubig. A zero-one law for densities of higher order. Manuscript, 2004.

308. Klaus Holzapfel, Sven Kosub, Moritz G. Maaß, and Hanjo Täubig. The complexity of detecting fixed-density clusters. In *Proceedings of the 5th Italian Conference on Algorithms and Complexity (CIAC'03)*, volume 2653 of *Lecture Notes in Computer Science*, pages 201–212. Springer-Verlag, 2003.

309. John E. Hopcroft and Robert E. Tarjan. Finding the triconnected components of a graph. Technical Report TR 72-140, CS Dept., Cornell University, Ithaca, N.Y., August 1972.

310. John E. Hopcroft and Robert E. Tarjan. Dividing a graph into triconnected components. *SIAM Journal on Computing*, 2(3):135–158, September 1973.

311. John E. Hopcroft and Robert E. Tarjan. Efficient algorithms for graph manipulation. *Communications of the ACM*, 16(6):372–378, June 1973.

312. John E. Hopcroft and Robert E. Tarjan. Dividing a graph into triconnected components. Technical Report TR 74-197, CS Dept., Cornell University, Ithaca, N.Y., February 1974.

313. John E. Hopcroft and Robert E. Tarjan. Efficient planarity testing. *Journal of the ACM*, 21(4):549–568, October 1974.

314. John E. Hopcroft and J.K. Wong. A linear time algorithm for isomorphism of planar graphs. In *Proceedings of the 6th Annual ACM Symposium on the Theory of Computing (STOC'74)*, pages 172–184, 1974.

315. Radu Horaud and Thomas Skordas. Stereo correspondence through feature grouping and maximal cliques. *IEEE Transactions on Pattern Analysis and Machine Intelligence*, 11(11):1168–1180, 1989.

316. Wen-Lian Hsu. $\mathcal{O}(MN)$ algorithms for the recognition and isomorphism problems on circular-arc graphs. *SIAM Journal on Computing*, 24:411–439, 1995.

317. T.C. Hu. Optimum communication spanning trees. *SIAM Journal on Computing*, 3:188–195, 1974.

318. Xiaohan Huang and Victor Y. Pan. Fast rectangular matrix multiplication and applications. *Journal of Complexity*, 14(2):257–299, 1998.

319. Charles H. Hubbell. In input-output approach to clique identification. *Sociometry*, 28:377–399, 1965.

320. Piotr Indyk and Jiří Matoušek. Low-distortion embeddings of finite metric spaces. In Jacob E. Goodman and Joseph O'Rourke, editors, *Handbook of Discrete and Computational Geometry*. Chapman & Hall/CRC, April 2004.

321. Alon Itai and Michael Rodeh. Finding a minimum circuit in a graph. *SIAM Journal on Computing*, 7(4):413–423, 1978.

322. Kenneth E. Iverson. *A Programming Language*. Wiley, 1962.

323. Matthew O. Jackson and Asher Wolinsky. A strategic model of social and economic networks. *Journal of Economic Theory*, 71:474–486, 1996.

324. Anil K. Jain and Richard C. Dubes. *Algorithms for clustering data*. Prentice Hall, 1988.

325. Anil K. Jain, M. N. Murty, and Patrick J. Flynn. Data clustering: a review. *ACM Computing Surveys*, 31(3):264–323, 1999.

326. Glen Jeh and Jennifer Widom. Scaling personalized web search. In *Proceedings of the 12th International World Wide Web Conference (WWW12)*, pages 271–279, Budapest, Hungary, 2003.

327. Hawoong Jeong, Sean P. Mason, Albert-László Barabási, and Zoltan N. Oltvai. Lethality and centrality in protein networks. *Nature*, 411, 2001. Brief communications.

328. Mark Jerrum. Large cliques elude the Metropolis process. *Randoms Structures and Algorithms*, 3(4):347–359, 1992.

329. Mark Jerrum and Alistair Sinclair. Fast uniform generation of regular graphs. *Theoretical Computer Science*, 73:91–100, 1990.

330. Tang Jian. An $O(2^{0.304n})$ algorithm for solving maximum independent set problem. *IEEE Transactions on Computers*, C-35(9):847–851, 1986.

331. Bin Jiang. I/O- and CPU-optimal recognition of strongly connected components. *Information Processing Letters*, 45(3):111–115, March 1993.

332. Cheng Jin, Qian Chen, and Sugih Jamin. Inet topology generator. Technical Report CSE-TR-433-00, EECS Department, University of Michigan, 2000.

333. David S. Johnson, Jan Karel Lenstra, and Alexander H. G. Rinnooy Kan. The complexity of the network design problem. *Networks*, 9:279–285, 1978.

334. David S. Johnson, Christos H. Papadimitriou, and Mihalis Yannakakis. On generating all maximal independent sets. *Information Processing Letters*, 27(3):119–123, 1988.

335. Ian T. Jolliffe. *Principal Component Analysis*. Springer-Verlag, 2002.

336. Camille Jordan. Sur les assemblages de lignes. *Journal für reine und angewandte Mathematik*, 70:185–190, 1869.

337. Ferenc Juhász. On the spectrum of a random graph. *Colloquia Mathematica Societatis János Bolyai*, 25:313–316, 1978.

338. Sepandar D. Kamvar, Taher H. Haveliwala, and Gene H. Golub. Adaptive methods for the computation of PageRank. Technical report, Stanford University, April 2003.

339. Sepandar D. Kamvar, Taher H. Haveliwala, Christopher D. Manning, and Gene H. Golub. Exploiting the block structure of the web for computing PageRank. Technical report, Stanford University, March 2003.

340. Sepandar D. Kamvar, Taher H. Haveliwala, Christopher D. Manning, and Gene H. Golub. Extrapolation methods for accelerating PageRank computations. In *Proceedings of the 12th International World Wide Web Conference (WWW12)*, pages 261–270, Budapest, Hungary, 2003.

341. Arkady Kanevsky, Roberto Tamassia, Guiseppe Di Battista, and Jianer Chen. On-line maintenance of the four-connected components of a graph. In *Proceedings of the 32nd Annual IEEE Symposium on Foundations of Computer Science (FOCS'91)*, pages 793–801, October 1991.

342. Ravi Kannan and V. Vinay. Analyzing the structure of large graphs. Manuscript, 1999.

343. David R. Karger and Matthew S. Levine. Finding maximum flows in undirected graphs seems easier than bipartite matching. In *Proceedings of the 30th Annual ACM Symposium on the Theory of Computing (STOC'98)*, pages 69–78, May 1998.

344. David R. Karger and Clifford Stein. An $\tilde{O}(n^2)$ algorithm for minimum cuts. In *Proceedings of the 25th Annual ACM Symposium on the Theory of Computing (STOC'93)*, pages 757–765, May 1993.

345. Richard M. Karp. Reducibility among combinatorial problems. In Raymond E. Miller and James W. Thatcher, editors, *Complexity of Computer Computations*, pages 85–103. Plenum Press, 1972.

346. Richard M. Karp. On the computational complexity of combinatorial problems. *Networks*, 5:45–68, 1975.

347. Richard M. Karp. Probabilistic analysis of some combinatorial search problems. In Joseph F. Traub, editor, *Algorithms and Complexity: New Directions and Recent Results*, pages 1–19. Academic Press, 1976.

348. George Karypis and Vipin Kumar. A fast and high quality multilevel scheme for partitioning irregular graphs. *SIAM Journal on Scientific Computing*, 20(1):359–392, 1998.

349. Alexander V. Karzanov. On finding maximum flows in networks with special structure and some applications. In *Matematicheskie Voprosy Upravleniya Proizvodstvom*, volume 5, pages 66–70. Moscow State University Press, 1973. (In Russian).

350. Alexander V. Karzanov. Determining the maximal flow in a network by the method of preflows. *Soviet Mathematics-Doklady*, 15(2):434–437, 1974.

351. Alexander V. Karzanov and Eugeniy A. Timofeev. Efficient algorithm for finding all minimal edge cuts of a nonoriented graph. *Cybernetics*, 22(2):156–162, 1986.

352. Leo Katz. A new status index derived from sociometric analysis. *Psychometrika*, 18(1):39–43, 1953.

353. Subhash Khot. Improved approximation algorithms for max clique, chromatic number and approximate graph coloring. In *Proceedings of the 42nd Annual IEEE Symposium on Foundations of Computer Science (FOCS'01)*, pages 600–609. IEEE Computer Society Press, 2001.

354. Subhash Khot. Ruling out PTAS for graph min-bisection, densest subgraph and bipartite clique. In *Proceedings of the 45th Annual IEEE Symposium on Foundations of Computer Science (FOCS'04)*, pages 136–145. IEEE Computer Society Press, 2004.

355. K. H. Kim and F. W. Roush. Group relationships and homomorphisms of boolean matrix semigroups. *Journal of Mathematical Psychology*, 28:448–452, 1984.

356. Valerie King, Satish Rao, and Robert E. Tarjan. A faster deterministic maximum flow algorithm. In *Proceedings of the 3rd Annual ACM-SIAM Symposium on Discrete Algorithms (SODA'92)*, pages 157–164, January 1992.

357. G. Kishi. On centrality functions of a graph. In N. Saito and T. Nishizeki, editors, *Proceedings of the 17th Symposium of Research Institute of Electrical Communication on Graph Theory and Algorithms*, volume 108 of *Lecture Notes in Computer Science*, pages 45–52, Sendai, Japan, October 1980. Springer.

358. G. Kishi and M. Takeuchi. On centrality functions of a non-directed graph. In *Proceedings of the 6th Colloquium on Microwave Communication*, Budapest, 1978.

359. Jon M. Kleinberg. Authoritative sources in a hyperlinked environment. *Journal of the ACM*, 46(5):604–632, 1999.

360. Jon M. Kleinberg. The small-world phenomenon: An algorithmic perspective. In *Proceedings of the 32nd Annual ACM Symposium on the Theory of Computing (STOC'00)*, May 2000.

361. Jon M. Kleinberg. An impossibility theorem for clustering. In *Proceedings of 15th Conference: Neiral Information Processing Systems, Advances in Neural Information Processing Systems*, 2002.

362. Daniel J. Kleitman. Methods for investigating connectivity of large graphs. *IEEE Transactions on Circuit Theory*, 16(2):232–233, May 1969.

363. Ton Kloks, Dieter Kratsch, and Haiko Müller. Finding and counting small induced subgraphs efficiently. *Information Processing Letters*, 74(3–4):115–121, 2000.

364. David Knoke and David L. Rogers. A blockmodel analysis of interorganizational networks. *Sociology and Social Research*, 64:28–52, 1979.

365. Donald E. Knuth. Two notes on notation. *American Mathematical Monthly*, 99:403–422, 1990.

366. Dénes Kőnig. Graphen und Matrizen. *Mat. Fiz. Lapok*, 38:116–119, 1931.

367. Avrachenkov Konstantin and Nelly Litvak. Decomposition of the Google PageRank and optimal linking strategy. Technical Report 5101, INRIA, Sophia Antipolis, France, January 2004.

368. Tamás Kővári, Vera T. Sós, and Pál Turán. On a problem of Zarankiewicz. *Colloquium Mathematicum*, 3:50–57, 1954.

369. Paul L. Krapivsky, Sidney Redner, and Francois Leyvraz. Connectivity of growing random networks. http://xxx.sissa.it/ps/cond-mat/0005139, September 2000.

370. Jan Kratochvíl. *Perfect Codes in General Graphs*. Academia Praha, 1991.

371. V. Krishnamoorthy, K. Thulasiraman, and M. N. S. Swamy. Incremental distance and diameter sequences of a graph: New measures of network performance. *IEEE Transactions on Computers*, 39(2):230–237, February 1990.

372. Joseph B. Kruskal. Multidimensional scaling by optimizing goodness of fit to a nonparametric hypothesis. *Psychometrika*, 29:1–27, March 1964.

373. Joseph B. Kruskal. Nonmetric multidimensional scaling: A numerical method. *Psychometrika*, 29:115–129, June 1964.

374. Luděk Kučera. Expected complexity of graph partitioning problems. *Discrete Applied Mathematics*, 57(2–3):193–212, 1995.

375. Ravi Kumar, Prabhakar Raghavan, Sridhar Rajagopalan, D. Sivakumar, Andrew S. Tomkins, and Eli Upfal. Stochastic models for the web graph. In *Proceedings of the 41st Annual IEEE Symposium on Foundations of Computer Science (FOCS'00)*, 2000.

376. Ravi Kumar, Prabhakar Raghavan, Sridhar Rajagopalan, and Andrew S. Tomkins. Trawling the web for emerging cyber-communities. *Computer Networks: The International Journal of Computer and Telecommunications Networking*, 31(11–16):1481–1493, 1999.

377. Johannes A. La Poutré, Jan van Leeuwen, and Mark H. Overmars. Maintenance of 2- and 3-connected components of graphs, Part I: 2- and 3-edge-connected components. Technical Report RUU-CS-90-26, Dept. of Computer Science, Utrecht University, July 1990.

378. Amy N. Langville and Carl D. Meyer. Deeper inside PageRank. Technical report, Department of Mathematics, North Carolina State University, Raleigh, NC, USA, March 2004. accepted by *Internet Mathematics*.

379. Amy N. Langville and Carl D. Meyer. A survey of eigenvector methods of web information retrieval. Technical report, Department of Mathematics, North Carolina State University, Raleigh, NC, USA, January 2004. accepted by *The SIAM Review*.

380. Luigi Laura, Stefano Leonardi, Stefano Millozzi, and Ulrich Meyer. Algorithms and experiments for the webgraph. In *Proceedings of the 11th Annual European Symposium on Algorithms (ESA'03)*, volume 2832 of *Lecture Notes in Computer Science*, 2003.

381. Eugene L. Lawler. Cutsets and partitions of hypergraphs. *Networks*, 3:275–285, 1973.

382. Eugene L. Lawler, Jan Karel Lenstra, and Alexander H. G. Rinnooy Kan. Generating all maximal independent sets: $\mathcal{NP}$-hardness and polynomial-time algorithms. *SIAM Journal on Computing*, 9(3):558–565, 1980.

383. Chris Pan-Chi Lee, Gene H. Golub, and Stefanos A. Zenios. A fast two-stage algorithm for computing PageRank. Technical Report SCCM-03-15, Stanford University, 2003.

384. Erich L. Lehmann. *Testing Statistical Hypotheses*. Springer Texts in Statistics. Springer-Verlag, 2nd edition, 1997.

385. Erich L. Lehmann and George Casella. *Theory of Point Estimation*. Springer Texts in Statistics. Springer-Verlag, 2nd edition, 1998.

386. L. Ya. Leifman. An efficient algorithm for partitioning an oriented graph into bicomponents. *Cybernetics*, 2(5):15–18, 1966.

387. Ronny Lempel and Shlomo Moran. The stochastic approach for link-structure analysis (SALSA) and the TKC effect. *Computer Networks: The International Journal of Computer and Telecommunications Networking*, 33:387–401, 2000. volume coincides with the Proceedings of the 9th international World Wide Web conference on Computer networks.

388. Ronny Lempel and Shlomo Moran. Rank-stability and rank-similarity of link-based web ranking algorithms in authority-connected graphs. *Information Retrieval, special issue on Advances in Mathematics/Formal Methods in Information Retrieval*, 2004. in press.

389. Thomas Lengauer. *Combinatorial Algorithms for Integrated Circuit Layout*. Wiley, 1990.

390. Linda Lesniak. Results on the edge-connectivity of graphs. *Discrete Mathematics*, 8:351–354, 1974.

391. Robert Levinson. Pattern associativity and the retrieval of semantic networks. *Computers & Mathematics with Applications*, 23(2):573–600, 1992.

392. Nathan Linial, László Lovász, and Avi Wigderson. A physical interpretation of graph connectivity and its algorithmic applications. In *Proceedings of the 27th Annual IEEE Symposium on Foundations of Computer Science (FOCS'86)*, pages 39–48, October 1986.

393. Nathan Linial, László Lovász, and Avi Wigderson. Rubber bands, convex embeddings and graph connectivity. *Combinatorica*, 8(1):91–102, 1988.

394. François Lorrain and Harrison C. White. Structural equivalence of individuals in social networks. *Journal of Mathematical Sociology*, 1:49–80, 1971.

395. Emmanuel Loukakis and Konstantinos-Klaudius Tsouros. A depth first search algorithm to generate the family of maximal independent sets of a graph lexicographically. *Computing*, 27:249–266, 1981.

396. László Lovász. Connectivity in digraphs. *Journal of Combinatorial Theory Series B*, 15(2):174–177, August 1973.

397. László Lovász. On the Shannon capacity of a graph. *IEEE Transactions on Information Theory*, 25:1–7, 1979.

398. Anna Lubiw. Some $\mathcal{NP}$-complete problems similar to graph isomorphism. *SIAM Journal on Computing*, 10:11–24, 1981.

399. Fabrizio Luccio and Mariagiovanna Sami. On the decomposition of networks in minimally interconnected subnetworks. *IEEE Transactions on Circuit Theory*, CT-16:184–188, 1969.

400. R. Duncan Luce. Connectivity and generalized cliques in sociometric group structure. *Psychometrika*, 15:169–190, 1950.

401. R. Duncan Luce and Albert Perry. A method of matrix analysis of group structure. *Psychometrika*, 14:95–116, 1949.

402. Eugene M. Luks. Isomorphism of graphs of bounded valence can be tested in polynomial time. *Journal of Computer and System Sciences*, 25:42–65, 1982.

403. Saunders Mac Lane. A structural characterization of planar combinatorial graphs. *Duke Mathematical Journal*, 3:460–472, 1937.

404. Wolfgang Mader. Ecken vom Grad n in minimalen n-fach zusammenhängenden Graphen. *Archiv der Mathematik*, 23:219–224, 1972.

405. Damien Magoni and Jean Jacques Pansiot. Analysis of the autonomous system network topology. *Computer Communication Review*, 31(3):26–37, July 2001.

406. Vishv M. Malhotra, M. Pramodh Kumar, and S. N. Maheshwari. An $\mathcal{O}(|V|^3)$ algorithm for finding maximum flows in networks. *Information Processing Letters*, 7(6):277–278, October 1978.

407. Yishay Mansour and Baruch Schieber. Finding the edge connectivity of directed graphs. *Journal of Algorithms*, 10(1):76–85, March 1989.

408. Maarten Marx and Michael Masuch. Regular equivalence and dynamic logic. *Social Networks*, 25:51–65, 2003.

409. Rudi Mathon. A note on the graph isomorphism counting problem. *Information Processing Letters*, 8(3):131–132, 1979.

410. David W. Matula. The cohesive strength of graphs. In *The Many Facets of Graph Theory, Proc.*, volume 110 of *Lecture Notes in Mathematics*, pages 215–221. Springer-Verlag, 1969.

411. David W. Matula. k-components, clusters, and slicings in graphs. *SIAM Journal on Applied Mathematics*, 22(3):459–480, May 1972.

412. David W. Matula. Graph theoretic techniques for cluster analysis algorithms. In J. Van Ryzin, editor, *Classification and clustering*, pages 95–129. Academic Press, 1977.

413. David W. Matula. Determining edge connectivity in $\mathcal{O}(nm)$. In *Proceedings of the 28th Annual IEEE Symposium on Foundations of Computer Science (FOCS'87)*, pages 249–251, October 1987.

414. James J. McGregor. Backtrack search algorithms and the maximal common subgraph problem. *Software - Practice and Experience*, 12(1):23–24, 1982.

415. Brendan D. McKay. Practical graph isomorphism. *Congressus Numerantium*, 30:45–87, 1981.

416. Brendan D. McKay and Nicholas C. Wormald. Uniform generation of random regular graphs of moderate degree. *Journal of Algorithms*, 11:52–67, 1990.

417. Alberto Medina, Anukool Lakhina, Ibrahim Matta, and John Byers. BRITE: An approach to universal topology generation. In *Proceedings of the International Symposium on Modeling, Analysis and Simulation of Computer and Telecommunication Systems (MASCOTS'01)*, 2001.

418. Alberto Medina, Ibrahim Matta, and John Byers. On the origin of power laws in Internet topologies. *Computer Communication Review*, 30(2), April 2000.

419. Karl Menger. Zur allgemeinen Kurventheorie. *Fundamenta Mathematicae*, 10:96–115, 1927.

420. Milena Mihail, Christos Gkantsidis, Amin Saberi, and Ellen W. Zegura. On the semantics of internet topologies. Technical Report GIT-CC-02-07, Georgia Institute of Technology, 2002.

421. Stanley Milgram. The small world problem. *Psychology Today*, 1:61, 1967.

422. Gary L. Miller and Vijaya Ramachandran. A new graph triconnectivity algorithm and its parallelization. *Combinatorica*, 12(1):53–76, 1992.

423. Ron Milo, Shai Shen-Orr, Shalev Itzkovitz, Nadav Kashtan, Dmitri Chklovskii, and Uri Alon. Network motifs: Simple building blocks of complex networks. *Science*, 298:824–827, October 2002.

424. J. Clyde Mitchell. Algorithms and network analysis: A test of some analytical procedures on Kapferer's tailor shop material. In Linton Clarke Freeman, Douglas R. White, and A. Kimbal Romney, editors, *Research Methods in Social Network Analysis*, pages 365–391. George Mason University Press, 1989.

425. Bojan Mohar. Isoperimetric numbers of graphs. *Journal of Combinatorial Theory Series B*, 47(3):274–291, 1989.

426. Bojan Mohar. Eigenvalues, diameter and mean distance in graphs. *Graphs and Combinatorics*, 7:53–64, 1991.

427. Bojan Mohar. The laplacian spectrum of graphs. In Yousef Alavi, Gary Chartrand, Ortrud R. Oellermann, and Allen J. Schwenk, editors, *Graph Theory, Combinatorics, and Applications*, pages 871–898. Wiley, 1991.

428. Bojan Mohar and Svatopluk Poljak. Eigenvalues in combinatorial optimization. In Richard A. Brualdi, Shmuel Friedland, and Victor Klee, editors, *Combinatorial and Graph-Theoretical Problems in Linear Algebra*, pages 107–151. Springer-Verlag, 1993.

429. Robert J. Mokken. Cliques, clubs, and clans. *Quality and Quantity*, 13:161–173, 1979.

430. Burkhard Möller. Zentralitäten in Graphen. Diplomarbeit, Fachbereich Informatik und Informationswissenschaft, Universität Konstanz, July 2002.

431. John W. Moon. On the diameter of a graph. *Michigan Mathematical Journal*, 12(3):349–351, 1965.

432. John W. Moon and L. Moser. On cliques in graphs. *Israel Journal of Mathematics*, 3:23–28, 1965.

433. Robert L. Moxley and Nancy F. Moxley. Determining Point-Centrality in Uncontrived Social Networks. *Sociometry*, 37:122–130, 1974.

434. N. C. Mullins, L. L. Hargens, P. K. Hecht, and Edward L. Kick. The group structure of cocitation clusters: A comparative study. *American Sociological Review*, 42:552–562, 1977.

435. Ian Munro. Efficient determination of the transitive closure of a directed graph. *Information Processing Letters*, 1(2):56–58, 1971.

436. Siegfried F. Nadel. *The Theory of Social Structure*. Cohen & West LTD, 1957.

437. Kai Nagel. Traffic networks. In Stefan Bornholdt and Heinz Georg Schuster, editors, *Handbook of Graphs and Networks: From the Genome to the Internet*. Wiley-VCH, 2002.

438. Walid Najjar and Jean-Luc Gaudiot. Network resilience: A measure of network fault tolerance. *IEEE Transactions on Computers*, 39(2):174–181, February 1990.

439. Georg L. Nemhause and Laurence A. Wolesy. *Integer and Combinatorial Optimization*. Wiley, 1988.

440. Jaroslav Nešetřil and Svatopluk Poljak. On the complexity of the subgraph problem. *Commentationes Mathematicae Universitatis Carolinae*, 26(2):415–419, 1985.

441. Mark E. J. Newman. Assortative mixing in networks. *Physical Review Letters*, 89(208701), 2002.

442. Mark E. J. Newman. Fast algorithm for detecting community structure in networks. arXiv cond-mat/0309508, September 2003.

443. Mark E. J. Newman. A measure of betweenness centrality based on random walks. arXiv cond-mat/0309045, 2003.

444. Mark E. J. Newman and Michelle Girvan. Mixing patterns and community structure in networks. In Romualdo Pastor-Satorras, Miguel Rubi, and Albert Diaz-Guilera, editors, *Statistical Mechanics of Complex Networks*, volume 625 of *Lecture Notes in Physics*, pages 66–87. Springer-Verlag, 2003.

445. Mark E. J. Newman and Michelle Girvan. Findind and evaluating community structure in networks. *Physical Review E*, 69(2):026113, 2004.

446. Mark E. J. Newman and Juyong Park. Why social networks are different from other types of networks. *Physical Review E*, 68(036122), 2003.

447. Mark E. J. Newman, Steven H. Strogatz, and Duncan J. Watts. Random graph models of social networks. *Proceedings of the National Academy of Science of the United States of America*, 99:2566–2572, 2002.

448. Mark E. J. Newman, Duncan J. Watts, and Steven H. Strogatz. Random graphs with arbitrary degree distributions and their applications. *Physical Review E*, 64:026118, 2001.

449. Andrew Y. Ng, Alice X. Zheng, and Micheal I. Jordan. Link analysis, eigenvectors and stability. In *Proceedings of the senventeenth international joint conference on artificial intelligence*, pages 903–910, Seattle, Washington, 2001.

450. Victor Nicholson, Chun Che Tsai, Marc A. Johnson, and Mary Naim. A subgraph isomorphism theorem for molecular graphs. In *Proceedings of The International Conference on Graph Theory and Topology in Chemistry*, pages 226–230, 1987.

451. U. J. Nieminen. On the Centrality in a Directed Graph. *Social Science Research*, 2:371–378, 1973.

452. National laboratory for applied network research routing data, 1999.

453. Krzysztof Nowicki and Tom A.B. Snijders. Estimation and prediction for stochastic blockstructures. *Journal of the American Statistical Association*, 96:1077–1087, 2001.

454. Esko Nuutila and Eljas Soisalon-Soininen. On finding the strongly connected components in a directed graph. *Information Processing Letters*, 49(1):9–14, January 1994.

455. Ortrud R. Oellermann. A note on the ℓ-connectivity function of a graph. *Congressus Numerantium*, 60:181–188, December 1987.

456. Ortrud R. Oellermann. On the l-connectivity of a graph. *Graphs and Combinatorics*, 3:285–291, 1987.

457. Maria G.R. Ortiz, Jose R.C. Hoyos, and Maria G.R. Lopez. The social networks of academic performance in a student context of poverty in mexico. *Social Networks*, 26(2):175–188, 2004.

458. Lawrence Page, Sergey Brin, Rajeev Motwani, and Terry Winograd. The PageRank citation ranking: Bringing order to the web. Manuscript, 1999.

459. Robert Paige and Robert E. Tarjan. Three partition refinement algorithms. *SIAM Journal on Computing*, 16(6):973–983, 1987.

460. Ignacio Palacios-Huerta and Oscar Volij. The measurement of intellectual influence. *Econometrica*, 2004. accepted for publication.

461. Christopher Palmer, Phillip Gibbons, and Christos Faloutsos. Fast approximation of the "neighbourhood" function for massive graphs. Technical Report CMUCS-01-122, Carnegie Mellon Uiversity, 2001.

462. Christopher Palmer, Georgos Siganos, Michalis Faloutsos, Christos Faloutsos, and Phillip Gibbons. The connectivity and fault-tolerance of the Internet topology. In *Workshop on Network-Related Data Management (NRDM 2001)*, 2001.

463. Gopal Pandurangan, Prabhakar Raghavan, and Eli Upfal. Using PageRank to characterize Web structure. In *Proceedings of the 8th Annual International Conference on Computing Combinatorics (COCOON'02)*, volume 2387 of *Lecture Notes in Computer Science*, pages 330–339, 2002.

464. Apostolos Papadopoulos and Yannis Manolopoulos. Structure-based similarity search with graph histograms. In *DEXA Workshop*, pages 174–178, 1999.

465. Britta Papendiek and Peter Recht. On maximal entries in the principal eigenvector of graphs. *Linear Algebra and its Applications*, 310:129–138, 2000.

466. Panos M. Pardalos and Jue Xue. The maximum clique problem. *Journal of Global Optimization*, 4:301–328, 1994.

467. Beresford N. Parlett. *The Symmetric Eigenvalue Problem*. SIAM, 1998.

468. Romualdo Pastor-Satorras, Alexei Vázquez, and Alessandro Vespignani. Dynamical and correlation properties of the internet. *Physical Review Letters*, 87(258701), 2001.

469. Romualdo Pastor-Satorras and Alessandro Vespignani. Epidemics and immunization in scale-free networks. In Stefan Bornholdt and Heinz Georg Schuster, editors, *Handbook of Graphs and Networks: From the Genome to the Internet*. Wiley-VCH, 2002.

470. Keith Paton. An algorithm for the blocks and cutnodes of a graph. *Communications of the ACM*, 14(7):468–475, July 1971.

471. Philippa Pattison. *Algebraic Models for Social Networks*. Cambridge University Press, 1993.

472. Philippa Pattison and Stanley Wasserman. Logit models and logistic regressions for social networks: II. Multivariate relations. *British Journal of Mathematical and Statistical Psychology*, 52:169–193, 1999.

473. Marvin C. Paull and Stephen H. Unger. Minimizing the number of states in incompletely specified sequential switching functions. *IRE Transaction on Electronic Computers*, EC-8:356–367, 1959.

474. Aleksandar Pekeč and Fred S. Roberts. The role assignment model nearly fits most social networks. *Mathematical Social Sciences*, 41:275–293, 2001.

475. Claudine Peyrat. Diameter vulnerability of graphs. *Discrete Applied Mathematics*, 9, 1984.

476. Steven Phillips and Jeffery Westbrook. On-line load balancing and network flow. *Algorithmica*, 21(3):245–261, 1998.

477. Jean-Claude Picard and Maurice Queyranne. A network flow solution to some nonlinear 0-1 programming problems, with an application to graph theory. *Networks*, 12:141–159, 1982.

478. Jean-Claude Picard and H. D. Ratliff. Minimum cuts and related problems. *Networks*, 5(4):357–370, 1975.

479. Gabriel Pinski and Francis Narin. Citation influence for journal aggregates of scientific publications: theory, with application to the literature of physics. *Information Processing & Management*, 12:297–312, 1976.

480. André Pönitz and Peter Tittmann. Computing network reliability in graphs of restricted pathwidth. http://www.peter.htwm.de/publications/Reliability.ps, 2001.

481. R. Poulin, M.-C. Boily, and B.R. Mâsse. Dynamical systems to define centrality in social networks. *Social Networks*, 22:187–220, 2000.

482. William H. Press, Saul A. Teukolsky, William T. Vetterling, and Brian P. Flannery. *Numerical Recipes in C*. Cambridge University Press, 1992.

483. C.H. Proctor and C. P. Loomis. Analysis of sociometric data. In Marie Jahoda, Morton Deutsch, and Stuart W. Cook, editors, *Research Methods in Social Relations*, pages 561–586. Dryden Press, 1951.

484. Paul W. Purdom, Jr. A transitive closure algorithm. Computer Sciences Technical Report #33, University of Wisconsin, July 1968.

485. Paul W. Purdom, Jr. A transitive closure algorithm. *BIT*, 10:76–94, 1970.

486. Pavlin Radoslavov, Hongsuda Tangmunarunkit, Haobo Yu, Ramesh Govindan, Scott Shenker, and Deborah Estrin. On characterizing network topologies and analyzing their impact on protocol design. Technical Report 00-731, Computer Science Department, University of Southern California, February 2000.

487. Rajeev Raman. Recent results on the single-source shortest paths problem. *ACM SIGACT News*, 28(2):81–87, 1997.

488. John W. Raymond, Eleanor J. Gardiner, and Peter Willet. RASCAL: Calculation of graph similarity using maximum common edge subgraphs. *The Computer Journal*, 45(6):631–644, 2002.

489. Ronald C. Read and Derek G. Corneil. The graph isomorphism disease. *Journal of Graph Theory*, 1:339–363, 1977.

490. John H. Reif. A topological approach to dynamic graph connectivity. *Information Processing Letters*, 25(1):65–70, 1987.

491. Franz Rendl and Henry Wolkowicz. A projection technique for partitioning the nodes of a graph. *Annals of Operations Research*, 58:155–180, 1995.

492. John A. Rice. *Mathematical Statistics and Data Analysis*. Duxbury Press, 2nd edition, 1995.

493. Fred S. Roberts and Li Sheng. $\mathcal{NP}$-completeness for 2-role assignability. Technical Report 8, Rutgers Center for Operation Research, 1997.

494. Garry Robins, Philippa Pattison, and Stanley Wasserman. Logit models and logistic regressions for social networks: III. Valued relations. *Psychometrika*, 64:371–394, 1999.

495. John Michael Robson. Algorithms for maximum independent sets. *Journal of Algorithms*, 7(3):425–440, 1986.

496. Liam Roditty and Uri Zwick. On dynamic shortest paths problems. In *Proceedings of the 12th Annual European Symposium on Algorithms (ESA'04)*, volume 3221 of *Lecture Notes in Computer Science*, pages 580–591, 2004.

497. Arnon S. Rosenthal. *Computing Reliability of Complex Systems*. PhD thesis, University of California, 1974.

498. Sheldon M. Ross. *Introduction to Probability Models*. Academic Press, 8th edition, 2003.

499. Britta Ruhnau. Eigenvector-centrality – a node-centrality? *Social Networks*, 22:357–365, 2000.

500. Gert Sabidussi. The centrality index of a graph. *Psychometrika*, 31:581–603, 1966.

501. Lee Douglas Sailer. Structural equivalence: Meaning and definition, computation and application. *Social Networks*, 1:73–90, 1978.

502. Thomas Schank and Dorothea Wagner. Approximating clustering-coefficient and transitivity. Technical Report 2004-9, Universität Karlsruhe, Fakultät für Informatik, 2004.

503. Claus P. Schnorr. Bottlenecks and edge connectivity in unsymmetrical networks. *SIAM Journal on Computing*, 8(2):265–274, May 1979.

504. Uwe Schöning. Graph isomorphism is in the low hierarchy. *Journal of Computer and System Sciences*, 37:312–323, 1988.

505. Alexander Schrijver. *Theory of linear and integer programming*. Wiley, 1986.

506. Alexander Schrijver. Paths and flows—a historical survey. *CWI Quarterly*, 6(3):169–183, September 1993.

507. Alexander Schrijver. *Combinatorial Optimization: Polyhedra and Efficiency*. Springer-Verlag, 2003.

508. Joseph E. Schwartz. An examination of CONCOR and related methods for blocking sociometric data. In D. R. Heise, editor, *Sociological Methodology 1977*, pages 255–282. Jossey Bass, 1977.

509. Jennifer A. Scott. An Arnoldi code for computing selected eigenvalues of sparse real unsymmetric matrices. *ACM Transactions on Mathematical Software*, 21:423–475, 1995.

510. John R. Seeley. The net of reciprocal influence. *Canadian Journal of Psychology*, III(4):234–240, 1949.

511. Stephen B. Seidman. Clique-like structures in directed networks. *Journal of Social and Biological Structures*, 3:43–54, 1980.

512. Stephen B. Seidman. Internal cohesion of LS sets in graphs. *Social Networks*, 5(2):97–107, 1983.

513. Stephen B. Seidman. Network structure and minimum degree. *Social Networks*, 5:269–287, 1983.

514. Stephen B. Seidman and Brian L. Foster. A graph-theoretic generalization of the clique concept. *Journal of Mathematical Sociology*, 6:139–154, 1978.

515. Stephen B. Seidman and Brian L. Foster. A note on the potential for genuine cross-fertilization between anthropology and mathematics. *Social Networks*, 172:65–72, 1978.

516. Ron Shamir, Roded Sharan, and Dekel Tsur. Cluster graph modification problems. In *Graph-Theoretic Concepts in Computer Science, 28th International Workshop, WG 2002*, volume 2573 of *Lecture Notes in Computer Science*, pages 379–390. Springer-Verlag, 2002.

517. Micha Sharir. A strong-connectivity algorithm and its applications in data flow analysis. *Computers & Mathematics with Applications*, 7(1):67–72, 1981.
518. Yossi Shiloach. An $\mathcal{O}(n \cdot I \log^2 I)$ maximum-flow algorithm. Technical Report STAN-CS-78-702, Computer Science Department, Stanford University, December 1978.
519. Alfonso Shimbel. Structural parameters of communication networks. *Bulletin of Mathematical Biophysics*, 15:501–507, 1953.
520. F. M. Sim and M. R. Schwartz. Does CONCOR find positions? Unpublished manuscript, 1979.
521. Alistair Sinclair. *Algorithms for Random Generation and Counting: A Markov Chain Approach.* Birkhäuser Verlag, 1993.
522. Brajendra K. Singh and Neelima Gupte. Congestion and Decongestion in a communication network. arXiv cond-mat/0404353, 2004.
523. Mohit Singh and Amitabha Tripathi. Order of a graph with given vertex and edge connectivity and minimum degree. *Electronic Notes in Discrete Mathematics*, 15, 2003.
524. Peter J. Slater. Maximin facility location. *Journal of National Bureau of Standards*, 79B:107–115, 1975.
525. Daniel D. Sleater and Robert E. Tarjan. A data structure for dynamic trees. *Journal of Computer and System Sciences*, 26(3):362–391, June 1983.
526. Giora Slutzki and Oscar Volij. Scoring of web pages and tournaments – axiomatizations. Technical report, Iowa State University, Ames, USA, February 2003.
527. Christian Smart and Peter J. Slater. Center, median and centroid subgraphs. *Networks*, 34:303–311, 1999.
528. Peter H. A. Sneath and Robert R. Sokal. *Numerical Taxonomy: The Principles and Practice of Numerical Classification.* W. H. Freeman and Company, 1973.
529. Tom A.B. Snijders. Markov chain monte carlo estimation of exponential random graph models. *Journal of Social Structure*, 3(2), April 2002.
530. Tom A.B. Snijders and Krzysztof Nowicki. Estimation and prediction of stochastic blockmodels for graphs with latent block structure. *Journal of Classification*, 14:75–100, 1997.
531. Anand Srivastav and Katja Wolf. Finding dense subgraphs with semidefinite programming. In *Proceedings of the 1st International Workshop on Approximatin Algorithms for Combinatorial Optimization (APPROX'98)*, volume 1444 of *Lecture Notes in Computer Science*, pages 181–191. Springer-Verlag, 1998.
532. Angelika Steger and Nicholas C. Wormald. Generating random regular graphs quickly. *Combinatorics, Probability and Computing*, 8:377–396, 1999.
533. Karen A. Stephenson and Marvin Zelen. Rethinking centrality: Methods and examples. *Social Networks*, 11:1–37, 1989.
534. Volker Stix. Finding all maximal cliques in dynamic graphs. *Computational Optimization and Applications*, 27(2):173–186, 2004.
535. Josef Stoer and Roland Bulirsch. *Introduction to Numerical Analysis.* Springer-Verlag, 1993.
536. Mechthild Stoer and Frank Wagner. A simple min-cut algorithm. *Journal of the ACM*, 44(4):585–591, 1997.
537. Sun Microsystems. *Sun Performance Library User's Guide.*
538. Melvin Tainiter. Statistical theory of connectivity I: Basic definitions and properties. *Discrete Mathematics*, 13(4):391–398, 1975.
539. Melvin Tainiter. A new deterministic network reliability measure. *Networks*, 6(3):191–204, 1976.
540. Hongsuda Tangmunarunkit, Ramesh Govindan, Sugih Jamin, Scott Shenker, and Walter Willinger. Network topologies, power laws, and hierarchy. Technical Report 01-746, Computer Science Department, University of Southern California, 2001.

464   Bibliography

541. Hongsuda Tangmunarunkit, Ramesh Govindan, Sugih Jamin, Scott Shenker, and Walter Willinger. Network topologies, power laws, and hierarchy. *ACM SIG-COMM Computer Communication Review*, 32(1):76, 2002.
542. Robert E. Tarjan. Depth-first search and linear graph algorithms. *SIAM Journal on Computing*, 1(2):146–160, June 1972.
543. Robert E. Tarjan. Finding a maximum clique. Technical Report 72-123, Department of Computer Science, Cornell University, Ithaca, NY, 1972.
544. Robert E. Tarjan. A note on finding the bridges of a graph. *Information Processing Letters*, 2(6):160–161, 1974.
545. Robert E. Tarjan and Anthony E. Trojanowski. Finding a maximum independent set. *SIAM Journal on Computing*, 6(3):537–546, 1977.
546. Mikkel Thorup. On RAM priority queues. In *Proceedings of the 7th Annual ACM-SIAM Symposium on Discrete Algorithms (SODA'96)*, pages 59–67, 1996.
547. Mikkel Thorup. Undirected single source shortest paths with positive integer weights in linear time. *Journal of the ACM*, 46(3):362–394, 1999.
548. Mikkel Thorup. On ram priority queues. *SIAM Journal on Computing*, 30(1):86–109, 2000.
549. Mikkel Thorup. Fully dynamic all-pairs shortest paths: Faster and allowing negative cycles. In *Proceedings of the 9th Scandinavian Workshop on Algorithm Theory (SWAT'04)*, volume 3111 of *Lecture Notes in Computer Science*, pages 384–396. Springer-Verlag, 2004.
550. Gottfried Tinhofer. On the generation of random graphs with given properties and known distribution. *Appl. Comput. Sci. Ber. Prakt. Inf.*, 13:265–296, 1979.
551. Po Tong and Eugene L. Lawler. A faster algorithm for finding edge-disjoint branchings. *Information Processing Letters*, 17(2):73–76, August 1983.
552. Miroslaw Truszczyński. Centers and centroids of unicyclic graphs. *Mathematica Slovaca*, 35:223–228, 1985.
553. Shuji Tsukiyama, Mikio Ide, Hiromu Ariyoshi, and Isao Shirakawa. A new algorithm for generating all the maximal independent sets. *SIAM Journal on Computing*, 6(3):505–517, 1977.
554. Pál Turán. On an extremal problem in graph theory. *Matematikai és Fizikai Lapok*, 48:436–452, 1941.
555. William T. Tutte. A theory of 3-connected graphs. *Indagationes Mathematicae*, 23:441–455, 1961.
556. William T. Tutte. *Connectivity in graphs*. Number 15 in Mathematical Expositions. University of Toronto Press, 1966.
557. Salil P. Vadhan. The complexity of counting in sparse, regular, and planar graphs. *SIAM Journal on Computing*, 31(2):398–427, 2001.
558. Thomas W. Valente and Robert K. Foreman. Integration and radiality: measuring the extent of an individual's connectedness and reachability in a network. *Social Networks*, 1:89–105, 1998.
559. Leslie G. Valiant. The complexity of computing the permanent. *Theoretical Computer Science*, 8:189–201, 1979.
560. Leslie G. Valiant. The complexity of enumeration and reliability problems. *SIAM Journal on Computing*, 8(3):410–421, 1979.
561. Edwin R. van Dam and Willem H. Haemers. Which graphs are determined by their spectrum? *Linear Algebra and its Applications*, 373:241–272, 2003.
562. René van den Brink and Robert P. Gilles. An axiomatic social power index for hierarchically structured populations of economic agents. In Robert P. Gilles and Picter H.M. Ruys, editors, *Imperfections and Behaviour in Economic Organizations*, pages 279–318. Kluwer Academic Publishers Group, 1994.
563. René van den Brink and Robert P. Gilles. Measuring domination in directed networks. *Social Networks*, 22(2):141–157, May 2000.

564. Stijn M. van Dongen. *Graph Clustering by Flow Simulation.* PhD thesis, University of Utrecht, 2000.

565. Santosh Vempala, Ravi Kannan, and Adrian Vetta. On clusterings - good, bad and spectral. In *Proceedings of the 41st Annual IEEE Symposium on Foundations of Computer Science (FOCS'00),* pages 367–378, 2000.

566. The Stanford WebBase Project. http://www-diglib.stanford.edu/ testbed/doc2/-WebBase/.

567. Yuchung J. Wang and George Y. Wong. Stochastic blockmodels for directed graphs. *Journal of the American Statistical Association,* 82:8–19, 1987.

568. Stephen Warshall. A theorem on boolean matrices. *Journal of the ACM,* 9(1):11–12, 1962.

569. Stanley Wasserman and Katherine Faust. *Social Network Analysis: Methods and Applications.* Cambridge University Press, 1994.

570. Stanley Wasserman and Philippa Pattison. Logit models and logistic regressions for social networks: I. An introduction to Markov graphs and p^*. *Psychometrika,* 60:401–426, 1996.

571. David S. Watkins. QR-like algorithms for eigenvalue problems. *Journal of Computational and Applied Mathematics,* 123:67–83, 2000.

572. Alison Watts. A dynamic model of network formation. *Games and Economic Behavior,* 34:331–341, 2001.

573. Duncan J. Watts and Steven H. Strogatz. Collective dynamics of "small-world" networks. *Nature,* 393:440–442, 1998.

574. Bernard M. Waxman. Routing of multipoint connections. *IEEE Journal on Selected Areas in Communications,* 6(9):1617–1622, 1988.

575. Alfred Weber. *Über den Standort der Industrien.* J. C. B. Mohr, Tübingen, 1909.

576. Douglas B. West. *Introduction to Graph Theory.* Prentice Hall, 2nd edition, 2001.

577. Jeffery Westbrook and Robert E. Tarjan. Maintaining bridge-connected and biconnected components on-line. *Algorithmica,* 7:433–464, 1992.

578. Douglas R. White and Stephen P. Borgatti. Betweenness Centrality Measures for Directed Graphs. *Social Networks,* 16:335–346, 1994.

579. Douglas R. White and Karl P. Reitz. Graph and semigroup homomorphisms on networks of relations. *Social Networks,* 5:193–234, 1983.

580. Scott White and Padhraic Smyth. Algorithms for estimating relative importance in networks. In *Proceedings of the 9th ACM SIGKDD International Conference on Knowledge Discovery and Data Mining (KDD'03),* 2003.

581. Hassler Whitney. Congruent graphs and the connectivity of graphs. *American Journal of Mathematics,* 54:150–168, 1932.

582. R. W. Whitty. Vertex-disjoint paths and edge-disjoint branchings in directed graphs. *Journal of Graph Theory,* 11(3):349–358, 1987.

583. Harry Wiener. Structural determination of paraffin boiling points. *Journal of the American Chemical Society,* 69:17–20, 1947.

584. Eugene P. Wigner. Characteristic vectors of bordered matrices with infinite dimensions. *Annals of Mathematics,* 62:548–564, 1955.

585. Eugene P. Wigner. On the distribution of the roots of certain symmetric matrices. *Annals of Mathematics,* 67:325–327, 1958.

586. Herbert S. Wilf. *generatingfunctionology.* pub-ap, 1994.

587. James H. Wilkinson. *The Algebraic Eigenvalue Problem.* Clarendon Press, 1965.

588. Thomas Williams and Colin Kelley. *Gnuplot documentation.*

589. Gerhard Winkler. *Image Analysis, Random Fields, and Markov Chain Monte Carlo Methods.* Springer-Verlag, 2nd edition, 2003.

590. Gerhard J. Woeginger. Exact algorithms for $\mathcal{NP}$-hard problems: A survey. In *Proceedings of the 5th International Workshop on Combinatorial Optimization (Aussois'2001),* volume 2570 of *Lecture Notes in Computer Science,* pages 185–207. Springer-Verlag, 2003.

591. Kesheng Wu and Hort Simon. Thick-restart Lanczos method for large symmetric eigenvalue problems. *SIAM Journal on Matrix Analysis and Applications*, 22(2):602–616, 2000.
592. Stefan Wuchty and Peter F. Stadler. Centers of complex networks. *Journal of Theoretical Biology*, 223:45–53, 2003.
593. Norman Zadeh. Theoretical efficiency of the Edmonds-Karp algorithm for computing maximal flows. *Journal of the ACM*, 19(1):184–192, 1972.
594. Bohdan Zelinka. Medians and peripherians tree. *Archivum Mathematicum (Brno)*, 4:87–95, 1968.
595. Uri Zwick. All pairs shortest paths using bridging sets and rectangular matrix multiplication. *Electronic Colloquium on Computational Complexity (ECCC)*, 60(7), 2000.

Index

Lecture Notes in Computer Science

For information about Vols. 1–3292

please contact your bookseller or Springer